Molecular Aspects of Monooxygenases and Bioactivation of Toxic Compounds

NATO ASI Series

Advanced Science Institutes Series

A series presenting the results of activities sponsored by the NATO Science Committee, which aims at the dissemination of advanced scientific and technological knowledge, with a view to strengthening links between scientific communities.

The series is published by an international board of publishers in conjunction with the NATO Scientific Affairs Division

A	**Life Sciences**	Plenum Publishing Corporation
B	**Physics**	New York and London
C	**Mathematical and Physical Sciences**	Kluwer Academic Publishers
D	**Behavioral and Social Sciences**	Dordrecht, Boston, and London
E	**Applied Sciences**	
F	**Computer and Systems Sciences**	Springer-Verlag
G	**Ecological Sciences**	Berlin, Heidelberg, New York, London,
H	**Cell Biology**	Paris, Tokyo, Hong Kong, and Barcelona
I	**Global Environmental Change**	

Recent Volumes in this Series

Volume 196—Sensory Abilities of Cetaceans: Laboratory and Field Evidence
edited by Jeanette A. Thomas and Ronald A. Kastelein

Volume 197—Sulfur-Centered Reactive Intermediates in Chemistry and Biology
edited by Chryssostomos Chatgilialoglu and Klaus-Dieter Asmus

Volume 198—Selective Activation of Drugs by Redox Processes
edited by G. E. Adams, A. Breccia, E. M. Fielden, and P. Wardman

Volume 199—Targeting of Drugs 2: Optimization Strategies
edited by Gregory Gregoriadis, Anthony C. Allison, and George Poste

Volume 200—The Neocortex: Ontogeny and Phylogeny
edited by Barbara L. Finlay, Giorgio Innocenti, and Henning Scheich

Volume 201—General and Applied Aspects of Halophilic
Microorganisms
edited by Francisco Rodriguez-Valera

Volume 202—Molecular Aspects of Monooxygenases and Bioactivation
of Toxic Compounds
edited by Emel Arınç, John B. Schenkman, and Ernest Hodgson

Series A: Life Sciences

Molecular Aspects of Monooxygenases and Bioactivation of Toxic Compounds

Edited by

Emel Arınç

Middle East Technical University
Ankara, Turkey

John B. Schenkman

University of Connecticut Health Center
Farmington, Connecticut

and

Ernest Hodgson

North Carolina State University
Raleigh, North Carolina

Plenum Press
New York and London
Published in cooperation with NATO Scientific Affairs Division

Proceedings of a NATO Advanced Study Institute on
Molecular Aspects of Monooxgenases and Bioactivation of
Toxic Compounds,
held August 27–September 7, 1989,
in Çeşme (Izmir), Turkey

QP
603
,M65
N37
1989

Library of Congress Cataloging-in-Publication Data

NATO Advanced Study Institute on Molecular Aspects of Monooxgenases
    and Bioactivation of Toxic Compounds (1989 : Çeşme, Turkey)
    Molecular aspects of monooxgenases and bioactivation of toxic
  compounds / edited by Emel Arınç, John B. Schenkman, and Ernest
  Hodgson.
        p.   cm. -- (NATO ASI series. Series A, Life sciences   v.
  202)
    "Proceedings of a NATO Advanced Study Institute on Molecular
  Aspects of Monooxgenases and Bioactivation of Toxic Compounds, held
  August 27-September 7, 1989 in Çeşme (Izmir), Turkey"--T.p. verso.
    "Published in cooperation with NATO Scientific Affairs Division."
    Includes bibliographical references and index.
    ISBN 0-306-43823-2
    1. Monooxygenases--Congresses. 2. Xenobiotics--Metabolic
  detoxication--Congresses. 3. Cytochrome P-450--Congresses.
  I. Arınç, Emel. II. Schenkman, John B. III. Hodgson, Ernest, 1932-
  . IV. North Atlantic Treaty Organization. Scientific Affairs
  Division. V. Title. VI. Series.
    [DNLM: 1. Biotransformation--congresses. 2. Oxygenases-
  -metabolism--congresses. 3. Xenobiotics--metabolism--congresses.
  4. Xenobiotics--toxicity--congresses.   QU 120 N2795m 1989]
  QP603.M65N37 1989
  615.9--dc20
  DNLM/DLC
  for Library of Congress   91-3008
                                                                CIP

© 1991 Plenum Press, New York
A Division of Plenum Publishing Corporation
233 Spring Street, New York, N.Y. 10013

Printed in the United States of America

PREFACE

This volume represents the proceedings of a NATO Advanced Study Institute on "Molecular Aspects of Monooxygenases and Bioactivation of Toxic Compounds" which was held in Çeşme (İzmir) Turkey, August 27 to September 7, 1989.

The Institute emphasized the potential dangers of drugs, pollutants, pesticides, carcinogens, and both nutrient and non-nutrient food additives and their interactions at the molecular level. These xenobiotics are metabolized in phase I reactions by monooxygenases, primarly cytochrome P450-isozyme dependent, followed by phase II conjugation enzymes. The versatility of these enzymes is unique in biochemistry since they catalyze the metabolism of an almost limitless number of compounds through a diverse array of reactions. Although these metabolic sequences are usually detoxication mechanisms, in some cases the reactive intermediates are more mutagenic, more carcinogenic and ultimately more toxic than the parent compound. The Institute was devoted to recent scientific progress in the understanding of the biochemical and molecular properties of these enzymes and factors that influence the metabolism of chemicals to toxic and non-toxic compounds.

Overview lectures focused on xenobiotic metabolism by cytochrome P450, the flavin-containing monooxygenase, cytochrome b5, epoxide hydrolase, glutathione S-transferase and UDP-glucuronosyl transferase. Several additional presentations stressed various aspects of cytochrome P450 mediated metabolism including: specific forms in the rabbit and insects; topology in membranes; posttranslational modification; control of constitutive forms in the rat; hormonal and xenobiotic regulation; prostanoid metabolism; endogenous substrate metabolism; interaction with estrogenic pesticides; pesticide bioactivation. The Institute also included talks on interactions between enzymes and on benzene metabolism and toxicity.

The Editors believe this book will prove useful both to beginners and to those with a long-time interest in xenobiotic metabolism, monooxygenases and bioactivation of toxic compounds.

E. Arınç
J. B. Schenkman
E. Hodgson

CONTENTS

Cytochrome P450-Dependent Monooxygenase : An Overview ...   1
   J. B. Schenkman

The Flavin-Containing Monooxygenase (EC 1.14.13.8) 11
   E. Hodgson and P. E. Levi

Multiple Forms of Rabbit Cytochrome P450 : Comparison
   of Chemical, Physical, Immunological and
   Biocatalytic Properties 23
   E. F. Johnson, T. Kronbach, A. S. Muerhoff,
   K. J. Griffin, U. R. Pendurthi and R. H. Tukey

Primary Structures and Regulation of Cytochrome P450
   Isozymes 2 (IIB) and 5 (IVB) and the Flavin-
   Containing Monooxygenase in Rabbit Liver and
   Lung 55
   R. M. Philpot, R. Gasser and M. P. Lawton

Insect Cytochrome P450 75
   E. Hodgson and R. Rose

I. Membrane Topology of Cytochromes P450 : Elements
   and Measurement by Spectroscopic Techniques 93
   A. Stier, V. Krüger, T. Eisbein and S. A. E. Finch

II. Membrane Topology of Cytochromes P450 : Oligomers
   and Cooperativity 115
   A. Stier, V. Krüger, T. Eisbein and S. A. E. Finch

NADPH-Dependent Cytochrome P450 Reductase 135
   A. Y. H. Lu

Essential Features of NADH Dependent Cytochrome b5
   Reductase and Cytochrome b5 of Liver and Lung
   Microsomes 149
   E. Arınç

Electron Transfer from Cytochrome b5 to Cytochrome
   P450 171
   C. Bonfils, J-L. Saldana, C. Balny and P. Maurel

Functional Aspects of Protein-Protein Interactions
   of Cytochrome P450, Cytochrome b5 and
   Cytochrome P450 Reductase 185
   J. B. Schenkman

Posttranslational Modification by Phosphorylation :
    Control of Cytochrome P450 and Associated
    Enzymes 199
    W. Pyerin and H. Taniguchi

Physiological and Pathophysiological Alterations in
    Rat Hepatic Cytochrome P450 233
    J. B. Schenkman, K. E. Thummel and L. V. Favreau

Ontogenesis of Liver Cytochromes P450 255
    C. Bonfils, J. Combalbert, T. Pineau,
    C. Larroque, R. Lange and P. Maurel

Mechanism of Steroid Hormome Action 267
    A. Berkenstam and J-Å Gustafsson

Xenobiotic Regulation of Cytochrome P450 Gene
    Expression 283
    M. Gillner, J. Bergman and J-Å Gustafsson

Prostanoid Metabolism and Biologically Active
    Product Formation 293
    D. Kupfer

Role of Cytochrome P450 in the Anabolism and
    Catabolism of Endobiotics 305
    H. Vanden Bossche, H. Moereels and P. A. J. Janssen

Interactions of Estrogenic Pesticides with
    Cytochrome P450 331
    D. Kupfer

Effects of Inhibitors on the P450-Dependent Metabolism
    of Endogenous Compounds in Fungi, Protozoa,
    Plants and Vertebrates 345
    H. Vanden Bossche, P. Marichal G. Willensens and
    P. A. J. Janssen

Mechanism of Metabolic Activation of Nitroimidozoles 365
    A. Y. H. Lu and P. G. Wislocki

Benzene Metabolism 375
    R. Snyder and S. P. Chatterjee

Mechanisms of Benzene Toxicity 387
    S. P. Chatterjee and R. Snyder

Hematotoxicity, Leukemogenecity and Carcinogenecity
    of Chronic Exposure to Benzene 415
    M. Aksoy

Epoxide Hydrolase Isoenzymes and Their Individual
    Contribution to the Control of Toxic
    Metabolites 435
    F. Oesch, L. Schladt, M. Knehr, J. Döhmer and
    H. Thomas

Role of the Well-Known Basic and Recently Discovered
      Acidic Glutathione S-Transferases in the
      Control of Genotoxic Metabolites 447
   F. Oesch, I. Gath, T. Igarashi, H. Glatt and
   H. Thomas

Characterisation and Regulation of UDP-Glucuronosyl –
      transferases 463
   B. Burchell

Molecular Cloning, Expression and Genetic Deficiencies
      of UDP-Glucuronosyltransferases 473
   B. Burchell

Contributors 489

Index 491

CYTOCHROME P450-DEPENDENT MONOOXYGENASE: AN OVERVIEW

John B. Schenkman

Department of Pharmacology
University of Connecticut Health Center
Farmington, CT   06032, USA

INTRODUCTION

In early studies on liver microsomes it was noticed that the heme content, measured by the pyridine hemochrome method, yielded values about double that of the content of the only known microsomal hemoprotein, cytochrome b_5 (1). Reduction of the microsomes with sodium dithionite yielded, in the presence of carbon monoxide a "carbon monoxide-binding pigment" which was rationalized as not to be a hemoprotein (1, 2). The reasons cited for the CO-binding pigment not to be a hemoprotein included a) the absence of a reduced minus oxidized difference spectrum in the Soret region, b) lack of α- and β-peaks for the reduced, CO-complex, c) displacement of the CO-reduced absorption peak much further to the red than other carbon monoxy-hemoproteins, and d) the CO-complex showed no photodissociation. P450 was coined as a tentative name for the CO-complex, which was first reported to be hemoprotein in nature, in 1962 (3). Within the next few years cytochrome P450 was demonstrated to be the terminal oxidase in a number of microsomal monooxygenation reactions (4, 5).

Cytochrome P450 is actually a family of different gene products, all with a single iron protoporphyrin IX prosthetic group containing a fifth ligand to the sulfur atom of a cysteine residue. A large number of cytochrome P450 forms have already been sequenced, all of which contained the cysteinyl group of the carboxy-terminal portion of the molecule in a region of highly conserved amino acids, and enough sequence similarity to crystallized, Pseudomonas P450 (P450CAM), the structure of which is known (6), to allow attempts to relate functional regions, e.g., heme binding, oxygen-binding, and substrate-binding domains (7, 8). Several of the subsequent chapters in this book discuss specific forms of cytochrome P450 and their structure, and the interaction of the cytochrome P450 with other microsomal electron transfer proteins.

Cytochrome P450 Catalyzed Reactions

Cytochrome P450 is a very versatile enzyme. It can catalyze a very wide range of different types of reactions, as pointed out by Gillette and by Brodie many years ago (9). These include a variety of reaction types (Figure 1), from oxidative deamination, desaturation of steroids, various heteroatom dealkylations and oxidations hydroxylations and a

1. DEAMINATION

AMPHETAMINE

2. DESATURATION

TESTOSTERONE

3. DEALKYLATIONS

A. NITROGEN

AMINOPYRINE   HCHO

B. OXYGEN

p-NITROANISOLE   HCHO

C. SULFUR

6-METHYLTHIO
PURINE   HCHO

4. HYDROXYLATIONS

A. ALIPHATIC

$$CH_3CH_2(CH_2)_{13}COOH$$

$$\longrightarrow HOCH_2CH_2(CH_2)_{13}COOH$$

B. AROMATIC

ANILINE

5. HETEROATOM OXIDATION

A. N-OXIDATION

SULFANILAMIDE

B. S-OXIDATION

CHLORPROMAZINE

C. O-OXIDATION

TESTOSTRONE

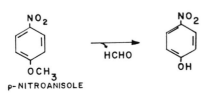

Figure 1

6. REDUCTIONS

A. DEHALOGENATION

$$CF_3CHClBr \longrightarrow CF_3CH_2Cl$$

HALOTHANE

B. AZO

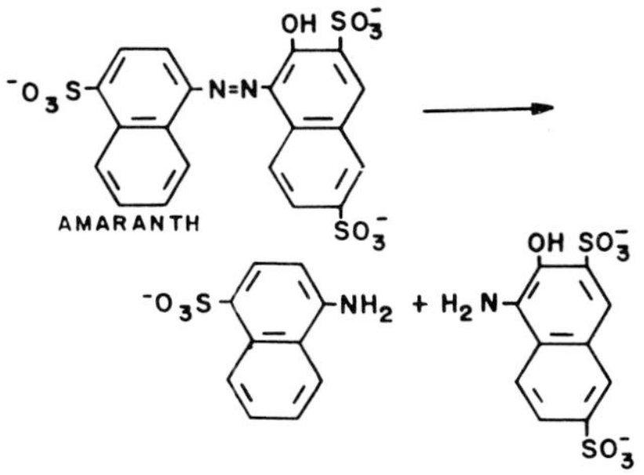

AMARANTH

C. NITRO

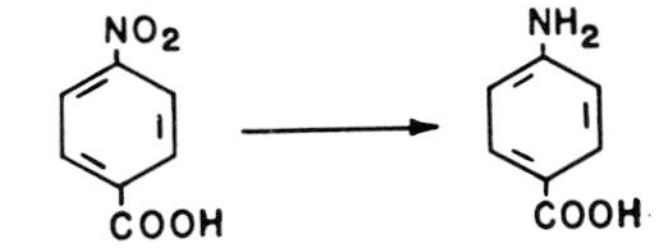

p-NITRO
BENZOATE

7. ESTER CLEAVAGE

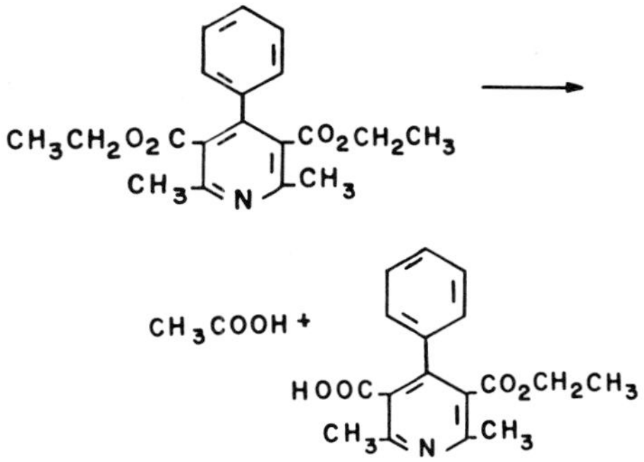

$CH_3COOH +$

number of reductive reactions, including reductive dehalogenation, azo reductive cleavage, and nitro group reduction, as well as ester group oxidative cleavage. The tremendous range of reactions catalyzed by cytochromes P450 is made possible, in part, by its very negative redox potential (E^0 = -350mV), plus its ability to reductively cleave molecular oxygen to active oxygen species. For those interested in the mechanisms of some such reactions, one might see studies by Parli and McMahon (10) on deaminations, by Cheng and Schenkman (11) on O-oxidation, by Guengerich (12) on ester oxidative cleavage and by Nagata, et al. (13) on steroid desaturation.

Although cytochrome P450 is classified as a monooxygenase, it does, in fact, also function as an NADPH oxidase (TPNH-oxidase (14)). As a monooxygenase it catalyzes the reaction in equation 1, where SH is the substrate, and SOH is the product, the hydroxylated compound.

$$NADPH + O_2 + SH + H^+ \longrightarrow SOH + H_2O + NADP^+ \qquad \text{eq. 1}$$

In the NADPH oxidase reaction, equation 2, the reducing equivalents of NADPH are used to partially reduce molecular oxygen to the level of

$$NADPH + O_2 + H^+ \longrightarrow H_2O_2 + NADP^+ \qquad \text{eq. 2}$$

hydrogen peroxide. Nordblom and Coon (15) showed with the reconstitued cytochrome P450 system a 1:1:1 stoichiometry could be obtained with NADPH:O_2:(H_2O_2 + SOH) measurements. A similar 1:1:1 stoichiometry of NADPH and oxygen consumption and product formation (SOH + H_2O_2) was obtained (16) with liver microsomes during aminopyrine demethylation. In more recent studies (17, 18), it was noted that with some reconstituted cytochrome P450 systems, with certain substrates an excess NADPH and O_2 consumption occurs in 2:1 stoichiometry, suggesting reduction of O_2 to the level of water (equation 3). Table 1 shows the stoichiometry breakdown for testosterone metabolism with one of the forms

$$2 \ NADPH + O_2 + 2H^+ \longrightarrow 2H_2O + 2 \ NADP^+ \qquad \text{eq. 3}$$

of cytochrome P450, RLM5 (gene IIC11). The turnover of the reconstituted system was measured by disappearance of enzymatically active NADPH and by disappearance of oxygen in the oxygen electrode. Testosterone consumption (AOH) was measured as total product formation from ^{14}C-testosterone on thin layer chromatography. This and other items measured are underlined in the actual stoichiometry. Using the total monooxygenase product sum (AOH) of 14 nmol/nmol P450/min, and adding appropriate amounts of oxygen and NADPH to fit the monooxygenase stoichiometry yields line 1. We can do the same in line 2, fitting the NADPH oxidase stoichiometry to the amount of H_2O_2 measured (13.9 nmol/nmol P450/min). In line 3 we fit the excess NADPH which was consumed and the excess O_2 consumed. Note that they fit a stoichiometry of 2:1, a value indicative of 4 electron reduction of dioxygen, i.e., reduction to the level of water. The data, then, indicates that cytochrome P450 can catalyze the oxidative metabolism of substrates, and in addition, reduce oxygen to H_2O_2 and to H_2O. This latter, believed to be a leaking of activated oxygen from the enzyme system, or an "uncoupling" of the system, releases hydrogen peroxide, which is potentially harmful to the cell, as well as fully reduced oxygen. While safer, the latter is, nevertheless, a waste of cellular reducing equivalents.

The P450 Cycle

The turnover of cytochrome P450 can be described in terms of a cycle

Table 1

Stoichiometry of Testosterone Oxidation

Monooxygenase $1\ NADPH + 1\ O_2 + 1\ AH \rightarrow 1\ AOH + 1\ H_2O + 1\ NADP^+$

Oxidase $1\ NADPH + 1\ O_2 \rightarrow 1\ NADP^+ + H_2O_2$

Oxidase $1\ NADPH + 1/2\ O_2 \rightarrow 1\ H_2O + 1\ NADP^+$

Turnover $\Sigma = NADPH + O_2 + AH \rightarrow AOH + H_2O + NADP^+ + H_2O_2$

RLM5 With Stoichiometry

Turnover $41.8\ NADPH + 35.9\ O_2 + 14\ AH \rightarrow 14\ AOH + H_2O + 41.8\ NADP^+ + 13.9\ H_2O_2$

Monooxygenase (eq. 1) $14\ NADPH + 14\ O_2 + 14\ AH \rightarrow 14\ AOH + 14\ H_2O + 14\ NADP^+$

Oxidase (eq. 2) $13.9\ NADPH + 13.9\ O_2 \rightarrow 13.9\ NADP^+ + 13.9\ H_2O_2$

Oxidase (eq. 3) $13.9\ NADPH + 7\ O_2 \rightarrow 14\ H_2O + 13.9\ NADP^+$

$41.8\ NADPH + 34.9\ O_2 + 14\ AH \rightarrow 14\ AOH + 28\ H_2O + 41.8\ NADP^+ + 13.9\ H_2O_2$

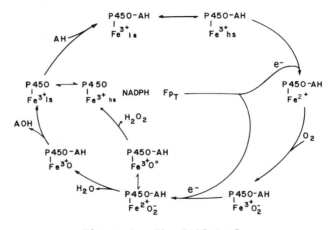

Figure 2.   The P450 Cycle

in which the enzyme undergoes a series of reductive and oxidative steps.
In order to function a source of reducing equivalents is required.   These
come from NADPH via a flavoprotein reductase, NADPH-cytochrome P450
reductase, abbreviated Fp_T in Figure 2.   Dr. Lu will discuss this
enzyme in his chapter.   The P450 cycle begins on the upper left side of
the cycle with cytochrome P450 as a ferric enzyme existing as an
equilibrium mixture of high and low spin enzyme.   Most of the forms of
cytochrome P450 found are primarily in the low spin state, but one or two
forms have been found, which exist primarily in the high spin
configuration.   [The spin state refers to the five iron valence
electrons.   In the high spin state the $3d^5$ electrons are in separate
orbitals, while in low spin P450 they are paired with only one unpaired
electron remaining for the ferric enzyme.   For further explanation see
ref. 19].   Substrate binding to cytochrome P450 frequently perturbs the
spin equilibrium, shifting it further to the high spin configuration.   In
difference spectra this is seen as a shift from absorption at about 420nm
(low spin absorption) to 390nm (high spin absorption), a shift termed
earlier the "type I spectral change".   The shift to high spin cytochrome
P450 has been shown to raise the redox potential of various forms of
cytochrome P450 from -350mV to as high as -178mV, and this correlates
well with a facilitation of reduction of the hemoprotein to the ferrous
state (19).   Since NADPH has a redox potential of about -360mV, and NADPH
cytochrome P450 reductase has redox potentials of -270 and -290mV for the
2- and 3-electron reduced states, respectively (20), raising the redox
potential of cytochrome P450, i.e., making it more positive, should favor
its reduction.

      Continuing around the cycle, oxygen binds to the reduced, or ferrous
hemoprotein, forming oxy-P450, and this is capable of accepting a second
electron.   This latter electron has been shown to come from P450
reductase, or, in the reconstituted system, from cytochrome b_5 (21).
The two electron reduced form can be called the peroxy-P450 state.   With
the bacterial P450CAM it has been shown that if input of the second
electron is delayed, superoxide anion can be released which yields, after
dismutation, hydrogen peroxide.   Hydrogen peroxide itself may be released
after input of the second electron, or, in the presence of some
monooxygenase substrate, water is released followed by the oxidation
product.   In a subsequent lecture chapter I have discussed the role of

cytochrome b_5 as an electron donor and effector of the system; binding of this hemoprotein can cause "metabolite switching", i.e., an increase in product formation at the expense of NADPH oxidase activity (H_2O_2 production). If we consider the release of H_2O_2 as being due to uncoupling of the monooxygenase reaction, then cytochrome b_5 serves to increase the coupling of the reaction. However this is accomplished, it requires the cytochrome b_5 be functionally capable of undergoing reduction (21). Table 2 shows the metabolic switching that occurs with cytochrome P450. In the table the percent of the NADPH consumed to yield monooxygenase product, H_2O_2 or 4 electron reduction product (H_2O), eq. 1, 2 and 3, respectively, are listed for three forms of cytochrome

Table 2

Stoichiometry of benzphetamine metabolism

| Reaction | % NADPH consumption | | | | | |
| | LM2 | | RLM5 | | RLM2 | |
	$-b_5$	$+b_5$	$-b_5$	$+b_5$	$-b_5$	$+b_5$
eq. 1	43	52	45	45	6	6
eq. 2	21	11	25	21	54	59
eq. 3	35	37	30	34	40	35

P450, rabbit LM2, rat RLM5 and rat RLM2. In the presence of cytochrome b_5 the extent of uncoupling of LM2 decreased from 21% of the NADPH yielding H_2O_2 to 11%. There was a corresponding rise in the proportion of reducing equivalents going to form the product of the monooxygenase reaction (eq. 1). In absolute terms, the actual rate of NADPH consumption was increased as were O_2 consumption and product formation. Only H_2O_2 formation decreased. This was not the case with RLM5 (18): with benzphetamine as the substrate the rate of H_2O_2 production by RLM5 was unchanged and a 20% increase was observed in consumption of NADPH and O_2, and in production of product. Thus, while cytochrome b_5 did speed up the monooxygenase reaction with RLM5, it did not cause much in the way of enhanced coupling with benzphetamine as substrate (Table 2). RLM2, interestingly, was unaffected by the presence of cytochrome b_5. Of special interest was the very high proportion of the NADPH going to form H_2O_2 with RLM2 (Table 2); over 50% of the reducing equivalents are wasted. If one includes the eq. 3 oxidase reaction, with RLM2 over 90% of the NADPH is wasted, as very little of the turnover of this enzyme yields product from benzphetamine.

Mechanism Of Oxidation

This portion of the P450 cycle still remains a black box. Suggestions are that the reaction subsequent to input of the second electron involves a disproportionation of the electrons between the atoms of dioxygen, e.g., a heterolytic cleavage.

$$Fe^{III}O_2^{=} \longrightarrow \left[Fe^{V}=O\right]^{3+} + O^{=} \qquad \text{eq. 4}$$

This reaction can be described as

$$Fe^{III}:\ddot{O}|\ddot{O}:^= \longrightarrow Fe^{V}:\ddot{O} + :\ddot{O}:^=$$

heterolytic   perferryl ion   water anion

A suggesting by Ullrich, et al. (22) provided a role for the cysteinyl sulfur heme ligand, that of stabilizing the ferric-complexed atomic

$$-SHFe^{III}O_2^={}^R_H \xrightarrow{H_2O} -S\cdots Fe^{III}O^-{}^R_H \xrightarrow{ROH} -S^-\cdots Fe^{III}{}^H_{O^-} \xrightarrow{RH} -SHFe^{III}{}^R_H \qquad eq.\ 5$$

oxygen. The cysteinyl ligand on P450 is unique and perhaps is responsible for the properties that permit substrate hydroxylation. Only P450, among hemoproteins is capable of substrate hydroxylation, as well as the many reactions shown in Figure 1. The perferryl ion is an extremely reactive species, and the suggestion was made (23) that perhaps it is stabilized by bonding to a pyrrole ring nitrogen, $Fe^{IV}O\text{-}por^{(+)}$.

An alternative mechanism for P450, exists, also utilizing the cysteinyl sulfur as a ligand of the heme, but calling for a homolytic cleavage mechanism for opening the peroxide oxygen-oxygen bond during substrate hydroxylation supported by peroxyacids (24). The mechanism, which can be related to 2 electron activated dioxygen bond cleavage, invokes a cleavage of the dioxygen bond by one electron reduction of the peroxide by the heme thiolate ion. In this proposal the cysteinyl sulfur is a participant in the redox reaction (eq. 6).

$$-S^-_{RH}Fe^{III}{}^H_O{-}O{-}\overset{O}{C}{-}R \longrightarrow -S^-Fe^{III}{}^H_O{-} + \cdot O{-}\overset{O}{C}{-}R \xrightarrow[RCOOH]{} -S^-Fe^{III}{}^H_O{-}{}_{R\cdot} \xrightarrow{ROH} -S^-Fe^{III} \qquad eq.\ 6$$

The dioxygen cleavage is homolytic,

$$S^-{-}Fe^{III}{}^H_O \!\!\diagup\!\! \ddot{O}{:}H \qquad\qquad \dot{S}\,{-}Fe^{III}{:}\ddot{O}{:}^- + :\dot{O}{:}\,H$$
Hydroxy radical

and the hydroxyl radical generated can abstract a hydrogen atom from the substrate (RH) in the active site in juxtaposition to it. The iron serves to stabilize the other half of the peroxide transferring an electron from the sulfur to yield iron bound hydroxide. This, perhaps, would react with the substrate radical (R·) in juxtaposition, with an electron then returning to the sulfur (eq. 6). While such a mechanism agrees with peroxyacid driven P450 reactions (24), it remains to be seen whether it might be a more general mechanism.

Multiple Forms of P450

At present at least 71 different genes for cytochrome P450 have been identified in different species (25), with 32 of these being sequenced in the rat. These include both mitochondrial and endoplasmic reticulum forms, and multiple, possibly allelic forms of some of these that have been deduced from cDNA but for which the protein have as yet not been isolated. Some of these forms are inducible, i.e., the levels are elevated by challenge of the animals with drugs or chemicals. Other forms have, as yet, not been found to be inducible. The latter forms, however, are not necessarily constant in level in the membranes. As discussed in my subsequent chapter of this book, such forms, too, are in a state of dynamic equilibrium, with levels responsive to homeostatic changes in the animal. Such forms have been shown to be responsive to developmental changes in vivo, sexual development, and pathophysiological conditions. In the following chapters, some of the different forms of cytochrome P450 will be discussed along with their characteristics.

Acknowledgements

Supported by US Public Heatlh Service grant from the NIH, GM26114. The help of Dr. Ingela Jansson in preparing this manuscript is gratefully acknowledged.

REFERENCES

1. M. Klingenberg, Pigments of rat liver microsomes, Arch. Biochem. Biophys., 75:376 (1958).
2. D. Garfinkel, Studies on pig liver microsomes. I. Enzymic and pigment composition of different microsomal fractions. Arch. Biochem. Biophys., 77:493 (1958).
3. T. Omura and R. Sato, A new cytochrome in liver microsomes, J. Biol. Chem., 237:PC1375 (1962).
4. R.W. Estabrook, D.Y. Cooper and O. Rosenthal, The light reversible carbon monoxide inhibition of the steroid C21-hydroxylase system of the adrenal cortex, Biochem. Z., 338:741 (1963).
5. D.Y. Cooper, S. Levin, S. Narasimhulu, O. Rosenthal and R.W. Estabrook, Photochemical action spectrum of the terminal oxidase of mixed function oxidase systems, Science, 147:400 (1965).
6. T.C. Poulos, B.C. Finzel and A.J. Howard, High-resolution crystal structure of cytochrome P450CAM, J. Mol. Biol., 195:687 (1987).
7. D.R. Nelson and H.W. Strobel, On the membrane topology of vertebrate cytochrome P450 proteins, J. Biol. Chem., 263:6038 (1986).
8. O. Gotoh and Y. Fujii-Kuriyama, Evolution, structure, and gene regulation of cytochrome P450, in: Frontiers in Biotransformation, Vol. 1, p. 195, K. Ruckpaul and H. Rein, eds., Akademie-Verlag, Berlin (1989).
9. J.R. Gillette, Oxidation and reduction by microsomal enzymes, in Proceedings of the First International Pharmacological Meeting, Vol. 6, p. 13, B.B. Brodie and E.G. Erdos, eds., The MacMillan Co., New York, 1962.
10. C.J. Parli and R.E. McMahon, The mechanism of microsomal deamination, Drug Metab. Dispos., 1:337 (1973).
11. K.-C. Cheng and J.B. Schenkman, Testosterone metabolism by cytochrome P450 isozymes RLM3 and RLM5 and by microsomes, J. Biol. Chem., 258:11738 (1983).
12. F.P. Guengerich, Oxidative cleavage of carboxylic esters by cytochrome P450, J. Biol. Chem., 262:8459 (1987).
13. K. Nagata, D.J. Liberato, J.R. Gillette and H.A. Sasame, An unusual metabolite of testosterone, 17β-hydroxy-4,6-androstadiene-3-one, Drug Metab. Dispos., 14:559 (1986).
14. J.R Gillette, B.B. Brodie and B.N. LaDu, The oxidation of drugs by liver microsomes. On the role of TPNH and oxygen, J. Pharm. Exp. Ther., 119:532 (1957).
15. G.D Nordblom and M.J. Coon, Hydrogen peroxide formation and stoichiometry of hydroxylation reactions catalyzed by highly purified liver microsomal cytochrome P450, Arch. Biochem. Biophys., 180:343 (1977).
16. I. Jansson and J.B. Schenkman, Stoichiometry of aminopyrine demethylation with and without NADH synergism, Drug Metab. Dispos., 9:461 (1981).

17.   L.D. Gorsky, D.R. Koop and M.J. Coon, On the stoichiometry of the oxidase and monooxygenase reactions catalyzed by liver microsomal cytochrome P450, J. Biol. Chem., 259:6812 (1984).

18.   I. Jansson and J.B. Schenkman, Influence of cytochrome b_5 on the stoichiometry of the different oxidative reactions catalyzed by liver microsomal cytochrome P450, Drug Metab. Dispos., 15:344 (1987).

19.   J.B. Schenkman, S.G. Sligar and D.L. Cinti, Substrate interaction with cytochrome P450, in Hepatic Cytochrome P450 monooxygenase system, p. 587, J.B. Schenkman and D. Kupfer, eds., Pergamon Press, N.Y. 1982.

20.   T. Iyanagi, N. Makino and H.S. Mason, Redox properties of the reduced NADP-cytochrome P450 and NAD-cytochrome b_5 reductases, Biochemistry, 13:1701 (1974).

21.   P.P. Tamburini and J.B. Schenkman, Purification to homogeneity and enzymological characterization of a functional covalent complex composed of cytochromes P450 isozyme 2 and b_5 from rabbit liver, Proc. Natl. Acad. Sci. USA, 84:11 (1987).

22.   V. Ullrich, H.H. Ruf and P. Wende, The structure and mechanism of cytochrome P450, Croatica Chem. Acta, 49:213 (1977).

23.   M. Imai, H. Shimada, Y. Watanabe, Y. Matsushimo-Hibaya, R. Makino, H. Koga, T. Horiuchi and Y. Ishimura, Uncoupling of the cytochrome P450 CAM monooxygenase reaction by a single mutation, threonine 252 to alanine or valine: A possible role of the hydroxy amino acid in oxygen activation, Proc. Natl. Acad. Sci. USA, 86:7823 (1989).

24.   R.E. White, S.G. Sligar and M.J. Coon, Evidence for a homolytic mechanism of peroxide oxygen-oxygen bond cleavage during substrate hydroxylation by cytochrome P450, J. Biol. Chem. 255:11108 (1980).

25.   D.W. Nebert, D.R. Nelson, M. Adesnik, M.J. Coon, R.W. Estabrook, F.J. Gonzalez, F.P. Guengerich, I.C. Gunsalus, E.F. Johnson, B. Kemper, W. Levin, I.R. Phillips, R. Sato and M.R. Waterman, The P450 superfamily: Updated listing of all genes and recommended nomenclature for the chromosomal loci, DNA, 8:1 (1989).

THE FLAVIN-CONTAINING MONOOXYGENASE (EC 1.14.13.8)

Ernest Hodgson and Patricia E. Levi

Department of Toxicology, Box 7633
North Carolina State University
Raleigh, NC   27695, U.S.A.

INTRODUCTION

The flavin-containing monooxygenase (EC 1.14.13.8) (FMO), originally described as an amine oxidase, was subsequently shown to be a versatile sulfur oxidase, the early studies being summarized by Ziegler (1). The FMO has been shown more recently to be a phosphorus oxidase (2,3). This enzyme and the cytochrome P450-dependent monooxygenase system are the two principal enzymes that catalyze the oxidation of lipophilic xenobiotics to electrophilic products capable of further metabolism, either to readily excretable conjugation products or to reactive intermediates with potential for adverse effects. Much of what is known about the substrate specificity of the FMO, is summarized in a recent review (4). Purification of pig liver FMO was accomplished some time ago (5) and the ability of the solubilized enzyme to catalyze the oxidation of the same wide variety of nucleophilic nitrogen, sulfur and phosphorus compounds as the membrane-bound enzyme has been established (1-5). The physiological role for this enzyme is not well known but may be involved in the maintenance of cellular thiol:disulfide ratios by the oxidation of cysteamine to cystamine (6).

More recently it has been demonstrated that the FMO exists in many species and in more than one organ and the FMO has been purified from a number of sources (7). Studies have been initiated on the molecular biology of FMO and will be discussed by Philpot et al. (this monograph). Much of the material in this chapter was previously reviewed by the authors (8).

SUBSTRATE SPECIFICITY

Among the sulfur compounds known to be oxidized at the sulfur atom are: thiols and disulfides; aminothiols; thioamides; thiocarbamides; thioacids; sulfides and sulfoxides. Nitrogen containing substrates include: tertiary

amines; secondary amines; hydrazines (4,9,10). The pulmonary
form of the enzyme can also oxidize primary amines (11,12).
Oxidations at phosphorus atoms include the formation of
phosphine oxides from either phosphines or phosphine sulfides
(2,3).

RELATIVE IMPORTANCE OF THE FMO AND P450 IN MICROSOMAL
OXIDATIONS IN DIFFERENT TISSUES

Many substrates, particularly sulfur-containing
substrates, are oxidized by both P450 and the FMO. A small
number of studies (13,14) have demonstrated the oxidation of
N,N-dimethylaniline by both P450 and the FMO in microsomes
from the same organ, however the relative contributions of
the two enzymes were not closely quantitated. In order to
study the relative contributions of these two enzymes toward
the oxidation of common substrates, we have developed methods
to measure each separately in the same microsomal
preparation. Two of these techniques have been particularly
useful. The first involves the inhibition of P450 activity
by the use of an antibody to NADPH cytochrome P450 reductase,
thus permitting the measurement of FMO activity alone. The
second utilizes the known heat lability of the FMO and
consists of heat treatment of the microsomal preparation (50C
for 1 min.) to inactivate the FMO, thus permitting
measurement of P450 alone, as P450 activity is unchanged by
the heat treatment (15,16).

The relative contributions of the two enzyme systems
with thiobenzamide as substrate are summarized in Table 1.
The FMO contribution is higher in female than in male mouse
liver and there is a difference in relative contribution
between liver and lung in the mouse but not in the rat.
Similar studies on the oxidation of the insecticide, phorate,
to phorate sulfoxide in the different tissues are shown in
Table 2 (17,18). In the livers of untreated animals P450 is
more important in the sulfoxidation of phorate (P450:FMO,
approximately 75:25); by contrast in the kidney and lung,
while the overall activity is low compared to the liver, the
relative contribtuion of the FMO is significantly higher.
This is particularly evident in renal microsomes from female
mice in which 90 percent of phorate sulfoxidation is due to
FMO.

Relative levels of activity are easily perturbed by
compounds or conditions that alter either the level of P450
or the level of the FMO. Thus pretreatment of mice with
phenobarbital significantly increases, not only the overall
rate of phorate oxidation, but also the relative contribution
of the P450 pathway (Table 2).

Piperonyl butoxide is a methylenedioxyphenyl insecticide
synergist with a biphasic effect on P450 activity in the
liver, first inhibition and subsequently induction. In mice
treated with piperonyl butoxide it can be demonstrated (Table
2) (18) that the proportion of phorate oxidation due to the
FMO first rises, during the P450 inhibition phase, and then
falls, during the P450 induction phase.

The effects of xenobiotics on the relative contributions

Table 1. Microsomal Oxidation of Thiobenzamide in Microsomes
        from Rat and Mouse Tissue

| Species | Sex | Tissues | Relative Activity (%) | |
			P-450	FMO
Mouse	M	Liver	50	50
Mouse-Pb[a]	M	Liver	65	35
Mouse	F	Liver	25	75
Mouse	M	Lung	20	80
Rat	M	Liver	35	65
Rat	M	Lung	40	60

[a]Mice pretreated with phenobarbital
  Modified from Tynes and Hodgson 1983 (15)

of the FMO and P450 appear to be mediated primarily via the
P450 component since the FMO does not appear to be
inducible by xenobiotics. FMO levels may, however, vary with
nutrition, diurnal rhythms, sex, pregnancy and
corticosteroids, such as dihydrocortisone, although the
effects appear to be both species and tissue dependent (see
1,4 for references). We have also studied (18) the effects
of hydrocortisone treatment on the metabolism of phorate and
thiobenzamide by the FMO of mouse liver and lung. The FMO
activity in the liver is increased for both substrates (+82%
for phorate, +52% for thiobenzamide) with only minor changes
in the lung (-15% for phorate, +20% for thiobenzamide). Such
alterations may assume toxicological importance when the
products from the two enzymes differ, particularly when one
metabolite is more toxic or more pharmacologically active
than the others.

CONTRIBUTION OF THE FMO TO COMPLEX METABOLIC PATHWAYS

        An interesting and relatively unexplored aspect of FMO
function is its contribution to complex metabolic pathways of
xenobiotics. It is clear that both purified enzymes and
intact microsomes must be used to elucidate the reactions and
products involved in complex metabolic pathways, especially
when several enzymes are involved and the oxidative pathways
involve both detoxication and activation reactions. We have
recently utilized both purified FMO and P450 isozymes to
examine in detail the oxidative pathways of phorate
metabolism (19,20) [Figure 1].

        Both P450 and FMO catalyze the initial sulfoxidation of
thioether-containing organophosphate insecticides such as
phorate and disulfoton to form the respective sulfoxides.
Subsequent oxidation reactions, however, such as formation of
the sulfone and oxidative desulfuration to the corresponding
oxons are catalyzed only by P450. Although both the FMO and
P450 catalyze the initial sulfoxidation, the products are
stereochemically different. The FMO forms the (-) phorate
sulfoxide while two of the P450 isozymes (P450 B2, a major

13

Table 2. Relative Contributions of P450 and FMO to the Microsomal Oxidation of Phorate in Microsomes from Several Mouse Tissues

| Tissue | Sex | Product formation (nmols phorate sulfoxide/mg protein/min) | | | | |
		Control	+AR[a]	+Heat	%FMO	%P450
Liver	M	12.7	2.8	9.3	21.7	78.3
Liver	F	14.4	3.7	11.7	24.0	76.1
Lung	M	3.3	1.9	-	59.1	41.3
Lung	F	5.7	3.1	-	54.0	46.0
Kidney	M	1.6	1.2	-	72.0	28.1
Kidney	F	2.0	1.8	-	90.0	10.0
Liver-PB[b]	M	69.7	10.1	59.6	14.3	85.5
Liver-PBO[c]						
2 hr	M	11.1	-	6.5	41.4	58.6
6 hr	M	19.4	16.3	-	16.0	84.0

[a]Antibody to P450 reductase
[b]Phenobarbital treated mice
[c]Piperonyl butoxide treated mice
Data from & Kinsler, Levi and Hodgson (17,18)

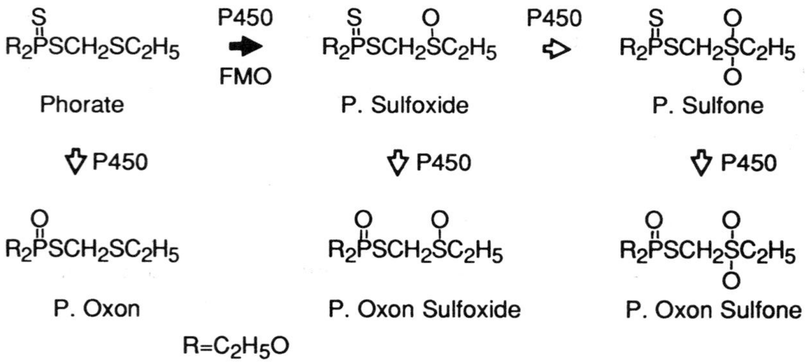

Figure 1. Oxidative metabolism of the insecticide phorate

constitutive form, and P450 PB, the principal form induced by phenobarbital) yield (+) phorate sulfoxide. The other three P-450 isozymes examined gave racemic mixtures.

Both (+) and (-) phorate sulfoxides are substrates for further oxidation by P450 to either the oxon or the sulfone. However, not only is (+) phorate sulfoxide the preferred substrate, but the percent of oxon sulfoxide relative to the percent phorate sulfone is higher with the (+) sulfoxide as substrate. It is interesting to note that the isozyme of P450 induced by phenobarbital (P450 PB) not only forms the

(+) phorate sulfoxide more rapidly but also produces the
highest percentage of oxon sulfoxide of any of the P450s.
Clearly environmental or physiological factors which increase
the level of this isozyme in vivo have the potential, by
preferentially increasing the activation product (the oxon,
an effective cholinesterase inhibitor), to enhance the
toxicity of this compound.

      Thioridazine is a phenothiazine neuroleptic that is
extensively metabolized after administration. Examination of
the chemical structure of thioridazine relative to known
substrates for the FMO and P450 isozymes indicate that it
should be a substrate for both of these monooxygenases. S-
oxidation is known probably to be the predominant route of
metabolism in man (21,22) producing the 2-sulfoxide, the 2-
sulfone, and the 5-sulfoxide, with the 2-sulfoxide and the 2-
sulfone having greater antipsychotic activity than the parent
compound (22,23) while the ring sulfoxides appear to be
largely responsible for the cardiotoxic side effects of
thioridazine (24). Northioridazine, the demethylation
product, is formed in significant quantities in rats, but not
in man. Additional minor metabolites include ring
hydroxylations and combinations of the above sulfoxides,
sulfones, demethylation products, and phenols.

      Preliminary results (Lembke, Mailman and Hodgson,
unpublished results) indicate the involvement of both P450
and FMO in the oxidation of thioridazine. Incubations with
partially purified mouse FMO (P450 free) and thioridazine in
which NADPH consumption was monitored gave an Km of 8.8µmol
and Vmax of 352 nmol/min/unit FMO, indicating that
thioridazine is, in fact, a good substrate for the FMO. Heat
pretreatment of microsomes caused a decrease in the amounts
of the 2-sulfoxide and northioridazine (an N-demethylation
product), indicating that these reactions may be mediated in
part by the FMO. On the other hand, the amount of the 5-
sulfoxide and 5-sulfone produced is increased in incubations
with heat treated microsomes. Such differential routes of
metabolism may assume pharmacological significance in the
case of activation of a prodrug or when one of the
metabolites is more toxic.

SPECIES DIFFERENCES

      The first comprehensive comparison of species and organs
employed an immunochemical method and provided much useful
qualitative information (25). The quantitative estimates,
however, particularly for extrahepatic tissues, are almost
certainly low due to lack of knowledge, at that time, of the
immunologically different forms of the enzyme. More detailed
comparisons of purified FMOs from different species are now
possible because of the development of newer purification
methods (7).

      A comparison of the substrate specificity of liver FMOs
of several species is shown in Table 3 and a comparison of
the purified enzymes from mouse and pig liver is shown in
Table 4. These enzymes are generally similar with some
relatively minor differences. Lung FMOs will be discussed in
more detail later but it may be noted that the liver FMOs

Table 3. Species Variation in the Oxidation of Sulfur-
containing Compounds by the FMO of Hepatic
Microsomes

| Substrate | Activity[a] | | | |
	Mouse	Rat	Rabbit	Pig
Thiourea	18.9	7.8	12.3	12.9(31.4)[b]
Phenylthiourea	9.0	4.4	7.7	9.3(26.1)
-Naphthylthiourea	7.3	4.0	6.6	7.9(21.6)
Methimazole	17.4	7.2	14.3	14.3(30.4)
2-Mercaptobenzimidazole	6.1	2.9	6.1	8.3(18.8)
Methylphenylsulfide	10.2	4.7	7.1	9.9(25.6)
Thioacetamide	18.6	7.5	11.6	13.4(31.0)
Cysteamine	17.8	8.1	11.6	13.6(30.9)
1-Butanethiol	9.2	7.1	7.2	7.8(20.7)
Trans-0-Dithiane-4,5-diol	4.8	2.5	6.9	7.3(10.9

[a]Activity: nmol NADPH/min/mg microsomal protein
[b]Values in parentheses are determined in presence of n-
octylamine.
Modified from Tynes & Hodgson (26)

from different species are broadly similar to each other but
differ from the pulmonary enzymes of the same or different
species.

One or more forms of FMO have been characterized in
microsomes from pig, mouse, rat, rabbit, guinea pig, hamster,
dog, human and Typanosoma cruzi. In addition one or more
forms of the enzyme have been purified from pig, mouse, rat,
rabbit, human and Trypanosoma cruzi (1,4,8,11,12,26-32).

ORGAN DIFFERENCES

Differences in catalytic activity between microsomes
from different organs have often been determined, and it is
clear that the activity in lung and kidney may be as high as
that of the liver. One such comparison is shown in Table 5.

The most extensive comparison between FMOs of different
organs is between those of the liver and the lung. The first
indication that these FMOs might differ from one another was
provided by Devereux et al. (34) who reported marked
differences in the effects of Hg^{2+} on partially purified FMO
preparations. Subsequently, differences in lung and liver
FMOs were suggested by the studies of Ohmiya and Mehendale
(35,36) on chlorpromazine and imipramine metabolism in the
lung of the rat and the rabbit. These and other compounds
are substrates for the liver, but not the lung, FMO.

More recently, our group (37) as well as Williams et al.
(38,39) purified an FMO from rabbit lung that was
catalytically and immunologically distinct from the liver
enzyme. The mouse and rabbit lung FMOs have a unique

Table 4. Comparison of FMO Purified from Mouse and Pig Liver

Characteristic	Mouse[a]	Pig[b]
Binding to Blue Agarose	+	+
Binding to Red Agarose	+	+
Binding to AMP Agarose	+	+
Binding to ADP Agarose	+	+
Molecular Weight	58,000[c]	58,000[c]
		64,000[b]
		56,000[c]
Spectral Maxima (oxidized)	383,453	382,450
pH Optimum	9.5	8.5
Km (M)(pH 7.6)		
Methimazole	10	5
NADPH	5	5
NADPH	86	>50

[a]Data from (25)
[b]Literature values
[c]By SDS-PAGE

ability for N-oxidation of primary aliphatic amines, including n-octylamine, a positive effector but not a substrate for the liver enzyme and also a compound often added to microsomal incubations to inhibit P450 (12,39). In the mouse lung, however, n-octylamine is both a substrate and a positive effector. The mouse and rabbit lung enzymes have a higher pH optimum, near 9.8, compared to that of the FMO from the liver which is approximately 8.8. Using antibodies raised in goats, Ouchterlony immunodiffusion analysis showed that the liver and lung proteins were immunochemically dissimilar (37). It has now become evident that there are several FMO isozymes with overlapping substrate specificities, and it is likely that the relative proportions of these isozymes vary in different tissues within and between species (29).

Although there have been no detailed studies of the FMOs from tissues other than liver and lung, apparent FMO activity has also been noted in kidney, bladder mucosa, testes, corpus luteum, thryoid, thymus, adrenal gland, placenta, aorta, lymph nodes, pancreas, small intestine and skin (1,10,25,40).

CELLULAR DISTRIBUTION

Recently, studies of the cellular localization of the FMO have been carried out (Overly, Lawton, Philpot, and Hodgson, unpublished results). An immunohistochemical method utilizing peroxidase labelled antibodies and diaminobenzidine revealed that in the rabbit lung the FMO is highly localized

Table 5. Comparative Rates of Sulfur Metabolism in
Selected Pesticide Substrates by the FMO of
Mouse Liver, Lung and Kidney Microsomes[a]

| | Activity (nmoles NADPH/min/mg) | | |
	Liver	Lung	Kidney
Phorate	6.47	9.98	5.44
Disulfoton	8.58	12.31	8.23
Fenthion	5.77	2.33	_b
Methyl Carbophenothion	3.76	2.86	-b
Croneton	3.31	2.00	_b
Aldicarb	1.65	1.13	1.79

[a] Data selected from (33)
[b] Not determined

in the non-ciliated bronchiolar epithelial (Clara) cells.
The lung FMO did not cross-react with the antibody to the
liver enzyme.

Similar immunohistochemical studies of FMO distribution
in the skin of mice and pigs (Levi, Inman, Venkatesh, Misra,
Monteiro-Riviere, and Hodgson,unpublished studies) reveal a
considerable concentration of this enzyme in epidermis,
sebaceous gland cells, and hair follicles. The presence of
FMO in skin has recently been confirmed by Western blot
analysis using an antibody to FMO purified from mouse liver
(Venkatesh, Levi and Hodgson, unpublished studies).

MOLECULAR BIOLOGY

The genes for the FMO from pig liver as well as those
from rabbit liver and lung have recently been cloned and
sequenced. These studies will be described by Philpot (this
monograph).

CONCLUSIONS

The pioneering work of Zeigler and associates
established that the FMO is a versatile nitrogen and sulfur
oxidase. The enzyme was first purified from pig liver, the
reaction mechanism was described, and the physiological role
of the enzyme investigated. Recently several laboratories,
including our own, have greatly extended the range of known
substrates as well as knowledge of the enzyme in several
species and organs. Purification of the FMO from other
species and organs has now established that these forms
differ in physical properties and substrate specificity.

There are a number of exciting prospects for the
immediate future. Investigation of the importance of the FMO
relative to other monooxygenases, particularly in complex
metabolic pathways for xenobiotic metabolism, has just begun.
Further studies particularly of the molecular biology of the
enzyme, are critical for an understanding of the role of this
enzyme in toxicological and pharmacological events.

REFERENCES

1. D. M. Zeigler, Microsomal flavin-containing monooxygenation of nucleophilic nitrogen and sulfur compounds, in: "Enzymatic Basis of Detoxication", W.B. Jakoby, ed., Academic Press, New York (1980).

2. N. P. Hajjar and E. Hodgson, "Flavin adenine dinucleotide-dependent monooxygenase as an activation enzyme, in: "Biological Reactive Intermediates - II, Part B", R. Snyder, D. V. Parke, J. J. Kocsis, D. J. Jollow, C. G. Gibson and C. M. Witmer, eds., Plenum Press, New York (1982).

3. B. P. Smyser and E. Hodgson, Metabolism of phosphorus-containing compounds by pig liver microsomal FAD-containing monooxygenase, Biochem. Pharacol. 34:1145-1150 (1985).

4. D. M. Zeigler, Flavin-containing monooxygenases: catalytic mechanism and substrate specificities, Drug Metab. Rev. 9:1-32 (1988).

5. D. M. Zeigler and L. L. Poulsen, Hepatic microsomal mixed-function amine oxidase, in: Methods in Enzymology", S. Fleisher and L. Packer, eds., New York (1978).

6   D. M. Zeigler and L. L. Poulsen, Protein disulfide bond synthesis: a possible intracellular mechanism, Trends Biochem. Sci. 2:79-82 (1977).

7. P. J. Sabourin, B. P. Smyser and E. Hodgson, Purification of the flavin-containing monooxygenase from mouse and pig liver microsomes, Int. J. Biochem. 16:713-720 (1984).

8. E. Hodgson and P. E. Levi, Species, organ and cellular variation in the flavin-containing monooxygenase, Drug Metabol. and Drug Interact. 6:219-233 (1989).

9. D. M. Zeigler, p. 297 in: "Microsomes and Drug Oxidations, J. Miners, D. J. Birkett, R. Drew and M. McManus, eds., Taylor and Francis, London (1988).

10. E. Hodgson, E. and P. E. Levi, The flavin-containing monooxygenase as a sulfur oxidase, in: "Metabolism of Xenobiotics", J.W. Gorrod, H. Oelschlager and J. Caldwell, eds., Taylor and Francis, London (1988).

11. D. E. Williams, S. E. Hale, A. S. Muerhoff and B. S. S. Masters, Rabbit lung flavin-containing monooxygenase. Purification, characterization, and induction during pregnancy, Mol. Pharmacol. 28:381-390 (1985).

12. R. E. Tynes, P. J. Sabourin, E. Hodgson and R. M. Philpot, Formation of hydrogen peroxide and n-hydroxylated amines catalyzed by pulmonary flavin-containing monooxygenases in the presence of primary alkylamines, Arch. Biochem. Biophys. 251:654-664 (1986).

13. P. Hlavica and M. Kehl, The role of cytochrome P-450 and mixed-function amine oxidase in the N-oxidation of N,N-dimethylaniline, Biochem. J. 164:487-496 (1977).

14. S. Hamill and D. Y. Cooper, The role of cytochrome P-450 in the dual pathways of N-demethylation of N,N-dimethylaniline by hepatic microsomes, Xenobiotica 14:139-149 (1984).

15. R. E. Tynes and E. Hodgson, Oxidation of thiobenzamide by the FAD-containing and cytochrome P-450-dependent monooxygenases of liver and lung microsomes, Biochem. Pharmacol. 32:3419-3428 (1983).

16. R. E. Tynes and E. Hodgson, The measurement of FAD-containing monooxygenase activity in microsomes containing cytochrome P-450, _Xenobiotica_ 14:515-520 (1984).

17. S. Kinsler, P. E. Levi and E. Hodgson, Hepatic and extrahepatic microsomal oxidation of phorate by the cytochrome P-450 and FAD-containing monooxygenase systems in the mouse, _Pestic. Biochem. Physiol._31:54-60 (1988).

18. S. Kinsler, P. E. Levi and E. Hodgson, Relative contributions of the cytochrome p-450 and flavin-containing monooxygenases to the microsomal oxidation of phorate following treatment of mice with phenobarbital, hydrocortisone, acetone, and piperonyl butoxide. In press.

19. P. E. Levi and E. Hodgson, Stereospecificity in the oxidation of phorate and phorate sulphoxide by purified FAD-containing monooxygenase and cytochrome P-450 isozymes, _Xenobiotica_ 18:29-39 (1988).

20. P. E. Levi and E. Hodgson, Metabolites resulting from oxidative and reductive processes, in: "Intermediary Xenobiotic Metabolism in Animals", D. J. Hutson and G. D. Paulson, eds., Taylor and Francis, London (1988).

21. P. W. Hale,Jr.and A. Poklis, Thioridazine-5-sulfoxide diastereoisomers in serum and urine from rat and man following chronic thioridazine administration, _J. Anal. Tox._ 9179-201 (1985).

22. C. C. Kilts, K. S. Patrick, G. R. Breese and R. B. Mailman, Simultaneous determination of thioridazine and its S-oxidized and N-demethylated metabolites using high performance liquid chromatography on radially compressed silica, _J. Chromatog._ 231:377-391 (1982).

23   C. D. Kilts, R. B. Mailman, E. Hodgson and G. R. Breese, Simultaneous determination of thioridazine and its sulfoxidized metabolites by HPLC use in clinical and preclinical metabolic studies, _Federation Proceedings_ 40:283 (1981).

24. P. W. Hale, Jr. and A. Poklis, Cardiotoxicity of thioridazine and two stereoisomeric forms of thioridazine-5-sulfoxide in the isolated perfused rat heart, Tox. Appl. Pharmacol. 86:44-55 (1986).

25. G. A. Dannan and F. P.Guengerich, Immunochemical comparison and quantitation of microsomal flavin-containing monooxygenases in varioius hog, mouse, rat, rabbit, dog and human tissues, _Mol. Pharmacol._ 22:787-794 (1982).

26. R. E. Tynes and E. Hodgson, Catalytic activity and substrate specificity of the flavin-containing monooxygenase in microsomal systems: characterization of the hepatic, pulmonary and renal enzymes of the mouse, rabbit and rat, _Arch. Biochem. Biophys._ 240:77-93 (1985).

27. P. J. Sabourin and E. Hodgson, Characterization of the purified microsomal FAD-containing monooxygenase from mouse and pig liver, _Chem. Biol. Interactions_ 51:125-139 (1984).

28. P. J. Sabourin, R. E. Tynes, B. P. Smyser and E. Hodgson, The FAD-containing monooxygenase of lung and liver tissue from rabbit, mouse and pig: species and tissue differences, in: "Biological Reactive Intermediates III", J. J. Kocsis, D. J. Jollow, C. M. Witmer, J. O. Nelson and R. Synder, eds., Plenum Press, New York (1986).

29. R. E. Tynes and R. M. Philpot, Tissue and species-dependent expression of multiple forms of mammalian microsomal flavin-containing monooxygenase. Mol. Pharmacol. 31:569-574 (1987).

30. M. E. McManus, I. Stupans, W. Burgess, J. A. Koenig, P. de la M Hall and D. J. Birkett, Flavin containing monooxygenase activity in human liver microsomes, Drug Metab. Disp. 15:256-261 (1987).

31. M. Agosin and G. T. Ankley, Conversion of N,N-dimethylaniline to N,N-dimethylaniline-N-oxide by a cytosolic flavin-containing enzyme from Trypanaosoma cruzi, Drug Metabol. Disp. 15:200-203 (1987).

32. B. P. Smyser, P. J. Sabourin and E. Hodgson, Oxidation of pesticides by purified microsomal FAD-containing monooxygenase from mouse and pig liver, Pestic. Biochem. Physiol. 24:368-374 (1985).

33. R. E. Tynes E. and Hodgson, Magnitude of involvement of the mammalian flavin-containing monooxygenase in the microsomal oxidation of pesticides, J. Agric. Food Chem. 33:471-479 (1985).

34. T. R. Devereux, R. M. Philpot and J. R. Fouts, The effects of Hg^{2+} on rabbit hepatic and pulmonary solubilized, partially purified N,N-dimethylaniline N-oxidases, Chem. Biol. Interact. 19:277-297 (1977).

35. Y. Ohmiya and H. M. Mehendale, Metabolism of chlorpromazine by pulmonary microsomal enzymes in the rat and rabbit, Biochem. Pharmacol. 31:157-162 (1982).

36. Y. Ohmiya and H. M. Mehendale, Species differences in pulmonary N-oxidation of chlorpromazine and imipramine, Pharmacology 28:289-295 (1984).

37. R. E. Tynes, P. J. Sabourin, and E. Hodgson, Identification of distinct hepatic and pulmonary forms of microsomal flavin-containing monooxygenase in the mouse and rabbit, Biochem. Biophys.Res. Commun. 126:1069-1075 (1985).

38. D. E. Williams, D. M. Ziegler, D. J. Nordin, S. E. Hale, and B. S. S. Masters, Rabbit lung flavin-containing monooxygenase is immunochemically and catalytically distinct from the liver enzyme, Biochem. Biophys. Res. Commun. 125:116-122 (1984).

39. L. L. Poulsen, K. Taylor, D. E. Williams, B. S. S. Masters, Substrate specificity of the rabbit lung flavin-containing monooxygenase for amines: oxidation products of primary alkylamines, Mol. Pharmacol. 30:680-685 (1986).

40. T. G. Osimitz and A. P. Kulkarni, Oxidative metabolism of xenobiotcs during pregnancy: significance of microsomal flavin-containing monooxygenase. Biochem. Biophys. Res. Commun. 4;1164-1171 (1982).

MULTIPLE FORMS OF RABBIT CYTOCHROME P450:   COMPARISON OF CHEMICAL

PHYSICAL, IMMUNOLOGICAL AND BIOCATALYTIC PROPERTIES

Eric F. Johnson, Thomas Kronbach, A. Scott Muerhoff, Keith J. Griffin, Usha R. Pendurthi* and Robert H. Tukey*

Scripps Clinic and Research Foundation
Department of Molecular and Experimental Medicine
La Jolla, California 92037, USA

*University of California San Diego
School of Medicine, Cancer Center T-012
La Jolla, California 92093, USA

INTRODUCTION

   The cytochrome P450 system of the rabbit is one of the most highly characterized of any species. The majority of the P450s characterized in the rabbit contribute to the metabolism of foreign compounds, where the multiple forms of P450 provide both functional diversity as well as redundancy of function. In addition, a variety of factors regulate the expression of the individual forms of P450. These can be ontogenetic factors leading to differential expression with age and tissue or environmental factors which induce specific P450s.

Nomenclature

     An early attempt to achieve a systematic nomenclature for the multiple forms of rabbit microsomal P450 was based on the relative electrophoretic mobilities exhibited by the purified proteins in the presence of sodium dodecyl sulfate (SDS PAGE). Many of the purified enzymes exhibit distinct electrophoretic mobilities when compared in this manner, and the procedure could be easily applied in any laboratory for the purpose of comparison.   Preparations were exchanged between laboratories, and a numerical designation was given to each enzyme. Eight distinct forms, 1, 2, 3a, 3b, 3c, 4, 5 and 6 were identified in this manner.

     The multiple forms of P450 exhibited a number of additional differences in their physical properties. These included distinct visible absorption spectra, amino acid compositions (1), N- and C- terminal amino acid sequences (2) and peptide maps (3). The latter clearly indicated that the different P450 proteins were likely to be encoded by distinct genes.

Molecular Aspects of Monooxygenases and Bioactivation of Toxic Compounds
Edited by E. Arınç *et al., Plenum Press, New York,*

TABLE 1

Nomenclature (4)

Electrophoretic	Sequence	Regulation In Liver
1(5)	IIC5(6)	Genetic Polymorphism (7,8)
2(9,10)	IIB4(11-14)	Induction Phenobarbital (15,16)
3a(17,18)	IIE1(19)	Induction Ethanol (20,21)
3b(22,23)	IIC3(24,25)	Genetic Polymorphism (26)
3c(27,28)	IIIA6(29,30)	Induction Rifampicin (31)
Pgw*(32,33)	IVA4(34)	Elevated During Pregnancy (35,36)
4(1,37-39)	IA2(40-42)	Induction TCDD (43)
5(44)	IVB1(45)	Induction Phenobarbital (46)
6(47)	IA1(40,41)	Induction TCDD (43)

---

*Not assigned an electrophoretic designation

Amino acid sequences are now available for 9 purified enzymes, Table 1. These sequences provide a basis for a systematic nomenclature based on likely evolutionary relationships between rabbit P450s and those of other mammalian species. P450s from different mammalian species that exhibit >65% amino acid sequence identity are thought to have arisen from the same gene in the last common ancestor of these species. This analysis has lead to the identification of 13 distinct classes of P450s that are generally conserved among mammals (4). These are designated by a Roman numeral and a letter. The P450s whose amino acid sequences exhibit greater than 40% identity share the same Roman numeral designation, whereas those that display greater than 65% identity share both Roman numeral and letter designations. P450s that are thought to have evolved from a single gene in the last common ancestor of mammals are designated by the same Roman numeral and letter. Because the number of P450s within these classes has increased since the radiation of mammals, each class often comprises several forms of P450 that are further designated by an Arabic numeral. The purified forms of P450 for which amino acid sequence data is available are listed in Table 1 and identified by each nomenclature. The corresponding gene is designated Cyp followed by an Arabic rather than a Roman numeral, a letter and an Arabic numeral for the same designation, i.e. the gene for P450IIIA6 is denoted Cyp3A6.

Orthologous enzymes in different mammalian species can share the same designation, eg. IVB1. It is often difficult, however, to identify orthologous enzymes by sequence comparisons because extensive duplication has occurred during the divergence of mammalian species. Thus, for many classes, each P450 is denoted by a unique designation. For instance, there are >5 class IIC P450 enzymes that have been isolated and characterized in the rat which are related to IIC3 and IIC5 of rabbit. Each exhibits distinct functional attributes and there is no indication that either rabbit enzyme is more closely related to one rat enzyme than to another. Thus, each is assigned a unique Arabic numeral that is not duplicated in another animal species.

In this review, we will discuss the enzymology of the rabbit class IIC, IIIA and IVA P450s. The discussion of each class will focus on specific aspects of the P450 system that are illustrated by work from our laboratories. We will discuss the induction of P450 enzymes and their activation by allosteric effectors with regard to class IIIA. The use of molecular biological approaches to the characterization of enzymes will be discussed using illustrations from class IIC and IVA. The occurrence and characterization of genetic polymorphisms affecting the expression of P450s in the rabbit will be discussed in relation to class IIC. In addition, we will discuss both the application of monoclonal antibodies to the characterization of P450 enzymes as well as the generation of new enzymes by the construction of chimeric proteins. Other chapters in this volume address the remaining classes of rabbit P450s. Dr. Philpot discusses the regulation and enzymology of classes IA, IIB, and IVB, whereas Dr. Bonfil discusses the regulation of class IIE.

CLASS IVA

The class IVA P450 enzymes are generally associated with the oxidation of primary rather than of secondary or tertiary carbon-hydrogen bonds. This reaction pathway, ω-hydroxylation, provides a catabolic pathway for fatty acids and for prostaglandins, and the products are further oxidized to dicarboxylic acids (48). In addition the ω-hydroxylation of arachidonic acid may generate vasoactive compounds that serve a physiologic regulatory function (49).

P450IVA4: Elevation during Pregnancy

A prostaglandin ω-hydroxylase is elevated in rabbit lung and liver during pregnancy (50). The P450 catalyzing this reaction was isolated from the lungs of pregnant rabbits or rabbits treated with progesterone and termed P450PGω (33) or P-2 (32), respectively. The purified protein catalyzes the ω-hydroxylation of prostaglandins E_1, E_2, A_1, A_2 and $F_{2\alpha}$ when reconstituted with reductase. The amino acid sequence predicted from a cDNA corresponding to this P450 (34) indicates that it is a member of the IVA gene family, and the enzyme is designated as P450IVA4 in a uniform system of P450 nomenclature (4). The abundance of mRNAs corresponding to this P450 increases during pregnancy in rabbit lung (34,35) or following treatment of male rabbits with progesterone (34). In contrast, the ω-hydroxylation of prostaglandins E_1, E_2, A_1, A_2 and $F_{2\alpha}$ is not elevated in rat lung or liver during pregnancy (50).

Class IVA Heterogeneity

As discussed by Dr. Gibson in another chapter of this volume, the prototypic activity for the rat liver class IVA enzymes is the ω-hydroxylation of lauric acid. Microsomes isolated from rabbit liver and kidney also catalyze this reaction (51-54), but preparations of the lung enzyme, P450IVA4, exhibit negligible lauric acid ω-hydroxylase activity (32,33). This suggested that class IVA P450s, distinct from P450IVA4, are also expressed in the rabbit. This prompted us to identify additional class IVA P450s by the characterization of cDNA clones. These were identified by screening a rabbit kidney cDNA library with a partial cDNA isolated in our laboratory for the rabbit lung P450IVA4 and subsequent sequence determination (55).

Six cDNAs encoding cytochrome P450 enzymes were cloned from a rabbit kidney cDNA library. Three of these cDNAs have been completely sequenced. The three clones exhibit >90% nucleotide sequence identity when the coding regions are aligned, but they diverge in their 3' untranslated regions suggesting that they are the products of distinct genes. The predicted amino acid sequences derived from each cDNA exhibit about 85% identity (55). This degree of sequence identity is also seen when each is compared

25

with the amino acid sequence of P450IVA4, the enzyme that catalyzes the ω-hydroxylation of prostaglandins and that is elevated during pregnancy and induced by progesterone in rabbit lung. The P450s encoded by the three cDNAs are designated as IVA5, IVA6 and IVA7 in the uniform system of nomenclature (4). The amino acid sequences of the 4 rabbit class IVA P450s are displayed in Fig. 1. The N-termini of these proteins appear to be processed because the N-terminus of purified P450IVA4 (34) corresponds to the fifth codon following the initiation methionine.

The capacity of the P450 encoded by each cDNA to metabolize lauric acid or prostaglandin A_1 was assessed following the expression of each in COS-1 cells by transfection with an expression vector harboring the cDNA. The expression of the protein was monitored using antibodies to P450IVA4 (35) which recognized a microsomal protein expressed in COS-1 cells transfected with each construct. Microsomes isolated from the transfected cells were assayed for their capacity to catalyze the ω-hydroxylation of lauric acid (55). These results indicate that IVA5, IVA6 and IVA7 exhibit similar laurate ω-hydroxylase activities (55), whereas IVA4 does not display laurate ω-hydroxylase activity when expressed in this manner. Only one of the three kidney enzymes, IVA7 ω-hydroxylates prostaglandin A_1 at a measurable rate. The specific activity was 100-fold lower, however, than that for lauric acid (55). Preparations of P450 isolated from kidney microsomes are reported to catalyze the ω-hydroxylation of prostaglandin A_1 (53). Thus, it is likely that additional class IVA P450s are expressed in the rabbit. It is possible that one or more of the three additional cDNAs that are currently being characterized in our laboratory will encode prostaglandin ω-hydroxylases.

Heterogeneity is Seen for Other Classes

The study described above indicates how the cloning of cDNAs for a purified P450 can lead to the identification of additional forms of P450. Additional cDNAs that encode P450s have been identified for 5 of the 9 purified forms of P450 bringing the total number of cDNAs with unique sequences to 20. The greatest complexity is seen for classes IVA and IIC. Class IIC will be discussed later. The microheterogeneity of classes IIB and IIE will be discussed in other chapters of this volume. This microheterogeneity is also seen in other species, but different numbers of enzymes are seen in each class (4).

It is likely that the genetic diversity seen for the P450 enzymes reflects the duplication and subsequent divergence of genes. Initially, this will produce multiple copies of the gene and can lead to an increase in the expression of the enzyme. In a sense, the multiple genes encoding class IVA P450s that catalyze the ω-hydroxylation of lauric acid provide this increase in capacity and a redundancy of function. Over time each gene will accumulate unique mutations, and this can lead to divergence in function and regulation. The differential regulation of P450IVA4 during pregnancy and its capacity to metabolize a distinct group of substrates from other class IVA enzymes are a likely manifestation of this process.

In many cases, the close sequence identity of P450s within a class is suggestive that gene duplication has occurred after the radiation of mammals. In addition, a given class may comprise different numbers of constituents in different species. This multiplicity and divergence may explain the many differences that occur between species in pathways of metabolism mediated by P450 enzymes.

```
              10                30                50
IVA5   MSVSALSPTRLPGSLSGLLQVAALLGLLLLLLLKAAQLYLRRQWLLRALQQFPCPPFHWLL
IVA6         N                G                H
IVA7        S       F   F    A                                    S
IVA4                       >                    H

              70                90               110
IVA5   GHSREFQMNQELQQILKWVEKFPRACPHWIGGNKVRVQLYDPDYMKVILGRSDPKSRGSY
IVA6      NGH     VM        S   R LW  SRAHLLI                   AQ
IVA7      PIDS    V  R      S   R LW  SELFLIC         T        A V
IVA4      ND    ER Q        G   W LS  A LLV         L          APRN

             130               150               170
IVA5   TFVAPWIGYGLLLLNGQPWFQHRRMLTPAFHYDILKPYVGLMVDSVQIMLDKWEQLVSQD
IVA6   R L              T                      A
IVA7   S L              E T                        V     L K ARK
IVA4   KLMT             D T                           R     I

             190               210               230
IVA5   SSLEVFQDISLMTLDTIMKCAFSYQGSVQLDSRNSQSYIQAVGDLNNLVFARVRNIFHQS
IVA6                      H              .              F S       V
IVA7   AP  IYEHV     E       H       E  T K      RE SD ALQ       V
IVA4        I  HV                        H       IN       Y A    V

             250               270               290
IVA5   DTIYRLSPEGRLSHRACQLAHEHTDRVIQQRKAQLQQEGELEKVRRKRRLDFLDVLLFAK
IVA6
IVA7   FL
IVA4   FL            F

             310               330               350
IVA5   MENGSSLSDQDLRAEVDTFMFEGHDTTASGVSWIFYALATHPEHQHRCREEIQGLLGDGA
IVA6                             I
IVA7                             I
IVA4

             370               390               410
IVA5   SITWEHLDQMPYTTMCIKEAMRLYPPVPAISRDLSSPVTFPDGRSLPKGFTVTLSIYGLH
IVA6                       L         GVG Q               VI      A
IVA7        K             L         GVGSK               III
IVA4                      L         SVT Q  K            VILF

             430               450               470
IVA5   HNPNVWPNPEVFDPSRFTPGSARHSHAFLPFSGGARNCIGKQFAMNELKVAVALTLVRFE
IVA6       K              A              P
IVA7       K              A              S
IVA4   Y   K Q          F A D   Y                        R

             490               510
IVA5   LLPDPTRIPKPTARLVLKSNNGIHLRLRKLQ
IVA6      K V DQKP      S          R
IVA7        V I IT      K          H
IVA4         I I V      K          H
```

Fig. 1.   Amino acid sequences of rabbit P450s IVA5, IVA6, IVA7 and IVA4. The sequence of IVA5 (55) is shown in its entirety, whereas only differences are shown where they occur for IVA6 (55), IVA7 (55), and IVA4 (34). The beginning of the sequence for IVA4 predicted from the cDNA is indicated by (>).

REDUNDANCY

Many reactions are catalyzed by more than one P450 enzyme. In the example discussed above, several closely related forms of P450 each catalyze the same reaction, lauric acid ω-hydroxylation. This redundancy of function is also seen between more distantly related forms of P450. Both P450s IIIA6 and IIC3 can contribute substantially to the 6β-hydroxylation of progesterone depending on their relative levels of expression. Induction by rifampicin of P450IIIA6 and/or activation of this enzyme by α-naphthoflavone can increase both the rate of 6β-hydroxylation and the relative contribution of P450IIIA6 to the overall rate. There is no evidence that P450IIC3 (P450 3b) is inducible. P450IIC3 is not fully expressed until 4-weeks of age and is genetically deficient in some strains of rabbits such as B/J. It should be noted that P450s IIIA6 and IIC3 differ greatly in the spectrum of substrates they metabolize. Thus, the redundancy of function between these two specific P450s is limited to a subset of the substrates each recognizes.

Another example of overlapping function is seen for the metabolism of benzo(a)pyrene. P450s IA1, IIIA6 and IIC5 each metabolize benzo(a)pyrene (56), and it is interesting to note that each is modulated by α-naphthoflavone. Both P450s IA1 and IIC5 are inhibited by α-naphthoflavone. This compound is a tightly bound, almost stoichiometric inhibitor of IA1, whereas it inhibits IIC5 at μM concentrations (57). As we will discuss later, α-naphthoflavone has the opposite affect on P450 IIIA6, stimulating the reaction it catalyzes (56). Thus, the overall effect of α-naphthoflavone on microsomal benzo(a)pyrene hydroxylation will reflect the relative expression of each enzyme which is dependent on induction, age, and genetic polymorphism.

REACTION PHENOTYPING

The characterization of the extent of participation by a specific form of P450 in a given reaction is called "reaction phenotyping." There are three basic elements of this process. One aspect is to demonstrate kinetic competence, ie. that the kinetic properties of the enzyme are consistent with those of the microsomal reaction. Reconstitution of the purified protein or expression of the enzyme from the cDNA can provide evidence that a given form of P450 catalyzes the reaction. Additional information such as Km can indicate whether the enzyme could catalyze the reaction observed with microsomes.

The second aspect is to correlate the expression of the enzymic reaction with the expression of the protein. If the protein is induced, is the enzyme activity induced? What is the correlation between enzyme expression and protein expression? These comparisons must also consider the contributions of other forms of P450 to the same reaction pathway.

The third approach is to employ selective inhibitors of P450 enzymes. The extent of inhibition seen for the microsomal reaction is used to estimate the relative participation of the P450 in the microsomal reaction.

The most widely used selective inhibitors of P450s are antibodies. Antibodies can provide greater selectivity toward different P450s than chemical inhibitors, and they inhibit extensively for a wide range of substrates and substrate-concentrations. Moreover, these antibodies can also be used to estimate the amount of P450 protein for comparison with enzyme activity.

Although antibodies can be highly discriminatory among P450s, polyclonal antibodies developed toward a member of a subclass will often react with other members of that subclass. For example, polyclonal antibodies to IA1 usually react to some extent with IA2. One means of improving the selectivity of the antibody is adsorption. In this procedure, the antibody is incubated with the cross-reactive protein to remove them from the serum. For instance, an anti-IA1 antibody can be incubated with P450IA2 immobilized by covalent attachment to an agarose matrix. This removes component antibodies that recognize both proteins. This approach has been used to generate specific antibodies to rat IA1 and IA2 (58).

Another approach is to clone the lymphocyte making selective, inhibitory antibodies by generating a hybridoma. The development of monoclonal antibodies is more labor intensive, but may be required when the cross-reactive proteins are not available in sufficient quantities for adsorption. We will discuss an application of this approach in the next section.

CLASS IIC P450s

P450 1 (P450IIC5)

Large variations occur among untreated rabbits in the liver microsomal 21-hydroxylation of progesterone. This is a major route for the catabolism of progesterone in rabbit liver. As a result, the mineralocorticoid deoxycorticosterone is produced in the liver and subsequently oxidized to acidic products that appear as major urinary metabolites of progesterone (59). The large variation in activity as well as the unusual regiospecificity of the 21-hydroxylase when compared to other hepatic P450 mediated steroid hydroxylations suggested that this activity might be catalyzed largely by a single P450 enzyme (7). Using classical techniques of protein separation we were able to isolate from microsomes exhibiting high 21-hydroxylase activity an electrophoretically homogeneous preparation of P450 that catalyzed 21-hydroxylation at high rates when compared to other purified forms of P450 and hepatic microsomes. The enzyme was designated as P450 1 based on a comparison of its relative electrophoretic mobility to that of other purified rabbit P450s (5), and it is denoted as P450IIC5 based on sequence comparisons (4).

These results suggested that differences in the microsomal concentration of P450 1 might underlie the variations in 21-hydroxylase activity. In order to test this hypothesis we developed a panel of monoclonal antibodies to P450 1.

P450 1 is part of complex and diverse family of homologous forms of P450

When we initiated the development of monoclonal antibodies to P450 1, the structural relationship of P450 1 to other purified forms of rabbit P450 was not available. We know now that P450 1 (P450IIC5) exhibits >65% amino acid identity with 6 other rabbit P450s, IIC1, IIC2, IIC3, IIC4, IIC14 and IIC15 (4). Only one, IIC3 or P450 3b had been purified (22). These enzymes are classified as a subfamily, IIC. The polymorphic S-mephenytoin hydroxylase of man (60,61) is also a class IIC P450 as are the principal human tolbutamide and hexobarbital hydroxylases (62,63). There have been more than 15 class IIC P450s described in man, rat and rabbit (4), and class IIC is the largest of the 11 classes, IA, IIA, IIB, IIC, IID, IIE, IIF, IIG, IIIA, IVA and IVB that include more than 70 mammalian P450 enzymes that metabolize xenobiotic substrates (4).

Characterization of Antibody Selectivity

The monoclonal antibodies developed to P450IIC5 were tested for their specificity by determining the reactivity toward other purified forms of P450. Two of the antibodies, 1G11 and 3C3, recognize P450 3b (P450IIC3) in addition to P450IIC5, but they differ because only 3C3 recognizes an antigen expressed in the kidney. The 2F5 antibody also recognizes an antigen in the kidney, but unlike the 3C3 antibody it does not react with P450 3b. This complex pattern of reactivity suggested that the four antibodies recognize distinct epitopes on P450IIC5 and that some epitopes were shared with other proteins which differed in their pattern of expression. However, only one of these proteins, IIC3, was identified among the purified forms of rabbit P450. A two-site immunoradiometric assay confirmed that each antibody recognizes a distinct epitope of IIC5 (64).

The reactivity of some of the antibodies toward additional proteins was also assessed by using the antibodies to isolate these enzymes from solubilized preparations of microsomes followed by electrophoretic analysis of the purified proteins using SDS PAGE, Fig. 2. As expected all of the antibodies sequestered P450IIC5 from liver microsomes and in addition the 1G11 and 3C3 antibodies sequestered P450IIC3. The two antibodies, 3C3 and 2F5, that react with an antigen expressed in kidney microsomes, sequestered a single electrophoretic species from solubilized kidney microsomes with a mobility intermediate between that of P450 3b and

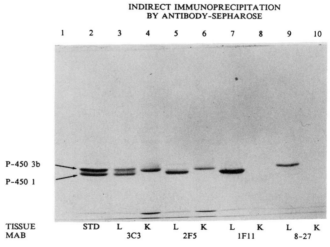

Fig. 2. Indirect immunoadsorbent precipitation of P450 isozymes using monoclonal antibodies linked to Sepharose. P450 isozymes were separated by polyacrylamide gel electrophoresis (7.5%) in the presence of SDS. Immunoprecipitations were conducted as described (64). Eluents were compared to purified P450 IIC5 and IIIC3 (5). The contents of lanes 1-10 are as follows: 1, tracking dye; 2, P450 IIC5 and IIC3; 3, 3C3 liver eluent; 4, 3C3 kidney eluent; 5, 2F5 liver eluent; 6, 2F5 kidney eluent; 7, 1F11 liver eluent; 8, 1F11 kidney eluent; 9, 8-27 liver eluent; 10, 8-27 kidney eluent. The 8-27 monoclonal antibody selectively immunoprecipitates P450 3b (65). Taken from Finlayson et al. (66) with permission.

P450 1 (66).   The two antibodies also sequestered a protein with a
similar mobility from liver microsomes (64). We designated this
cytochrome as P450 K.

Identification of P450 K as P450IIC2

        Using immunocytochemical techniques, the protein recognized by the
2F5 and 3C3 antibodies in kidney, P450 K (64), was localized to the S_2 and
S_1 segments of the proximal tubule in the renal cortex (66). In addition,
the treatment of rabbits with PB, but not TCDD, was found to induce the
level of P450 K in the kidney as judged by Western blotting, Fig. 3. The
2F5 antibody inhibited 50% of the $(\omega-1)$-hydroxylation of lauric acid in
kidney microsomes. This indicates that P450 K participates in the lauric
acid 11-hydroxylase activity (66).

2F5 IMMUNOBLOT

PB vs Untreated

Kidney Microsomes

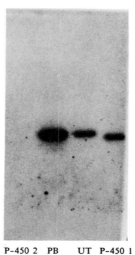

P-450 2   PB   UT P-450 1

[3pmol] [10μg] [10μg] [3pmol]

Fig. 3. Immunoblot of kidney microsomes isolated from untreated and
        phenobarbital-treated rabbits as well as purified P450s IIC5 and
        IIB4.   Following electrophoretic transfer, the nitrocellulose
        sheet was incubated with 1% BSA for 30 min at 37°C to reduce
        nonspecific binding.   The blot was then allowed to react with 3 μg
        of ^{125}I-2F5 monoclonal antibody (5.415 X 10_6 cpm/μg) for 2 h at
        room temperature with constant agitation.   Unbound antibody was
        then removed with several washes and the blot was visualized by
        autoradiography.   Track: 1, purified P450IIB4 (3 pmol); 2, kidney
        microsomes from a rabbit treated with phenobarbital (10 μg); 3,
        kidney microsomes from an untreated rabbit (10 μg); 4, purified
        P450IIC5 (3 pmol).   Taken from Finlayson et al. (66) with
        permission.

A cDNA, pP450PBC2 had been characterized by Leighton et al. (25), that codes for a P450, IIC2, that had not been isolated and characterized. Its sequence similarity to P450IIC5 as well as the observation that the cDNA hybridizes to phenobarbital-inducible mRNA in kidney (67) prompted us to examine whether this cDNA encodes P450 K. For this purpose, the P450 encoded by pP450PBC2 was synthesized using an in vitro transcription/translation system. This in vitro translation product was indirectly immunoprecipitated by the 2F5 monoclonal antibody and SDS-polyacrylamide gel electrophoresis revealed that the translated protein product exhibits the same relative electrophoretic mobility as P450 K (68). The N-terminal amino acid sequence was determined for P450 K immunopurified from rabbit kidney using the monoclonal antibody 2F5 (68), and it was found to agree with the predicted amino acid sequence derived from the Cyp2C2 gene (69). These results indicate that P450 K is encoded by the Cyp2C2 gene (68). Imai (70) subsequently expressed this enzyme from the cDNA in yeast and confirmed its capacity to catalyze the (ω-1)-hydroxylation of lauric acid. We have obtained similar results following the expression of the protein from the cDNA in COS-1 cells.

Southern blots of genomic DNA (6) and the cloning of the gene for P450IIC5 and other closely related genes suggest that additional P450s that are more closely related to P450IIC5 than IIC3 and IIC2 are expressed in liver. These results do not indicate how many of these gene-like sequences are expressed in liver, but the potential clearly exists for the occurrence in rabbit liver of two or more forms of P450 closely related to P450IIC5 that may exhibit the same electrophoretic mobility as P450IIC5.

One of these is P450IIC4. The predicted amino acid sequence of P450IIC4 differs at only 24 of 487 amino acids from that of P450IIC5 (71). We therefore synthesized P450IIC4 in vitro using rabbit reticulocyte lysate primed with RNA transcribed from the coding sequence of the cDNA using a bacteriophage T7 RNA polymerase/promoter system and examined its reactivity with the monoclonal antibodies, Fig. 4. The nascent P450IIC4 reacts with two monoclonal antibodies that recognize P450IIC5 and exhibits the same relative electrophoretic mobility as P450IIC5. In contrast, a third monoclonal antibody recognizing P450IIC5, 1F11, did not recognize the P450IIC4 synthesized in vitro (71). We have subsequently expressed IIC4 in COS cells and have shown that the 1F11 antibody does not recognize the functional protein.

These results indicate that the 1F11 antibody shows a high degree of specificity toward IIC5 and that it can discriminate IIC5 from closely related proteins such as IIC4. In contrast, the other monoclonal antibodies display distinct patterns of cross-reactivity toward other class IIC proteins. Given the genetic complexity of this family, it becomes increasingly difficult to characterize the specificity of these antibodies. The 1F11 antibody may recognize additional proteins. However, as we will demonstrate, the 1F11 antibody has proven useful for the characterization of the expression of P450IIC5. In addition, this antibody extensively inhibits liver microsomal progesterone 21-hydroxylase activity, >95% (64), indicating that P450IIC5 accounts for the high hepatic 21-hydroxylase activity exhibited by some rabbits.

GENETIC POLYMORPHISM

Two genetic polymorphisms have been described which affect the expression of P450 enzymes in the rabbit. One governs the expression of P450IIC5 and the other affects a form of P450IIC3.

P450IIC5

As we discussed earlier, P450IIC5 catalyzes the 21-hydroxylation of progesterone, and it accounts for most of the activity exhibited by liver microsomes. The distribution of liver microsomal 21-hydroxylase activities among rabbits is bimodal with the two groups, 21H and 21L, exhibiting a 10-fold difference in mean progesterone 21-hydroxylase activities (7). Serial liver biopsies of outbred rabbits indicate that the 21H and 21L phenotypes are traits of individual animals rather than a transient elevation of 21-hydroxylase activity (72). In addition, the inbred strain of rabbits, III/J, exhibits only the 21L phenotype. The offspring of matings between this inbred strain and rabbits identified as 21H by liver biopsy were analyzed in order to characterize the inheritance of the 21H and 21L phenotypes (8).

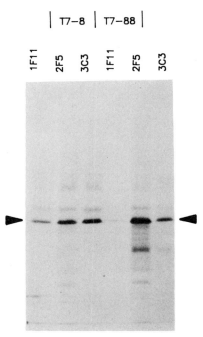

Fig. 4. Synthesis of the protein products of the cDNAs encoding IIC4 (p1-88) and IIC5 (p1-8). The coding and 3' untranslated portions of p1-8 and p1-88 were inserted behind the T7 promoter in the plasmid pT7-1. Capped mRNAs were synthesized from pT7-8 and pT7-88 using T7 RNA polymerase following the linearization of the plasmid by digestion with EcoRV thereby terminating the transcription unit in the 3' untranslated portion of the insert. The RNAs were then translated in a rabbit reticulocyte lysate system, and the protein products were immunopurified using one of three monoclonal antibodies directed to P450IIC5, 1F11, 2F5, or 3C3, that were linked to Sepharose. The immunopurified in vitro translation products were eluted from the antibody-Sepharose and analyzed by SDS-polyacrylamide (7.5%) gel electrophoresis and autoradiography. The rabbit reticulocyte lysate was supplemented with [^{35}S]methionine. The arrowheads indicate the relative mobility of purified P450IIC5. Taken from Johnson et al. (71) with permission.

Both the 21H and 21L phenotypes were evident among male and female offspring in roughly equal numbers. Backcrosses between 21L progeny and III/J rabbits exhibit only the 21L phenotype, whereas 21H offspring yield both 21H and 21L progeny when backcrossed to the 21L inbred strain III/J. These results are consistent with autosomal dominant inheritance of the 21H phenotype (8).

Analysis of Southern blots of genomic DNA digested with the restriction endonuclease Kpn I reveals 20, 13 and 9 kb fragments that hybridize with a probe derived from the 3' untranslated region of the P450IIC5 cDNA, Fig. 5. The 13 kb band is not observed for strain III/J or 21L progeny of strain III/J crossed with 21H rabbits, but it is detected for both 21H fathers and 21H progeny indicating that the genetically determined difference of 21-hydroxylase expression is linked to allelic forms of P450IIC5 (8).

Heritable Differences in the Expression P450IIC5

The difference in 21-hydroxylase activity between 21H and 21L rabbits reflects differences in the microsomal concentration of P450IIC5. The 1F11 antibody was used in a two-site immunoradiometric assay to quantify the microsomal concentration of P450IIC5. The microsomal concentration of P450IIC5 was found to vary from <0.02 to >0.4 nmol of P450IIC5 per mg of microsomal protein. In addition, the results of this assay indicate that the microsomal concentration of P450IIC5 is highly correlated with the microsomal 21-hydroxylase activity, Fig. 6. Scatchard analysis of the binding of the 1F11 antibody to microsomal samples revealed that the differences reflect a variation in the maximal extent of binding rather than in the dissociation constant governing the binding reaction. Thus, the variations in 21-hydroxylase activity reflect

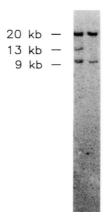

20 kb —
13 kb —
9 kb —

Phenotype	13 kb Frg	
	+	−
21H	13	0
21L	0	11

Fig. 5. Restriction fragment length polymorphism. The autoradiographic image of a Southern blot of rabbit genomic DNA is shown. Each sample was digested with the enzyme KpnI. Following electrophoresis and transfer to nitrocellulose, the filter-bound DNA was probed with a ^{32}P-labeled portion of the P450IIC5 cDNA corresponding to the 3'-untranslated region. Two patterns which differ in regard to the presence or absence of the 13-kb band are seen. The frequency of occurrence of each pattern for 21H and 21L offspring of 21H NZW matings with III/J rabbits is shown next to the autoradiographic image. Taken from Johnson et al. (8) with permission.

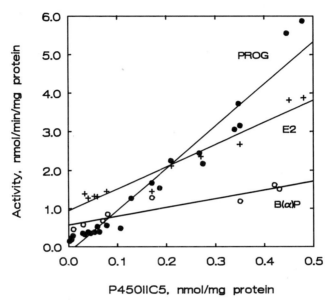

Fig. 6. Relation between the expression of P450 IIC5 and the activity of
liver microsomes for the 2-hydroxylation of 17β-estradiol (+)
(57), benzo(a)pyrene hydroxylation (o) (56), and progesterone 21-
hydroxylation (•) (73). Each point represents the value for a
preparation of microsomes from a single animal. The samples were
chosen to provide a uniform distribution of microsomal
concentrations of P450IIC5, and the distribution shown does not
reflect the natural, bimodal distribution of P450IIC5
concentrations. Linear regression lines are shown. The
y-intercepts reflect the contributions of other P450s to each
activity.

differences in the microsomal concentration of P450IIC5. Moreover, the *in situ* turnover number for the microsomal enzyme is very similar to that obtained for the reconstituted, purified enzyme (73).

Heritable Differences in IIC5 Concentration Affect the Metabolism of Other Substrates

The differences among outbred rabbits in the microsomal concentration of P450IIC5 also affect the metabolism of both estradiol and benzo(a)pyrene, Fig. 6. The purified cytochrome catalyzes both reactions with a relatively high turnover number when reconstituted with reductase (56,57). Microsomes from untreated rabbits containing low concentrations of P450IIC5 generally exhibit an estradiol 2-hydroxylase activity characterized by a high Km, >50 μM for estradiol, whereas those containing high concentrations of this enzyme exhibit a Km of ca. 1 μM. In addition, the high-affinity estradiol 2-hydroxylase is inhibited by the 1F11 antibody (57).

The rate of benzo(a)pyrene metabolism exhibited by microsomes prepared from different, untreated rabbits varies from 0.4 to 1.2 nmol/min/mg microsomal protein. Linear regression analysis indicates that this variation is correlated to differences in the microsomal concentration of P450 1. The *in situ* turnover number of 2.3 nmol/min/nmol P450 1 determined from this analysis agrees well with that determined for the reconstituted enzyme. Moreover, the 1F11 antibody inhibits this activity to a constant basal level (56) that corresponds to the intercept seen in Fig. 6.

The Abundance of IIC5 mRNAs is Genetically Determined

The gene encoding IIC5 has been cloned from 21L III/J rabbits. The sequences of the exons are identical to that of the IIC5 cDNA cloned from a 21H rabbit. The difference in protein concentration appears to reflect a difference in the abundance of mRNAs encoding IIC5 in the 21H and 21L animals. Because of the close nucleotide similarity between IIC4 and IIC5, the abundance of specific transcripts was determined using S1 nuclease protection assays. These results indicate that the abundance of mRNAs encoding IIC5 is much lower in 21L when compared to 21H rabbits, whereas mRNAs corresponding IIC4 are similar in abundance for the two phenotypes (71). The variation in abundance of IIC5 mRNAs is also evident for Northern blots probed with an oligonucleotide probe selected to discriminate IIC5 from other class IIC mRNAs. Primer extension analysis using this oligonucleotide indicates that transcription of the Cyp2C5 gene is initiated at the same site in both 21H and 21L rabbits although the abundance of the transcripts differs. In addition, sequence analysis of IIC5 cDNA generated from the low abundance mRNAs isolated from 21L III/J rabbits and amplified by polymerase chain reaction indicates that the transcript is properly spliced to encode IIC5. Thus, the allelic difference between 21L and 21H Cyp2C5 genes remains uncertain, but it could lead to a difference in gene transcription. This is currently under investigation.

Interstrain Differences in P450IIC3

The characterization of a genetic polymorphism affecting P450 3b (IIC3) (26) was one of the first examples of a genetic polymorphism affecting the expression of a P450 enzyme that was characterized at the molecular level. The particularly interesting aspect of this genetic difference is that it leads to a selective loss of only one of several reactions catalyzed by P450 3b, the 6β-hydroxylation of progesterone. The capacity of this enzyme to effect biphenyl hydroxylation and aminopyrine N-dimethylation was unaffected. As a result inbred strains of rabbits which express this alternate form of P450 3b exhibit much lower levels of liver microsomal 6β-hydroxylase activity (74). Preparations of P450 3b from animals exhibiting the two phenotypes differ by the presence or absence of a single tryptic peptide suggesting that they differ very little in structure. Kinetic studies suggest that animals, including inbred strains expressing the 6β-hydroxylase form of P450 3b, express both forms of the enzyme indicating that the P450 3b catalyzing this reaction is actually one of two or more closely related forms of P450 and that the genetic polymorphism affects the expression of one of these forms (26,74). The second form of P450 3b which catalyzes the 16α-hydroxylation of progesterone is expressed in all of these strains (74). It exhibits allosteric activation by a variety of steroids of which 5β-pregnane-3β,20α-diol has the greatest effect increasing the efficiency of the enzyme by 4-fold (75).

P450s IIC4 and IIC5 are 21-Hydroxylases with Different Kinetic Properties

As we indicated earlier P450s IIC4 and IIC5 differ by only 24 of 487 amino acids. Yet the capacity of the 1F11 monoclonal antibody to extensively inhibit progesterone 21-hydroxylase activity as well as the dramatic affect of genetic differences on the 21-hydroxylase activity of liver microsomes suggest that only IIC5 participates extensively in the observed activity. The catalytic properties of the two enzymes have been characterized following their expression by transfection of COS cells with expression vectors harboring the coding regions of each cDNA (76). Both P450s were shown to catalyze this reaction, however P450IIC5 expressed in this manner exhibits a more than 10-fold lower apparent Km for progesterone 21-hydroxylation than P450IIC4, Fig. 7. Thus, P450IIC5 differs from P450IIC4 in its immunoreactivity with the 1F11 antibody and its catalytic properties although their amino acid sequences are highly similar (≈95%).

Differences between P450s exhibiting extensive amino acid sequence identity is seen for P450s in other species as well. Two mouse class IIA proteins exhibit only 11 differences of amino acid sequence yet one catalyzes the 7-hydroxylation of coumarin and the other catalyzes the 15β-hydroxylation of testosterone (77) but not visa versa.

HYBRID ENZYMES

Although the three dimensional structure of a soluble bacterial P450 (P450cam) has been solved (78), little is known about the structures of the membrane-bound mammalian P450s. Thus, the structural features of these enzymes that determine their multisubstrate specificities are unknown. The generation of hybrid enzymes by the site-directed mutagenesis or the splicing together of portions of cDNAs encoding different P450s provides a means of identifying whether specific segments of the amino acid sequence determine the substrate selectivity of P450s. We have used this approach to identify which segments of the amino acid sequence contain differences leading to the 10-fold difference in apparent Km between IIC5 and IIC4.

Chimeric cDNAs were constructed and expressed in COS-1 cells which code for hybrids between these enzymes. The hybrid enzymes were assayed for catalytic activity and compared to the parental proteins by assaying them at low concentrations of substrate so that we could discriminate between low and high Km forms. This approach is outlined in Fig. 8. It was found that when a segment of IIC5 was substituted into IIC4, it conferred the higher affinity (lower Km) to P450IIC4. Sequential reduction of the length of the exchanged segments led to a hybrid enzyme with a Km similar

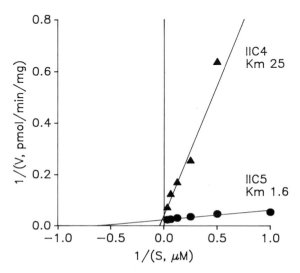

Fig. 7. Double-reciprocal plots of the substrate-dependent formation of 21-hydroxyprogesterone by microsomes from COS-1 cells transfected with pCMVIIC5 (•; Km = 1.6 μM, Vmax = 42 pmol/min^{-1}/mg^{-1}) or pCMVIIC4 (▲ Km = 25 μM, Vmax = 25 pmol/min^{-1}/mg^{-1}). Lines show the estimates obtained by nonlinear minimization of residuals from untransformed values assuming Michaelis-Menten kinetics. Taken from Kronbach et al. (76) with permission.

to that of IIC5. The protein is derived largely from P450IIC4 and contains 3 amino acid residues derived from P450IIC5 clustered between positions 113 to 118. This suggests that this region is part of a substrate binding domain (76). It is also interesting to note that this region is highly variable among class IIC P450s, and this variation may therefore contribute to the distinct substrate selectivities of the different class IIC P450s. Moreover, it is a region which shows variation within other classes, such as the IVA enzymes described earlier.

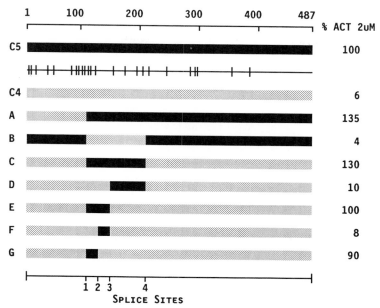

Fig. 8. Structure and relative progesterone 21-hydroxylase activities of
P450IIC5, P450IIC4, and the hybrid proteins expressed in COS-1
cells.   Progesterone 21-hydroxylase activity of hybrids between
P450IIC4 and P450IIC5 was determined for cultured cells incubated
with 2 μM progesterone in order to distinguish high and low Km
enzymes.   The left column identifies each hybrid.   Hybrids are
represented by solid and stippled segments corresponding to each
of the parent proteins P450IIC5 (C5) or P450IIC4 (C4),
respectively.   Numbers at the bottom refer to the following splice
sites used for the constructions: 1, StyI restriction site in
P450IIC4, a silent mutation introduced this site into P450IIC5; 2,
BspHI; 3, a DraIII restriction site corresponding to a site in
P450IIC5 was introduced into P450IIC4, this yields a Leu[162] -> Ser
point mutation with no increased 21-hydroxylase activity; 4, StyI.
The four splice sites correspond to codons 110, 128, 162 and 210.
The distribution of the 25 differences of amino acid sequence
exhibited by IIC4 and IIC5 is shown by vertical lines between the
parent proteins.   Left column shows the progesterone 21-
hydroxylase activity of cells transfected with chimeric genes or
the parent plasmids expressed as a percentage of the activity (%
ACT) of P450IIC5 (3-6 nmol per plate per 60 min).   The initial
concentration of progesterone was chosen to maximize differences
between low Km and high Km enzymes.   The results of two or more
experiments are summarized.   Taken from Kronbach et al. (76) with
permission.

39

This region maps by alignment of amino acid sequences to a residue of P450cam, Tyr 96, which has been implicated in substrate binding (79) suggesting that these segments of the primary structure serve a similar functional role in these only distantly related proteins. This segment of the amino acid sequence is adjacent to one of two highly conserved regions of these otherwise distantly related P450s. This suggests that some features of the tertiary structure of P450 cam that determine the substrate binding pocket may also be conserved in P450 1 (76).

Lindberg and Negishi (77) have also identified determinants of the different substrate specificities of two mouse class IIA P450s. Reciprocal substitution of three residues at positions 117, 209, and 365 from one enzyme to the other switched the substrate selectivity of the two enzymes, a steroid 15α-hydroxylase and a coumarin 7-hydroxylase exhibiting 11 differences of amino acid sequence. The substitution at 209 is critical for steroid hydroxylase activity whereas all three substitutions affect the level of coumarin hydroxylase activity. The substitution at 117 aligns with the segment defined in our studies (76). On the other hand, the substitution at 209 maps to helix E of P450 cam (80,81). This helix does not contact the substrate binding pocket in the structure of P450 cam (82). Thus, mapping of residue 209 to this topological component of P450 cam may not be correct, or the changes at 209 could have an affect on the substrate binding site from a distance or effect the docking of the substrate.

It is not clear what features of the topological organization of P450 cam are conserved in the structures of the mammalian microsomal enzymes. Analyses of secondary structure predictions for large numbers of mammalian microsomal P450s (>52) suggest that 10 of the 14 α-helices, a β-structure related to heme ligation and 6 of 7 β-pair structures found in P450 cam are generally predicted for the mammalian P450s (80,81). However, the mammalian microsomal P450s are longer, >487 residues compared to 414 for P450 cam (83), and it is not known how these insertions are accommodated within the basic framework seen for P450 cam. These insertions are most likely to occur at the N-terminus, between helices D and E, and between helices J and K (81). In addition, many of the substrates such as steroids that are metabolized by the mammalian microsomal P450s are larger than camphor, the substrate of P450 cam, and this probably requires a larger substrate binding pocket.

The greatest similarity in the aligned residues between P450 cam and the mammalian P450s is seen for structural elements that are likely to form the heme binding pocket (80-82). This is not surprising because the basic chemistry of oxygen reduction and of the reaction of the resulting oxidant with the organic substrate is likely to be conserved, whereas structural elements that contribute to the binding of the organic substrate are more likely to diverge in order to accommodate the many different substrates metabolized by the mammalian enzymes. Thus, helices C, D, I, and L as well as the β-bulge preceding helix L which contains the cysteine residue that binds to the axial position of the heme are likely to be conserved in all P450s (84). This may account for our results indicating that residues adjacent to helix pair C-D contribute to substrate binding as is seen in P450 cam (76).

The structures of the mammalian enzymes must also include features that are not required in P450 cam that determine their microsomal localization and interaction with the flavoprotein, P450 reductase. The latter supplies reducing equivalents from NADPH for the reduction of oxygen. In contrast, P450cam is soluble and interacts with an iron-sulfur protein which shuttles electrons from an NADH-dependent flavoprotein reductase (85). Each of the mammalian P450s exhibits a hydrophobic N-terminus which serves as a membrane anchor (86-88). Although it is likely that the bulk of the protein is cytosolic (89), it remains uncertain whether additional segments of the amino acid sequence are involved in membrane anchorage (87,88).

We are now characterizing chimeras constructed between these two enzymes, IIC4 and IIC5, and two, more distantly related enzymes, IIC1 and IIC2, (72% identity with IIC5). Preliminary experiments indicate that the region defined previously for IIC5 confers the progesterone 21-hydroxylase activity of IIC5 to IIC1 but not to IIC2 indicating that differences between IIC1 and IIC2 in their carboxyterminal portions contribute to P450 substrate specificity. Work is in progress to localize the critical segment of the sequence for these differences in substrate selectivity.

ENZYME INDUCTION

As indicated in Table 1, several of the purified P450 enzymes can be induced by foreign compounds. In most cases these foreign compounds induce enzymes of the same class in rodents. The principal mechanism of induction appears to be an increase in gene transcription. This has been shown experimentally by nuclear run-on assays in the rat for the induction of class IIB enzymes by phenobarbital (90,91), of class IA enzymes by 2,3,7,8-tetrachlorodibenzo-p-dioxin (92,93), and of class IVA enzymes by clofibrate (94). The induction of the corresponding forms of P450 in the rabbit is likely to reflect the same mechanism although it has not been verified experimentally. In contrast, the induction of class IIE enzymes in either rat (95) or rabbit (96) by ethanol or heterocyclic compounds such as pyrazole or imidazole appears to reflect a stabilization of the enzyme to turnover because the accumulation of protein is not preceded by an elevation of mRNA concentration.

CLASS IIIA

The induction of class IIIA P450s by different compounds can occur through distinct mechanisms. An increase in gene transcription has been demonstrated for the induction of the rat class IIIA enzymes by dexamethasone (97). In contrast, a study (98) of the induction of P450IIIA6 by troleandromycin in rabbits did not detect any significant changes in the transcriptional activity of the corresponding gene in response to the antibiotic. The antibiotics erythromycin and troleandromycin have been reported to increase the half-life of class IIIA P450s (99) suggesting that protein stabilization or other post-transcriptional events could underlie the induction of class IIIA P450s by troleandromycin in rabbits (98).

The spectrum of inducing agents for a class of P450 may differ between species. Rifampicin is an antibiotic that induces class IIIA enzymes in man (100,101) and in rabbits (31), but rats are refractory (102). In rabbits, elevations of progesterone 6β-hydroxylation and erythromycin N-demethylation are associated with the induction of P450IIIA6. Thus, the rabbit is an excellent animal model for the investigation of the mechanism of the induction of class IIIA P450s by rifampicin.

Recent work in our laboratories demonstrates that in immature
rabbits, rifampicin transiently activates the transcription of the Cyp3A6
gene which encodes P450IIIA6 (30). The maximum increase in transcription
occurred at 12 hr following administration of rifampicin as judged by
transcription run-on assays using isolated nuclei, Fig. 9. Experiments
using a cDNA to P450IIIA6 to probe Northern blots and slot blots indicate
that mRNAs corresponding to P450IIIA6 accumulate during the period of
increased transcription and persist at 18hr when the rate of transcription
has returned to basal levels, Fig. 9. P450IIIA6 protein accumulates in

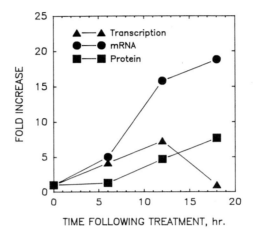

Fig. 9. Time course of induction. Changes with time in the rate of
transcription of the Cyp3A6 gene (▲), the concentration of
P450IIIA6-related RNA (●), and the microsomal concentration of
P450IIIA6 protein (■) are shown following treatment with
rifampicin. The amounts of specific transcripts generated by
nuclear run-on experiments were determined by densitometry of
autoradiographs. The relative abundance of RNAs hybridizing with
the P450IIIA6 cDNA were estimated by slot blot analysis.
Immunoblotting was used to characterize relative changes in the
microsomal concentration of P450IIIA6 protein. Taken from Potenza
et al. (30) with permission.

liver microsomes over this period. At 24 hr., a >10-fold increase in
microsomal P450IIIA6 protein is detected by immunoblotting using a
monoclonal antibody to P450IIIA6. The time course and magnitude of the
changes elicited by rifampicin are similar to those effected by
dexamethasone. However, in contrast to dexamethasone, rifampicin does not
induce tyrosine amino transferase, a "classical" glucocorticoid responsive
enzyme. Thus, the initial event in the induction of P450IIIA6 by
rifampicin is an increase in the transcription of the Cyp3A6 gene (30).

Direct activation of P450 enzymes provides another mechanism regulating the activity of P450 enzymes. Several flavonoids have been shown to stimulate the metabolism of the carcinogens benzo(a)pyrene and aflatoxin B1 by cytochrome P450 monooxygenases in man (103-107), and rabbit (56,106,107). This effect can be determined directly for the reaction catalyzed _in vitro_ and does not require the synthesis of

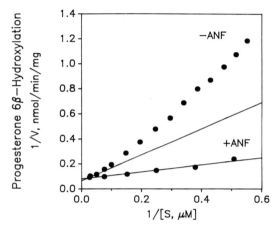

Fig. 10. Lineweaver-Burk analysis of the 6β-hydroxylation of progesterone as catalyzed by microsomes obtained from RIF-B/J rabbits in the absence (-ANF) and presence (+ANF) of 5 μM α-naphthoflavone. A linear regression line (r^2 = 0.95) is shown for the results determined in the presence of α-naphthoflavone. The line shown for the results obtained with substrate alone has been drawn as a tangent to the points obtained at the highest substrate concentrations in order to illustrate the curvilinearity of the experimental points. Taken from Schwab et al. (110) with permission.

additional protein. Early studies indicated that benzo[a]pyrene metabolism catalyzed by rabbit P450IIIA6 (P450 3c) is activated by α-naphthoflavone (56,108). P450 3c is induced in rabbit liver by rifampicin (102), dexamethasone (109) and troleandromycin (109), and we have shown that rifampicin greatly induces a microsomal benzo(a)pyrene hydroxylase that is stimulated by α-naphthoflavone (56). The progesterone 6β-hydroxylase induced by rifampicin in B/J rabbits is also activated by α-naphthoflavone (110). This strain is deficient in the expression of the

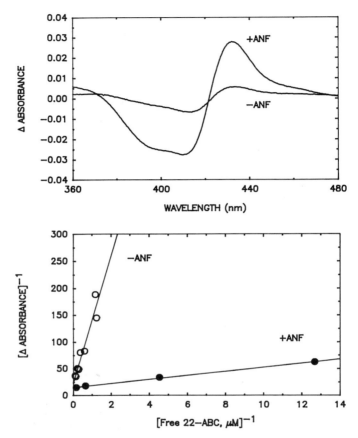

Fig. 11. The effect of α-naphthoflavone on the binding of 22-ABC to cytochrome P450IIIA6. Upper panel, type II difference spectra elicited by 3 μM 22-ABC for 1.9 μM cytochrome P450IIIA6 in the presence (+ANF) or absence (-ANF) of 10 μM α-naphthoflavone. Lower panel, double reciprocal plot of the spectral changes elicited by 22-ABC in the presence (•) or absence (○) of 10 μM α-naphthoflavone. The concentration of free 22-ABC was estimated by subtracting the amount bound from the total concentration of 22-ABC. Taken from Johnson et al. (111) with permission.

IIC3 enzyme which also catalyzes 6β-hydroxylation (74). The substrate-concentration dependence of the rate of 6β-hydroxylation in the presence or absence of α-naphthoflavone is shown in Fig. 10. In the absence of the effector, a curvilinear double-reciprocal plot, suggestive of substrate activation, is seen. In the presence of α-naphthoflavone, a linear plot is obtained which exhibits a lower apparent Km, although the non-linearity of the plot in the absence of effector confounds a simple comparison. Similar alterations of the kinetic properties of human liver microsomes were elicited by α-naphthoflavone indicating that the activation of these class IIIA enzymes by α-naphthoflavone is conserved in man and rabbit (110).

In order to determine whether α-naphthoflavone enhances the affinity of the enzyme for the substrate (lower Km), we examined the binding of a substrate analog, 22-amino-23,24-bisnor-5-cholen-3β-ol (22-ABC) to purified P450IIIA6 and liver microsomes prepared from rifampicin-treated B/J rabbits. 22-ABC binds to purified P450IIIA6 producing a Type II spectral change reflecting the coordination of the amine with the heme iron of the protein, Fig. 11. The magnitude of this spectral change is increased in the presence of α-naphthoflavone at less than saturating concentrations of 22-ABC. This reflects a change in the apparent dissociation constant for 22-ABC as shown in Fig. 11. In the absence of allosteric effectors, the binding is characterized by a Ks of 5 μM. In the presence of α-naphthoflavone, the Ks decreases to 0.8 μM (111).

The magnitude of the type II spectral change elicited by 22-ABC for microsomes prepared from rifampicin treated B/J rabbits is greater than that observed for microsomes from untreated rabbits. For microsomes from rifampicin-treated rabbits, the apparent binding constant for 22-ABC was decreased 5-fold in the presence of α-naphthoflavone indicating that α-naphthoflavone affects the properties of P450IIIA6 in situ in a manner similar to that of the purified enzyme (111).

In contrast, pregnenolone reverses the effects of α-naphthoflavone on the binding of 22-ABC, but does not alter the binding of 22-ABC to P450IIIA6 in the absence of α-naphthoflavone (111). This suggests that α-naphthoflavone and pregnenolone compete for binding to P450IIIA6 at a site distinct from that of 22-ABC. The apparent 1:1 stoichiometry for the binding of 22-ABC with P450IIIA6 is not altered by α-naphthoflavone indicating that these effects do not arise through the interaction of P450 monomers through a half-of-sites reactivity mechanism.

SUMMARY

The rabbit has proven to be an excellent species for the characterization of cytochrome P450 multiplicity. It provided a source for the first purified preparations of the multiple forms of P450. The differences and similarities exhibited by the rabbit, rat and human P450 systems, the most widely characterized, provide a basis for developing the general features of this system while the differences exhibited by these species continue to reinforce our understanding of how evolution has shaped these enzymes in each species. The rabbit has proven to be a model system for the characterization of modes of regulation for P450 enzymes such as enzyme activation by flavonoids and induction by rifampicin which occur in man but which do not occur in rat. Moreover, the larger organ sizes when compared to the rat have permitted more extensive studies of the P450 enzymes expressed in non-hepatic tissues and in fetal and neonatal life. These aspects of the rabbit P450 system are discussed in other chapters of this volume.

ACKNOWLEDGEMENT

The work of the authors laboratories is supported by USPHS Grants HD04445 (EFJ), GM31001 (EFJ), GM36590 (RHT), and CA37139 (RHT).

REFERENCES

1. D. A. Haugen, and M. J. Coon, Properties of electrophoretically homogeneous phenobarbital-inducible and β-naphthoflavone-inducible forms of liver microsomal cytochrome P-450, J.Biol.Chem. 251:7929 (1976).

2. D. A. Haugen, L. G. Armes, K. T. Yasunobu, and M. J. Coon, Amino-terminal sequence of phenobarbital-inducible cytochrome P-450 from rabbit liver microsomes: Similarity to hydrophobic amino-terminal segments of preproteins, Biochem.Biophys.Res.Commun. 77:967 (1977).

3. E. F. Johnson, M. C. Zounes, and U. Muller-Eberhard, Characterization of three forms of rabbit microsomal cytochrome P-450 by peptide mapping utilizing limited proteolysis in sodium dodecyl sulfate and analysis by gel electrophoresis, Arch.Biochem.Biophys. 192:282 (1979).

4. D. W. Nebert, D. R. Nelson, M. Adesnik, M. J. Coon, R. W. Estabrook, F. J. Gonzalez, F. P. Guengerich, I. C. Gunsalus, E. F. Johnson, B. Kemper, W. Levin, I. R. Phillips, R. Sato, and M. R. Waterman, The P450 superfamily: Updated listing of all genes and recommended nomenclature of the chromosomal loci, DNA 8:1 (1989).

5. H. H. Dieter, U. Muller-Eberhard, and E. F. Johnson, Identification of rabbit microsomal cytochrome P-450 isozyme, form 1, as a hepatic progesterone 21-hydroxylase, Biochem.Biophys.Res.Commun. 105:515 (1982).

6. R. H. Tukey, S. T. Okino, H. J. Barnes, K. J. Griffin, and E. F. Johnson, Multiple gene-like sequences related to the rabbit hepatic progesterone 21-hydroxylase cytochrome P-450 1, J.Biol.Chem. 260:13347 (1985).

7. H. H. Dieter, U. Muller-Eberhard, and E. F. Johnson, Rabbit hepatic progesterone 21-hydroxylase exhibits a bimodal distribution of activity, Science 217:741 (1982).

8. E. F. Johnson, M. Finlayson, C. M. Hujsak, U. R. Pendurthi, and R. H. Tukey, Genetic contributions to the variation among rabbits of liver microsomal deoxycorticosterone synthesis, Arch.Biochem.Biophys. 273:273 (1989).

9. T. A. van der Hoeven, D. A. Haugen, and M. J. Coon, Cytochrome P-450 purified to apparent homogeneity from phenobarbital-induced rabbit liver microsomes: Catalytic activity and other properties, Biochem.Biophys.Res.Commun. 60:569 (1974).

10. Y. Imai, and R. Sato, A gel-electrophoretically homogeneous preparation of cytochrome P-450 from liver microsomes of phenobarbital-pretreated rabbits, Biochem.Biophys.Res.Commun. 60:8 (1974).

11. G. E. Tarr, S. D. Black, V. S. Fujita, and M. J. Coon, Complete amino acid sequence and predicted membrane topology of phenobarbital-induced cytochrome P-450 (isozyme 2) from rabbit liver microsomes, Proc.Natl.Acad.Sci.USA 80:6552 (1983).

12. R. Gasser, M. Negishi, and R. M. Philpot, Primary structures of multiple forms of cytochrome P-450 isozyme 2 derived from rabbit pulmonary and hepatic cDNAs, Mol.Pharmacol. 33:22 (1988).

13. F. S. Heinemann, and J. Ozols, The complete amino acid sequence of rabbit phenobarbital-induced liver microsomal cytochrome P-450, J.Biol.Chem. 258:4195 (1983).

14. M. Komori, Y. Imai, S. Tsunasawa, and R. Sato, Microheterogeneity in the major phenobarbital-inducible forms of rabbit liver microsomal cytochrome P-450 as revealed by nucleotide sequencing of cloned cDNAs, Biochemistry 27:73 (1988).

15. D. A. Haugen, T. A. van der Hoeven, and M. J. Coon, Purified liver microsomal cytochrome P-450. Separation and characterization of multiple forms, J.Biol.Chem. 250:3567 (1975).

16. G. E. Schwab, R. L. Norman, U. Muller-Eberhard, and E. F. Johnson, Identification of the form of cytochrome P-450 induced in neonatal rabbit liver microsomes by phenobarbital, Mol.Pharmacol. 17:218 (1980).

17. D. R. Koop, E. T. Morgan, G. E. Tarr, and M. J. Coon, Purification and characterization of a unique isozyme of cytochrome P-450 from liver microsomes of ethanol-treated rabbits, J.Biol.Chem. 257:8472 (1982).

18. M. Ingelman-Sundberg, and A. -L. Hagbjork, On the significance of the cytochrome P-450-dependent hydroxyl radical-mediated oxygenation mechanism, Xenobiotica 12:673 (1982).

19. S. C. Khani, P. G. Zaphiropoulos, V. S. Fujita, T. D. Porter, D. R. Koop, and M. J. Coon, cDNA and derived amino acid sequence of ethanol-inducible rabbit liver cytochrome P-450 isozyme 3a (P-450 alc), Proc.Natl.Acad.Sci.USA 84:638 (1987).

20. D. R. Koop, G. D. Nordblom, and M. J. Coon, Immunochemical evidence for a role of cytochrome P-450 in liver microsomal ethanol oxidation, Arch.Biochem.Biophys. 235:228 (1984).

21. M. Ingelman-Sundberg, and H. Jornvall, Induction of the ethanol-inducible form of rabbit liver microsomal cytochrome P-450 by inhibitors of alcohol dehydrogenase, Biochem.Biophys.Res.Commun. 124:375 (1984).

22. E. F. Johnson, Isolation and characterization of a constitutive form of rabbit liver microsomal cytochrome P-450, J.Biol.Chem. 255:304 (1980).

23. D. R. Koop, and M. J. Coon, Purification and properties of P-450LM3b, a constitutive form of cytochrome P-450, from rabbit liver microsomes, Biochem.Biophys.Res.Commun. 91:1075 (1979).

24. J. Ozols, F. S. Heinemann, and E. F. Johnson, The complete amino acid sequence of a constitutive form of liver microsomal cytochrome P-450, J.Biol.Chem. 260:5427 (1985).

25. J. K. Leighton, B. A. DeBrunner-Vossbrinck, and B. Kemper, Isolation and sequence analysis of three cloned cDNAs for rabbit liver proteins that are related to rabbit cytochrome P-450 (form 2), the major phenobarbital-inducible form, Biochemistry 23:204 (1984).

26. H. H. Dieter, and E. F. Johnson, Functional and structural polymorphism of rabbit microsomal cytochrome P-450 form 3b, J.Biol.Chem. 257:9315 (1982).

27. N. Miki, T. Sugiyama, T. Yamano, and Y. Miyake, Characterization of a highly purified form of cytochrome P-450 B1, Biochem.Int. 3:217 (1981).

28. D. R. Koop, A. V. Persson, and M. J. Coon, Properties of electrophoretically homogeneous constitutive forms of liver microsomal cytochrome P-450, J.Biol.Chem. 256:10704 (1981).

29. C. Dalet, P. Clair, M. Daujat, P. Fort, J. -M. Blanchard, and P. Maurel, Complete sequence of cytochrome P450 3c cDNA and presence of two mRNA species with 3' untranslated regions of different lengths, DNA 7:39 (1988).

30. C. L. Potenza, U. R. Pendurthi, D. K. Strom, R. H. Tukey, K. J. Griffin, G. E. Schwab, and E. F. Johnson, Regulation of the rabbit cytochrome P-450 3c gene: Age dependent expression and transcriptional activation by rifampicin, J.Biol.Chem. 264:16222 (1989).

31. R. Lange, C. Larroque, C. Balny, and P. Maurel, Isolation and partial characterization of a rifampicin induced rabbit liver microsomal cytochrome P-450, Biochem.Biophys.Res.Commun. 126:833 (1985).

32. S. Yamamoto, E. Kusunose, K. Ogita, M. Kaku, K. Ichihara, and M. Kusunose, Isolation of cytochrome P-450 highly active in prostaglandin ω-hydroxylation from lung microsomes of rabbits treated with progesterone, J.Biochem. 96:593 (1984).

33. D. E. Williams, S. E. Hale, R. T. Okita, and B. S. S. Masters, A prostaglandin ω-hydroxylase cytochrome P-450 (P-450 PG-ω) purified from lungs of pregnant rabbits, J.Biol.Chem. 259:14600 (1984).

34. S. Matsubara, S. Yamamoto, K. Sogawa, N. Yokotani, Y. Fujii-Kuriyama, M. Haniu, J. E. Shively, O. Gotoh, E. Kusunose, and M. Kusunose, cDNA cloning and inducible expression during pregnancy of the mRNA for rabbit pulmonary prostaglandin ω-hydroxylase (cytochrome P-450$_{p-2}$), J.Biol.Chem. 262:13366 (1987).

35. A. S. Muerhoff, D. E. Williams, M. T. Leithauser, V. E. Jackson, M. R. Waterman, and B. S. S. Masters, Regulation of the induction of a cytochrome P-450 prostaglandin ω-hydroxylase by pregnancy in rabbit lung, Proc.Natl.Acad.Sci. 84:7911 (1987).

36. Y. Kikuta, E. Kusunose, S. Matsubara, Y. Funae, S. Imaoka, I. Kubota, and M. Kusunose, Purification and characterization of hepatic microsomal prostaglandin omega-hydroxylase cytochrome P-450 from pregnant rabbits, J.Biochem. 106:468 (1989).

37. E. F. Johnson, and U. Muller-Eberhard, Purification of the major cytochrome P-450 of liver microsomes from rabbits treated with 2,3,7,8-tetrachlorodibenzo-p-dioxin (TCDD), Biochem.Biophys.Res.Commun. 76:652 (1977).

38. J. C. Kawalek, W. Levin, D. Ryan, P. E. Thomas, and A. Y. H. Lu, Purification of liver microsomal cytochrome P-448 from 3-methylcholanthrene-treated rabbits, Mol.Pharmacol. 11:874 (1975).

39. C. Hashimoto, and Y. Imai, Purification of a substrate complex of cytochrome P-450 from liver microsomes of 3-methylcholanthrene-treated rabbits, Biochem.Biophys.Res.Commun. 68:821 (1976).

40. S. T. Okino, L. C. Quattrochi, H. J. Barnes, S. Osanto, K. J. Griffin, E. F. Johnson, and R. H. Tukey, Cloning and characterization of cDNAs encoding 2,3,7,8-tetrachlorodibenzo-p-dioxin-inducible rabbit mRNAs for cytochrome P-450 isozymes 4 and 6, Proc.Natl.Acad.Sci.USA 82:5310 (1985).

41. N. Kagawa, K. Mihara, and R. Sato, Structural analysis of cloned cDNAs for polycyclic hydrocarbon-inducible forms of rabbit liver microsomal cytochrome P-450, J.Biochem. 101:1471 (1987).

42. J. Ozols, Complete amino acid sequence of a cytochrome P-450 isolated from β-naphthoflavone-induced rabbit liver microsomes. Comparison with phenobarbital-induced and constitutive isozymes and identification of invariant residues, J.Biol.Chem. 261:3965 (1986).

43. R. L. Norman, E. F. Johnson, and U. Muller-Eberhard, Identification of the major cytochrome P-450 form transplacentally induced in neonatal rabbits by 2,3,7,8-tetrachlorodibenzo-p-dioxin, J.Biol.Chem. 253:8640 (1978).

44. R. M. Philpot, and E. Arinc, Separation and purification of two forms of hepatic cytochrome P-450 from untreated rabbits, Mol.Pharmacol. 12:483 (1976).

45. R. Gasser, and R. M. Philpot, Primary structures of cytochrome P-450 isozyme 5 from rabbit and rat and regulation of species-dependent expression and induction in lung and liver: Identification of cytochrome P-450 gene subfamily IVB, Mol.Pharm. 35:617 (1989).

46. Z. Parandoosh, V. S. Fujita, M. J. Coon, and R. M. Philpot, Cytochrome P-450 isozymes 2 and 5 in rabbit lung and liver. Comparisons of structure and inducibility, Drug Metab.Dispos. 15:59 (1987).

47. E. F. Johnson, and U. Muller-Eberhard, Multiple forms of cytochrome P-450: Resolution and purification of rabbit liver aryl hydrocarbon hydroxylase, Biochem.Biophys.Res.Commun. 76:644 (1977).

48. D. Kupfer, Endogenous substrates of monooxygenases: Fatty acids and prostaglandins, Pharmacol.Ther. 11:469 (1980).

49. M. L. Schwartzman, J. R. Falck, P. Yadagiri, and B. Escalante, Metabolism of 20-hydroxyeicosatetraenoic acid by cyclooxygenase, J.Biol.Chem. 264:11658 (1989).

50. W. S. Powell, ω-Oxidation of prostaglandins by lung and liver microsomes. Changes in enzyme activity induced by pregnancy, pseudopregnancy, and progesterone treatment, J.Biol.Chem. 253:6711 (1978).

51. A. Y. H. Lu, K. W. Junk, and M. J. Coon, Resolution of the cytochrome P-450-containing ω-hydroxylation system of liver microsomes into three components, J.Biol.Chem. 244:3714 (1969).

52. E. Kusunose, K. Ogita, K. Ichihara, and M. Kusunose, Effect of cytochrome b5 on fatty acid ω- and (ω-1)-hydroxylation catalyzed by partially purified cytochrome P-450 from rabbit kidney cortex microsomes, J.Biochem. 90:1069 (1981).

53. S. Yamamoto, E. Kusunose, M. Kaku, K. Ichihara, and M. Kusunose, Effect of peroxisomal proliferators on microsomal prostaglandin A ω-hydroxylase, J.Biochem. 100:1449 (1986).

54. M. Kusunose, E. Kusunose, K. Ichihara, K. Ogita, M. Kaku, and S. Yamamoto, Cytochrome P-450-linked prostaglandin ω-hydroxylase, in: "Advances in Prostaglandin, Thromboxane, and Leukotriene Research," O. Hayaishi, and S. Yamamoto, eds., Raven Press, New York (1985).

55. E. F. Johnson, D. W. Walker, K. J. Griffin, J. E. Clark, R. T. Okita, A. S. Muerhoff, and B. S. Masters, Cloning and expression of three rabbit kidney cDNAs encoding lauric acid ω-hydroxylases, Biochemistry 29:873 (1990).

56. J. L. Raucy, and E. F. Johnson, Variations among untreated rabbits in benzo(a)pyrene metabolism and its modulation by 7,8-benzoflavone, Mol.Pharmacol. 27:296 (1985).

57. G. E. Schwab, and E. F. Johnson, Variation in hepatic microsomal cytochrome P-450 1 concentration among untreated rabbits alters the efficiency of estradiol hydroxylation, Arch.Biochem.Biophys. 237:17 (1985).

58. P. E. Thomas, D. Korzeniowski, D. Ryan, and W. Levin, Preparation of monospecific antibodies against two forms of rat liver cytochrome P-450 and quantitation of these antigens in microsomes, Arch.Biochem.Biophys. 192:524 (1979).

59. I. R. Senciall, G. Bullock, and S. Rahal, Progesterone C21 hydroxylation and steroid carboxylic acid biosynthesis in the rabbit. In vitro studies with endocrine, metabolic, and potential target tissues, Can.J.Biochem.Cell Biol. 61:722 (1983).

60. C. Ged, D. R. Umbenhauer, T. M. Bellew, R. W. Bork, P. K. Srivastava, N. Shinriki, R. S. Lloyd, and F. P. Guengerich, Characterization of cDNAs, mRNAs, and proteins related to human liver microsomal cytochrome P-450 (S)-Mephenytoin 4'-hydroxylase, Biochemistry 27:6929 (1988).

61. T. Yasumori, N. Murayama, Y. Yamazoe, A. Abe, Y. Nogi, T. Fukasawa, and R. Kato, Expression of a human P-450IIC gene in yeast cells using galactose-inducible expression system, Mol.Pharm. 35:443 (1989).

62. G. R. Wilkinson, F. P. Guengerich, and R. A. Branch, Genetic polymorphism of S-mephenytoin hydroxylation, Pharmacol.Ther. 43:53 (1989).

63. W. R. Brian, P. K. Srivastava, D. R. Umbenhauer, R. S. Lloyd, and F. P. Guengerich, Expression of a human liver cytochrome P-450 protein with tolbutamide hydroxylase activity in Saccharomyces cerevisiae, Biochemistry 28:4993 (1989).

64. I. Reubi, K. J. Griffin, J. L. Raucy, and E. F. Johnson, Three monoclonal antibodies to rabbit microsomal cytochrome P-450 1 recognize distinct epitopes that are shared to different degrees among other electrophoretic types of cytochrome P-450, J.Biol.Chem. 259:5887 (1984).

65. I. Reubi, K. J. Griffin, J. Raucy, and E. F. Johnson, Use of a monoclonal antibody specific for rabbit microsomal cytochrome P-450 3b to characterize the participation of this cytochrome in the microsomal 6β- and 16α-hydroxylation of progesterone, Biochemistry 23:4598 (1984).

66. M. J. Finlayson, J. H. Dees, B. S. S. Masters, and E. F. Johnson, Differential expression of cytochrome P-450 1 and related forms in rabbit liver and kidney, Arch.Biochem.Biophys. 252:113 (1986).

67. J. K. Leighton, and B. Kemper, Differential induction and tissue-specific expression of closely related members of the phenobarbital-inducible rabbit cytochrome P-450 gene family, J.Biol.Chem. 259:11165 (1984).

68. M. J. Finlayson, B. Kemper, N. Browne, and E. F. Johnson, Evidence that rabbit cytochrome P-450 K is encoded by the plasmid pP-450 PBc2, Biochem.Biophys.Res.Commun. 141:728 (1986).

69. S. Govind, P. A. Bell, and B. Kemper, Structure of genes in the cytochrome P-450PBc subfamily. Conservation of intron locations in the phenobarbital-inducible family, DNA 5:371 (1986).

70. Y. Imai, Characterization of rabbit liver cytochrome P-450 (Laurate ω-1 hydroxylase) synthesized in transformed yeast cells, J.Biochem. 103:143 (1988).

71. E. F. Johnson, H. J. Barnes, K. J. Griffin, S. Okino, and R. H. Tukey, Characterization of a second gene product related to rabbit cytochrome P-450 1, J.Biol.Chem. 262:5918 (1987).

72. L. Ghizzoni, U. Muller-Eberhard, H. H. Liem, M. New, M. Finlayson, and E. F. Johnson, Characterization of variations in rabbit hepatic progesterone 21-hydroxylase activity by serial biopsy, Biochem.Biophys.Res.Commun. 130:43 (1985).

73. E. F. Johnson, and K. J. Griffin, Variations in hepatic progesterone 21-hydroxylase activity reflect differences in the microsomal concentration of rabbit cytochrome P-450 1, Arch.Biochem.Biophys. 237:55 (1985).

74. G. E. Schwab, and E. F. Johnson, Two catalytically distinct subforms of P-450 3b as obtained from inbred rabbits, Biochemistry 24:7222 (1985).

75. E. F. Johnson, G. E. Schwab, and H. H. Dieter, Allosteric regulation of the 16α-hydroxylation of progesterone as catalyzed by rabbit microsomal cytochrome P-450 3b, J.Biol.Chem. 258:2785 (1983).

76. T. Kronbach, T. M. Larabee, and E. F. Johnson, Hybrid cytochromes P-450 identify a substrate binding domain in P-450IIC5 and P-450IIC4, Proc.Natl.Acad.Sci.USA 86:8262 (1989).

77. R. L. P. Lindberg, and M. Negishi, Alteration of mouse cytochrome P450coh substrate specificity by mutation of a single amino-acid residue, Nature 339:632 (1989).

78. T. L. Poulos, B. C. Finzel, and A. J. Howard, High-resolution crystal structure of cytochrome P450cam, J.Mol.Biol. 195:687 (1987).

79. W. M. Atkins, and S. G. Sligar, The roles of active site hydrogen bonding in cytochrome P-450cam as revealed by site-directed mutagenesis, J.Biol.Chem. 263:18842 (1988).

80. D. R. Nelson, and H. W. Strobel, Secondary structure prediction of 52 membrane-bound cytochromes P450 shows a strong structural similarity to P450$_{cam}$, Biochemistry 28:656 (1989).

81. R. J. Edwards, B. P. Murray, A. R. Boobis, and D. S. Davies, Identification and location of α-helices in mammalian cytochrome P450, Biochemistry 28:3762 (1989).

82. T. L. Poulos, Site-directed mutagenesis: Reversing enzyme specificity, Nature 339:580 (1989).

83. B. P. Unger, I. C. Gunsalus, and S. G. Sligar, Nucleotide sequence of the pseudomonas putida cytochrome P-450cam gene and its expression in Escherichia coli, J.Biol.Chem. 261:1158 (1986).

84. T. L. Poulos, Cytochrome P450: Molecular architecture, mechanism, and prospects for rational inhibitor design, Pharmac.Res. 5:67 (1989).

85. P. S. Stayton, T. L. Poulos, and S. G. Sligar, Putidaredoxin competitively inhibits cytochrome b5-cytochrome P-450cam association: A proposed molecular model for a cytochrome P-450cam electron-transfer complex, Biochemistry 28:8201 (1989).

86. G. Vergeres, K. H. Winterhalter, and C. Richter, Identification of the membrane anchor of microsomal rat liver cytochrome P-450, Biochemistry 28:3650 (1989).

87. S. Monier, P. Van Luc, G. Kreibich, D. D. Sabatini, and M. Adesnik, Signals for the incorporation and orientation of cytochrome P450 in the endoplasmic reticulum membrane, J.Cell.Biol. 107:457 (1988).

88. E. Szczesna-Skorupa, N. Browne, D. Mead, and B. Kemper, Positive charges at the NH$_2$ terminus convert the membrane-anchor signal peptide of cytochrome P-450 to a secretory signal peptide, Proc.Natl.Acad.Sci.USA 85:738 (1988).

89. C. A. Brown, and S. D. Black, Membrane topology of mammalian cytochromes P-450 from liver endoplasmic reticulum. Determination by trypsinolysis of phenobarbital-treated micrsomes, J.Biol.Chem. 264:4442 (1989).

90. J. P. Hardwick, F. J. Gonzalez, and C. B. Kasper, Transcriptional regulation of rat liver epoxide hydratase, NADPH-cytochrome P-450 oxidoreductase, and cytochrome P-450b genes by phenobarbital, J.Biol.Chem. 258:8081 (1983).

91. S. F. Pike, E. A. Shephard, B. R. Rabin, and I. R. Phillips, Induction of cytochrome P-450 by phenobarbital is mediated at the level of transcription, Biochem.Pharmacol. 34:2489 (1985).

92. D. I. Israel, and J. P. Whitlock Jr., Regulation cytochrome P1-450 gene transcription by 2,3,7,8-tetrachlorodibenzo-p-dioxin in wild type and variant mouse hepatoma cells, J.Biol.Chem. 259:5400 (1984).

93. F. J. Gonzalez, R. H. Tukey, and D. W. Nebert, Structural gene products of the Ah locus. Transcriptional regulation of cytochrome P1-450 and P3-450 mRNA levels by 3-methylcholanthrene, Mol.Pharmacol. 26:117 (1984).

94. J. P. Hardwick, B-J. Song, E. Huberman, and F. J. Gonzalez, Isolation, complementary DNA sequence, and regulation of rat hepatic lauric acid ω-hydroxylase (cytochrome P-450LAω), J.Biol.Chem. 262:801 (1987).

95. B-J. Song, H. V. Gelboin, S-S. Park, C. S. Yang, and F. J. Gonzalez, Complementary DNA and protein sequences of ethanol-inducible rat and human cytochrome P-450s, J.Biol.Chem. 261:16689 (1986).

96. T. D. Porter, S. C. Khani, and M. J. Coon, Induction and tissue-specific expression of rabbit cytochrome P450IIE1 and IIE2 genes, Mol.Pharm. 36:61 (1989).

97. D. L. Simmons, P. McQuiddy, and C. B. Kasper, Induction of the hepatic mixed-function oxidase system by synthetic glucocorticoids, J.Biol.Chem. 262:326 (1987).

98. C. Dalet, J. M. Blanchard, P. Guzelian, J. Barwick, H. Hartle, and P. Maurel, Cloning of a cDNA coding for P-450 LM3c from rabbit liver microsomes and regulation of its expression, Nucleic Acids Res. 14:5999 (1986).

99. P. B. Watkins, S. A. Wrighton, E. G. Schuetz, P. Maurel, and P. S. Guzelian, Macrolide antibiotics inhibit the degradation of the glucocorticoid-responsive cytochrome P-450p in rat hepatocytes in vivo and in primary monolayer culture, J.Biol.Chem. 261:6264 (1986).

100. J. Combalbert, I. Fabre, G. Fabre, I. Dalet, J. Derancourt, J. P. Cano, and P. Maurel, Metabolism of cyclosporin A. IV. Purification and identification of the rifampicin-inducible human liver cytochrome P-450 (cyclosporin A oxidase) as a product of P450IIIA gene subfamily, Drug Metab.Dispos. 17:197 (1989).

101. C. Ged, J. M. Rouillon, L. Pichard, J. Combalbert, N. Bressot, P. Bories, H. Michel, P. Beaune, and P. Maurel, The increase in urinary excretion of 6β-hydroxycortisol as a marker of human hepatic cytochrome P450IIIA induction, Br.J.Clin.Pharmacol. 28:373 (1989).

102. S. A. Wrighton, E. G. Schuetz, P. B. Watkins, P. Maurel, J. Barwick, B. S. Bailey, H. T. Hartle, B. Young, and P. Guzelian, Demonstration in multiple species of inducible hepatic cytochromes P-450 and their mRNAs related to the glucocorticoid-inducible cytochrome P-450 of the rat, Mol.Pharmacol. 28:312 (1985).

103. J. Kapitulnik, P. J. Poppers, M. K. Buening, J. G. Fortner, and A. H. Conney, Activation of monooxygenases in human liver by 7,8-benzoflavone, Clin.Pharmacol.Ther. 22:475 (1977).

104. M. K. Buening, J. G. Fortner, A. Kappas, and A. H. Conney, 7,8-benzoflavone stimulates the metabolic activation of aflatoxin B1 to mutagens by human liver, Biochem.Biophys.Res.Commun. 82:348 (1978).

105. M. K. Buening, R. L. Chang, M-T. Huang, J. G. Fortner, A. W. Wood, and A. H. Conney, Activation and inhibition of benzo(a)pyrene and aflotoxin B1 metabolism in human liver microsomes by naturally occurring flavonoids, Cancer Res. 41:67 (1981).

106. D. R. Thakker, W. Levin, M. Buening, H. Yagi, R. E. Lehr, A. W. Wood, A. H. Conney, and D. M. Jerina, Species-specific enhancement by 7,8-benzoflavone of hepatic microsomal metabolism of benzo[e]pyrene 9,10-dihydrodiol to bay-region diol epoxides, Cancer Res. 41:1389 (1981).

107. M-T. Huang, R. L. Chang, J. G. Fortner, and A. H. Conney, Studies on the mechanism of activation of microsomal benzo(a)pyrene hydroxylation by flavonoids, J.Biol.Chem. 256:6829 (1981).

108. M-T. Huang, E. F. Johnson, U. Muller-Eberhard, D. R. Koop, M. J. Coon, and A. H. Conney, Specificity in the activation and inhibition by flavonoids of benzo(a)pyrene hydroxylation by cytochrome P-450 isozymes from rabbit liver microsomes, J.Biol.Chem. 256:10897 (1981).

109. P. B. Watkins, S. A. Wrighton, P. Maurel, E. G. Schuetz, G. Mendez-Picon, G. A. Parker, and P. S. Guzelian, Identification of an inducible form of cytochrome P-450 in human liver, Proc.Natl.Acad.Sci.USA 82:6310 (1985).

110. G. E. Schwab, J. L. Raucy, and E. F. Johnson, Modulation of rabbit and human hepatic cytochrome P450-catalyzed steroid hydroxylations by α-naphthoflavone, Mol.Pharmacol. 33:493 (1988).

111. E. F. Johnson, G. E. Schwab, and L. E. Vickery, Positive effectors of the binding of an active site-directed amino steroid to rabbit cytochrome P-450 3c, J.Biol.Chem. 263:17672 (1988).

PRIMARY STRUCTURES AND REGULATION OF CYTOCHROME P-450 ISOZYMES 2 (IIB) AND 5 (IVB) AND THE FLAVIN-CONTAINING MONOOXYGENASE IN RABBIT LIVER AND LUNG

Richard M. Philpot, Rodolfo Gasser, Michael P. Lawton

Laboratory of Cellular and Molecular Pharmacology
National Institute of Environmental Health Sciences
Research Triangle Park, North Carolina 27709, U.S.A.

INTRODUCTION

Cytochrome P-450 (P-450) isozymes 2 and 5, which belong to P-450 gene subfamilies IIB and IVB according to the nomenclature devised by Nebert et al. (1), comprise over 90% of the P-450 in rabbit lung and are the major drug-metabolizing enzymes in that tissue. Results of immunochemical, catalytic, and biochemical studies (2-5) suggest that orthologs of isozymes 2 and 5 also account for the majority of the pulmonary P-450 in other species (rats, mice, guinea pigs, hamsters, and monkeys). A third drug-metabolizing enzyme present in rabbit lung at a relatively high concentration is the flavin-containing monooxygenase (FMO). This enzyme, which catalyzes the oxidation of sulfur, phosphorous, and nitrogen in a wide variety of compounds (6, 7), has also been detected in lungs from a number of species (8).

Examination of the catalytic, immunochemical and structural properties of purified and microsomal P-450 indicated that highly similar, if not identical, forms of isozymes 2 and 5 are expressed in rabbit lung and liver (9-11). However, regulation of these isozymes in the two tissues was found to be quite different; the concentrations of both increase markedly in liver, but not in lung, following treatment of rabbits with phenobarbital (PB). Although isozyme 2 is generally thought of as the "PB-inducible" P-450 in rabbits, increases in the hepatic content of isozyme 5 by PB are equally significant (12). Taken together these results suggest the possibility that isozymes 2 and 5 in lung and liver are each products of related, but distinctly different, genes. If this were true, their primary structures would likely be somewhat different and their substrate specificities might not be exactly the same.

In contrast to the similarities between P-450 isozymes 2 and 5 expressed in liver and lung, the FMO enzymes expressed in the two tissues differ with respect to a number of immunochemical, physical, and catalytic properties. The enzymes exhibit no immunochemical cross-reactivity (13-14), have different apparent monomeric molecular weights (8), and have substrate specificities that overlap but are not identical. For example, n-octylamine is a substrate only for the lung FMO (15), and imipramine only for the liver FMO (13). While these differences could result from alternative gene splicing or post-translational modifications, a more likely explanation is that the two enzymes are products of distinct genes.

Molecular Aspects of Monooxygenases and Bioactivation of Toxic Compounds
Edited by E. Arınç *et al., Plenum Press, New York,*

55

The precise relationship between the FMO enzymes expressed in liver and lung has now been examined with the techniques of molecular biology. Similar work has been carried out with P-450 isozymes 2 and 5, and uniquely different results have been obtained for each of the three enzymes studied.

MATERIALS AND METHODS

Animals, Tissues, and Subcellular Preparations

Adult, male New Zealand White rabbits were used. Animals were killed by suffocation with CO_2 and the tissues were removed as quickly as possible, frozen in liquid nitrogen, and stored at -70°C. Pulmonary and hepatic mRNA was purified by a modification (16) of the methods of Chirgwin et al. (17) and Glisin et al. (18), and genomic DNA was purified by standard procedures (19, 20).

Construction and Screening of cDNA Libraries

Liver cDNA libraries were constructed with mRNA from untreated rabbits (Lv library), from rabbits treated with PB once for 12h (Lv-1 library), and from rabbits treated with PB daily for 4 days (Lv-4 library). Lung libraries were constructed with mRNA from untreated rabbits (Lv library). All libraries were made with lambda gt11 or lambda ZAP (Stratagene, La Jolla, CA). Initial screening for isozyme 2 was with cDNA for a homologous murine P-450, for isozyme 5 with antibodies, for liver FMO with the cDNA for the FMO from pig liver (21), and for lung FMO with antibodies. Subsequent screenings were carried out with cDNAs isolated from the rabbit libraries.

Nucleotide Sequencing

Insert fragments formed by sonication (22) or restriction (23) were subcloned into M13 and sequenced by the dideoxy chain termination method (24, 25). Some sequencing was also done by making use of the T3 and T7 promoter regions flanking some inserts (26). Sequence data were analyzed with software from the University of Wisconsin (27), and gap alignments were constructed with the algorithm developed by Wilbur and Lipman (28).

Analysis of mRNA and Genomic DNA

Purified mRNA was separated by electrophoresis in agarose gels with methylmercury and transferred overnight to nylon membranes. Northern analysis was carried out with standard procedures using cDNAs labeled by the random primer method (29). Following hybridization, blots were washed twice for 15 min at room temperature in 0.1xSSC containing 0.1% SDS and then subjected to autoradiography. Genomic DNA was assessed by the method of Southern (30).

Peptide Sequences of FMO Purified from Rabbit Lung

The FMO was purified from rabbit lung by the method of Tynes et al. (14), alkylated, and digested with trypsin. Peptides were separated on a 300A C-18 column by elution with a acetonitrile-water gradient. The flow rate was 0.9 ml/min, and peptide elution was monitored at 214 and 254 nm. Fractions were collected by hand, and sequencing was done with a gas phase instrument (Applied Biosystems 407A).

RESULTS AND DISCUSSION

Structure and Regulation of P-450 Isozyme 2

The first library screened for cDNAs encoding isozyme 2 was Lv-1. Over 50 clones were isolated initially and the largest was sequenced. This clone (Lv-1:B2-1) contained 2077 bases with an open reading frame of 1473 bases, a 3'-flanking region of 572 bases and a 5'-flanking region of 32 bases. The derived amino acid sequence (for complete sequences see 16) contained the same number of residues (491) as reported by Tarr et al. (31) for purified isozyme 2, but different amino acids were present at eleven positions. Two of these differences, at positions 174 and 370, coincided with the only Bst EII restriction sites present in the cDNA. With the sequence reported for the protein, the Bst EII site at position 174 would not exist. Therefore, we subjected an additional 26 clones from the Lv-1 library to restriction analysis with Bst EII. Two sites were found in seventeen of the clones, and these clones (B2 clones) produced patterns identical to those formed by Lv-1:B2-1 upon further restriction with a number of enzymes. However, each of the other nine clones contained only a single Bst EII restriction site. Restriction of these "B1" clones with additional enzymes gave uniform results. Because the single Bst EII site present in the B1 clones turned out to be associated with residue 174, it was clear that B1 cDNA did not encode the protein for which the sequence was known. In fact, the sequence derived from Lv-1:B1-1 differed from that published for the protein at six positions and differed from the sequence derived for B2 at 15 positions. The insert for clone Lv-1:B1-1 contained 2063 bases of which 1473 were in an open reading frame, 564 in the 3'-flanking region and 26 in the 5'-flanking region. A sequence of 1850 bases from a second B1 clone was identical to that of Lv-1:B1-1, and a sequence of 720 bases from a second B2 clone was identical to that of Lv-1:B2-1. Further analysis of clones from the Lv-1 library gave no evidence for isozyme 2-encoding cDNAs other than those described as B1 or B2.

The second library screened for cDNAs encoding isozyme 2 was the lung library (Lg library), which was screened with Lv-1:B2-1. Twenty-two clones were isolated, eight of which contained more than 1.5 kb. None of these eight clones was altered by restriction with Bst EII, and the sequence derived from the longest (Lg:B0-1; 2044 bases) was identical to the sequence published for the protein. This clone had an open reading frame of 1473 bases, a 3'-flanking region of 556 bases and a 5'-flanking region of 15 bases. The base sequence differed from that of Lv-1:B2-1 at 86 positions and from that of Lv-1:B1-1 at 56 positions. Examination of pulmonary clones between 1.0 and 1.5 kb revealed several with characteristics similar to the B1 clones from the Lv-1 library. The sequences (1091 and 1074 bases) of two of these clones (Lg:B1-1 and Lg:B1-2) were the same and were identical to the sequence of Lv-1:B1-1. The sequence of Lg:B0-1 was confirmed when an identical sequence was obtained from a second clone (Lg:B0-2).

Liver clones with the characteristics of B0 clones from lung were isolated from the Lv-4 library. One clone (Lv-4:B0-1) was shown to have a base sequence identical to that of Lg:B0-1. Finding this sequence for a liver clone was expected since the protein that had been sequenced was purified from livers of rabbits treated with phenobarbital (31). A summary of the sequencing experiments carried out with clones encoding forms of isozyme 2 is presented in Table 1, base differences and amino acids differences are summarized in Table 2, and the amino acid substitutions are shown in Table 3. Complete sequences of B0, B1, and B2 are presented in reference 16.

Table 1. Cloning and Sequencing of cDNAs Encoding the B0, B1, and B2
Forms of Isozyme 2

Form of Isozyme	Library	Clone Desigination	Number of Bases
BO	Lg	Lg:BO-1	2044
BO	Lg	Lg:BO-2	860
BO	Lv-4	Lv-4:BO-1	1735
B1	Lv-1	Lv-1:B1-1	2063
B1	Lv-1	Lv-1:B1-2	1850
B1	Lg	Lg:B1-1	1091
B1	Lg	Lg:B1-2	1074
B2	Lv-1	Lv-1:B2-1	2077
B2	Lv-1	Lv-1:B2-1	720

Table 2. Differences in Nucleotide Sequences and Derived Amino Acid
Sequences for cDNAs Encoding Forms of Isozyme 2

	Number of Differences		
	Nucleotide Sequence		Amino Acid Sequence
Comparison	Total	Coding Region	
Form B0 vs. Form B1	56	19	6
Form B0 vs. Form B2	86	31	11
Form B1 vs. Form B2	82	31	15

Table 3. Amino Acid Substitutions in the Primary Structures of Isozyme 2
Variants

Variant	Residue Number															
	035	039	057	114	120	174	248	286	290	294	314	363	367	370	417	420
BO	SER	VAL	ARG	ILE	ARG	ile	SER	GLN	LEU	SER	MET	ILE	VAL	THR	ASN	LEU
B1	SER	ile	gln	ILE	ARG	VAL	SER	GLN	ile	SER	leu	ILE	VAL	THR	ASN	met
B2	pro	VAL	ARG	phe	his	VAL	thr	arg	LEU	thr	MET	val	ala	met	asp	LEU

Analysis of total mRNA from liver and lung of untreated rabbits and rabbits treated with PB for species encoding forms of isozyme 2 gave results that were consistent with those obtained by analysis of microsomal protein. Levels of hepatic mRNA that hybridized with Lv-1:B2-1 increased about 10-times following treatment with PB, whereas levels of pulmonary mRNA remained unchanged. In untreated rabbits, however, more specific mRNA was detected in samples from lung than in samples from liver. In all cases single bands of mRNA (~2.4 kb) were detected. It is of interest that all of the mRNA detected in pulmonary samples was present in the polyadenylated fraction, in contrast to what has been reported for the orthologous mRNA in rats (32). Our analysis of rat pulmonary mRNA was also positive only with the polyadenylated fraction. We believe that this difference is caused by shearing of the polyadenylated tails of the mRNA by the isolation procedure used by Omiencinski (32). Our results (16) show clearly that extreme care must be taken in the isolation of intact mRNA from lung.

In order to evaluate the expression of mRNA specific for the B0, B1, and B2 forms of isozyme 2, oligonucleotides able to differentiate among these forms were obtained. Synthesis of specific oligonucleotides was facilitated by a highly variable area present in the 3' noncoding regions of the cDNAs. This variability results from a deletion of 19 bases in B0 and 11 bases in B1, both relative to B2. The alignment of this area and the oligonucleotides synthesized are shown in Table 4. These probes were used to evaluate mRNA from livers and lungs of both untreated rabbits and rabbits treated with PB. Several conclusions can be drawn from the results of these experiments. First, B0 and B1 appear to be allelic variants of the same gene. This conclusion is based on finding animals containing B0 without B1, B1 without B0, or both B0 and B1. In addition, identical B0/B1 phenotypes were found when pulmonary and hepatic samples from individual animals were examined. Second, B2 mRNA was found to be expressed in liver but not in lung. Third, expression of B0 appears to be coupled with expression of B1; B2 is never observed in animals homozygous for B0. Fourth, mRNA for B0, B1, and B2 can be increased in liver by treatment of rabbits with PB. Fifth, PB has no effect on the level of mRNA for either B0 or B1 in lung. Expression of the three forms of isozyme 2 is shown in Table 5.

It is interesting that expression of the forms of isozyme 2 exhibits differences in tissue specificity. This suggests that the regulatory

Table 4. The 3'-Flanking Sequences used for the Synthesis of Oligonucleotide Probes (Complementary Sequences are Underlined)
Specific for the B0, B1 and B2 Forms of Isozyme 2

Variant	Stop	Alignment of 3'-Flanking Regions[a]
B0	ATG	AGGGGCCACAGG<u>GACCC</u>------------------<u>TGGTCATTAGGTGTC</u>
B1	===	==================<u>=CCCAGCCCA</u>----------===G=========
B2	===	===============A=====CC==<u>ACCGCACCGCCCT</u>=GTCG=========

[a]In the alignment the dashes (-) indicate apparent deletions relative to the B2 sequence and the equal signs (=) designate identical bases.

portions of the genes for B0/B1 and B2 must differ in some significant manner even though the gene products are remarkably similar. This appears to be analogous to the case for the b and e forms of P-450 IIB in rat (3, 32-34). Also, we have recently found (Ryan and Philpot, unpublished) that the hamster has a form of IIB P-450 that is expressed in liver but not in lung. Thus, this aspect of the regulation of forms of IIB P-450 likely originated prior to speciation, and the similarity of the gene products is probably controlled by some recombination mechanism. Additional research is required in order to calrify this point. It should also be mentioned that B0, B1, and B2 do not account for the total of the IIB P-450 in the rabbits we work with. We have characterized one additional form (Bx) from kidney, and have genomic evidence that additional gene products and allelic variants exist.

Structure and Regulation of Cytochrome P-450 Isozyme 5

The initial cDNA clones encoding isozyme 5 were isolated from a Lv-1 library by detection of the fused β-galactosidase-antigen products with antibodies to isozyme 5. The largest of these clones was then used to screen the lung library from which a number of clones were detected and isolated. The largest of these was then used to isolate clones from the Lv-4 library, from a rat lung library (Rat Lg), and from a control rabbit liver library (Lv-C). Several clones were then selected for sequencing as detailed in Table 6. All five clones (Lg:1, Lg:2, Lv-1:1, Lv-4:1, Lv-C:1) from rabbit libraries had identical sequences for their areas of overlap. Clones Lg:1, Lv-4:1, and Lv-C:1 contained open reading frames of 1518 bases that encoded for identical proteins of 506 amino acids. Identification of this sequence as that of isozyme 5 was based on the exact match of the N-terminal sequence with that of the protein (Table 7). Comparison of the derived sequence and that of rat isozyme 5 with sequences for other isozymes of P450 showed less than 30% identity with all except members of gene subfamily IVA (Table 8). The 54% identity between rabbit isozyme 5 and rabbit isozyme p2 places isozyme 5 in gene family IV, subfamily B. Clearly, the sequence obtained for rat isozyme 5, which is 87% identical to rabbit isozyme 5, indicates the rabbit and rat isozymes are orthologues. The extent of this identity is shown by a direct comparison of the amino acid sequences (Table 9). The two members of gene subfamily IVA and the two members of gene subfamily IVB have a 16 residue peptide in common. The significance of this peptide, which is shown in Table 10, is not known. Complete nucleotide sequences for the cDNAs encoding isozyme 5 from rabbit and rat are given in reference 35.

Analysis of mRNA samples from liver and lung of rabbits showed that the induction of isozyme 5 by PB is associated with an increase in specific mRNA content. This finding demonstrates that the inductive process is similar to that involved with the induction of isozyme 2. As with isozyme 2, PB administration has no effect on the level of mRNA for isozyme 5 in the lung. Results with analysis of isozyme 5 protein and activity have suggested that the pattern of regulation for isozyme 5 in rabbits is not observed with other species (2, 5). These results are summarized in Table 11. Western blot analysis of pulmonary microsomal samples shows that antibodies to rabbit isozyme 5 detect a single band of protein with all species examined. The greatest intensity of staining is seen with samples from rabbit, although relatively high levels of protein are detected with guinea pig and hamster. Weak, but clearly positive staining is observed with samples from rat and mouse. In contrast to

Table 5. Expression of mRNA Encoding B0, B1, and B2 Forms of Isozyme 2 in Lungs and Livers of Untreated Rabbits and Rabbits Treated with PB

Phenotype	Expression		Induction	
	Liver	Lung	Liver	Lung
1	B0	B0	B0	-
2	B0/B1/B2	B0/B1	B0/B1/B2	-
3	B1/B2	B1	B1/B2	-

Table 6. Cloning and Sequencing of cDNAs Encoding Isozyme 5

Library	Species	Tissue	Treatment	Clone	Number of Bases
Lg	Rabbit	Lung	Control	Lg:1	1764
Lg	Rabbit	Lung	Control	Lg:2	1240
Lv-1	Rabbit	Liver	PB	Lv-1:1	930
Lv-4	Rabbit	Liver	PB	Lv-4:1	2034
Lv-C	Rabbit	Liver	Control	Lv-C:1	1762
Rat Lg	Rat	Lung	Control	Rat Lg:1	1930

Table 7. Identity of Derived Amino Acid Sequence with that Published for the N-Terminal Region of the Protein

1. Nucleotide Sequence   ACG ATG CTC GGC TTC CTC TCC CGC CTG GGC CTG TGG
2. Derived Sequence   Met Leu Gly Phe Leu Ser Arg Leu Gly Leu Trp
3. Protein Sequence (11)   Met Leu Gly Phe Leu Ser Arg Leu Gly Leu Trp

    1. GCT TCC GGA CTG ATC TTG ATC CTA GGC TTC CTC AAG CTC CTC
    2. Ala Ser Gly Leu Ile Leu Ile Leu Gly Phe Leu Lys Leu Leu
    3. Ala Ser Gly Leu Ile Leu Ile Leu Gly Phe

Table 8. Comparison (% Identity) of Rabbit and Rat Isozyme 5 Sequences with Sequences of other P-450 Isozymes

RAT	RABBIT					
	IA (6)	IIB (2)	IIE (3a)	IIIA (3c)	IVA (p2)	IVB (5)
IA (c)	73	<30	<30	<30	<30	<30
IIB (b)	<30	76	51	<30	<30	<30
IIE (j)	<30	47	82	<30	<30	<30
IIIA (PCN1)	<30	<30	<30	70	<30	<30
IVA (LAω)	<30	<30	<30	<30	74	53
IVB (5)	<30	<30	<30	<30	52	87

results obtained with pulmonary samples, no staining is detected with
hepatic microsomal samples from rat, mouse, or guinea pig. Hepatic
samples from hamster stain in the same manner as pulmonary samples, but no
increase in staining is observed following treatment with PB. These
findings are reflected by experiments in which the effect of the antibodies to rabbit isozyme 5 on the metabolism of 2-aminofluorene was
examined. As is the case with rabbits, N-hydroxylation of 2-aminofluorene
catalyzed by pulmonary microsomal preparations is markedly inhibited with
every species examined. In contrast, significant inhibition of hepatic
activity is observed only in the case of rabbits and hamsters.

Table 9. Amino Acid Sequences of P450 Isozyme 5 from Rabbit and Rat

```
RABBIT   ML-GFLS-RL---GLWASGLILILGFLKLLRLLLRRQRLARAMDSFPGPPTHWLFGHALE    55
RAT      =VLN===PS=SRL=====VV==MVIV===FS======K=====================    60

RABBIT   IGKTGSLDKVVTWTQQFPYAHPLWVGQFIGFLNIYEPDYAKAVYSRGDPKAPDVYDFFLQ   115
RAT      ===L=======S=A====H=====F===V================A============    120

RABBIT   WIGKGLLVLDGPKWFQHRKLLTPGFHYDVLKPYVAIFADSTRIMLEKWEKKACEGKSFDI   175
RAT      ============================E===M==D======S=N=====           180

RABBIT   FSDVGHMLDTLMKCDTFGKGDSGLNHRDSSYYVAVSELTLLMQQRIDSFQYHNDFIYWLT   235
RAT      =C====================G===N===L===D======================    240

RABBIT   PHGRRFLRACRAAHDHTDRVIRQRKAALQDEKEREKIQNRRHLDFLDILLDVRGESGVOL   295
RAT      ===========KI======E============K===Q============G==D===IK=    300

RABBIT   SDTDLRAEVDTFMFEGHDTTTSGISWKLYCMALYPEHQQRCREEVREILGDQDSFQWEDL   355
RAT      ==AE===========================L=====G=============D==       360

RABBIT   AKMTYLTMCMKECFRLYPPVPQVYRQLSKPVSFVDGRSLPAGSLISLHIYALHRNSDVWP   415
RAT      ==========================N===T===================T===      420

RABBIT   DPEVFDPLRFSPENSSGRHPYAFIPFSAGPRNCIGQQFAMNEMKVVTALCLLRFEFSVDP   475
RAT      ==============AA====F==M=======================L==          480

RABBIT   LRLPIKLPQLVLRSKNGIHLYLKPLGPKAEK                              506
RAT      SKM===V===I=============ASRSG=                               511
```


Table 10. Common Peptide for P450 Gene Subfamilies IVA and IVB

Species	Gene Family	#	Amino Acid Sequence	#
Rabbit	IVB	293	QLSDTDLRAEVDTFMFEGHDTTTSGISW	321
Rabbit	IVA	300	S===Q==================A====S	328
Rat	IVB	299	K===AE==================S	326
Rat	IVA	304	S===K==================A==V==	331

aData taken from 35, 36 and 37.

62

The tissue and species distribution of isozyme 5 was further examined by Northern analysis of mRNA preparations by hybridization with cDNA encoding rabbit isozyme 5. The results, described in Table 12, were in complete agreement with what is known concerning expression and distribution of the protein recognized by antibodies to isozyme 5 (see Table 11). Therefore, a number of lines of evidence show that the hepatic distribution and regulation of isozyme 5 is quite variable across species lines. In the rabbit isozyme 5 is expressed in the liver and is highly induced by

Table 11. Immunological Detection of Homologues of Rabbit Isozyme 5 in Other Species

| Species | Western Blot[a] | | | Inhibiton of Activity[b] | | |
| | Liver | | | Liver | | |
	Control	PB	Lung	Control	PB	Lung
Rabbit	++	+++++ +++++	++++	60%	80%	80%
Rat	-	-	+	<10%	<10%	80%
Guinea Pig	-	-	+++	<10%	<10%	80%
Mouse	-	-	+	<10%	<10%	75%
Hamster	++	++	++	40%	40%	85%

[a]Relative intensity of blot staining.
[b]Percent inhibition of the N-hydroxylation of 2-aminofluorene.

Table 12. Hybridization of mRNA with cDNA Encoding Rabbit Isozyme 5

Species	Tissue	Treatment	Staining Intensity
Rabbit	Lung	None	++++++
Rabbit	Liver	None	++
Rabbit	Liver	PB	+++++++++++++++
Rat	Lung	None	+
Rat	Liver	None	-
Rat	Liver	PB	-
Hamster	Lung	None	++
Hamster	Liver	None	++
Hamster	Liver	PB	++
Guinea Pig	Lung	None	++++
Guinea Pig	Liver	None	-
Guinea Pig	Liver	PB	-

PB. In the hamster isozyme 5 is expressed in the liver but PB has no effect on its concentration. On the other hand, in the rat, mouse, and guinea pig no isozyme 5 is detected in liver. Pulmonary expression of isozyme 5 is observed with all species examined.

Structure and Regulation of the Flavin-Containing Monooxygenase in Pig and Rabbit

Background. While comparisons of lung and liver have generally supported the conclusion that isozymes of P-450 expressed in the two tissues are the same, this has not been the case for the FMO. As early as 1977 it was noted that partially-purified preparations of FMO from rabbit lung and liver had different characteristics (43). In particular, the lung enzyme was found to be stimulated by concentrations of Hg^{2+} that inhibited the hepatic enzyme. Later, it was observed that imipramine, which undergoes FMO-catalyzed N-oxidation in hepatic microsomal incubations, is not metabolized by the pulmonary microsomal fraction (12). Characterization of purified FMO from rabbit (13, 14) and mouse (14) provided the first direct evidence for differences between the lung and liver enzymes. The lung and liver enzymes show no immunochemical cross-reactivity (13, 14), have somewhat different substrate specificities (13, 15), and respond differently to temperature and ionic detergents (13, 14).

Because there is little information about the structure of the FMO proteins, the reasons for the marked differences between the lung and liver enzymes are not clear. However, it is clear that the existence of a "pulmonary" enzyme is not confined to the rabbit (8). Antibodies to the rabbit lung enzyme detect protein in pulmonary microsomal preparations from rabbit, rat, mouse, guinea pig, and hamster (Table 13). In contrast, the same antibodies do not detect hepatic microsomal proteins in any species examined. Tissue-specific expression of the "hepatic" enzyme is not as highly regulated (Table 13). Antibodies to the pig liver enzyme detect proteins in hepatic samples from all species and in pulmonary samples from all species except rabbit. Kidney proteins are detected in all species by the hepatic antibodies, but only in rabbit, mouse and hamster by antibodies to the pulmonary enzyme. The reactivities of antibodies to the mouse liver enzyme are qualitatively the same as those of antibodies to the pig liver enzyme (8).

Pig liver enzyme. Investigations of the structure of FMO proteins began with the enzyme from pig liver. Antibodies to the pig liver FMO were used to screen a library constructed from pig liver mRNA. Several clones were isolated and the largest (A; ~1400 bases) was labeled and used as a probe for further screening. A clone containing an insert of ~2000 bases (B) was isolated, fragmented by restriction with PstI and RsaI, subcloned, and sequenced. Clone B contained an open reading frame of 1596 bases, with 5'- and 3'-flanking regions of 56 and 418 bases, respectively. Clone A was also sequenced following fragmentation by sonication. The sequence of A was identical to that of B for the overlapping area of 1374 bases of which 956 were in the coding region. The base sequences for the pig liver FMO are presented in reference 21, and the derived amino acid sequence is shown in Table 14. Positive identification of the cDNA as encoding the pig liver FMO was made on the basis of the amino acid composition and the sequence of a peptide obtained from the protein (Table 15).

Analysis of mRNA from pig liver, lung and kidney shows that species encoding the liver FMO are present in all three tissues. This is consistent with the finding of protein for the liver FMO in microsomal samples from the same tissues. However, a comparison of the mRNA and protein amounts showed an apparent discrepancy. Whereas the liver contained about

64

5-times more FMO protein than did the lung and kidney, the kidney appeared to contain the most mRNA (21). Southern analysis of genomic DNA with a 5' fragment (1-388) of the pig liver cDNA gave evidence for a single gene. The following restriction enzymes resulted in the hybridization of single bands of DNA with the cDNA probe: XhoI, 14 kb; ClaI, 8 kb; BamHI, 24 kb; KpnI, 23 kb; PstI, 4.5 kb; EcoRI, 20 kb. A second, partially-related gene was detected under conditions of low stringency.

Rabbit Liver Enzyme. The cDNA clone encoding the pig liver FMO provided an excellent probe for screening a library constructed from hepatic mRNA of rabbit. Six clones were detected with this probe, and one was fragmented by a combination of restriction and sonication and sequenced. The complete sequence contained 2046 bases with an open reading frame of 1605 bases and flanking regions of 48 (5') and 393 (3') bases. The sequence of this clone (hFMO-1) is presented in reference 38, and the derived amino acid sequence is shown in Table 14. The derived sequence of the rabbit liver FMO shows 87% identity with the sequence of the pig liver FMO. The two liver enzymes have two large areas of absolute identity, residues 1-69 and residues 328-380. One notable difference between the two enzymes is the gap of three residues in the sequence of the pig liver protein relative to the rabbit liver protein. This gap aligns with residues 317-319 of the rabbit liver sequence. As a consequence of this difference, the rabbit liver enzyme contains 535 residues rather than the 532 determined for the pig liver enzyme. Analysis of mRNA from rabbit liver, lung and kidney for species encoding the pig liver FMO gave results consistent with previous findings for distribution of the protein. Single bands of mRNA (~2.6 kb) were detected with the hepatic and kidney samples and no mRNA was detected with the pulmonary sample. In contrast to the findings for mRNA encoding the pig liver FMO, the relative amounts of mRNA encoding the rabbit liver FMO in liver and kidney were similar to the relative amounts of protein detected. Analysis of genomic DNA by Southern blotting suggested that the rabbit liver FMO is encoded by a single gene.

Rabbit lung enzyme. Polyclonal antibodies to the rabbit pulmonary FMO were used to screen a library constructed with mRNA from rabbit lung. Seven clones were detected and isolated from the 50,000 plaques screened. The longest clone insert (pFMO-1) was sequenced following fragmentation with HaeIII, PstI, BalI, AluI, HincII, RsaI, and ScaI. The insert contained 2611 bases with an open reading frame of 1605 bases and flanking regions of 82 (5') and 924 (3') bases. The complete base sequence of the

Table 13. Detection of Hepatic, Pulmonary, and Renal Microsomal Proteins by Antibodies to the Rabbit Lung and Pig Liver Flavin-Containing Monooxygenases

Species	Antibody to Lung FMO			Antibody to Liver FMO		
	Liver	Lung	Kidney	Liver	Lung	Kidney
Rabbit	-	+	+	+	-	+
Rat	-	+	-	+	+	+
Mouse	-	+	+	+	+	+
Hamster	-	+	+	+	+	+
Guinea Pig	-	+	-	+	+	+

Table 14. Primary Sequences of Rabbit Lung and Liver and Pig Liver
           Flavin-Containing Monooxygenases

Enzyme	Sequence	
Rabbit Lung	MAKKVAVIGAGVSGLISLKCCVDEGLEPTCFERTEDIGGLWRFKENVEDG	50
Rabbit Liver	===R==IV======A=I===LE===========SD=L======T=H==E=	50
Pig Liver	===+==+========+=+===++==============++=+======+=+==+=	50
Rabbit Lung	RASIYQSVITNTSKEMSCFSDFPMPEDFPNFLHNSKLLEYFRIFAKKFDL	100
Rabbit Liver	===L=K==VS=SC=====Y====F===Y==YVP==QF=D=LKMY=DR=S=	100
Pig Liver	===+=+==++=++=====+P===+====+==+++==H+===+=++=NQ=N=	100
Rabbit Lung	LKYIQFQTTVISVKKRPDFASSFQWEVVTQSNSKQQSAVFDAVMVCSGHH	150
Rabbit Liver	==S===K===F=IT=CQ==NV========LHEG==E==I=======T=FL	150
Pig Liver	==C===+=K=C==+=HE==+TT===D===+C++==+=======+=++	150
Rabbit Lung	ILPNIPLKSFPGIEKFKGQYFHSRQYKHPAGLEGKRILVIGIGNSASDIA	200
Rabbit Liver	TN=HL==GC====KT===============DIFKD==V==V=M===GT===	200
Pig Liver	++=Y===D======N+=================+++++=S+==+=+===++===	200
Rabbit Lung	VELSKKAAQVYISTRKGSWVMSRISEDGYPWDMVFHTRFSSMLRNVLPRM	250
Rabbit Liver	==A=HV==KK==FL==TG=A==I==VFDS======T===QNFI==S==TP	250
Pig Liver	==+=+L=++=++==++=+==+=++++======M==++=F==+===++	250
Rabbit Lung	IVKWMMEQQMNRWFNHEAYGLAPENKYLMKEPVLNDDLPSRILYGTIKVK	300
Rabbit Liver	==T=LVAKK==S====A====V=KDRIQL=======E==G==IT=KVFIR	300
Pig Liver	==N=+I+++==+====+====I==+++++R======+==+==++=++L+=	300
Rabbit Lung	RRVKELTESAAIFEDG---TVEEDIDVIVFATGYTFAFPFLEESLVKIED	350
Rabbit Liver	PSI==VK=NSVV=GNAHNTPS==P==================D==V==V==	347
Pig Liver	+++==++=++++=NSS---+E==+==I===============+==+==+==	347
Rabbit Lung	NMVSLYKYMFPPQLEKSTFACLGLIQPLGSIFPTVELQARWATRVFKGLC	400
Rabbit Liver	GQA=====I==AH=Q=P=L-VI===K====ML==G=T====YTVQ====VI	397
Pig Liver	+++==+====+==++=+=+=+==++=====L+==+D+=====+==+==+N	397
Rabbit Lung	SLPSKETMMADIIKRNENRIALFGESLSQKLQTNYIDYLDELALEIGAKP	450
Rabbit Liver	K==PTSV=IKEVNE=K==KHNG==LCYCKA==AD==T=I=D=LTS=N===	447
Pig Liver	+==+S++=+Q===T=+===+PS+==++++++==S+==A=+===++Y=D==	447
Rabbit Lung	DLVSFLFKDPKLAVKLYFGPCNSYQYRLVGPGQWEGARNAIFTQKQRILK	500
Rabbit Liver	N=F=L=LT==L==LTMF===YSP==F==T===K=K======M==WD=TF=	497
Pig Liver	+M+=+=++==H==++I+====T+==+==+===+=========+==++=++=	497
Rabbit Lung	PLKTRTLKASSNFPVSFLLKFLGLFALVLAFLFQLQWF	535
Rabbit Liver	VT===IVQE==SPFE=L=KL=AV=AL==SV==IF=	535
Pig Liver	++===++=+=P+++A=+=++=SF=++==AI=Q++=	532

[a]Equal sign (=) indicates same amino acid as in the rabbit lung sequence,
and the plus sign (+) indicates the same amino acid as in the rabbit liver
sequence.

Table 15. Positive Identification of the cDNA Clone Encoding the Pig Liver Flavin-Containing Monooxygenase[a]

Amino Acid Composition (Residues/Molecule)

Source	A	R	D	C	N+ E	G	H	I	L	K	M	F	P	S	T	W	Ÿ	V
Protein	31	21	52	09	52	37	10	29	58	39	10	37	42	38	33	09	19	44
Derived	30	19	47	12	45	33	09	30	51	36	12	37	34	37	31	09	19	41

Peptide Sequence Comparison

Source	308	Sequence	320
Protein		ASN SER VAL VAL PHE ASN SER SER PRO GLU	
Derived		GLU ASN SER VAL VAL PHE ASN SER SER PRO GLU GLU GLU	

[a]Data taken from 21 and 45.

rabbit pulmonary FMO cDNA is presented in reference 38, and the derived amino acid sequence is shown in Table 14. The sequenced clone terminated (3') with 10 adenosine nucleotides, but no consensus polyadenylation signal was present 10-30 bases upstream. However, within a 36-base region 385 bases from the 3' terminus three putative polyadenylation signals (AATAAA) were found. Analysis of additional clones yielded two variations at the 3' end: first, clones 2, 4, and 6 showed a single adenosine deletion in the area where clone 1 contained 7 consecutive adenosines (bases 2109-2155); second, clone 7 terminated in CAAAA which might represent the beginning of a poly(A) tail for a transcript shorter than that of clone 1. At their 5' ends clones 2 and 4 also differed in sequence from clone 1. Both clones have the same two differences, substitutions of T for G at position 440 and G for C at position 485. These substitutions both result in amino acid changes; serine for alanine at position 120 and glutamine for glutamic acid at position 136. Additional evidence from an expression system indicates that clones 1 and 2 encode for allelic variants and account for the different phenotypes detected by protein mobility on polyacrylamide gels containing SDS (8). Comparison of the primary structures of the rabbit pulmonary and hepatic FMO isozymes (Table 14) shows 56% identity. Evidence that the pulmonary cDNA did in fact encode the pulmonary FMO came from matching peptide sequences with sequences derived from the cDNA (Table 16). In all, the sequences of four peptides, obtained from purified pulmonary FMO by tryptic digestion, were determined. These sequences, which contained a total of 31 amino acids, matched perfectly with sequences derived from the cDNA.

Multiple bands of mRNA were detected in samples from lung and kidney by hybrization with cDNA encoding the pulmonary FMO. Four distinct bands (2.4, 2.6, 4.8, and 6.0 kb) were resolved. The relative intensities of these bands were constant among individual samples and gave no evidence of being associated with the expression of different proteins. Likely the band at 2.4 kb is represented by clone 7 and the band at 2.6 kb by clone 1. The origin of these bands is under investigation, and preliminary evidence suggests that each contains the same coding information and that the four populations arise by variable 3' processing. The possibility of a single gene encoding for the pulmonary FMO is reinforced by results of Southern analysis; single bands of DNA were detected by hybridization with a 3' fragment (bases 1368-2611) after restriction with NcoI, HindIII, EcoRI, or PstI.

Comparison of FMO enzymes from pig and rabbit. The comparison of primary sequences for the pig hepatic and the rabbit hepatic and pulmonary FMO enzymes shown in Table 14 provides a great deal of information. First, the overall similarity of the liver enzymes (87% identity) is seen to be much greater than the similarity between the liver and lung enzymes from rabbit (56%). This difference indicates that the evolutionary divergence of the pulmonary and hepatic isozymes actually began prior to speciation. It should be noted that in several instances the pig liver and rabbit lung enzymes share have some common characteristics that set them apart from the rabbit liver enzyme. For example, the three residue gap in the pig liver sequence relative to the rabbit liver sequence is also found in the sequence of the rabbit lung enzyme, and the same amino acid substitutions are found in the pig liver and rabbit lung enzymes at positions 88, 92, 113, 139, 159, 242, 274, 300, 391, 392, 394, 441, 471, 485, and 508 of the rabbit liver sequence. Second, nearly identical pyrophosphate-binding regions are present in each enzyme. The FAD region (GAGVSG) at residues 9-14 and the NADP region (GxGxxG/A) at residues 191-196 very closely resemble similar binding sites present in a number of enzymes from a wide variety of sources (Table 17). It is of interest that with each of these enzymes the FAD and NADP sites are separated by about 170 residues. Third, the primary sequence data yields information about the hydropathy of the enzymes. Although the primary sequences are significantly different, the liver and lung enzymes have predicted membrane-associated peptides that are extremely similar. The liver enzyme does, however, contain a hydrophilic peptide (centered at residue 461) not present in the pulmonary enzyme.

The differences between the primary structures of the rabbit liver and lung FMO enzymes demonstrates that they are the products of different genes and not derived from the same genomic information by alternative

Table 16. Peptide Sequences from Purified Rabbit Lung Flavin-Containing Monooxygenase and Derived from cDNA Nucleotide Sequences

Source of Sequence	#	Sequence
Protein	032	THR GLU ASP ILE GLY GLY LEU TRP ARG
cDNA		GLU ARG THR GLU ASP ILE GLY GLY LEU TRP ARG PHE LYS
Protein	213	LYS GLY SER TRP VAL MET SER ARG
cDNA		THR ARG LYS GLY SER TRP VAL MET SER ARG ILE SER GLU
Protein	362	SER THR PHE
cDNA		GLU LYS SER THR PHE ALA CYS LEU GLY LEU ILE GLN PRO
Protein	473	LEU VAL GLY PRO GLY GLN TRP GLU GLY ALA ARG
cDNA		TYR ARG LEU VAL GLY PRO GLY GLN TRP GLU GLY ALA ARG

splicing or some other mechanism. The patterns obtained by Southern analysis of genomic DNA indicate that each enzyme may not be encoded for by more than one gene, although one or more related genes (<70% identity) may be present. Clearly, the multiple populations of mRNA associated with the pulmonary enzyme are not related to multiple genes, and the proteins of different mobilities are likely allelic variants.

The problem of multiple pulmonary proteins associated with the pulmonary FMO is now being investigated in guinea pigs. Three protein phenotypes have been identified and cDNA libraries have been constructed with mRNA from homozygous animals. Preliminary evidence indicates that the proteins expressed in guinea pig lung are highly identical (~85%) to the rabbit pulmonary enzyme, and that the protein phenotypes reflect the expression of allelic variants.

SUMMARY AND CONCLUSIONS

The tissue and species expression of various drug-metabolizing enzymes is just beginning to be understood in detail. These details include structure, substrate specificity, and regulation, all of which are important determinants in tissue- and species-specific aspects of the metabolism of xenobiotics. The importance of this information is directly related to the fact that most toxic chemicals, including carcinogens and mutagens, have little biological activity without first being modified by a process that has come to be known as "metabolic activation." For such chemicals it is understood that metabolism is a requisite step in the pathway from exposure to toxicity. What is not so well understood is the relationship between metabolism and the species-, tissue- and cell-specific effects of many toxicants. The lung is a good example of a tissue that exhibits species- and cell-specific toxic responses to various chemicals. In order to understand whether or not the selective effects of these toxic agents are related to a tissue-specific aspect of their metabolism, the enzyme systems involved must be understood completely. As a contribution to this understanding we have examined several drug-metabolizing enzymes expressed in liver and lung. This tissue comparison has yielded different answers for each enzyme, P-450 isozymes 2 and 5 and the FMO (Table 18).

Table 17. Pyrophosphate-Binding Regions of the Flavin-Containing Monooxygenases and Some Other FAD Enzymes

Enzyme	#	FAD Region	#	NADP Region
		******		******
Liver FMO	006	AVIGAGVSGLIS	188	LVVGMGNSGTDIA
Lung FMO	006	AIVGAGVSGLIS	188	LVIGIGNSASDIA
TMO	042	AIVGAGISCLVA	235	GSLGIGSGGFLPV
MRase	102	AVIGSGGAAMAA	274	AVIGSSVVALELA
ADR	010	CVVGSGPAGFYT	150	VILGQGNVALDVA
GRase	024	LVIGGGSGGLAS	191	VIVGAGYIAVEMA
		******		******

aThe enzymes are: FMO, flavin-containing monooxygenase (21, 38); TMO, tryptophan 2-monooxygenase (39); MRase, mercurate reductase (40); ADR, adrenodoxin oxidoreductase (41); GRase, glutathione reductase (42).

The comparison of isozyme 5 expressed in lung and liver provides the least complicated case. Identical forms of isozyme 5 are expressed in both tissues with the concentration somewhat greater in lung than liver. Sequencing of cDNAs from libraries of both tissue indicates that a single gene produce is expressed. However, Southern analysis of genomic DNA indicates the possibility of more than one gene. Additional work is required in order to clarify this point. Although the same isozyme 5 is expressed in liver and lung, regulation in the two tissues is clearly different, induction by PB occurs in liver but not lung. This effect in liver is species-specific and is not observed with rats, mice, guinea pigs, or hamsters. In fact, homologs of isozyme 5 are not expressed at all in livers of rats, mice or guinea pigs. Therefore, in most species isozyme 5 can be considered an extrahepatic P-450.

The situation with isozyme 2 is clearly more complicated than with isozyme 5. First of all, multiple forms of the enzyme are expressed. These forms, which appear to include products of different genes as well as allelic variants, are all expressed in liver but at least one is not expressed in lung. As is the case for isozyme 5, induction by PB occurs in liver and is observed with all three forms of the enzyme. Potentially, the tissue-specific distribution of forms of isozyme 2 could result in tissue-specific differences in the metabolism of certain toxicants. Attempts to determine whether or not this is true are underway. The various forms of isozyme 2 will be expressed in cell systems and their substrate specificities examined. This information will then be applied to investigations of microsomal preparations phenotyped by results of Northern analysis. Also, recent evidence indicates that additional forms of isozyme 2 exist. These forms need to be characterized with respect to structure, distribution and regulation.

Finally, the FMO presents a picture that is unique with respect to that for either isozyme 5 or isozyme 2. With the FMO the enzymes expressed in liver and lung are products of distinct genes. These genes are related, but only distantly. The FMO expressed in rabbit lung is 56% identical to the FMO expressed in rabbit liver, a relationship that is not nearly as extensive as that between the hepatic FMOs from rabbit and pig (87% identity). It is already known that the liver and lung enzymes have somewhat different substrate specificities, and their activities are now being further investigated in expression systems. As with isozyme 5, species differences in tissue distribution are observed. The "liver" enzyme is expressed in liver of all species examined and in lungs of all species except rabbits. In contrast, the "lung" enzyme, which is detected in all lungs, is not detected in any liver samples.

Table 18. Summary of the Expression and Regulation of P450 Isozymes 2 and 5 and the FMO in Rabbit Liver and Lung

Enzyme	Tissue Expression	Content	Induction
Isozyme 2-B0	Lung and Liver	Lung>Liver	Liver only
Isozyme 2-B1	Lung and Liver	Lung>Liver	Liver only
Isozyme 2-B2	Liver only		Liver only
Isozyme 5	Lung and Liver	Lung>Liver	Liver only
Pulmonary FMO	Lung and Kidney	Lung>Kidney	None
Hepatic FMO	Liver and Kidney	Liver>Kidney	None

Our results show that the patterns of expression and regulation of drug-metabolizing enzymes in lung and liver vary considerably with respect to the specific enzyme and species in question. For the enzymes studied, qualitative uniformity is observed across species lines for pulmonary expression. In contrast, hepatic expression is highly species-dependent.

REFERENCES

1. D. W. Nebert, D. R. Nelson, M. Adesnik, M. J. Coon, R. W. Estabrook, F. J. Gonzalez, F. P. Guengerich, I. C. Gunsalus, E. F. Johnson, B. Kemper, W. Levin, I. R. Phillips, R. Sato, and M. R. Waterman, The P450 superfamily: Updated listing of all genes and recommended nomenclature for the chromosomal loci, DNA 8:1 (1989).
2. R. R. Vanderslice, B. A. Domin, G. Carver, and R. M. Philpot, Species-dependent expression and induction of homologues of rabbit cytochrome P-450 isozyme 5 in liver and lung, Mol. Pharmacol. 31:320 (1987).
3. B. A. Domin, C. J. Serabjit-Singh, R. R. Vanderslice, T. R. Devereux, J. R. Fouts, J. R. Bend, and R. M. Philpot, Tissue and cellular differences in the expression of cytochrome P-450 isozymes, in: "Proceedings of IUPHAR 9th International Congress of Pharmacology," W. Paton, J. Mitchell, and P. Turner, eds., Macmillan Press Ltd., London (1984).
4. R. M. Philpot, B. A. Domin, T. R. Devereux, C. Harris, M. W. Anderson, J. R. Fouts, and J. R. Bend, Cytochrome P-450 monooxygenase systems of the lung: Relationships to pulmonary toxicity, in: "Microsomes and Drug Oxidations, Proceedings of the 6th International Symposium," A. R. Bobbis, J. Caldwell, F. DeMateis, and C. R. Elcombe, eds., Taylor and Francis Ltd., London (1985).
5. R. Gasser and R. M. Philpot, Structure and function of homologous pulmonary and hepatic isozymes of cytochrome P-450, in: "Microsomes and Drug Oxidations, Proceedings of the 7th International Symposium," J. Minors, D. J. Birkett, R. Drew, and M. McManus, eds., Taylor and Francis Ltd., London (1988).
6. D. M. Ziegler, Flavin-containing monooxygenases: Catalytic mechanisms and substrate specificities, Drug Metab. Rev. 6:1 (1988).
7. B. P. Smyser and E. Hodgson, Metabolism of phosphorous-containing compounds by pig liver microsomal FAD-containing monooxygenase, Biochem. Pharmacol. 34:1145 (1985).
8. R. E. Tynes and R. M. Philpot, Tissue- and species-dependent expression of multiple forms of mammalian microsomal flavin-containing monooxygenase, Mol. Pharmacol. 31:569 (1987).
9. C. J. Serabjit-Singh, C. R. Wolf, and R. M. Philpot, The rabbit pulmonary monooxygenase system: Immunochemical and biochemical characterization of the enzyme components, J. Biol. Chem. 254:9901 (1979).
10. S. R. Slaughter, C. R. Wolf, J. P. Marciniszyn, and R. M. Philpot, The rabbit pulmonary monooxygenase system: Partial structural characterization of the cytochrome P-450 components and comparison to the hepatic cytochrome P-450, J. Biol. Chem. 256:2499 (1981).
11. Z. Parandoosh, V. S. Fujita, M. J. Coon, and R. M. Philpot, Cytochrome P-450 isozymes 2 and 5 in rabbit lung and liver: Comparisons of structure and inducibility, Drug Metab. Dispo. 15:59 (1987).
12. I. G. C. Robertson, C. J. Serabjit-Singh, J. E. Croft, and R. M. Philpot, The relationship between increases in the hepatic content of cytochrome P-450, form 5, and in the metabolism of

aromatic amines to mutagenic products following treatment of rabbits with phenobarbital, Mol. Pharmacol. 24:156 (1983).

13. Williams, D. M. Ziegler, D. J. Nordin, S. E. Hale, and B. S. S. Masters, Rabbit lung flavin-containing monooxygenase is immuno-chemically and catalytically distinct from the liver enzyme, Biochem. Biophys. Res. Commun. 125:116 (1984).

14. R. E. Tynes, P. J. Sabourin, and E. Hodgson, Identification of distinct hepatic and pulmonary forms of microsomal flavin-containing monooxygenase in the mouse and rabbit, Biochem. Biophys. Res. Commun. 126:1069 (1985).

15. R. E. Tynes, P. J. Sabourin, E. Hodgson, and R. M. Philpot, Formation of hydrogen peroxide and N-hydroxylated amines catalyzed by pulmonary flavin-containing monooxygenases in the presence of primary alkylamines, Arch. Biochem. Biophys. 251:654 (1986).

16. R. Gasser, M. Negishi, and R. M. Philpot, Primary sequences of multiple forms of cytochrome P-450 isozyme 2 derived from rabbit pulmonary and hepatic cDNAs, Mol. Pharmacol. 32:22 (1988).

17. J. M. Chirgwin, A. E. Przybyla, R. J. Macdonald, and W. J. Rutter, Isolation of biologically active ribonucleic acid from sources enriched in ribonuclease, Biochemistry 18:5294 (1979).

18. V. Glisin, R. Crkvenjakov, and C. Byus, Ribonucleic acid isolated by cesium chloride centrifugation, Biochemistry 13:2633 (1974).

19. M. Gross-Bellard, P. Oudet, and P. Chambon, Isolation of high-molecular-weight DNA from mammalian cells, Eur. J. Biochem. 36:32 (1973).

20. P. J. Enrietto, L. N., Payne, and M. J. Hayman, A recovered avian myelocytomaosis virus that induces lymphomas in chickens: Patho-genic properties and their molecular basis, Cell 35:369 (1983).

21. R. Gasser, R. E. Tynes, M. P. Lawton, K. K. Korsmeyer, D. M. Ziegler, and R. M. Philpot, The flavin-containing monooxygenase expressed in pig liver: Primary sequence, distribution and evidence for a single gene, Biochemistry 29:119 (1990).

22. P. L. Deininger, Random subcloning of sonicated DNA: Application to shotgun DNA sequence analysis, Anal. Biochem. 129:216 (1983).

23. J. Messing, New M13 vectors for cloning, Methods Enzymol. 57:20 (1983).

24. F. Sanger, S. Nicklen, and A. R. Coulson, DNA sequencing with chain terminating inhibitors. Proc. Natl. Acad. Sci. USA 74:5463 (1977).

25. M. D. Biggin, T. J. Gibson, and G. F. Hong, Buffer gradient gels and [^{35}S] label as an aid to rapid DNA sequence determination, Proc. Natl. Acad. Sci. USA 80:3963 (1983).

26. R. Kraft, J. Tardiff, K. S. Krauter, and L. A. Leinwand, Using mineprep plamid DNA for sequencing double stranded templates with sequenase, BioTechniques 6:544 (1988).

27. J. Devereux, P. Haeberli, and O. Smithies, A comprehensive set of sequence analysis programs for the VAX, Nucleic Acids Res. 12:387 (1984).

28. W. J. Wilbur and D. J. Lipman, Rapid similarity searches of nucleic acid and protein data banks, Proc. Natl. Acad. Sci. USA 80:726 (1983).

29. A. Feinberg and B. Vogelstein, A technique for radiolabeling DNA restriction endonuclease fragments to high specific activity, Anal. Biochem. 132:6 (1983).

30. E. M. Southern, Detection of specific DNA fragments separated by gel electrophoresis, J. Mol. Biol. 98:503 (1975).

31. G. E. Tarr, S. D. Black, V. S. Fujita, and M. J. Coon, Complete amino acid sequence and predicted membrane topology of pheno-barbital-induced cytochrome P-450 (isozyme 2) from rabbit liver microsomes, Proc. Natl. Acad. Sci. USA 80:1169 (1983).

32. C. J. Omiecinski, Tissue-specific expression of rat mRNAs homologous to cytochrome P-450b and P-450e, Nucleic Acids Res. 14:1525 (1986).

33. F. G. Walz, Jr., G. P. Vlasuk, C. J. Omiecinski, E. Bresnick, P. E. Thomas, D. E. Ryan, and W. Levin, Multiple, immunoidentical forms of phenobarbital-induced rat liver cytochrome P-450 are encoded by different mRNAs, J. Biol. Chem. 257:4023 (1982).

34. A. Rampersaud and F. G. Walz, Jr., At least six forms of extremely homologous cytochrome P-450 are encoded at two closely linked genetic loci, Proc. Natl. Acad. Sci. USA 80:6542 (1983).

35. R. Gasser and R. M. Philpot, Primary structures of cytochrome P-450 isozyme 5 from rabbit and rat and regulation of species-dependent expression and induction in lung and liver: Identification of cytochrome P-450 gene subfamily IVB, Mol. Pharmacol. 35:617 (1989).

36. J. P. Hardwick, B.-J. Song, E. Huberman, and F. J. Gonzalez, Isolation, complementary DNA sequence, and regulation of rat hepatic lauric acid ω-hydroxylase (cytochrome P-450LAω): Identification of a new cytochrome P-450 gene family. J. Biol. Chem. 262:810 (1987).

37. S. Matsubara, S. Yamamoto, K. Sogawa, N. Yokotne, Y. Fujii-Kuriyama, M. Haniu, J. E. Shively, O. Gotoh, E. Kusunose, and M. Kusunuse, cDNA cloning and inducible expression during pregnancy of the mRNA for rabbit pulmonary prostaglandin ω-hydroxylase (cytochrome P-450 p-2), J. Biol. Chem. 262:13366 (1987).

38. M. P. Lawton, R. Gasser, R. E. Tynes, E. Hodgson, and R. M. Philpot, The flavin-containing monooxygenase enzymes expressed in rabbit liver and lung are products of related but distinctly different genes, J. Biol. Chem. 265:5855 (1990).

39. T. Yamada, C. J. Palm, B. Brooks, and T. Kosuge, Nucleotide sequences of the Pseudomonas savastanoi indoleacetic acid genes show homology with Agrobacterium tumefaciens T-DNA, Proc. Natl. Acad. Sci. USA 82:6522 (1985).

40. N. L. Brown, S. J. Ford, R. D. Pridmore, and D. C. Fritzinger, Nucleotide sequence of a gene from the Pseudomonas transposon Tn501 encoding mercuric reductase, Biochemistry 22:4089 (1983).

41. Y. Nonaka, H. Murakami, Y. Yabusaki, S. Kuramitsu, H. Kagamiyama, T. Yamano, and M. Okamoto, Molecular cloning and sequence analysis of full-length cDNA for mRNA of adrenodoxin oxidoreductase from bovine adrenal cortex, Biochem. Biophys. Res. Commun. 145:1239 (1987).

42. R. L. Krauth-Siegel, R. Blatterspiel, M. Saleh, E. Schiltz, R. H. Schirmer, and R. Untucht-Grau, Glutathione reductase from human erythrocytes. The sequences of the NADPH domain and of the interface domain, Eur. J. Biochem. 121:259 (1982).

43. T. R. Devereux, R. M. Philpot, and J. R. Fouts, The effect of $Hg2^+$ on rabbit hepatic and pulmonary microsomal and purified N,N-dimethylaniline N-oxidases, Chemico-Biol. Interact. 18:277 (1977).

44. Y. Ohmiya and H. M. Mehendale, Species differences in the pulmonary N-oxidation of chlorpromazine and imipramine, Pharmacology 22:289 (1984).

45. L. L. Poulsen and D. M. Ziegler, The liver microsomal FAD-containing monooxygenase: Spectral characterization and kinetic studies, J. Biol. Chem., 254:6449 (1979).

INSECT CYTOCHROME P450

E. Hodgson and Randy Rose

Department of Toxicology, Box 7633
North Carolina State University
Raleigh, North Carolina 27695 U.S.A.

INTRODUCTION

Cytochrome P450 (P450) dependent monooxygenase systems occur in most if not all organisms, from bacteria to mammals and higher plants. In insects, the biochemical functions ascribed to this enzyme system include the metabolism of a wide variety of endogenous substrates such as hormones and the oxidation of xenobiotics such as pesticides and secondary plant metabolites. Insect monooxygenase systems appear to be involved in many different physiological processes with roles in growth and development, insecticide resistance and in the interaction of phytophagous insects with multiple host plants. A number of reviews have appeared on these topics in recent years (1-5). Due to it's importance in the study of the biochemistry and genetics of insecticide resistance the house fly (Musca domestica) has been a useful model species for the study of insect monooxygenases. The purification and characterization of multiple cytochrome P450 isozymes from this species, as well as their role in insecticide resistance is discussed. Recently with the cloning and sequencing of a P450 isozyme from the housefly (6,7) a beginning has been made in the study of the molecular biology of insect cytochrome P450. Heliothis spp are also useful model species since their polyphagy has led to the study of the role of P450 in insect-plant interactions.

DISTRIBUTION AND GENERAL PROPERTIES OF INSECT MONOOXYGENASES

P450 was first demonstrated in insects by Ray (8,9) and is now known to be distributed in many tissues and species, the highest concentrations being generally found in the midgut, fat body and Malpighian tubules. At the sub-cellular level, insect monooxygenases are found in both the endoplasmic reticulum (10,11), (the microsomal monooxygenases), and in mitochondria (12). As in mammals, insect microsomal monooxygenase systems consist of two components, the flavoprotein NADPH-cytochrome P450 reductase and the hemeprotein, cytochrome P450, the latter existing as multiple isozymes displaying overlapping substrate specificities. The mechanisms of action appear to be similar to those seen in other animals. Cytochrome b_5 is present in insect microsomes and, as in mammals and fish (13,14), appears to be involved synergistically in the oxidation of a number of substrates eg. benzo(a)pyrene and lauric acid (15,16) possibly supplying one electron from NADH via cytochrome b_5 reductase. Spectral interactions between insect P450 and monooxygenase substrates and inhibitors are much the same as those seen in mammalian systems. A characteristic type I binding spectrum is often difficult to demonstrate but, when present, has an absorption maximum at 385-390 nm and a minimum at 420 nm. It is seen with the oxidized form of the cytochrome and substrates while type II binding spectra, with a maximum at 420-425 nm and a minimum at 390-410 nm, are seen with many nitrogen containing inhibitors. The reduced cytochrome shows the well known carbon monooxide difference spectrum and a type III spectrum with ethyl isocyanide and methylenedioxyphenyl compounds consisting of two pH dependant peaks at 430 nm and 455 nm. Spectral studies with insect microsomes have been considered in detail in recent reviews (3,17).

METABOLIC REACTIONS

Metabolism of Hormones and Hormone Analogues

P450-dependent monooxygenases play an important role in the synthesis and degradation of insect hormones. The terpenoid juvenile hormones (JHs), which suppress molting, and the steroid ecdysones, which promote it, have both been studied. The final step of JH-III (methyl,10,11-epoxy-3,7,11-trimethyl-trans-2,6-dodecadienoate), synthesis, an epoxidation, occurs in the corpora allata of several insects. This reaction has been shown to be due to a P450 dependent monooxygenase in microsomes from the corpora allata of both Blaberus giganticus (18) and Locusta micratoria (19). The anti-JHs, precocenes I and II (7-methoxy-2,2-dimethylchromene and 6,7-dimethoxy-2,2-dimethylchromene) which cause precocious molts and destroy the corpora allata are thought to act as "suicide substrates" for a form of P450 (20) found in this endocrine organ. Oxidative degradation of JH I (E,E,cis-methyl-10,11-epoxy-7-ethyl-3,7-dimethyl-2,6-tridecadienoate) has been demonstrated in housefly and blowfly microsomes (21) and JH I acts as a competitive inhibitor of several P450 dependant oxidations in house fly microsomes (22). Moreover, JH analogues of the alkyl 3,7,11-trimethyl-2,4-dodecadienoate type are easily oxidized by fly microsomes (23,24). However, despite this facile oxidation of JHs in vitro, much of the metabolism in vivo appears to be due to epoxide hydrolase and esterase (25,26). Oxidative metabolism of JH III could not be demonstrated in house fly microsomes (16) and only the JH dihydrodiol, JH acid and JH acid dihydrodiol, characteristic of epoxide hydrolase and esterase attack, were recovered.

Ecdysteroids that play a major role in insect development and metamorphosis are a group of 8 or 9 closely related C27 steroids whose synthesis and degradation appears to be carried out via a number of mitochondrial and microsomal monooxygenases present in different tissues (27). The best studied of these reactions is ecdysone 20-hydroxylase which was localized in fat body mitochondria from Manduca sexta (27,28). However, this monooxygenase has also been shown to be microsomal in the fat body and Malpighian tubules of the locust Locusta migratoria (29,30) and to have a dual location, in both mitochondria and microsomes, from Manduca sexta midgut (12) and from Drosophila melanogaster larvae (31).

A role has also been suggested for monooxgenases in insect pheromone biosynthesis and degradation since many of these hormones are steroid or terpenoid in nature (32,33). White et al (34,35) have produced evidence for the role of a monooxygenase in the synthesis of a terpenoid pheromone from alpha-pinene in the bark beetle Dendroctonus terebrans and, more recently, Ahmad et al (36) have produced evidence for the metabolism of sex pheromones by a P450 dependent monooxygenase in the house fly.

Metabolism of Fatty Acids

Insect monooxygenases are involved in the oxidative metabolism of fatty acids in much the same way as in mammals. Lauric acid has been shown to be a substrate for monooxygenase in house fly microsomes (16,37) showing both omega and omega minus one hydroxylase activity.These reactions are higher in susceptible than in insecticide resistant strains and appear to be stimulated when NADH is present in addition to NADPH, suggesting the involvement of cytochrome b_5.

METABOLISM OF XENOBIOTICS

Because of the importance of insect monooxygenases in the metabolism of pesticides and their participation in insecticide resistance the majority of published studies of xenobiotic metabolism have involved the oxidation of insecticides (See reference 3 for a detailed review). From those studies, it is apparent that insect monooxygenases, like those of mammals and other animal groups, catalyze a wide variety of oxidations resulting in either detoxification or activation. These reactions have been classified (38), as desulfuration and ester cleavage, epoxidation, aliphatic hydroxylation, aromatic ring hydroxylation, N-, 0- and S-dealkylation, sulfoxidation, dehydrogenation and dioxole ring cleavage. An important activation catalyzed preferentially by insect monooxygenases is the oxidative desulfuration of many phosphorothioate insecticides such as parathion to their oxons, resulting in inhibition of acetylcholinesterase and the toxicity associated with these compounds in both insects and vertebrates.

FUNCTIONAL ROLES OF INSECT MONOOXYGENASES

Growth and Development

Dramatic variations are seen in the levels of P450 during the developmental cycle of most insect species (3,39). In general P450 is undetectable in the egg and then rises and falls in each larval instar, being at a maximum in the intermolt period and a minimum during molting. P450 is undetectable in pupae (40,41). In adult insects, P450 concentrations increase after pupation to a maximum at 7-9 days post-emergence and then slowly decline (42).

Changes in monooxygenase levels are clearly an important feature of insect growth and development. Monooxygenases are intimately involved in the synthesis and degradation of insect growth hormones but are also regulated by these same hormones. Synthetic JH and ecdysterone have been shown to induce heptachlor epoxidation in house fly microsomes (43), probably via a transcriptional mechanism.

Insect-Plant Interactions

The importance of microsomal monooxygenases in plant feeding was first demonstrated by Krieger et al (44), who, on examination of aldrin epoxidase activity in mid-gut preparations from 55 species of lepidopterous larvae, were able to show a greater than 10-fold higher activity in polyphagous compared to monophagous species. Oligophagous insects showed intermediate values. This suggested that a relationship exists between monooxygenase activity and adaptation to a wide variety of food plants. Much of the work on insect monooxygenase, food plant interactions was reviewed by Brattsten (32,33).

Such a relationship between monooxygenase activity and diet has also been shown in other animal groups. In birds, fish-eating seabirds and raptors which have restricted diets, show lower relative monooxygenase activities than do herbivorous and omnivorous species (45,46). Adaptation to plant secondary products encountered in the diet has also been suggested as the basis for the rapid evolution of the debrisoquine hydroxylases (P450 family IID) in mammals (47).

In recent years, many investigations have focused on the dietary induction of insect P450 due to the importance of this phenomenon in insect/host plant adaptation and the cross resistance to pesticides that often accompanies such induction. This area has been the subject of detailed reviews (4,32). In experiments in which lepidopterous larvae were fed either artificial diets or on various food plants, peppermint was found to be a strong inducer of both P450 and aldrin epoxidase activity (48,49). Monoterpines such as pulegone, menthone and menthol were shown to be the active components (50,51). In the tobacco budworm (Heliothis virescens), Riskallah et al (52,53) have shown that P450 content is increased 2-3 fold in mid-gut microsomes from animals fed on wild tomato leaves compared to an artificial diet. This induction was accompanied by a more than 4-fold increase in tolerance to diazinon toxicity. Diazinon was degraded faster in induced animals which displayed increases in desulfuration, side chain hydroxylation and oxidative dearylation of the pesticide. The active component in tomato leaves previously identified as 2-tridecanone (54) induced 3 protein bands in the 48,000-58,000 M.W. region on SDS-PAGE, the most prominent at M.W. 50,000. Purification and characterization of this tridecanone inducible P450 isozyme is in progress.

Although many studies have amply documented that induction by allelochemicals is accompanied by increases in metabolic capacity, few studies have been conducted with insects possessing metabolic resistance to these allelochemicals. The recent development of tobacco budworm strains with heritable resistance to the allelochemicals, 2-tridecanone, nicotine and quercetin has allowed for recent comparisons of monooxygenase enzymes between these strains (Rose and Hodgson - unpublished results).

A comparison of P450 content between strains resistant to 2-tridecanone (WAK-T) and nicotine (WAK-N) demonstrated a 2-fold increase above levels of P450 found in the susceptible strain (WAK) from which they were derived (Table 1). In order to further characterize allelochemical resistance, a variety of monooxygenase substrates were examined in the various strains. The substrates selected for these studies represent several different reaction pathways, including O-demethylation (p-nitroanisole and methoxyresorufin), N-demethylation (benzphetamine), aryl and alkyl hydroxylation (benzo(a)pyrene and lauric acid, respectively), and sulfoxidation (phorate). These studies were performed on progeny of parents which had been reared in the absence of selection pressure in order to avoid the possibility that parental treatment might bias the results.

The increases in P450 content observed in the WAK-N strain were associated with increases in metabolism of several substrates including p-nitroanisole (2x, benzphetamine (2x), phorate (2x), and benzo(a)pyrene (3x). Larvae of the WAK-T strain also had increased metabolism towards benzo(a)pyrene (2x) and phorate (2x). Since the substrates induced by 2-tridecanone resistance are a subset of those induced by nicotine resistance, one might expect that nicotine resistant larvae might also be tolerant to 2-tridecanone. Preliminary data suggest that this is the case, however, larvae resistant to 2-tridecanone have no advantage when exposeed to nicotine. These results, combined with the general lack of cross resistance between these strains and the quercetin resistant strain suggest that different isozymes of P450 are involved in the resistance mechanisms.

Although the P450 content of the quercetin-resistant strain (BC-Q) was only slightly elevated, significant increases in activity were observed in this strain towards benzo(a)pyrene (3x), p-nitroanisole (2x) and phorate (2x) relative to the strain from which it was derived (NRA). These results suggest that while specific isozymes may have increased as a result of resistance to quercetin, other isozymes contributing to overall P450 content are repressed by this resistance.

Comparisons of metabolic activity among strains following allelochemical exposure should show both the effect of previous selection and induction. Larvae of the control strain exposed to 2-tridecanone (Wak on T) have essentially the same metabolic profile as 2-tridecanone resistant larvae reared on 2-tridecanone (Wak-T on T) (Table 2). Although P450 content of the Wak-T and Wak-T on T strains are similar, there are significant changes in the metabolic profile for these two conditions. Wak-T on T larvae have increased p-

Table 1. A Comparison of Resistant and Susceptible Strains[a]
of the Tobacco Budworm with Respect to Various
Substrates[b] of Monooxygenase Enzymes

Strain	P450	PNA	BP	MRR	B(a)P	Phorate	LA
NRA.	.255	0.70	1.61	.024	0.52	2.11	2.39
BC-Q	.369	1.43	2.25	.020	1.51	4.23	3.55
WAK	.300	1.85	2.16	.030	0.60	7.42	4.01
WAK-T	.603	2.04	4.56	.052	2.01	12.07	4.86
WAK-N	.621	3.47	7.95	.074	2.70	11.54	5.55

[a]The quercetin resistant strain (BC-Q) is derived from the
susceptible (NRA) strain, while the 2-tridecanone (WAK-T)
and nicotine (WAK-N) strains are derived from the
susceptible (WAK) strain.
[b]P450 content in nmoles/mg protein. Substrates are p-
nitroanisole (PNA), benzphetamine (BP), methoxyresorufin
(MRR), benzo(a)pyrene (B(a)P), phorate, and lauric acid
(LA). Activities are expressed in nmole/min/mgp with the
exception of MRR and·B(a)P which are in fluorescent
units/min/mg protein.

Table 2. The Effect of Induction by 2-tridecanone and
Nicotine on Substrate[a] Specificity in Resistant and
Susceptible Strains[b] of the Tobacco Budworm

Strain	P450	PNA	BP	MRR	B(a)P	Phorate	LA
WAK	.300	1.85	2.16	.030	0.60	7.42	4.01
WAK on T	.526	10.00	11.41	.074	10.60	14.65	4.85
WAK on N	.344	3.38	3.86	.046	1.84	8.09	5.42
WAK-T	.603	2.04	4.56	.052	2.01	12.07	4.86
WAK-T on T	.524	6.32	12.97	.044	9.03	11.29	6.81
WAK-N	.621	3.47	7.95	.074	2.70	11.54	5.55
WAK-N on N	.944	7.18	11.83	.150	7.09	18.11	6.45

[a]P450 content in nmoles/mg protein. Substrates are p-
nitroanisole (PNA), benzphetamine (BP), methoxyresorufin
(MRR), benzo(a)pyrene (B(a)P), phorate, and lauric acid (LA).
Activities are expressed in nmole/min/mgp with exception of
MRR and B(a)P which are in fluorescent units/min/mg protein.
[b]Larvae of the susceptible (WAK) strain were exposed during
the last instar on diet containing either 2-tridecanone (on
T) or nicotine (on N). Similar designations are identified
for the 2-tridecanone resistant (WAK-T) and nicotine
resistant (WAK-N) strains.

nitroanisole O-demethylase (3X) and benzphetamine N-demethylase activity (3X) relative to Wak-T on control diets. Yet neither of these two substrates were induced appreciably above control values by resistance alone. With respect to the substrates which were induced by resistance, sulfoxidation of phorate showed no increase as a result of induction by 2-tridecanone while benzo(a)pyrene was induced 4.5-fold.

Larvae of the Wak strain reared on nicotine (Wak on N), did not differ in P450 content from their control (Wak). The lack of induction by nicotine was accompanied by oxidative activity similar to that of control larvae In contrast, larvae resistant to nicotine (Wak-N) possess levels of enzyme activity which were similar to those of the Wak-T strain. In addition, Wak-N larvae reared on nicotine (Wak-N on N) have the greatest cytochrome P450 content of all the strains and conditions examined. Wak-N on N larvae showed a 2X increase above that of Wak-N larvae in oxidation of all substrates with the exception of lauric acid. Incidentally, hydroxylation of lauric acid was not increased significantly for any strain or condition examined. These results suggest that patterns of substrate specificity vary widely, depending upon whether larvae are resistant or susceptible, as well as upon the previous exposure of the larvae to the allelochemicals mediating resistance.

Resistance and Tolerance to Insecticides

Resistance is defined as an increase in the ability of an organism to survive a toxicant due to genetic selection exerted on previous generations by the toxicant or, in the case of cross-resistance, by another toxicant. Resistance increases from one generation under selection pressure to the next until a maximum is reached. Tolerance reflects the ability of one unselected population of an organism to survive exposure to a toxicant more effectively than another unselected population.

Many behavioral, physiological and biochemical mechanisms have been implicated in insect resistance and tolerance to pesticides (3,5,55). Changes in oxidative metabolism appears to be one of the major biochemical mechanisms in the resistance of insects to insecticides of several different chemical classes. The most important early evidence for the importance of insect monooxygenases in resistance came from studies of pesticide synergists such as the methylenedioxyphenyl compounds, piperonyl butoxide and sesamex. Eldefrawi et al (56) observed that a carbaryl-sesamex combination could reverse resistance and it was shown that sesamex inhibited monooxygenase activity both in vivo (57) and in vitro (58,59). Since then the interaction between synergists and insect P450 has been examined in detail (See refs. 2,3). Following activation, probably to a carbene, (60), the methylenedioxyphenyl compounds appear to interact directly with the heme iron of P450 forming a type III binding complex similar to that seen with ethyl isocyanide, producing a non-competitive inhibition of monooxygenase activity. Many studies have demonstrated pronounced increases in monooxygenase activity in resistant insect strains (See ref.

3), yet in the most intensively studied species, the house fly, a no greater than 2-fold increase in P450 content has been observed between resistant and susceptible strains. This suggests that complex changes in the pattern of multiple P450 isozymes probably accompany the appearance of resistance in those species where increases in oxidation are an important factor.

The housefly has many advantages as a model species for the investigation of insect P450. Many resistant and susceptible strains are available, some with visible mutant markers, thus permitting analysis of changes in the monooxygenase system to be made in conjunction with investigations of biochemical genetics (61). Comparative studies have indicated the existence of a qualitatively different P450 isozyme composition in microsomes derived from resistant compared to susceptible house fly strains. Microsomes from resistant strains generally show a hypochromic shift in the lambda max of the reduced CO-difference spectra from 450 to 448-449 nm (62,63,64). Resistant strain microsomes show significant type I binding spectra and a lowered 455 nm peak in ethyl isocyanide type III binding spectra compared to the reduced CO-difference spectrum. In addition they show a single 390 nm trough in the n-octylamine type II spectrum in contrast to microsomes from susceptible strains which show a double trough at 410 and 394 nm (17,65).

Two or more semi-dominant genes for the high oxidase activity associated with pesticide resistant house fly strains have been associated with chromosomes II and V (61,66) and these loci are always linked to the presence of high type I binding spectra in resistant strain microsomes. However, from studies of P450 multiplicity (64,67-73), it now appears probable that these genes control the synthesis of different forms of the cytochrome, rather than the physical characteristics or catalytic properties of a single form.

Many of the early studies used a single substrate, such as aldrin, to assay monooxygenase activity, however, it is now known that there are many isozymes of monooxygenases in a single organism with different substrate specificities. Therefore it is necessary to use a variety of substrates to accurately characterize monooxygenase activity. In one instance, O-dealkylation of methoxy- and ethoxyresorufin were highly elevated in a pyrethroid resistant strain of houseflies while much smaller elevations of activities were observed for four other monooxygenase substrates (74). These results suggest that pyrethroid resistance in houseflies may be related to an isozyme of P450 which might be characterized by resorufin substrates.

PURIFICATION AND PROPERTIES

The purification of insect P450 has proven difficult and has lagged far behind similar studies of mammalian P450. One of the major reasons for this has been the difficulty in obtaining sufficient material. Individual insects are small and dissection of single organs is generally impractical. Furthermore, the eye pigment xanthommatin is a powerful monooxygenase inhibitor (75), necessitating the separation of

82

heads from the rest of the body for the successful preparation of active microsomes. For spectral studies it is also necessary to remove the thoraces since otherwise fragmented sarcosomes contaminate the microsomal fraction. An additional problem is the instability of insect P450 compared to that from mammals.

The first attempts at purification were those of Capdevila et al (70) and Agosin (76) using microsomes from the resistant Fc strain of house fly. Two distinct P450 containing fractions were separated but with only low specific contents (0.8 nmol/mg in the best case). In other early studies Schonbrod and Terriere (77) were able to separate two fractions of P450 from the microsomes of both a susceptible (NAIDM, CSMA) and a resistant (Rutgers) strain of houseflies. Yu and Terriere (78) separated 6 distinct P450 containing fractions from both susceptible and resistant strain microsomes. However, although SDS-PAGE showed the presence of several heme-containing protein bands of M.W. between 40 and 60,000, specific contents up to 2.7 nmol/mg were still very low compared to purified mammalian P450 preparations.

A more successful purification of house fly P450 was that of Capdevila et al (68,69) who characterized three isozymes with specific contents of 3.5, 13.9 and 7.5 nmol/mg with lambda max values of 450, 452, and 453 nm respectively from the susceptible NAIDM strain. Molecular weights for the 3 P450 isozymes were 48,000, 43,000 and 53,000. Fisher and Mayer (79) reported the purification of a P450 from of somewhat higher M.W. from phenobarbital induced Rutgers strain house fly microsomes. A 60,000 M.W. isozyme was purified to a specific content of 10 nmol/mg. Kodama et al (80) also found M.W.s of 59,000-60,000 for 4 P450 isozymes partially characterized from abdominal microsomes of the susceptible CSMA house fly strain to specific contents of 2-7 nmol/mg.

One limitation of the above studies has been the small yield of purified material for subsequent reconstitution studies or for the raising of anti-sera. In an attempt to overcome this problem we have been using a bulk preparation of fly abdomens prepared by shaking together flies and rubber stoppers frozen with liquid nitrogen followed by sieving and partitioning of the separated heads, thorax and abdomens, (16). Solubilization of microsomes thus prepared from the resistant Rutgers strain house fly in 0.8% sodium cholate was followed by chromatography on phenyl sepharose 4B and DEAE-cellulose (DE-52). A fraction which passed through the DE-52 column contained three forms of P450 and was stable for long periods at $-20^{\circ}C$. This allowed accumulation of sufficient P450 for further purification on CM-Sepharose, hydroxylapatite and an affinity column of rat cytochrome b_5 bound to Sepharose 4B. Four P450 isozymes were characterized. The fraction of highest specific content contained predominently one protein of M.W. 49,000 and a lambda max of 452 nm with a specific content of 6-8 nmol/mg.

From the above studies, it appears that the house fly monooxygenase system is quite complex, consisting of at least six forms of P450 and showing differences in isozyme composition between microsomes prepared from susceptible and resistant strains. This appears to be in contrast to the situation in Drosophila, the only other dipteran system to have been characterized in any detail. Here only 2-3 P450 forms have been described. Naquira et al (81) and Agosin (82), reported three different P450 fractions with specific contents of 8-12 nmol/mg, distinct spectral and catalytic properties and M.W. in the 51,000-54,000 range. More recently, Waters et al (83), Waters and Nix (84) and Sundseth et al (85) have reported two P450 isozymes in Drosophila - A with M.W. 59,000 and B with M.W. 56,000. P450A appears to be common to all Drosophila strains while the appearance of P450B is associated with resistance to DDT and malathion and with high microsomal nitrosodimethylamine demethylase activity (83).

House fly P450 reductase is considerably more labile than that of mammals (3,16,86), readily losing a 5000-7000 M.W. fragment to yield a partly degraded protein capable of reducing cytochrome c but not P450. Insect P450 reductase does not bind to 2',5', ADP-Sepharose affinity columns normally used to purify the mammalian reductase but can be purified on NADP-agarose columns (16,84,86).

Insect cytochrome b5 has been purified and characterized in microsomes from the mediterranian fruit fly Ceratitis capitata by Megias et al (87) and substantially purified from house fly microsomes as a by-product of P450 purificaton (16). It appears to have similar structure and properties to mammalian cytochrome b5.

Reconstitution of monooxygenase activity using purified insect P450 has met with varying degrees of success. Agosin (82) reported relatively good aldrin epoxidase and 7-ethoxycoumarin 0-deethylase activity in fractions of P450 from Drosophila microsomes using rat P450 reductase to reconstitute. P450I had 14 nmol./mg/min aldrin epoxidase activity while P450II had similar activity towards 7-ethoxycoumarin. However, in most other studies (16,77,78,85,88) only a small fraction of the activity seen in microsomes could be recovered in reconstitutions after microsomal solubilization and separation on one or more columns. At present the reasons for this are not readily apparent. Optimal conditions for the reconstitution of insect monooxygenase activity may vary considerably from those used with mammalian systems. Microsomes from dipterous insects show a different lipid composition from those of mammalian species or of other insects with little phosphatidylcholine (PC) and a predominance of phosphatidylethanolamine (PE). Ronis et al (16) found p-nitroanisole 0-demethylase activity was higher in reconstituted systems containing PE than in those containing PC. In addition, other components may be necessary for activity, Klotz et al (14) found that cytochrome b5 when included in reconstitutions of some fish P450s gave much higher activity than reductase and P450 alone. Cytochrome b5 also appears to play a role in some insect monooxygenations such as benzo(a)pyrene hydroxylase and lauric acid hydroxylase (15,16,37) and this aspect requires further investigation. Purified insect P450 shows rapid reduction but only very slow binding of carbon monoxide (Agosin, 82, Ronis, unpublished observations) which may be one reason why preparations of P450 which appear pure on SDS-PAGE have lower specific contents than equivalent mammalian preparations. It has been suggested that this reflects tight binding of endogenous substrates such as ecdysteroids to the P450 during the purification process (82).

The use of molecular probes against the insect monooxygenase system will be necessary for more detailed studies of the role of P450 isozymes in pesticide resistance and to define the place of insect P450s in the evolution of the P450 multigene superfamily. Recently Sundseth et al (88,89), and Sundseth (91) have generated monoclonal antibodies against Drosophila P450s A and B. Feyereisen and co-workers (6,7) have raised a polyclonal antibody against a purified house fly P450 and used it to obtain a cDNA clone. Sequence analysis showed little similarity to sequences of known mammalian P450 isozymes and this P450 has been assigned to a separate gene family - P450 family VI.

REFERENCES

1. C. F. Wilkinson, Role of mixed function oxidases in pesticide resistance, in: "Pest Resistance to Pesticides: Challenge and Prospects", G. P. Georghiou and M. Saito, eds., Plenum Press, NY (1983).
2. E. Hodgson, The significance of cytochrome P450 in insects, Insect Biochem. 13:237-246 (1983).
3. E. Hodgson, Microsomal monooxygenases, in: "Complete Insect Physiology, Biochemistry and Pharmacology" V.11, G.A. Kerkut and L. I. Gilbert, eds., Pergamon Press, Oxford (1985).
4. L. C. Terriere, Induction of detoxication enzymes in insects, Ann. Rev. Ent. 29:71-88 (1984).
5. M. J. J. Ronis and E. Hodgson, Cytochrome P450 monooxygenases in insects, Xenobiotica 19:1077-1092 (1989)
6. D. W. Nebert, D. R. Nelson and R. Feyereisen, Evolution of the cytochrome P450 genes, Xenobiotica 19:1149-1160 (1989).
7. R. Feyereisen, J. F. Koener, D. E. Farnsworth and D. W. Nebert, Isolation and sequence of cDNA encoding a cytochrome P-450 from an insecticide-resistant strain of the house fly, Musca domestica, Proc. Natl. Acad. Sci. 86:1465-1469 (1989).
8. J. W. Ray, Pest Infestation Research, Her Majesty's Stationary Office, London (1965).
9. J. W. Ray, The epoxidation of aldrin by housefly microsomes and its inhibition by carbon monoxide, Biochem. Pharmacol. 16:99-107 (1967).
10. J. D. Cassidy, E. Smith and E. Hodgson, An ultrastructural analysis of microsomal preparations from Musca domestica and Proenia eridania, J. Insect. Physiol. 15:1573-1578 (1969).
11. M. D. Gilbert and C. F. Wilkinson, Microsomal Oxidases in the honey bee, Apis mellifera, Pest. Biochem. Physiol. 4:56-66 (1974).
12. G. F. Weirich, J. A. Svoboda and M. J. Thompson, Ecdysone 20-monooxygenase in mitochondria and microsomes of Manduca sexta Midgut: Is dual localization real?, Arch. Insect. Biochem. Physiol. 2:385-396 (1985).
13. M. D. Gorsky and M. J. Coon, Effects of conditions for

reconstitution with cytochrome b5 on the formation of products in cytochrome P450 catalysed reactions, Drug Metab. Disp. 14:89-96 (1986).

14. A. Klotz, J. J. Stegman, B. Woodin, E. Snowberger, P. Thomas, and C. Walsh, Cytochrome P450 isozymes from the marine teleost Stenotomus chrysops: Their roles in steroid hydroxylation and the role of cytochrome b5, Arch. Biochem. Biophys. 249:326-338 (1986).

15. A. J. Baars, J. A. Zijlstra, E. Vogel and D. D. Breimer, The occurrence of cytochrome P450 and aryl hydrocarbon hydroxylase activity in Drosophila melanogaster microsomes and the importance of the metabolizing capacity for the screening of carcinogenic and mutagenic properties of foreign compounds, Mutation Rsch. 44:257-268 (1977).

16. M. J. J. Ronis, W. C. Dauterman and E. Hodgson, Characterization of multiple forms of cytochrome P450 from an insecticide resistant strain of housefly, Musca domestica, Pest. Physiol. Biochem. 32:74-90 (1988).

17. E. Hodgson and A. P. Kulkarni, Characterization of cytochrome P450 in studies of resistance, in: "Pest Resistance to Pesticides: Challenges and Prospects", G. P. Georghiou and M. Saito, eds., Plenum Press, New York (1983).

18. B. D. Hammock, NADPH dependant epoxidation of methyl farnesoate to juvenile hormone in the cockroach Blaberus giganticus, Pest. Sci. 17:323-328 (1975).

19. R. Feyereisen, G. E. Pratt and A. F. Hamnett, Enzymic synthesis of juvenile hormone in locust corpora allata: Evidence for a microsomal cytochrome P450-linked methyl fornesoate epoxidase, European J. Biochem. 188:231-238 (1981).

20. G. E. Pratt, R. C. Jennings, A. F. Hamnett and G. T. Brooks, Lethal metabolism of precosene I to a reactive epoxide by locust corpora allata, Nature 284:320-323 (1980).

21. S. J. Yu and L. C. Terriere, Metabolism of juvenile hormone I by microsomal oxidase, esterase and epoxidase hydratase of Musca Domestica and some comparisons with Phormia regina and Sarcophaga bullata, Pest. Biochem. Physiol. 9:237-246 (1978).

22. C. W. Fisher and R. T. Mayer, Characterization of house fly microsomal mixed function oxidases: Inhibition of juvenile hormone I and piperonyl butoxide, Toxicology 24:15-31 (1982).

23. S. J. Yu and L. C. Terriere, Esterase and oxidase activity of housefly microsomes against juvenile hormone analogues containing branched chain ester groups and its induction by phenobarbital, J. Agri. Food Chem. 25:1333-1336 (1977).

24. S. J. Yu and L. C. Terriere, Metabolism of ^{14}C-hydropene (ethyl,3,7,11-trimethyl-2,4-dodecadieneoate), by microsomal oxidases and esterases of three species of diptera, J. Agri. Food Chem. 25:1076-1080 (1977).

25. G. T. Brooks, Insect epoxide hydratase inhibition by juvenile hormone analogues and metabolic inhibitors, Nature 245:382-384 (1973).

26. G. T. Brooks, Epoxidase hydratase as a modifier of

biotransformation and biological activity, Gen.Pharmacol. 8:221-226 (1977).

27. S. L. Smith, W. E. Bollenbacker and L. I. Gilbert, Studies on the biosynthesis of ecdysone and 20-hydroxyecdysone in the Tobacco Hornworm, Manduca sexta, in: "Progress in Ecdysterone Research", A. J. Hoffmann, ed. Elsevier, Holland (1980).

28. S. L. Smith, W. E. Bollenbacker, D. Y. Cooper, H. Schleyer, J. J. Wielgus and L. I. Gilbert, Ecdysone 20-monooxygenases: Characterization of an insect cytochrome P450 dependant steroid hydroxylase, Mol. Cell. Endocrin. 15:111-133 (1979).

29. R. Feyereisen, Cytochrome P450 et metabolisme de l'ecdysone, Societie Zooligique de France Bulletin 102:310-311 (1977).

30. R. Feyereisen and F. Durst, Ecdysterone biosynthesis: A microsomal cytochrome P450-linked ecdysone 20-monooxygenase from tissues of the African migratory locust, European J. Biochem. 88:37-47 (1979)

31. M. J. Mitchell and S. L. Smith, Characterization of ecdysone 20-monooxygenase activity in wandering stage larvae of Drosophila melanogaster, Insect Biochem. 16:525-537 (1986).

32. L. B. Brattsten, Biochemical defense mechanisms in herbivores against plant allochemicals, in: "Herbivores: Their Interactions with Secondary Plant Metabolites", G. A. Rosenthal and D. H. Janzen, eds., Academic Press, New York (1979).

33. L. B. Brattsten, Ecological significance of mixed-function oxidations, Drug Metab. Rev. 10:35-58 (1979).

34. R. A. White, Jr., R. T. Franklin and M. Agosin, Conversion of alpha-pinene to alpha-pinene oxide by rat liver and the bark beetle Dendroctonus terebrans, Pest. Biochem. Physiol. 10:233-242 (1979).

35. R. A. White, Jr. M. Agosin, R. T. Franklin and J. W. Webb, Bark Beetle pheromones: Evidence for physiological synthesis mechanisms and their ecological implications, Zeitschrift fur Angewante Entomologie 90:255-274 (1980).

36. S. Ahmad, K. E. Kirkland and G. J. Bloomquist, Evidence for a sex pheromone metabolizing cytochrome P450 monooxygenase in the housefly, Archives of Insect Physiol. and Biochem. 6:121-130 (1987).

37. S. E. Clarke, C. J. Brealey and G. G. Gibson, Cytochrome P450 in the housefly: Induction, substrate specificity and comparison to three rat hepatic isozymes, Xenobiotica 19:1175-1180 (1989).

38. A. P. Kulkarni and E. Hodgson, Multiplicity of cytochrome P450 in microsomal membranes from the housefly, Musca Domestica, Biochem. Biophys. Acta 632:573-588 (1980).

39. C. F. Wilkinson and L. B. Brattsten, Microsomal drug metabolizing enzymes in insects, Drug Metab. Rev. 1:153-228 (1972).

40. R. I. Krieger and C. F. Wilkinson, Microsomal mixed function oxidases in insects, I. Localization and properties of an enzyme system effecting aldrin epoxidation in larvae of the Southern Armyworm, (Prodenia eridana), Biochem. Pharmacol. 18:1403-1415 (1969).

41. R. S. Anderson, Aryl hydrocarbon hydroxylase in an insect, Spodoptera eridania (Cramer), _Comp. Biochem. Physiol._ 59:87-93 (1978).

42. R. D. Schonbrod, W. W. Philleo and L. C. Terriere, Hydroylation as a factor in resistance in houseflies and blowflies, _J. Econ. Entomol._ 58:74-77 (1965).

43. S. J. Yu and L. C. Terriere, Hormonal modification of microsomal oxidase activity in the housefly, _Life Sci._ 10:1179-1185 (1971).

44. R. I. Kreiger, P. P. Feeny and C. F. Wilkinson, Detoxication enzymes in the guts of caterpillars: An evolutionary answer to plant defences, _Science_ 172:579-581 (1971).

45. C. H. Walker, Species variation in some microsomal drug metabolizing enzymes, _Prog. Drug Metab._ 5:113-163 (1980).

46. C. H. Walker, I. Newton, S. Hallam, and M. J. J. Ronis, Activities and toxicological significance of the hepatic microsomal enzymes of the Kestral and Sparrowhawk, _Compar. Biochem. Physiol._ 86C:359-363 (1986).

47. R. C. Skoda, F. J. Gonzalez and D. A. Meyer, Identification of mutant alleles of the P450DB1 gene associated with deficient metabolism of Debrisoquine, in: "Cytochrome P450 Biochemistry and Biophysics", I. Schuster, ed., Taylor and Francis, UK (in press).

48. R. E. Berry, S. J. Yu and L. C. Terriere, Influence of host plants on insecticide metabolism and management of variegated cutworm, _J. Econ. Entomol._ 73:771-774 (1980).

49. D. E. Farnsworth, R. E. Berry, S. J. Yu and L. C. Terriere, Aldrin epoxidase activity and cytochrome P450 content of microsomes prepared from alfalfa and cabbage looper larvae fed various plant diets, _Pest. Biochem. Physiol._ 15:158-165 (1871).

50. A. F. Moldenke, R. E. Berry and L. C. Terriere, Cytochrome P450 in insects V. Monoterpene induction of cytochrome P450 and associated Monooxygenase activities in the larvae of the variegated cutworm Peridroma saucia, _J. Compar. Physiol. Biochem._ 74C:365-371 (1983).

51. M. F. Christian and S. J. Yu, Cytochrome P450-dependent monooxygenase activity in the Velvetbean caterpillar, Anticarsia gemmatalis Hubner, _Comp. Biochem. Physiol._ 83C:23-27 (1986).

52. M. Riskallah, W. C. Dauterman and E. Hodgson, Host plant induction of microsomal monooxygenase activity in relation to diazinon metabolism and toxicity in larvae of the Tobacco Budworm, Heliothis virescens, _Pest. Physiol. Biochem._ 25:233-247 (1986).

53. M. Riskallah, W. C. Dauterman and E. Hodgson, Nutritional effects on the induction of cytochrome P450 and glutathione-transferase in larvae of the Tobacco Budworm, Heliothis virescens, _Insect Biochem._ 16:491-505 (1986b).

54. G. G. Kennedy, 2-Tridecanone, tomatoes and Heliothis zea, potential incompatibility of plant antibiosis

with chemical insect control. <u>Entomologia exp. appl</u>. 35:305 (1984).

55. L. B. Brattsten, C. W. Holyoke Jr., J. R. Leeper and K. F. Raffa, Insecticide resistance: Challenge to pest management and basic research, <u>Science</u> 231:1225-1260 (1986).

56. M. E. Eldefrawi, R. Miskus and V. Sutcher, Methylenedioxyphenyl derivatives as synergists for carbamate insecticideson susceptible, DDT- and parathion-resistant houseflies, <u>J. Econ. Entomol.</u> 53:231-234 (1960).

57. Y. P. Sun and E. R. Johnson, Synergistic and antagonistic actions of insecticide-synergist combinations and their mode of action, <u>J. Agri. Food Chem.</u> 8:261-266 (1960).

58. E. Hodgson and J. E. Casida, Biological oxidation of N,N, dialkyl carbamates, <u>Biochem. Biophys. Acta</u> 42:184-186 (1975).

59. E. Hodgson and J. E. Casida, Metabolism of N,N, dialkyl carbamates and related compounds by rat liver, <u>Biochem. Pharmacol.</u> 8:179-191 (1961).

60. A. R. Dahl and E. Hodgson, The interaction of aliphatic analogues of methylenedioxyphenyl compounds with cytochromes P450 and P-420, <u>Chem. Biol. Interact.</u> 27:163-175 (1979).

61. F. W. Plapp, Jr., Biochemical genetics of insecticide resistance, <u>Ann. Rev. Entomol.</u> 21:179-197 (1976).

62. A. S. Perry and A. J. Bucknor, Studies on microsomal cytochrome P450 resistant and susceptible houseflies, <u>Life Sci.</u> 9:335-350 (1970).

63. A. S. Perry, W. E. Dale and A. J. Bucknor, Induction and repression of microsomal mixed-function oxidases and cytochrome P450 in resistant and susceptible houseflies, <u>Pest. Biochem. Physiol.</u> 1:131-142 (1971).

64. R. M. Philpot and E. Hodgson, Differences in the cytochrome P450s from resistant and susceptible houseflies, <u>Chem.-Biol. Interact.</u> 4:399-408 (1972).

65. E. Hodgson, L. G. Tate, A. P. Kulkarni and F. W. Flapp, Jr., Microsomal cytochrome P450: Characterization and possible role in insecticide resistance in Musca domestica, <u>J. Agri. Food Chem.</u> 22:360-366 (1974).

66. F. W. Plapp, Jr., The genetic basis of insecticide resistance in the house fly: Evidence that a single locus plays a major role in metabolic resistance to insecticides, <u>Pest. Biochem. Physiol.</u> 22:194-201 (1984).

67. J. Capdevila, A. Morello, A. S. Perry and M. Agosin, Effect of phenobarbital and naphthalene on some of the components of the electron transport system and the hydroxylating activity of housefly microsomes, <u>Biochemistry</u> 12:1445-1451 (1973).

68. J. Capdevila, A. S. Perry, A. Morello and M. Agosin, Some spectral properties of cytochrome P450 from microsomes isolated from control, phenobarbital and naphthalene treated houseflies, <u>Biochem. Biophys. Acta.</u> 314:93-103 (1973).

69. J. Capdevila and M. Agosin, Multiple forms of housefly cytochrome P450, in: "Microsomes and Drug

Oxidations", V. Ullrich, ed., Pergamon Press, Oxford (1977).

70. J. Capdevila, N. Ahmad and M. Agosin, Soluble cytochrome P450 from housefly microsomes. Partial purification and characterization of two hemoprotein forms, J. Biol. Chem. 250:1048-1060 (1975).

71. L. G. Tate, F. W. Plapp, Jr. and E. Hodgson, Cytochrome P450 difference spectra of microsomes from several different insecticide-resistant and susceptible strains of housefly, Musca domestica, Chem.-Biol. Interact. 6:237-247 (1973).

72. L. G. Tate, F. W. Plapp, Jr. and E. Hodgson, Genetics of cytochrome P450 in two insecticide resistant strains of the housefly Musca domestica, Biochem. Genetics 11:49-63 (1974).

73. L. C. Terriere and S. J. Yu, Cytochrome P450 in insects 2. Multiple forms in the flesh fly (Sarcophaga bullata, Parker) and the blow fly (Phormia regina Meigen), Pest. Biochem. Physiol. 12:249-256 (1979).

74. S. S. T. Lee and J. G. Scott, Microsomal cytochrome P450 monooxygenases in the house fly (Musca domestica L.): Biochemical changes associated with pyrethroid resistance and phenobarbital induction, Pest. Biochem. Physiol. 35:1-10 (1989).

75. T. G. Wilson and E. Hodgson, Mechanisms of microsomal mixed-function oxidase inhibitor from the housefly, Musca domestica, Pest. Biochem. Physiol. 2:64-71 (1972).

76. M. Agosin, Insect cytochrome P450, Mol. and Cell. Biochem. 2:33-44 (1976).

77. R. D. Schonbrod and L. C. Terriere, Solubilization and separation of two forms of microsomal cytochrome P-450 from the housefly Musca domestica, Biochem. Biophys. Rsch. Commun. 64:829-835 (1975).

78. S. J. Yu and L. C. Terriere, Enzyme induction in the housefly, the specificity of cyclodiene insecticides, Pest. Biochem. Physiol. 2:184-190 (1972).

79. C. W. Fisher and R. T. Mayer, Partial purification and characterization of phenobarbital-induced housefly cytochrome P450, Arch. Insect Biochem. Physiol. 1:127-138 (1984).

80. O. Kodama, E. Hodgson and W. C. Dauterman, Unpublished observations (1985).

81. C. Naquira, R. A. White and M. Agosin, Multiple forms of Drosophila cytochrome P450, in: "Biochemistry, Biophysics and Regulation of Ctyochrome P450", J. A. Gustafsson, J. Carlstedt-Duke, A. Mode and J. Rafer, eds., Elsevier (1980).

82. M. Agosin, Multiple forms of insect cytochrome P450: Role in insecticide resistance, in: "Cytochrome P-450, Biochemistry, Biophysics and Environmental Implications", E. Hietanen, M. Laitinen and O. Hanninen, eds., Elsevier (1982).

83. L. C. Waters, S. I. Simms and C. E. Nix, Natural variation in the expression of cytochrome P450 and dimethylnitrosamine demthylase in Drosophila, Biochem. Biophys. Rsch. Commun. 123:907-913 (1984).

84. L. C. Waters and C. E. Nix, Regulation of insecticide

resistance related cytochrome P450 expression in Drosophila melanogaster, _Pest. Biochem. Physiol._ 30:214-227 (1988).

85. S. Sundseth, C. E. Nix and L. C. Waters, Isolation of insecticide resistance-related forms of cytochrome P-450 from Drosophila melanogaster, _Biochem. J._ 265:213-217 (1990).

86. D. R. Vincent, A. F. Moldenke and L. C. Terriere, NADPH-reductase from the housefly, Musca domestica, _Insect Biochem._ 13:559-566.

87. A. Megias, A. Saborido and A. M. Municio, Cytochrome b from the insect Ceratitis capitata, _Biochem. Biophys. Acta_ 872:116-124 (1986).

88. A. F. Moldenke, D. R. Vincent, D. E. Farnsworth and L. C. Terriere, Cytochrome P450 in insects 4. Reconstitution of cytochrome P450 dependant monooxygenase activity in the housefly, _Pest. Physiol. Biochem._ 21:358-368 (1984).

89. S. Sundseth, S. J. Kennel and L. C. Waters, Monoclonal antibodies to Drosophila cytochrome P450s, _Fed. Proc._ 46:2142 (1987).

90. S. Sundseth, S. J. Kennel and L. C. Waters, Correlation of cytochrome P450 expression and insecticide resistance in Drosophila, _FASEB Journal_ 2:A1012 (1988).

91. S. Sundseth, Cytochrome P450 related to insecticide resistance in Drosophila melanogaster, Ph.D. dissertation, University of Tennessee (1988).

I. MEMBRANE TOPOLOGY OF CYTOCHROMES P-450: ELEMENTS AND

MEASUREMENT BY SPECTROSCOPIC TECHNIQUES

Stier, A., Krüger, V., Eisbein, T., and Finch, S.A.E.

Max-Planck-Institut für biophysikalische Chemie
Abteilung Spektroskopie, D-3400 Göttingen, F.R.G.

INTRODUCTION

The main concern of these lectures is:
Do liver microsomal cytochromes P-450 have an oligomeric structure and is
cooperativity involved in their function?
Elements of topology of membrane proteins and the spectroscopic techniques
used to investigate this toplogy are described. Oligomeric structure is an
important element of membrane topology and information on it can be
obtained from measurements of rotational diffusion of the proteins.
Emphasis is therefore laid on photoselection techniques (time-dependent
polarized delayed fluorescence and phosphorescence and time-dependent
absorption dichroism).

ELEMENTS OF MEMBRANE TOPOLOGY OF PROTEINS

We use topology here as a term to deal with a collection of physical
elements related to the interaction of membrane proteins with a
phospholipid bilayer. We prefer topology as a more profound term
describing general principles of this interaction to the term topography
used more commonly (1) to denote a "cartographic" description of a
specific protein. Topology investigates causes and consequences of this
interaction and therefore contributes more to an understanding of the
relationship between structure and function.

The topology of membrane proteins results from an interaction of
structural elements of proteins and phospholipid membranes (see tables 1
and 2).

Table 1. Scales of protein structural elements

Linderstroem-Lang (2)	actual
primary	amino acid sequence
secondary	α, β, reverse turn, "coil"
supersecondary	β-α-β, "greek key", β-meander
tertiary	tertiary, domain, globular
quaternary	oligomer, aggregate

Molecular Aspects of Monooxygenases and Bioactivation of Toxic Compounds
Edited by E. Arınç *et al., Plenum Press, New York,*

Table 2. Elements of lipid membrane structure

gross:	sidedness, hydrophobic interior, charged surface polymorphism of phospholipid classes
molecular:	headgroup orientation, length, kinks and unsaturation of fatty acids, conical or cylindrical overall shape of phospholipids
dynamics:	rotational and lateral mobility, chain flexibility

The sidedness and the hydrophobic interior of a phospholipid membrane are dominating elements. Hydrophobic elements of secondary protein structure can enter or span a membrane, but compact globular domains with hydrophilic or charged surfaces cannot.

A description of elements of membrane topology may start with the fact that all membranes are asymmetric. The asymmetry is a correlate of the vectorial nature of membrane functions: in a plasma cellular membrane, signals are transduced from receptor proteins on the extracellular side to effector proteins on the cytoplasmic side or molecules traverse the barrier with the help of transporter proteins. In the inner mitochondrial membrane, the lateral and transverse organization of proteins combines electron transfer with vectorial proton transport.

A basis to differentiate topological classes of membrane proteins is the number of times the polypeptides spans the membrane (1). The polypeptide chain of monotopic proteins does not penetrate the lipid membrane. Cytochrome b5 is considered as a monotopic protein being anchored in the membrane matrix by a hydrophobic hairpin loop. Microsomal cytochromes P-450 may be a larger class of monotopic proteins, but it is not yet settled whether the amino terminal end is located on the cytoplasmic or on the luminal side of the endoplasmic reticulum (see part II). With the latter location the cytochromes P-450 would have to be classified sensu strictu as bitopic proteins. However, they would be very different from other bitopic proteins like the epidermal growth factor receptor, which has functional domains on both sides of the membrane, on the extracellular side for hormone binding and on the cytoplasmic side the phosphokinase activity. Transporters are examples of polytopic membrane proteins with more than four polypeptide crossings of the membrane. Thus, essential elements of membrane protein topology are membrane spanning polypeptides stretches. An alpha helix of about 20 residues oriented perpendicular to the membrane can span a lipid bilayer. These helical elements may be hydrophobic in the case of receptors and amphipatic in the case of channels or porters. Interestingly, non-helical structures can also be membrane spanning elements; for example, a β strand of the adenine nucleotide translocater of the inner mitochondrial membrane. For a more detailed description of these elements of membrane topology see (1).

All these monotopic, bitopic and polytopic proteins fulfill the definition of integral membrane proteins as their release from the membrane requires disruption of the phospholipid structure by detergents. This does not exclude additional ionic binding between domains of the protein located outside the membrane and the charged headgroups of the phospholipids. Evidence for this kind of additional binding of microsomal cytochromes P-450 and its functional meaning will be given in part II.

94

It is apparent that the two-dimensionality of the membrane limits the variability of the overall shape of integral membrane proteins. Two basic types may be distinguished (Fig.1): either cylinders span the membrane and protrude more or less from the surface of the lipid bilayer on one or both sides, or the hydrophilic bulk of the protein is anchored within the membrane by a conical hydrophobic part which invades the bilayer interior to some depth. In the simplest case the anchor may be a membrane spanning helix as it is assumed for cytochrome b5. Some of the proteins of the second type demand a higher degree of complementarity of the lipid matrix: for example, conically shaped phospholipids like phosphatidylethanolamine would better adapt these structures to the membrane than cylindrical lipids like lecithin (3). All of the proteins of the mixed function oxygenase system, microsomal cytochromes P-450, NADPH- cytochrome P-450 reductase including cytochrome b5 and presumably NADH- cytochrome b5 reductase are of the second type.

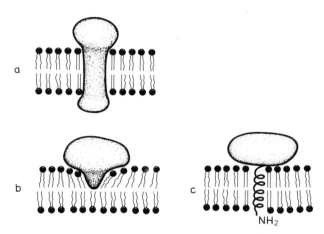

Figure 1. Three typical models of membrane topology of proteins

A special feature of biological membrane structure is the high mole fraction of proteins. A number of theories explain the observation of formation of protein aggregates by specific interaction between proteins and phospholipids (4). A topological element of special interest in relation to the function and regulation of membrane proteins is the supramolecular structure of an oligomeric state of defined size. This quaternary structure differs from the aforementioned aggregates because it is stabilized by specific interactions between the subunits or protomers (the latter term is used for the identical subunits of homooligomers): the geometric complementarity of parts of the surfaces of these subunits permits the formation of hydrogen bonds, salt bridges and/or hydrophobic contacts between subunit interfaces.

A feature of oligomer structure as an element of topology is lateral organization in the plane of the membrane. This reduces the dimensions of spatial organization to two, and so limits the number of necessary contact faces between the subunits. Lateral organization simplifies the symmetry rules in homooligomers which consist of identical subunits (5,6). There is only one symmetry element left, a line perpendicular to the surface of the membrane, and the degree of rotation around this axis generating congruency of structure is a function of the number of protomers in an oligomer. A hexamer of 6-fold rotational symmetry exhibits congruency after a rotation of 60 degrees around the symmetry axis. This arrangement necessitates heterologous binding between the protomers, which means that the two binding faces are located in different areas of the surface of the protomers. The symmetrical lateral organization of protomers in a membrane does not allow their isologous interaction. As symmetry of structure is not only important for the assembly mechanism of oligomeric proteins, but also for their cooperativity and allosteric behaviour, homooligomer structure can have important functional implications.

A dynamic topological element is given by the possible existence of different conformational states. Ligand induced conformational changes can alter membrane topology by affecting protein complementarity to dissociate an oligomer or by causing conformational change of monomeric proteins to induce oligomerisation.

While detailed consideration of protein-lipid interactions (7,8) is far outside the scope of this overview, one point of interest, the influence of a phospholipid membrane on the fluctuations between protein conformational states (9,10) has not been referred to in the literature. This influence may be of significance as "functionally important motions" apparently depend strongly on solvent. The aspect of ionic interaction of cytochromes P-450 and membrane phospholipids will be discussed in the chapter on membrane topology of cytochromes P-450 of part II.

OLIGOMERIC STATES IN FUNCTIONS OF MEMBRANE PROTEINS

The reasons for the existence of protein oligomers are manifold (5), and include molecular evolution of proteins, economy of genetic information, fidelity of genetic expression, allosteric and cooperative control (11) and the organisation of multifunctional multiple enzymes systems. There are two major types of multienzyme organization. One consists of large supramolecular structures of several enzymes catalyzing a sequence of reactions (12). Compartmentation of this kind provides control of substrate flux through a metabolic pathway, and cooperativity of the enzymes involved in this metabolic channeling is an important part of the physiological regulation of metabolism. The other type of multienzyme organization is embodied in a membrane. The peculiar dynamic properties of membrane constituents (7) which include lateral and rotational diffusion of lipids (13,14,15) and proteins (16), allow rapid, selective interaction of the protein partners in the two-dimensional 'compartment' of the membrane. This permits rapid ligand-induced association of proteins to functional complexes. The lifetime of the oligomer formed depends on its biological processing during execution of its function. In the simplest case, release of the ligand dissociates the oligomer. Heterooligomer formation is increasingly recognized as an important step in cellular regulation. The plasma membrane of the cell is a compartment across which hormone and transmitter signals are transduced by interaction of a number of proteins forming transient or longlived heterocomplexes. In a similar way different receptors 'crosstalk' (17) during signal transduction. For instance: a generalised concept for growth factor receptors proposes a signal transduction mechanism by their oligomerization. Binding of the hormone at the extracellular site induces oligomer formation thereby making mutual phosphorylation of the cytoplasmic sites of the protomer possible. In this way the allosteric effect of hormone binding crosses the membrane barrier and results in activation of the kinase function (18). Ligand-induced allosteric control of dual functions of preexisting oligomers is realized, for example, in acetylcholine-induced opening of an ion channel. Transmitter binding in the acetylcholine receptor is linked to control of opening of an ion channel by an allosteric interaction between the acetylcholine binding sites and the ion channel located in different protein segments (19). This involves homotropic cooperative interaction between two acetylcholine binding sites and heterotropic cooperativity between them, the channel, and various allosteric binding sites for noncompetitive blockers.

The two-dimensionality of the inner mitochondrial membrane channels electron transfer in the respiratory chain (20). In a similar manner the endoplasmic reticulum membrane channels electron transfer to cytochromes P-450 system. A preexisting oligomeric structure of cytochromes P-450 may have advantages for this function of the system over a monomeric state by increasing the probability of the transfer (see part II). Moreover, the cooperativity associated with the oligomeric state allows efficient control of the mixed function oxidation.

RECENT PROGRESS IN METHODOLOGY FOR INVESTIGATION OF MEMBRANE TOPOLOGY

Good progress in methodology to study membrane topology has been achieved in the last decade, as can be seen from a comparison of tables 3 and 4. The first table summarizes the methods applied to liver microsomal membranes up to 1977. Since critical evaluation of the application of these techniques up to 1977 appears to have only historical value it is only mentioned here that

Table 3. Summary of the methods to study membrane topology of liver microsomal membranes up to 1977 (21)

Enzyme topology in the lateral plane

Subfractionation of microsomes
Normal and freeze-fracture electron microscopy
Crosslinking of membrane proteins
Crosslinking of fatty acids of phospholipids to proteins

Enzyme topology in the transverse plane

Accessibility to proteases
Accessibility to antibodies, lectins and other macromolecules with specific binding sites
Accessibility to non-penetrating substances including substrates and effectors of enzyme activity
Involvement in vectorial processes
Reconstitution of membranes from isolated components

Table 4. Compilation of groups of current methods useful for
studies of protein membrane topology

Computational prediction of secondary and tertiary
structures from primary structure

Vectorial labelling (1) which includes surface
labelling of either side of a sealed membrane
preparation with hydrophilic agents and labelling the
interior of the membrane with hydrophobic agents
such as small radioactive, fluorescent or spin-
labelled molecules.
Surface mapping with monoclonal antibodies prepared
against peptide fragments of membrane proteins.

Genetic engineering synthesis of fusion proteins and
of proteins modified by site-directed mutagenesis

Target inactivation analysis of quaternary structure

Spectroscopic techniques for measurement of protein
mobility:
time resolved X-ray diffraction
ESR spectroscopy
NMR spectroscopy
Fluorescence, phosphorescence and absorption
dichroism spectroscopy

numerous studies came to the erroneous conclusion that the
bulk of a cytochrome P-450 molecule is embedded in the
hydrophobic part of the membrane (e.g. in 22) .
Numerous more advanced techniques have been developed during the last
decade for the study of membrane protein topology (see table 4). No single
technique suffices to present a complete picture and the various
techniques differ greatly in the informational quality and extent of their
contribution to a consensus model. As protein structure is the principal
element of membrane topology all techniques to study primary, secondary,
tertiary or quaternary protein stucture can be applied to membrane
topology.

The membrane spanning segments of proteins are important elements of
membrane topology. Rules have been established to compute such secondary
structures from the amino acid sequence of soluble proteins (23), but a
recent critical evaluation disclosed that methods of computational
prediction of secondary structure from folding motifs established in
soluble proteins are unsuccessful when applied to 16 integral membrane
proteins (24). This list can be supplemented by cytochrome P-450 (25).
More knowledge of rules governing the secondary structure of membrane
proteins is necessary for future predictions. On the other hand
computational prediction of the tertiary structure of proteins by
comparison of the primary structure with the amino acid sequence of a
protein whose tertiary structure is known from X-ray crystallography can
be helpful if these proteins belong to the same class. It is well known
that the motif of spatial organisation of helices is well conserved in the
evolution of globins and classes of cytochromes (26,27).

All techniques which map protein surfaces by their accessibility to different classes of agents are suitable for membrane proteins as they give information on the topological elements of part of the protein structure transverse to the membrane including sidedness of location (vectorial techniques) (1). The vectorial techniques include immunological, chemical and enzymic methods. Major parts of proteins can be localized from quantitation of antibody binding or visualized by electron microscopy. Details of the surface structure of membrane proteins can be localized using monoclonal antibodies against specific protein epitopes (28). These sequence directed antibodies can be raised against peptides prepared by tryptic digestion or chemical synthesis. In situ proteolysis gives information on the sidedness of proteins and the degree of their extension from the surface of the membrane. Location of specific amino acids can be accomplished by side chain specific reagents or enzymic phosphorylation (of serine, for example)(29).

Modification of protein structure by genetic engineering techniques can give important and detailed information on protein topology if a three-dimensional model based on the crystal structure of at least one member of the corresponding class of proteins is available. The exchange of one amino acid by site directed mutagenesis, if followed by considerable functional changes can give distinct structural information (30). In a multienzyme system this can be about the nature of the interaction site with another protein, e.g. cytochrome P-450 with the cytochrome P-450 reductase or cytochrome b5. The production of chimeric oligomers serves a similar purpose (31). Comparison of their enzymic properties with the properties of the separate components may yield information on mechanisms of heterooligomer formation and lifetime. This information can be confirmed by chemical crosslinking experiments in situ.

TARGET INACTIVATION ANALYSIS

High energy electron- or X-ray irradiation can be applied to determinate protein size. The target volume theory relates inactivation sensitivity to molecular weight (32) - the larger the target the more likely it will be hit. The high energy dissipated explains that the function of a molecule is completely destroyed by a single hit. One hit inactivation of an oligomeric enzyme could be caused by massive transfer of energy between subunits, or be due to a tight functional coupling of the subunits. The latter explanation applies to the radiation inactivation of the glutamate dehydrogenase hexamer (33). Six cytochromes P-450 in liver microsomes from Arochlor 1454 induced rats were found to be monomeric and cytochrome P-450-PCN-E trimeric by this technique (34). The authors considered these results inconclusive regarding the state of aggregation, as purified cytochrome P-450-PB-B, NADPH cytochrome P-450 reductase and epoxide hydrolase in solution also appeared monomeric when subjected to target technique whereas hydrodynamic and gel exclusion chromatography results indicated that the three proteins were oligomers. Apparently, like in glutamate dehydrogenase, the transfer of damaging energy between the subunits is negligible. The "inactivation" parameter in the case of the cytochromes P-450 was spectral analysis of the heme and quantitative immunoelectrophoresis. So far no experiments have been done with the target inactivation analysis of enzymic functions of cytochromes P-450. Such an analysis may yield indications of an oligomeric state if the subunits are functionally coupled to a reasonable degree.

Such studies of lateral topology of membrane bound multienzyme systems can be complemented by techniques which more directly investigate the dynamics of protein-protein interaction.

TIME SCALES OF SPECTROSCOPIC TECHNIQUES RELEVANT TO STUDIES OF PROTEIN
MEMBRANE TOPOLOGY

Several spectroscopic techniques have been developed to measure the
mobility of membrane proteins which is intimately related to their
membrane topology. In a wider sense some of these techniques are
noninvasive and suitable for studies of dynamic processes.

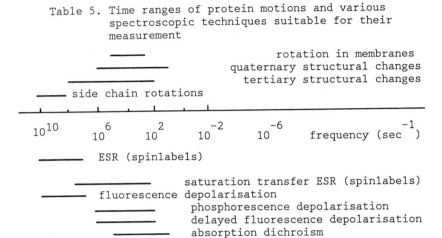

Table 5. Time ranges of protein motions and various
spectroscopic techniques suitable for their
measurement

From a comparison of ranges of rates of protein motions in membranes with
the temporal resolution of different spectroscopic techniques it can be
seen that by combined use of these techniques all types of motion in and
of proteins can be studied. Saturation transfer ESR, NMR and triplet
emission, depolarization, absorption dichroism cover the time range of
membrane protein rotation, the parameter most relevant to topology. This
parameter can give information on the size, shape and conformation of
membrane proteins, their aggregation state and, most important, their
various functional states. However, the techniques mentioned differ
greatly in the ease with which experimental results may be interpreted.

HIGH INTENSITY X-RAY DIFFRACTION

Under favourable conditions the three-dimensional structure of
isolated membrane proteins can be determined in great detail by X-ray
crystallography (35). General application of this technique is limited by
the difficulties of crystallizing membrane proteins and in situ
investigation of their topology by this technique is excluded because most
membrane preparations are not crystalline. However, time-resolved
macromolecular crystallography using high intensity X-rays from a
synchrotron for diffraction or scattering measurements, allows the
investigation of wide range of conformational changes (36). In principle,

time resolution is limited by the duration of the X-ray pulse which is typically > 140 psec. The higher the energy, the better the time resolution. At present, exposures in the millisecond range are needed. With a new generation of instruments, in laboratories in Grenoble, Argonne, Cornell and Stanford the microsecond range may be approached. Conformational changes are provoked by temperature or pressure jump, photochemical activation or ligand diffusion to permit examination of the association - dissociation kinetics of protein oligomers in the membrane.

SATURATION TRANSFER ESR SPECTROSCOPY (ST-ESR)

Conventional ESR of spin labelled proteins is only sensitive to submicrosecond rotational motions. Saturation transfer ESR can measure rotational motion in the microsecond to millisecond time range (37,38). This technique is similar in principle to polarized optical spectroscopy, particularly to polarized ground state depletion. The technology for time-resolved saturation transfer has been developed recently but awaits application to membrane proteins. Instead, steady state methods have so far been used in ST-ESR, so information about motion has to be retrieved from line shape analysis with the drawback that the time-resolution is poor in comparison to the optical technique. With the latter, actual correlation times can be determined directly from the data and certain model-independent conclusions can be drawn, in ST-ESR detailed analysis usually requires some assumptions about the type of motion taking place. The poor time resolution does not allow detection of multiple motional components nor detection of residual anisotropy due to the presence of large protein clusters with restricted motion. The difficulty of line shape analysis can be circumvented by comparison of the experimental spectra with spectra of some model probes, a procedure which usually does not improve the safety of interpretations. This is clearly demonstrated in an example in which the authors (39) calculated a rotational correlation time 480 μsec for cytochromes P-450 in phenobarbital treated rabbit liver microsomes five times greater than the time determined by optical spectroscopy. Thus, conclusions about the size of P-450 clusters based on ST-ESR measurements so far are not reliable.

PROSPECTS FOR NMR SPECTROSCOPY

NMR spectroscopy has been successfully applied to determination of the secondary and tertiary structures of small proteins and intramolecular motions in larger ones, including membrane proteins (40). It is the most versatile technique to study structure and dynamics of organic molecules. Its two-dimensional version is the only technique, except X-ray crystallography, by which the complete tertiary structures of proteins have been determined (41). To date, solution structures of around 50 small proteins or protein domains have been investigated. Tertiary structures which have been determined include basic pankreatic trypsin inhibitor (42), glucagon, an epidermal growth factor and amylase inhibitor (40). However, even with the advance of high resolution instrumentation the full capability of the technique is limited to proteins below a molecular weight of 20,000. With increasing molecular weight crowding of signals and broadening due to decreasing rotational mobility results in overlapping bands. Even so the structures of aromatic side chains possessing a higher motional freedom can be observed. Chemical exchange of a specific group between two magnetically nonequivalent environments can be exploited in the time range between 10 μsec and 1 sec. Thus NMR is outstanding in respect of the range of motional frequencies which can be measured.

Its application to membrane proteins is limited because of the orientational order in the anisotropic environment of the biomembrane, which results in broad, featureless lines due to extensive dipolar coupling between protons. Protein motion in membranes is too slow to average dipolar broadening due to anisotropy of dipolar, quadrupolar and chemical shift interactions. Some of the difficulties can be circumvented by solid state NMR techniques (43). Experimentally, methods like magic angle spinning and multiple pulse techniques have been developed for averaging anisotropic interaction. These techniques restore high resolution NMR spectra and permit detailed investigation of structural and dynamic properties of proteins. In this way the rotational diffusion of gramicidin A, a membrane channel, C-13 substituted on one amino acid position has been investigated in oriented lipid multilayers by observation of the chemical shift anisotropy dependence on sample orientation (44).

NMR may become particularly useful for characterization of protein surfaces: it has considerable potential for studying the nature of protein contacts in oligomer formation in solution and in membranes.

PHOTOSELECTION FOR MEASUREMENTS OF ROTATIONAL DIFFUSION OF MEMBRANE PROTEINS

For measurement of rotational diffusion in a population of molecules which are randomly oriented in space it is necessary to select a number of molecules with a particular orientation and observe their time-dependent randomisation. It is the transition of a subpopulation from anisotropism to isotropism which is investigated and for this it is necessary that the molecules become endowed with a vectorial property which is fixed in their geometry. Furthermore, the process of endowment needs to be rapid in comparison to the rate of motion. The dipole moments of absorption and emission of fluorescing molecules are properties ideally suited as they are geometrically fixed in a molecule. By excitation with a short pulse of linearly polarized light a subpopulation of molecules with aborption dipole moments nearly parallel to the exciting light are photoselected, their motional reorientation being followed by observation of the intensity decay of polarized light emitted from them. The kinetics of this decay are modulated by rotational motion, so its rate should be in the range of the rate of motion to be measured. Rates of rotational motion of proteins in solution are in the order of nanoseconds, whereas proteins in membranes rotate in the microsecond to millisecond range. Consequently, prompt fluorescence, with lifetimes in the nanosecond range, cannot be used to measure rotation of membrane proteins. On the other hand, emissions from longlived triplet states such as delayed fluorescence or phosphorescence, decay with the required rate. Unfortunately, the number of light-activated, long-lived intrinsic chromophores of membrane proteins is limited to rhodopsin, components of the photosynthetic system and some flavoproteins. It is therefore necessary to label other proteins with dyes having a reasonable quantum yield of delayed fluorescence or phosphorescence. Both types of triplet state emission have been used to measure rotational diffusion of membrane proteins (45,46,47,48,49). As an alternative to direct observation of polarized emission from the transient excited triplet state, repletion of the ground state population after pulsed depletion by excitation can be measured by monitoring polarized prompt fluorescence.

Rotational diffusion of membrane bound cytochromes can also be investigated with some limitations (see below) by absorption dichroism of the triplet state (50,51). In the photochemical version, the reassociation rate of the carbon monoxide complex dissociated by a pulse of polarized light falls into the desired range (52,53) and can be used as the optical absorption signal which is modulated by rotational diffusion (54,55,56).

ENERGETICS OF TRIPLET STATES IN ABSORPTION AND EMISSION (57)

A Jablonski energy state diagram helps to explain the experimental problems of photoselection using triplet probes (Fig.2). Absorption spectra display the energy of light necessary to excite a molecule from the ground state S_0 to the vibrational levels of the higher energy state S_1. After radiationless conversion to the lowest S_1 state, part of the energy is released as fluorescence by transition to the different vibrational levels of the S_0 state.

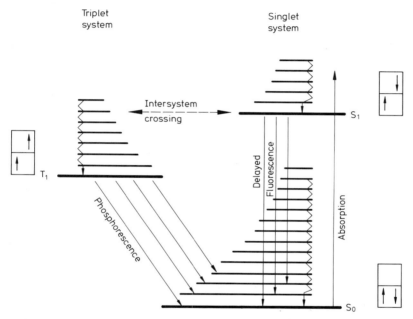

Figure 2. Jablonski diagram of singlet and triplet energy states

Under favourable conditions (e.g. a heavy atom effect favouring spin orbital coupling) the energy of excitation can cross to the triplet system in which, unlike the singlet system, the electron spins are unpaired. Part of it is released as phosphorescence after rapid internal conversion from higher vibrational levels to the lowest T_1 level. Another part undergoes intersystem crossing from the triplet system to the singlet system and is emitted as fluorescence as so-called delayed E-type fluorescence, because the decay is determined by the longer lifetime of the triplet state and not as in the case of the prompt fluorescence, by the lifetime of the singlet state S_1. Thus, we can learn two things from this Jablonski diagram: phosphorescence and delayed fluorescence have the same rate of emission decay in the micro- to millisecond range typical of triplet state (prompt fluorescence decays in nanoseconds), and the energy of the phosphorescence is lower than the energy of the delayed fluorescence. That

is, delayed fluorescence appears at shorter wavelengths than
phosphorescence and has the same spectral distribution as the prompt
fluorescence. In some dyes the yield of phosphorescence is higher than the
yield of delayed fluorescence, but lower than the yield of prompt
fluorescence by a factor of about 10000. Triplet yield is considerably
higher in a dye in which the Br-atoms of eosin are replaced by iodine.
This has considerable experimental advantages as the reasonably high yield
of delayed fluorecence from diiodofluorescein can be distinguished from
its strong prompt fluorescence by time gating the photomultiplier.
Separation of phosphorescence from prompt fluorescence can be done by the
straightforward optical filtering, making the use of phosphorescence as
the signal to be modulated by motion appear to be more simple. However,
phosphorescence has a drawback compared to delayed fluorescence in another
respect. In prompt and delayed fluorescence the orientation of the dipole
moment of emission is nearly parallel to the dipole moment of absorption,
whereas in phosphorecence there is a considerable angle between them. This
diminishes the value of the anisotropy, the critical parameter for the
evaluation of the rotational diffusion (see below).

METHODS AND INSTRUMENTATION FOR MEASURING POLARIZED EMISSION OF DELAYED
FLUORESCENCE MODULATED BY ROTATIONAL MOTION (48,49)

   In most cases an extrinsic dye label has to be introduced into the
proteins. For this purpose the protein has to be purified to homogeneity
and special care has to be taken to label the molecule at one site only.
For the evaluation of the results it may be helpful if different sites of
a protein are labelled selectively in separate experiments, for example,
an amino terminal group by diiodofluorescein isothiocyante and a thiol
group by diiodofluorescein iodoacateamide. The labelled protein has to be
reconstituted into phospholipid vesicle membranes.

   In our laboratory the nanosecond pulse of a nitrogen pumped dye laser
is used for photoselection. The exciting light is linearly polarized by a
Glan-Thompson prism. Polarized light emitted from the sample passes
through an analyser which alternates between orientations parallel and
perpendicular to the plane of the exciting light once per second. The
light emitted from the sample passes through an interference filter which
transmits only the emitted light and is photoncounted (Ortec system with a
time resolution of 100 MHz - photon counting has distinct advantages over
analogue techniques in quantitation of weak signals). To separate
longlived delayed fluorescence from shortlived prompt fluorescence the
potential of the first dynode, triggered by each laser pulse, is set for
5-7 µsec to the potential of the photocathode, whereby the release of
electrons from the latter and their acceleration is diminished. This time-
gating, together with the relatively high yield of the delayed
fluorescence of the extrinsic dye, avoided any disruption of the kinetics
of delayed fluorescence decay. Counts from individual photoselection
pulses are collected in a homebuilt digital buffer with a channel
resolution > 0.2 µsec. Counts corresponding to the two orientations of the
analysator are separately accumulated in a multichannel analyser. Data
processing (calculation of the anisotropy from the two experimental curves
of decay of delayed fluorescence parallel and perpendicular to the plane
of polarsation of the exciting light) (Fig. 3) and evaluation (statistical
analysis of the kinetics of anisotropy decay) are done with an LSI 11/73
computer. Multiexponential functions are fitted to the experimental
anisotropy decay curve by a program (simplex) which weights the
experimental data and minimizes the residual sum of squares of deviation
from the fit. Quality of fit is demonstrated by graphical presentation of
residuals and the autocorrelation function in Fig. 3.

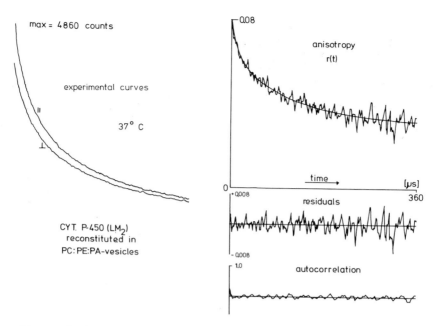

Figure 3. Decay of delayed fluorescence perpendicular and
parallel to the plane of the polarized excitation
and anisotropy decay of purified cytochrome
P-450 LM2 , labelled at cystein 152 with
diiodofluorescein iodoaceteamide and reconstituted
into phospholipid membrane vesicles (PC =
phosphatidylcholine; PE = phosphatidylethanolamine; PA = phosphatidic acid 2:2:0.12 w:w, from
egg). The continuous line represents a fitted bi-
exponential function. Goodness of the fit is
demonstrated by the residuals and the autocorrelation function

COMPARISON OF THE ADVANTAGES AND DISADVANTAGES OF PHOTOSELECTION
TECHNIQUES APPLIED TO THE MEASUREMENT OF CYTOCHROMES P-450

The photoselection techniques differ greatly in their general
applicability to cytochromes P-450 and the range of problems they can
approach. Major differences exist between absorption dichroism on the one
hand and delayed fluorescence and phosphorescence on the other, minor ones
between the two "triplet" techniques. These use extrinsic triplet probes
and are therefore more invasive. They require more sample preparation: the
cytochromes have to be purified, labelled with a dye while taking many
precautions - some of them have been mentioned above - and reconstituted
into phospholipid membranes. Absorption dichroism uses as an intrinsic
probe the heme of cytochrome P-450 which makes only the preparation of
microsomes or mitochondria necessary, but restricts investigation to the
reduced carbon-monoxide liganded cytochrome (54,55,56). Investigation of

microsomes is further complicated by the multiplicity of cytochromes P-450 which differ in their rotational mobility (see part II) and also in the rate of reassociation of the photochemically split CO-complex (53). Therefore unequivocal interpretation of experiments demands investigation of purified cytochromes P-450.

The main technical difference is that the sensitivity of absorption dichroism is several orders of magnitude less than that of triplet techniques with the consequence that measuring times are longer, time resolution and signal to noise ratio poor, which excludes statistical data evaluation in most cases.

The quality of experimental curves observed in our laboratory using delayed fluorescence (see Fig.3), which allowed kinetic evaluation of the anisotropy decay, was achieved by several experimental factors: 1) high yield from delayed fluorescence of diiodofluorescein used as the extrinsic dye-label, 2) optimal anisotropy of delayed fluorescence in contrast to phosphorescence due to parallel dipole moments of absorption and emission, 3) signal detection by photoncounting and 4) use of a fast homebuilt digital buffer allowing counting of multiple photons per photoselection pulse.

KINETICS OF MOTIONAL MODULATION OF THE EMISSION DECAY: THEORY (58,59)

In principle, the mode and rate of motion reveals the molecular structure of a particle. Depending on their shape, the motion of macromolecules can be rather complex. The two-dimensionality of biological membranes simplifies matters by restricting the motional freedom of membrane-bound protein to rotation about an axis normal to the membrane, upon which a wobbling motion can be superimposed (see Fig. 4.).

Rotational motion modulates the kinetics of polarized emission. From the time-dependent decay of intensity I(t) of emission polarized parallel II and perpendicular $\perp$ to the polarized excitation, an anisotropy function r(t) can be calculated

$$r(t) = \frac{I_{II}(t) - I_{\perp}(t)}{I_{II}(t) + 2 I_{\perp}(t)} \tag{1}$$

The emission anisotropy of a rigid ellipsoid (compare Fig.4) in isotropic media can be described by

$$(t) = \sum_{i=1}^{5} \beta_i \exp(-t/\Phi_i) \tag{2}$$

Practically, the anisotropy decay reduces to three exponentials as the rotational correlation times in two pairs nearly coincide.

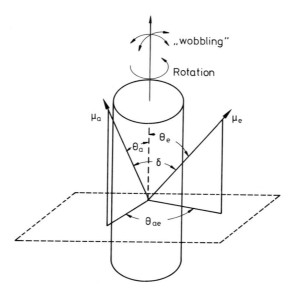

Figure 4. Model of a fluorescing cylindrical molecule
presenting the case of nonparallel transition
moments of absorption μ_a and emission μ_e

The prefactors β_i of the exponents are trigonometric functions of the
angles between the transition moments of absorption μ_a and emission μ_e and
the axis of symmetry of the ellipsoid (see figure 4).

The rotational correlation times Φ are related to the diffusion constants
of the rotation about the axis perpendicular to ($D_\perp$) and parallel (D_{II}) to
the axis of symmetry.

$$\Phi_1^{-1} = 5D_\perp + D_{II}, \qquad \Phi_2^{-1} = 2D_\perp + 4D_{II}, \qquad \Phi_3^{-1} = 6D_\perp \quad (3)$$

In anisotropic media such as membranes the modes of rotational freedom are
reduced. Two limiting cases can be differentiated (Fig. 4):
1. The uniaxial rotation of a chromophore about the normal to
   the membrane.
2. A wobbling diffusion in which the axis of symmetry of the
   molecules performs wobbling within a cone of half angle Θ_0
   about the membrane normal.

The uniaxial rotation can be described by equation (2) in which the
diffusion coefficients $D_\perp$ (see equations 3) for rotation about the in
plane axis disappear; with $d_\perp = 0$ we get

$$\Phi_1^{-1} = D_{II}, \quad \Phi_2^{-1} = 4D_{II}, \quad \Phi_3 = 0 \qquad\qquad (4)$$

and equation (2) converts to

$$r(t)/r_0 = A_1 \exp(-D_{II}t) + A_2 \exp(-4D_{II}t) + A_3 \qquad (5)$$

Thus a bisexponential anisotropy decay with a ratio of the rotational correlation times of 1:4 is characteristic for strict uniaxial rotation in a membrane.

The initial anisotropy at $t = 0$ depends only the angle δ between the transition moments of absorption μ_a and emission μ_e.

$$r(t=0) = r_0 = \frac{2}{5} \frac{(3\cos^2\delta - 1)}{2} \qquad (6)$$

r_0 can have values between -0.2 ($\delta = 90°$) and 0.4 ($\delta = 0°$). The maximum value can only be achieved if the transition moments μ_a and μ_e are parallel and angles Θ_a and Θ_e coincide with angle Θ (see Fig. 4) In this case the factors A_x depend only on the angle Θ between the axis of rotation and the orientation of the coincident transition moments (see equation 7). This applies to delayed fluorescence but not to phosphorescence.

$$
\begin{aligned}
A_1 &= 3\sin^2\Theta\cos^2\Theta \\
A_2 &= 3/4\sin^4\Theta \\
A_3 &= 1/4(3\cos^2\Theta - 1)^2
\end{aligned}
\qquad (7)
$$

For uniaxial rotation the angle Θ can be evaluated (in cases of delayed fluorescence and absorption dichroism) from

$$r_\infty/r_0 = 1/4(3\cos^2\Theta - 1)^2 \qquad (8)$$

or from
$$A_1/A_2 = 4 \cot^2\Theta \qquad (9)$$

$A_3 = r_\infty/r_0$ contains the rest anisotropy r_∞: it is typical for the limitation of the degrees of motional freedom of a membrane protein that the depolarization of emission due to motion is imperfect. r_∞ is a measure for this limitation. For isotropic rotation the rest anisotropy r_∞ vanishes.

Wobbling rotation can be kinetically described by

$$r(t) = [r(0) - r_\infty] \cdot \exp(t/\Phi) + r_\infty \qquad (10)$$

in which the apparent relaxation time Φ is related to a wobbling diffusion constant D_W by

$$\Phi = \langle\sigma\rangle/D_W \qquad (11)$$

$\langle\sigma\rangle$ is a relaxation time described by

$$\langle\sigma\rangle = \sum_{i=1}^{\infty} r_i\sigma_i/[r_0 - r_\infty] \qquad (12)$$

with σ_i being a function of the half angle Θ_0 of the wobbling cone. D_W can be approximatively determined from the initial slope of $r(t)$.

$$D_W \approx (1 - r_\infty/r_0)/6\Phi \qquad (13)$$

and Θ_0 from

$$_\infty/r_0 = [1/2\ \cos\Theta_0(1+\cos\Theta_0)]^2. \tag{14}$$

In general wobbling is superimposed on uniaxial rotation yielding a complex anisotropy decay kinetic.

In phosphorescence the transition moments of absorption μ_a and emission μ_e are often not parallel with the consequence of an increase of anisotropy with time for the cases

$$\Theta_{ae} > 45°, \Theta_{ae} > 90°, \Theta_a, \Theta_e > 54.7°.$$

USEFUL PARAMETERS OF MEMBRANE TOPOLOGY OF CYTOCHROME P-450 WHICH CAN BE RETRIEVED FROM THE ANISOTROPY DECAY KINETICS IN CASE OF UNIAXIAL ROTATION

The dimensions of a membrane spanning protein approximated by a cylinder can be determined from

$$\Phi_{II} = \frac{1}{D_{II}} = \frac{4V\eta}{kT} = \frac{4\pi^2 h\eta}{kT} \tag{15}$$

Φ_{II} = rotational correlation time about the membrane normal
D_{II} = rotational diffusion coefficent
η = viscosity of the membrane
V = volume of a cylinder
a = radius of a cylindrial protein
h = height of the protein cylinder in the membrane
k = Boltzmann-constant
T = temperature in °K

For a cylindrical protein, partially immersed in the membrane, equation (15) has to be modified (49,60) to

$$\Phi_{rot} = \frac{4\pi\eta a^3}{kT}\left[\frac{h}{a} + \frac{8}{3\pi}\right] \tag{16}$$

Protein conformational changes can become visible by a change of orientation of the chromophore according to equations (8) and (9).

Part of large clusters of proteins f_{im} can be detected by deviation of the factor $A_3 = (r_\infty/r_0)_{exp}$ in equation (5) from the theoretical value according to

$$(r_\infty/r_0)_{exp} = f_{im} + (1-f_{im})\ (r_\infty/r_0)_{theo} \tag{17}$$

The orientation of the heme relative to the plane of the membrane has been determined by photoselection experiments using absorption dichroism (56) from equation (8)

REFERENCES

1. M.L. Jennings, Topography of membrane proteins,
    Annu. Rev. Biochem. 58: 999 (1989).
2. K.U. Linderstrøm-Lang and J.A.Schellman, Protein structure and enzyme
    activity, in: "The Enzymes", P.D. Boyer, H. Lardy, and K. Myrbäck,
    eds., Academic Press, New York (1959).
3. A. Stier and S.A.E. Finch, Rotational diffusion of homo- and
    heterooligomers of cytochrome P-450: the functional significance of
    cooperativity and the membrane structure,
    in: "Frontiers in Biotransformation", Vol. 1, K. Ruckpaul and H.
    Rein, eds., Taylor & Francis, London, submitted.
4. J.R. Abney and J.C. Owicki, Theories of protein-lipid and
    protein-protein interactions in membranes, in: "Progress in Protein-
    Lipid Interactions", Vol. 1, A. Watts and J.J.H.M. De Pont, eds.,
    Elsevier Science Publishers, Amsterdam (1985).
5. I. Klotz, D.W. Darnall, and N.R. Langerman, Quaternary
    structure of proteins, in: "The Proteins", Vol. 1,
    H. Neurath and R.L. Hill, eds, Academic Press,
    New York (1975).
6. D.E. Koshland Jr., G. Nemethy, and D. Filmer, Comparison of
    experimental binding data and theoretical models in proteins
    containing subunits, Biochemistry 5: 365 (1966).
7. D. Marsh, Molecular mobility in membranes, in: "Physical
    Properties of Biological Membranes and Their Functional
    Implications", C. Hidalgo, ed., Plenum Publishing Company,
    New York (1988).
8. H. Sandermann Jr., Cooperativity of lipid-protein
    interactions, in: "Progress in Protein-Lipid Interactions"
    Vol. 2, A. Watts and J.J.H.M. De Pont, eds., Elsevier
    Science Publishers, Amsterdam (1986).
9. H. Frauenfelder, F. Parak, and R.D. Young, Conformational
    substates in proteins, Annu. Rev. Biophys. Biophys. Chem. 17: 451
    (1988).
10. C. Jung, F. Marlow, O. Ristau, S. Falsett. I.C. Gunsalus, and H.
    Frauenfelder, Accessibility and dynamics of the active site in
    bacterial cytochrome P-450, in: "Cytochrome P-450: Biochemistry and
    Biophysics", I. Schuster , ed. Taylor & Francis, London (1989).
11. J. Monod, J. Wyman, and J.-P. Changeaux, On the nature of
    allosteric transitions: A plausible model , J. Mol. Biol. 12: 88
    (1965).
12. G.R. Welch (ed.) "Organized Multienzyme Systems: Catalytic
    Properties", Academic Press Inc., Orlando (1985).
13. E. Sackmann and H. Träuble, Studies of the crystalline-liquid
    crystalline phase transition of lipid model membranes. II. Analysis
    of the electron spin resonance spectra of steroid labels
    incorporated into lipid membranes, J. Am. Chem. Soc. 94: 4491
    (1972).
14. H. Träuble and E. Sackmann, Studies of the crystalline-liquid
    crystalline phase transition of lipid model membranes. III.
    Structure of a steroid-lecithin system below and above the lipid
    phase transition, J. Am. Chem. Soc. 94: 4499 (1972).
15. C.J. Scandella, P. Devaux, and H.M. McConnell, Rapid lateral
    diffusion of phospholipids in rabbit sarcoplasmic reticulum, Proc.
    Natl. Acad. Sci USA 69: 2056 (1972).
16. M. Edidin, Rotational and translational diffusion in
    membranes, Annu. Rev. Biophys. Bioeng. 3: 179 (1974).
17. M.D. Hollenberg, Examples of homospecific and heterospecific receptor
    regulation, Trends Pharmacol. Sci. 6: 242 (1985).

18. J. Schlessinger, Signal transduction by allosteric receptor oligomerization, *Trends Biochem. Sci.* 13: 443 (1988).
19. J.-P. Changeux, A. Devillers-Thiery, and P. Chemouilli, Acetylcholine receptor: An allosteric protein, *Science* 225: 1355 (1984).
20. C.R. Hackenbrock, B. Chazotte, and S.S. Gupte, The random collision model and a critical assessment of diffusion and collision in mitochondrial electron transport, *J. Bioenerg. Biomembr.* 18: 331 (1986).
21. J.W. DePierre and L. Ernster, Enzyme topology of intracellular membranes, *Ann. Rev. Biochem.* 46: 201 (1977).
22. M. Ingelman-Sundberg, Cytochrome P-450 organization and membrane interaction,, *in*: "Cytochrome P-450", P.R. Ortiz de Montellano, ed., Plenum Press, New York (1986).
23. E. Schulz, A critical evaluation of methods for prediction of protein secondary structures, *Annu. Rev. Biophys. Biophys. Chem.* 17: 1 (1988).
24. B.A. Wallace, M. Cascio, and D.L. Mielke, Evaluation of methods for the prediction of membrane protein secondary structures, *Proc. Natl. Acad. Sci. USA* 83: 9423 (1986).
25. G.E. Tarr, S.D. Black, V.S. Fujita, and M.J. Coon, Complete amino acid sequence and predicted membrane topology of phenobarbital-induced cytochrome P-450 (isozyme 2) from rabbit liver microsomes, *Proc.Natl. Acad. Sci. USA* 80: 6552 (1983).
26. A. M. Lesk and C. Chothia, How different amino acid sequences determine similar protein structure: the structure and evolutionary dynamics of the globins, *J. Mol. Biol.* 136: 225 (1980).
27. C. Chothia and A.M. Lesk, Helix movements and the reconstruction of the haem pocket during the evolution of the cytochrome c family, *J. Mol. Biol.* 182: 151 (1985).
28. C. De Lemos-Chiarandini, A.B. Frey, D.D. Sabatini and G. Kreibich, Determination of the membrane topology of the phenobarbital-inducible rat liver cytochrome P-450 isoenzyme PB-4 using site-specific antibodies, *J. Cell Biol.* 104: 209 (1987).
29. R. Müller, W.E. Schmidt, and A. Stier, The site of cyclic AMP-dependent protein kinase catalyzed phosphorylation of cytochrome P-450 LM2. *FEBS Lett.* 187: 21 (1985).
30. H. Furuya, T. Shimizu, K. Hirano, M. Hatano, Y. Fujii-Kuriyama, R. Raag, and T.L. Poulos, Site-directed mutageneses of rat liver cytochrome P-450$_d$: catalytic activities toward benzphetamine and 7-ethoxycoumarin, *Biochemistry*: 28: 6848 (1989).
31. J. Murakami, Y. Yabusaki, T. Sakaki, M. Shibata, and H. Ohkawa, A genetically engineered P-450 monooxygenase: Construction of the functional fused enzyme between rat cytochrome P-450c and NADPH-cytochrome P-450 reductase, *DNA* 6: 189 (1987).
32. E.S. Kempner and W. Schlegel, Size determination of enzymes by radiation inactivation, *Analyt.Biochem.* 92: 2 (1979).
33. E.S. Kempner and J.H. Miller, Radiation inactivation of glutamate dehydrogenase hexamer: lack of energy transfer between subunits, *Science* 222: 586 (1983).
34. F.P. Guengerich, P.F. Churchill, C.Y. Jung and S. Fleischer, Target inactivation analysis applied to determination of rat liver proteins in the purified state and in microsomal membranes, *Biochim. Biophys. Acta* 915: 246 (1987).
35. J.Deisenhofer, O. Epp, K. Miki, R. Huber, and H. Michel, structure of the protein subunits in the photosynthetic reaction centre of Rhodopseudomonas viridis at 3 Å resoluton, *Nature* 318: 618 (1985).

36. K. Moffat, Time-resolved macromolecular crystallography, Annu. Rev. Biophys. Biophys. Chem. 18: 309 (1989).

37. P. Fajer and D. Marsh, Sensitivity of saturation transfer ESR spectra to anisotropic rotation. Application to membrane systems, J. Magn. Reson. 51: 446 (1983).

38. D.D. Thomas, T.M. Eads, V.A. Barnett, K.M. Lindahl, D.A.Momont, and T.C. Squier, Saturation transfer EPR and triplet anisotropy: Complementary techniques for the study of microsecond rotational dynamics, in: "Spectroscopy and the Dynamics of Molecular Biological Systems", P.M. Bayley, R.E. Dale, eds., Academic Press, London, (1985).

39. D. Schwarz, J. Pirrwitz, and K. Ruckpaul, Rotational diffusion of cytochrome P-450 in the microsomal membrane - evidence for a clusterlike organization from saturation transfer electron paramagnetic resonance spectroscopy, Archives Biochem.Biophys. 216: 322 (1982).

40. R.M. Cooke and I.D. Campbell, Protein structure determination by nuclear magnetic resonance, Bio Essays 8: 52 (1988)

41. A. Bax, Two-dimensional NMR and protein structure, Annu. Rev. Biochem. 58: 223 (1989).

42. K. Wüthrich, "NMR of proteins and nucleic acids", Wiley, New York (1986).

43. S.O. Smith and R.G. Griffin, High-resolution solid-state NMR of proteins, Annu Rev. Phys. Chem. 39: 511 (1988).

44. R. Smith, D.E. Thomas, F. Separovic, A.R. Atkins, and B.A. Cornell, Determination of the structure of a membrane-incorporated ion channel, Biophys. J. 56: 307 (1989).

45. T.M. Jovin and W.L.C. Vaz, Rotational and translational diffusion in membranes measured by fluorescence and phosphorescence methods, Methods.Enzymol. 172: 471 (1989).

46. R. Greinert, H. Staerk, A. Stier, and A. Weller, E-type delayed fluorescence depolarization, a technique to probe rotational motion in the microsecond range. J. Biochem. Biophys. Methods. 1: 77 (1979).

47. R. Greinert and A. Stier, Rotational diffusion of cytochrome P-450 in a reconstituted system measured by depolarization of delayed fluorescence, in: "Biochemistry, Biophysics and Regulation of Cytochrome P-450", J.Å. Gustafsson, J. Carlstedt Duke, A. Mode, and J. Rafter, eds., Biomedical Press, Elsevier/North-Holland (1980).

48. R. Greinert, S.A.E. Finch, and A. Stier, Cytochrome P-450 rotamers control mixed-function oxygenation in reconstituted membranes. Rotational diffusion studied by delayed fluorescence depolarization. Xenobiotica 12: 717 (1982a).

49. R. Greinert, S.A.E. Finch, and A. Stier, Conformation and rotational diffusion of cytochrome P-450 changed by substrate binding. Biosci. Rep. 2: 991 (1982b).

50. R.J. Cherry, Transient dichroism of bacteriorhodopsin, Methods Enzymol. 88: 248 (1982).

51. W.L.C. Vaz, R.H. Austi, and H. Vogel, The rotational diffusion of cytochrome b5 in lipid bilayer membranes, Biophys.J.26: 415 (1979)

52. P. Roesen and A. Stier, Kinetics of CO and O2 complexes of rabbit liver microsomal cytochrome P450, Biochem. Biophys.Res. Commun. 51: 603 (1973).

53. F. Mitani, T. Iizuka, H. Shimada, R. Ueno, and Y. Ishimura, Flash photolysis studies on the CO complexes of ferrous cytochrome P-450scc and cytochrome P-45011β, J. Biol. Chem. 260: 12042 (1985).

54. C. Richter, K.H. Winterhalter, and R.J. Cherry, Rotational diffusion of cytochrome P-450 in rat liver microsomes, FEBS Lett. 102: 151 (1979).

55. S. Kawato, J. Gut, R.J. Cherry, K.H. Winterhalter, and C. Richter, Rotation of cytochrome P-450. I. Invetigation of protein-protein interaction of cytochrome P-450 in phospholipid vesicles and liver microsomes, J. Biol. Chem. 257: 7023 (1982).

56. J. Gut, C. Richter, R.J. Cherry, K.H. Winterhalter, and S. Kawato, Rotation of cytochrome P-450. II. Specific interactions of cytochrome P-450 with NADPH-cytochrome P-450 reductase in phospholipid vesicles, J. Biol. Chem. 257: 7030 (1982).

57. C.A. Parker, " Photoluminescence of Solutions", Elsevier, Amsterdam (1968).

58. S. Kawato and K. Kinosita, Time-dependent absorption anisotropy and rotational diffusion of proteins in membranes, Biophys.J. 36: 277 (1981).

59. K. Kinosita Jr., A. Ikegami, and S. Kawato, On the wobbling-in-cone analysis of fluorescence anisotropy decay, Biophys. J. 37: 461 (1982).

60. B.D. Hughes, B.A. Pailthorpe, L.T. White, and W.H. Sawyer, Extraction of membrane microvisosity from translational and rotational diffusion coefficients, Biophys. J. 37: 673 (1982).

II. MEMBRANE TOPOLOGY OF CYTOCHROMES P-450: OLIGOMERS AND

COOPERATIVITY

A. Stier, V. Kruger, T. Eisbein and S.A.E. Finch

Max-Planck-Institut für biophysikalische Chemie
Abteilung Spektroskopie, D-3400 Göttingen, F.R.G.

INTRODUCTION

The main concern of this contribution is whether liver microsomal cytochromes P-450 have an oligomeric structure and whether cooperativity of the oligomeric subunits has functional significance.

The liver microsomal cytochromes P-450 are the main components of a multienzyme system involved in the phase I of biotransformation of endogenous and exogenous substrates. They interact in the mixed function oxygenation (MFO) of a wide variety of substrates with three other membrane-bound proteins, NADPH cytochrome P-450 reductase, cytochrome b5 reductase and cytochrome b5. Additionally there may be a functional linkage to epoxide hydratase and/or UDP-glucoronyltransferase, two other microsomal enzymes of phase II of biotransformation, by channelling of metabolites. As multiple cytochromes P-450 differing in their enzymic behaviour coexist in the microsomal membrane it appears that such a system, comprising a large number of proteins handling a practically unlimited number of exogenous substrates and products of different degrees of toxicity as well as endogenous substances of different physiological importance, requires some sort of molecular organisation and regulation if it is to function properly.

Supramolecular structure as an important element of topology of proteins in the membrane (compare our description in part I) can have a critical influence at a number of points of MFO: (1) electron transfer, (2) substrate-gating of electron transfer, (3) product release, (4) transfer of metabolites to the phase II enzymes and (5) recycling oxygenation. We may ask, for example, how an electron is transferred to the right isoform – among the many coexisting in the membrane – which is specific for the substrate in question? Are electrons transferred in the absence of substrate? Does a mechanism exist which controls electron flow and avoids futile electron transfer to oxygen with production of superoxide anion, hydrogen peroxide or even water? Instead of asking the question 'to which end is the system regulated?' we may ask: 'do certain elements of control exist which are characteristic for well known principles of enzymic regulation?' We will follow the second approach and may find that we can from principles of cooperative control ask new questions about some details of functioning of the system.

The measurement of the rotational diffusion of membrane-bound proteins by photoselection techniques as described in part I can give conclusive (decisive) information on membrane protein topology,

Molecular Aspects of Monooxygenases and Bioactivation of Toxic Compounds
Edited by E. Arınç *et al.*, *Plenum Press, New York*,

particularly about oligomeric structure. In this part we describe results obtained on membrane topology of cytochromes P-450 LM2 and LM4 from which, in combination with results obtained with other useful techniques, we elaborate a model of membrane topology of cytochromes P-450. Rotational diffusion studies also give information about protein interactions between components of the MFO system and between cytochromes P-450 and phase II enzymes. The verification of an oligomeric structure of cytochromes P-450 has significant consequences for the kinetics of their reduction, substrate binding and product release as well as for regulation of the system. In addition, the application of the concepts of cooperativity should be very useful for an understanding of the function of the system.

THE CYTOCHROME P-450 ROTAMER IN THE MEMBRANE IS AN OLIGOMER

Rotational diffusion of cytochromes P-450 using photoselection techniques has been measured in reconstituted and liver microsomal membranes. Evidence for the existence of P-450 oligomers of a defined size comes mainly from photoselection experiments performed on purified cytochrome P-450 LM2 using motional modulation of delayed fluorescence emitted from a diiodofluorescein marker (1,2,3). The drawbacks of these experiments in which part of the cytochrome P-450 molecules investigated are marked by the reporter dye and the cytochromes have to be reconstituted in artificial membranes is by far outweighed by a number of positive points which are missing in measurements of the rotational behaviour of cytochromes P-450 in liver microsomal membranes by absorption dichroism (4,5,6) (see part I): a) The higher sensitivity allows statistical analysis of the kinetics of the anisotropy decay curve, a precondition for the evaluation of the mode of rotational diffusion; only from a strict uniaxial rotation can conclusions on the size of the rotamer be drawn. b) The motion of only one species of P-450 is monitored. This is important as different cytochromes P-450 differ in rate and mode of motion. c) Motional behaviour of cytochromes P-450 in different functional states can be investigated. Regarding the artificial nature of the reconstituted system it is noteworthy that cytochrome P-450 LM2 labelled with marker dye at cysteine 152 exhibited exactly the same rotational behaviour as cytochrome P-450 LM2 marked at the N-terminus (7).

For consideration of the membrane topology differentiation of protein rotational modes may be reduced to the operational terms uniaxial rotation about the normal to the membrane and non-uniaxial rotation or complex motion (e.g. wobbling compare figure 1). Only in the case of uniaxial rotation can a correlation time of rotation be determined from which conclusions about the size of the rotamer can be drawn. In this case the kinetics of the anisotropy decay is bisexponential with a 1:4 ratio of the decay times (8). If this kinetic is not observed, wobbling rotation has to be assumed. In the wobbling case the presence of a membrane spanning cylindrical protein or of a disc-like structure firmly attached to the surface of the membrane structure may be excluded.

The anisotropy decay of reconstituted cytochrome P-450 LM2 follows the kinetics which theory predicts for uniaxial rotation (see table 1): a ratio ∅1/∅2 of 4 was observed, which was independent of temperature, binding of type I and type II substrates, reduction by dithionite and the lipid composition of the membrane. Rotation became faster with increasing temperature, whereby the change was greater between 10 and 20°C than between 25 and 35°C (table 1).

motion

mode   kinetics

$$r(t) = \sum_{i=1}^{n} A_i \ \exp\left(-\frac{t}{\varnothing_i}\right)$$

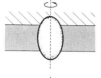

uniaxial   bisexpotential
$\varnothing_1 : \varnothing_2 = 4$

uniaxial   bisexpotential
$\varnothing_1 : \varnothing_2 = 4$

nutation   complex
wobbling

Fig. 1. Protein membrane topology, mode of rotation and
kinetics of anisotropy decay

Table 1. Effects of temperature, ligands, reduction and NADPH
cytochrome P-450 reductase (RED) on rate and mode of
motion of cytochrome P-450 LM2 (LM2) reconstituted in
phospholipid membranes (phosphatidylcholine :
phosphatidylethanolamine : phosphatidic acid =
2:1:0.12 (w/w, from egg) lipid : protein = 10:1
(w/w)) (1,2,3,9)

	rotational correlation time (μs)	mode	temp (°C)
LM2	193	uniaxial	8
LM2	133	"	16
LM2	110	"	24
LM2	126	"	32
LM2	129	"	37
LM2 + benzphetamine	204	"	37
LM2 + benzphetamine + dithionite	51	"	37
LM2 + metyrapone	81	"	37
LM2	114	"	30
LM2 + RED (1+1)	73	wobbling	30
LM2 + RED (1+1) + NADPH	84	uniaxial	30
LM2 + RED (5+1)	107	"	30
LM2 + RED (5+1) + NADPH	65	"	36

Lipid composition greatly affected the mode and rate of motion. Reconstitution with egg lecithin as the only lipid yielded an apparent kinetics of 2 isotropic rotations greatly varying in relaxation times indicating 2 populations of differently sized lipoprotein complexes. The presence of phosphatidylethanolamine (50% of total phospholipid) resulted in uniaxial rotation. Apparently, for incorporation of cytochrome P-450 LM2 with the 'right' membrane topology, phospholipids with a negative headgroup charge as distinct from 'neutral' lecithin (PC) are required.

The degree of lipid unsaturation also influences the rate of rotation: in a mixture of microsomal lipids, which are relatively unsaturated, the rate was twice as great as in the corresponding mixture of egg lipids.

How can we delimit the size of the cytochrome P-450 rotamer from the rotational correlation time measured? Using the only known tertiary structure of cytochrome P-450 (10) – the structure of cytochrome P-450 CAM (figure 2a)– as a lead (for justification see below) and the evidence indicating that the bulk of the protein is located outside the membrane, we may construct three different rotamers approaching a cylindrical shape. We also assume an extension of the N-terminus of the cytochrome P-450 CAM by a peptide of 60 amino acids including 22 hydrophobic amino acids near the N-terminus (11,12). We adapt a theoretical treatment (13) to the present case:

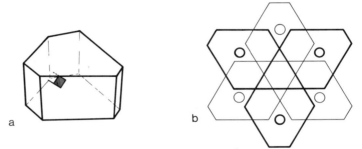

a   b

Fig. 2. a) The shape of cytochrome P-450 CAM (10). The intramolecular location of the heme group (shaded square with appending vinyl side chains) is shown. b) Schematic drawing of the hexamer of 32 dihedral symmetry of cytochrome P-450 LM2 found in solution of cytochrome P-450 LM2 (14). As the subunit of the two superimposed trimers the shape of the cytochrome P-450 CAM was assumed. Circles indicatethe site where the hydrophobic N-terminal sequence is appended to the bulk molecule. It is assumed that these hydrophobic structures of one trimer extend to the surfaces of the other trimer thereby stabilizing an hexamer.

For a cylindrical protein, partially immersing into the membrane equation (1) relates

$$\Phi_{rot} = \frac{4\pi\eta a^3}{kT} \left[\frac{h}{a} + \frac{8}{3\pi} \right] \qquad (1)$$

Φrot = rotational correlation time about the membrane normal
to η = viscosity of the membrane
   a = radius of a cylindrical protein
   h = height of the protein cylinder in the membrane and
   T = temperature in °K
  (k = Boltzmann constant)

This relationship has been calculated for 3 cylinders,
1) for a monomer taking the geometrical centre of the quasi prism as the
rotational axis and the height of the prism as the length of the cylinder
(30 Å) (table 2),
2) for a pentamer (figure 3a) taking a) the radius of its greatest
extension or b) the radius of a quasi cylinder delimited by the
hydrophobic N-terminal helices (table 3),
3) for a hexamer (figure 3b) taking radii as described for the pentamer
(table 4).

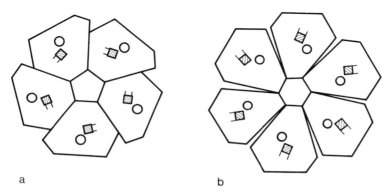

a   b

Fig. 3. a) pentamer, b) hexamer constructed from subunits of
the shape of cytochrome P-450 CAM as drawn in figure 2a, The
intramolecular location of the heme group is indicated by a
shaded square with appended vinyl side chains. Circles
indicate the site where the hydrophobic N-terminal sequence is appended to
the bulk molecule.

In all cases the rotational correlation times were calculated for
different degrees of depth of immersion, h, a membrane viscosity of 1.4
poise (for a critical examination of the viscosities of membranes see
(15)) and a temperature of 310°K.

Table 2. Rotational correlation times Φ_{rot} of a cylinder
         approximated from a monomer model (see text)
         immersing to different extents (h) into a
         phospholipid membrane.

a [Å]	h [Å]	Φ_{rot} [μs]
30	10	13.1
	20	16.8
	30	20.5

Table 3. Rotational correlation times Φ_{rot} of cylinders of
different radii (a) approximated from a pentamer
model (see text) immersing to different extents (h)
into a phospholipid membrane.

a [Å]	h [Å]	Φ_{rot} [µs]
60	10	90.1
	20	105.0
	30	119.8
	40	134.6
47	10	45.3
	20	54.4
	30	63.5
	40	72.7

Table 4. Rotational correlation times Φ_{rot} of cylinders of
different radii (a) approximated from a hexamer model
(see text) immersing to different extents (h) into a
phospholipid membrane.

a [Å]	h [Å]	Φ_{rot} [µs]
69	10	134.2
	20	153.8
	30	173.3
	40	192.8
43	10	35.3
	20	42.9
	30	50.5
	40	58.1

Inspection of table 2 immediately reveals that a cytochrome P-450 monomer
totally immersed in the membrane would have a rotational correlation time
of about 20 µs, much shorter than the experimentally determined
correlation time of 129 µs. Assumption of a pentamer of the shape shown in
figure 3 immersed more than 30 Å deep into the membrane gives a rotational
rate similar to the experimental value (table 3), as does a hexamer with a
hydrophobic basis immersing to 10 Å in the membrane (table 4). On the
other hand, five or six N-terminal hydrophobic sequences combined in a
cylinder of 47 or 43 Å diameter have to span the whole membrane in order
to explain the measured correlation time (tables 3 and 4). Thus it is
clear from rotational diffusion measurements that a defined oligomer
structure is the important topological element of cytochrome P-450 in the
membrane. This model can be complemented by a number of published
experimental results.

ELEMENTS OF MEMBRANE TOPOLOGY OF CYTOCHROMES P-450

As mentioned above, a model of the membrane topology of cytochromes
P-450 can be elaborated on the basis of the known tertiary structure of
cytochrome P-450 CAM (10). A comparison of the primary structure of more
than 50 cytochromes P-450 indicates that major motifs of tertiary
structure are conserved in evolution (12,16). This is consistent with the
results from structural investigation of globins and other cytochromes

which confirm the evolutionary conservation of their general tertiary
structure (17,18). The shape of the monomeric cytochrome P-450 is
approximately a trigonal prism (10) having a side length of some 60 Å in
the plane of the membrane and a height perpendicular to the membrane of
about 30 Å (figure 2a). This monomer must be a subunit of an oligomer
since a monomeric cytochrome P-450 whose bulk consists of several membrane
spanning helices can be excluded because it is inconsistent with both our
knowledge of the protein's surface structure and the rotational
diffusional arguments given above. A disc-like oligomer with circular
symmetry would complement the prism shape of the monomer (figure 3). A
feature of oligomer structure as an element of topology is lateral
organization in the plane of the membrane. This reduces the dimensions of
spatial organization to two, and so limits the number of necessary contact
faces between the subunits. Lateral organization simplifies the symmetry
rules in homooligomers which consist of identical subunits (19,20). There
is only one symmetry axis left, a line perpendicular to the surface of the
membrane. A hexamer has 6-fold, a pentamer 5-fold rotational symmetry.
This arrangement necessitates heterologous binding between the protomers,
which means that the two binding faces are located in different areas of
the surface of the protomers. In an oligomer of cyclic symmetry (hexamer
or pentamer) two of the five surfaces of the prismlike protomer are
interprotein contacts, one interfaces with the phospholipid membrane and
two are open sides which are accessible to encounters with partners in
function and with analytical reagents. (see table 5).

Open Sides

      Clearly, the surface structures of these open sides of the membrane
bound oligomer are more easily mapped due to their accessibility. As the
open sides in the oligomer account roughly for only 2/5 of the total
surface of the monomer surface mapping by antibodies should be more
effective in solubilized cytochrome P-450 monomers than in cytochrome P-
450 in situ.

        Table 5. Surface structures of cytochrome P-450 LM2
                 (cytochrome P-450b) accessible to functional and
                 analytical contacts

        Binding site for P-450 reductase (21,22)
        Binding site for cytochrome b5 (23,24)
        Serine 128 phosphorylated by cAMP dependent phosphokinase (25)
        Cysteine 152 (26)
        Lysines 139,144,251,384 (22)
        Monoclonal antibodies (27)

Of 16 antibodies raised against 15 peptides comprising 40% of the primary
sequence of cytochrome P-450b (which has a primary structure very similar
to cytochrome P-450 LM2) 13 bound to the purified cytochrome in an ELISA,
while only 5 bound to a similar extent to microsomes (27). The open sides
should contain the 5 peptides by which these antibodies have been raised
and in addition various lysines (see table 5), some of them in positions
which are presumably involved in interaction with the reductase (22), Ser
128 which can be enzymically phosphorylated (25), and the fast-labelling
Cys 152 (26) which is the binding site of the diiodofluorescein
iodoacetamide fluorescence marker introduced for measuring rotational
diffusion (2).

In fact, computer modelling (using SYBYR on an Evans & Sutherland computer) of the cytochrome P-450 CAM structure (10) in which most amino acids have been replaced by sequences from cytochrome P-450 LM2 based on the alignment of the primary structure of 38 cytochromes P-450 (12) showed that all these residues are easily accessible on the sides left open if the hexamer is attached to the membrane.

Membrane Contacts

For the membrane side, hydrophobic and electrostatic contacts have to be considered. The secondary structure of the hydrophobic contacts of the N-terminal sequence of < 29 amino acids may be either a membrane spanning helix ending with the methionine on the luminal side or a β-sheet like structure. This uncertainty is in principle a matter of the unsolved question of which rules can be applied to predict secondary structure elements in membranes (28,29) (see part I). In detail, the hydrophobic N-terminal sequence of < 29 amino acids may be considered as an insertion signal and stop transfer sequence of a type II plasma membrane protein (11). This is consistent with the finding that the N-terminal methionine is accessible to fluorescein isothiocyanate binding from the cytoplasmic side (30,31) and binding of antibodies against the N-terminus was not observed on the luminal side (32,27). However, others (33) found that the N-terminal sequence 1-21 of purified rat liver microsomal cytochrome P-450IIB1 reconstituted into liposomes spans the membrane. The binding stretch – a loop or a β-sheet like structure – may be supplemented by other membrane contacts mediated by hydrophobic sequences accessible to epitope mapping in purified cytochrome P-450b but not in microsome-bound cytochrome P-450 (27). Some of these contacts may be involved in the interprotomer binding of the only defined oligomer structure in solution of cytochrome P-450 LM2 so far known (14): by computer-aided image evaluation of electron micrographs a hexamer of 32 dihedral symmetry (see figure 2b) could be visualized. These hydrophobic structure may constitute a compact domain which anchors the protein but presumably does not penetrate the membrane. Within this domain the opening of the substrate/product channel may be located.

A conspicious accumulation of basic amino acids near the base of the model structure facing the membrane suggests an electrostatic interaction with the phosphate group of phospholipid head groups as an additional membrane attachment mode. Cations, in particular magnesium, may be in competition with these lysine residues of the protein for the binding to the phosphate groups. These electrostatic interactions may explain the influence of magnesium and ionic strength on the rotational rate of cytochrome P-450 (34). An interesting regulatory aspect is the possibility that binding of protein to the membrane may change the orientation of the head groups normally oriented parallel to the membrane surface (35). In this orientation the membrane surface is electrically neutral in spite of the large dipole moments of the choline and ethanolamine head groups. The reorientation therefore induces an electric field. Charged membranes attract and orient proteins and in addition may change their conformation depending on the strength of the electric field and the size of the dipole moment of the protein (36,37). Cytochromes P-450 have a number of surface regions crowding positive or negative charges. This charge separation may sum up to a large dipole moment in the larger structure of an oligomer, increasing the protein's sensitivity to electro-conformational coupling (38). Local fields may add to the control of electron transfer to the heme group as this transfer must be mediated through electron accepting side chains; the structure of cytochrome P-450 CAM reveals that a direct contact transfer to the heme of the P-450 cytochrome is impossible (10).

Interprotomer Contacts

Protomer interfaces may contain both electric and hydrophobic contacts. In addition to the N-terminal sequence and one of the two other sequences constituting the hydrophobic anchor, at least five other sequences are not accessible to antibodies in microsome-bound cytochrome P-450b (27), though they are in purified cytochrome P-450b. As the peptides used for raising the antibodies in this investigation include only 40% of the proteins primary sequence even more structural elements may exist which are hidden in the membrane-bound oligomer but not in the purified monomer.

What evidence points to the existence of an oligomer structure of cytochromes P-450 in liver microsomal membranes? That available from experiments using polarized absorption dichroism which are limited to the investigation of the rotation of the reduced carbon monoxide liganded cytochrome, allow, from the correlation times observed, 63 µs (37°C), 120 µs (20°C) (5), the conclusion that at least the major part of cytochrome P-450 in this membrane is not in a monomeric state – with the reservation in mind that rotation of multiple cytochromes P-450 differing in rotational behaviour are measured. A supportive link to this investigation is the correlation time of 208 µs observed at 8°C by photoselection for a cytochrome P-450 LM2 labelled with diiodofluorescein and incorporated in phenobarbital-induced rabbit liver microsomal membranes (39). The "reconstitution" was accomplished by mixing the labelled purified cytochrome with cholate solubilized microsomes and slow removal of the detergent by subsequent dialysis.

Is the oligomer structure unique to cytochrome P-450 LM2? This is improbable from the similarity of the primary structure of many microsomal cytochromes P-450 and the similarity of their physico-chemical properties which becomes apparent in part by their chromatographic behaviour and in part by their ability to form defined oligomers in solution in the presence of detergents (40,41). Cytochrome P-450 LM4 (IA2) exhibits wobbling rotation when measured in reconstituted phospholipid membranes by the photoselection technique using delayed fluorescence. This mode of motion prohibits estimation of the dimensions of the rotamer (9) but the anisotropy decay parameters are similar to those of a subform isolated from cytochrome P-450 LM2 by cation exchange high pressure liquid chromatography which was found wobbling at 37°C (7). Interestingly, the latter protein at lower temperature rotated strictly uniaxially with the same correlation time as another subfraction which showed uniaxial rotation behaviour at all temperatures. Thus, apparently cytochrome P-450 LM4 has similar size to LM2, but due to the special features of its membrane contact side – like one cytochrome P-450 LM2 isoform – has additional degrees of freedom of motion, particularly when at higher temperature a more extensive vertical mobility of the phospholipid molecules is possible.

Subunit Interaction in Mixed Function Oxygenation

The rate but not the mode of motion of cytochrome P-450 LM2 differs greatly depending on the functional state, ligand binding and reduction (3,9) (compare table 1). The binding of type II substrate or dithionite reduction makes the cytochrome rotate faster, but never to the extent that we have to assume a dissociation of the oligomer to monomers. The considerable effect of dithionite in accelerating the rate of rotation may be explained by additional artificial effects of this reducing agent on the surface charge and therefore on the electrostatic membrane protein interaction. We cannot exclude a dissociative rearrangement of the oligomer but it has to be considered that small changes of the radius of the oligomer have large effects on the rotational diffusion. (compare eq. 1 and tables 2,3,4). Anyhow, the changes of the rotational rate are a manifestation of significant conformational changes of the protein. They also become visible in a change of orientation of the marker dyes in relation to the axis of uniaxial rotation (3). Both sites used for labelling the protein, the cysteine 152 and the N-terminal of methionine, are affected (7).

The dependence of rotational relaxation time on benzphetamine concentration is sigmoidal (42). A halfmaximal effect is seen at $2 \cdot 10^{-6}$ M benzphetamine. This concentration differs greatly from the spectral binding constant ($Km = 3.5 \cdot 10^{-5}$ M) which is a measure of spin state changes, and the Michaelis Menten constant ($Km = 1 \cdot 10^{-5}$ M). The implication is that there is a cooperativity of subunit interaction on ligand binding at a high affinity allosteric binding site. In this context it is interesting that benzphetamine-binding changes only the spin state of an oligomeric but not that of a monomeric cytochrome P-450 LM2 preparation as seen by Raman resonance spectroscopy (43,44). As in hemoglobin, electronic changes of the heme are intimately interlinked with stereo-chemical displacements of subunit interfaces.

Another example of externally induced subunit interaction is enzymic reduction. Cytochrome P-450 LM2 when reconstituted at a molar ratio of 1:1 with NADPH cytochrome P-450 reductase in phospholipid vesicles forms a complex which exhibits wobbling rotation with a much shorter rotational relaxation time than cytochrome P-450 reconstituted alone, which rotates uniaxially (9,45). When a surplus of cytochrome P-450 is reconstituted with reductase (at a ratio of 5:1 or 10:1), uniaxial rotation with a rate similar to the case in the absence of reductase is observed. This strongly suggests the formation of a complex with a 1:1 stoichiometry. However, the mutual affinity of the two proteins in the heterocomplex is not great enough to compete with the formation of a homooligomer of cytochrome P-450 when this protein is abundant. Enzymic reduction of the cytochrome in the 1:1 complex dissociates it with concomitant restoration of the cytochrome P-450 oligomer, as shown by the appearance of a rotational mobility typical of the cytochrome's oligomer in the absence of reductase (9) (see table 1). Apparently, transition to the reduced state results in a conformation less prone to interact with the reductase. The existence of two redox states differing in their conformation is corroborated by the finding that dithionite reduction of cytochrome P-450 LM2 significantly enhances the rate of rotational motion without changing its uniaxial mode. Obviously, there are two important steps of mixed function oxidation, substrate binding and reduction, which involve considerable changes of the cytochrome's rotational diffusion reflecting a change in membrane topology due to a change of conformation. This must - in an oligomer - have significant consequences for the kinetics of these processes as conformational changes of the protomers must change their interprotomer binding properties. In conclusion, transition between different functional states of cytochromes P-450 in the membrane involves changes in the protein's conformation and quaternary structure. This necessarily implies subunit interaction and therefore cooperativity.

SOME PRINCIPLES OF COOPERATIVITY

A functional implication of oligomer structure of enzymes is cooperativity in binding substrates and allosteric effectors (19,46,47).

Monomeric enzymes can display apparent cooperativity involving slow conformational changes (hysteresis) but this is at the expense of catalytic efficiency (48). In contrast, cooperative subunit interation in oligomeric enzymes can enhance catalytic activity as substrate binding to the catalytic site of one subunit changes the substrate affinity of the others by transmission of a conformational change from the liganded subunit to the other protomers. The homotropic allosteric effect of substrate binding in a positively cooperative system is to make the enzyme more sensitive to changes in substrate concentration, in the extreme case an all-or-none response. Negative cooperativity decreases the sensitivity of the enzyme to changes in substrate concentration (49). In the same way, the cooperativity of oligomeric enzymes enhances their sensitivity to the heterotropic allosteric effects of effectors binding to non-catalytic sites. Such agents can be effective in regulating enzyme reactions as activators or inhibitors. Increased sensitivity of enzyme activity to changes in substrate or allosteric effector concentration is due to a narrowing of the range between low and saturating binding and occurs at the expense of a decrease in ligand affinity. The degree of sensitivity enhancement depends on the tightness of subunit coupling - two models describe the limits of the range. In tight coupling (46) symmetry of protein subunit interaction stabilizes the conformational states. As a consequence all protomers are in the same conformational state and transition to another is concerted. This model of tight coupling allows only for positive cooperativity. The sequential model (19), implying induced fit of ligand binding, assumes loose coupling of the subunits and allows for positive, negative and mixed cooperativity. Different subunit conformations can coexist in an oligomer, and different proteins may form a heterooligomer. This model also applies to the case in which different subunits serve different functions, manifest in the aspartate transcarbamylase of Escherischia coli in which three regulatory and three catalytic subunits combine to a hexamer (50).

Of particular interest in the context of an oligomeric state of cytochrome P-450 is the fact that the loose coupling model allows negative cooperativity and explains the phenomenon that substrate binding to one subunit decreases the affinity of subsequent substrate binding to the others. The degree of negative cooperativity is sensitive to pH, temperature and ligand structure. In its extreme only half of the sites display enzymic activity (49).

The functional aspects of the oligomeric structure of liver microsomal cytochromes P-450 have received some attention regarding allosteric control, less regarding cooperativity. The list of inhibitors of MFO, some of them categorized as non-competitive and uncompetitive, is long (51). Interestingly, the list of allosteric activators includes endogenous factors sought in the soluble supernatant of liver microsomal preparations (52). An example underlining the significance of direct activation of MFO in vivo has been given for flavone stimulation of metabolism of zoxazolamine in (53). A positive cooperativity of 6-β-hydroxylation of progesterone in membrane reconstituted purified cytochrome P-4503c was found (54,55) which raises the question of allosteric control of certain cytochromes P-450 by endogenous compounds. The finding that this positive cooperativity could only be observed with membrane reconstituted cytochrome but not in a preparation dispersed with dilaurylPC (56) suggests that membrane incorporation supports the formation of an oligomeric structure capable of positively cooperative interaction of the subunits.

To date, multiplicity may have obscured experimental verification of cooperativity in preparations in vitro, like liver microsomes, by an overlapping of the kinetic parameters of several cytochrome P-450 isozymes. Future studies in vitro using preparations of cytochromes P-450 purified to homogeneity, reconstituted into a membrane of the proper phospholipid composition and incubated with the 'right' substrate under other appropriate conditions (for example ionic milieu, Mg^{2+} ions, pH and perhaps endogenous effectors from cytoplasm) may yield more conclusive results.

PHYSIOLOGICAL SIGNIFICANCE OF COOPERATIVITY IN CYTOCHROMES P-450

Positive cooperativity of substrate binding and MFO may regulate the enzymic process, in its extreme in a switchlike on/off manner. This would avoid reduction of oxygen bound to cytochrome P-450 in the absence of substrate and protect cells against energetically wasteful production of noxious reduced oxygen species, but it would only be partially effective because uncoupling of the stoichiometry of product formation, NADPH oxidation and O_2 consumption may be also a matter of unfavourable substrate orientation at the active site (57). It is interesting in this context that two studies demonstrate that the steady state content of reduced cytochrome P-450 in the isolated perfused liver in the absence of exogenous substrates is very low (58,59).

A critical step in cooperative regulation of MFO is the reduction of the cytochromes by NADPH cytochrome P-450 reductase. Spin state, substrate binding, membrane topology and diffusibility of the redox partners determine the kinetics of the transfer of one electron under anaerobic conditions. These factors have received different weighting in explanations of the apparent biphasicity of the kinetics of enzymic cytochrome P-450 reduction observed in liver microsomes and reconstituted phospholipid membrane systems. The observed biphasicity of reduction in liver microsomes may be multiphasic (60). A number of kinetic models have been elaborated which imply spin equilibria (61,62) or assume random collisions between freely diffusing reaction partners (63), or reduction in a preexisting rosette-like cluster of the partners (64). Some of these models allow for an oligomeric structure of the cytochromes but do not include cooperativity of the protomers. Most relevant to this aspect are the findings that dithionite reduction of non-induced and phenobarbital-induced rat and rabbit liver microsomes, and of cytochrome P-450 LM2 both reconstituted into membrane vesicles and as oligomers in solution in the absence of NADPH cytochrome P-450 reductase is 'biphasic' while that of cytochrome P-450 LM2 monomers in solution is not. In the latter case monoexponential kinetics are observed (65), a strong indication that an oligomer is the structural correlate of the observed multiplicity of reduction kinetics. This lends support to the idea that subunit interaction is involved in reduction, a prerequisite for cooperative interaction.

The multiphasicity of cytochrome P-450 reduction can be explained by negative cooperativity of the protomers of an oligomer during transition from the oxidized to the reduced state conformation. Negative cooperativity is a feature of protein complexes which undergo transitions between function-dependent conformations. An example revealing this aspect of negative cooperativity is the behaviour of the tetrameric single strand DNA binding protein of Escherischia coli as a function of NaCl and $MgCl_2$ concentration. At low salt concentration extreme negative cooperativity stabilizes the binding mode in which only two subunits of the tetramer bind to DNA whereas at high salt concentration all four subunits interact with positive cooperativity. This enables the protein to serve two functions by switching from the state of 'half of the sites reactivity' used for repair and recombination processes to the positively cooperative binding mode used transiently during DNA replication (66).In case of cytochrome P-450 a number of functional states are involved in the substrate binding, acceptance of the first and second electrons from NADPH cytochrome P-450 reductase and/or cytochrome b5 respectively, oxygen binding and product release phases of MFO. Electron transfer involves transient complex formation of the substrate-tagged cytochrome oligomer with the reductase: negative cooperativity would facilitate dissociation of this complex at the expense of half of the sites' catalytic activity. This is another aspect of the problem of functional and nonfunctional complexes of the liver microsomal hydroxylase system (67).

Another feature of negative cooperativity in enzyme systems is desensitization of activity to substrate concentration, damping oscillations of product release. Negative and positive cooperativity phenomena have been observed in the binding of insulin to its receptor in cells and recently in receptor preparations purified to homogeneity (68). A model of three states of the receptor differing in their ligand affinities has been proposed. Transition from the unliganded state of intermediate affinity to the state of lower affinity induced by insulin proceeds with negative cooperativity and to the high affinity state with positive cooperativity (69). The physiological role of the negative cooperativity may be to buffer against high insulin secretions and to reduce oscillations in insulin action (70).

In the case of cytochrome P-450 this could have physiological significance by avoiding accumulation of toxic amounts of products like arene-epoxides or phenols biotransformed from exogenous substances, or of membrane perturbing cholesterol metabolites. This would improve metabolic coupling between phases I and II of detoxication: product release from cytochrome P-450 is one step in this and little is known about its mechanism and kinetics. Release from isozymes with low substrate specificity may be slow if the physico-chemical properties of the products are not much different from those of the substrates. The tendency to recycling oxygenation in such cases would be counteracted by negative cooperativity of the binding sites in an oligomer favouring product release. Moreover, the substrate binding and/or catalytic activity of one protomer would be improved at the expense of ligand binding energy of others and binding of substances with high affinity would decrease binding of others with lower affinity. In this way a preference for oxygenation of endogenous substrates in competition with exogenous substrates may be established.

We conclude therefore, that cooperativity would allow more pronounced regulatory influence of effectors (including cytochrome b5) on the cytochrome P-450 system. It is intriguing to think of effectors which switch the system between positive and negative cooperativity in analogy with the DNA single strand binding protein.

SUMMARY

Results from experiments on rotational diffusion obtained by photoselection techniques, particularly by measurement of time-dependent depolarization of delayed fluorescence clearly indicate that the purified cytochrome P-450 LM2 reconstituted in membrane vesicles, two subforms of this protein and cytochrome P-450 LM4 form homooligomers of about the size of hexamers. The shape of the protomer was deduced from cytochrome P-450 CAM, whose tertiary structure is well known, by computer modelling mainly on the basis of published comparative analyses of primary structure of

multiple cytochromes P-450 and from surface mapping data. A hydrophobic
non-penetrating part of cytochrome P-450 anchors the protomer in the
membrane. Elements of membrane phospholipid structure like the cones of
phosphatidylethanolamine molecules complementary to the shape of the
protein's anchor may adapt it to the membrane, electrostatic interactions
between protein side chains and phospholipid groups hold the oligomer in
position. Complexes between unliganded cytochrome P-450 LM2 and NADPH
cytochrome P-450 reductase are only observed in the presence of an
unphysiologically high reductase content. Interpretation of the kinetics
of substrate binding, product formation and reduction of cytochromes P-450
must take into consideration the oligomeric structure. Positive
cooperativity of substrate binding may regulate electron transfer to
cytochromes P-450, negative cooperativity of binding of cytochrome P-450
reductase may facilitate the dissociation of a transient heterocomplex
after reduction and negative cooperativity of product binding may
facilitate product release.

ACKNOWLEDGEMENTS

R. Greinert, H. Taniguchi, H. Garda, P. Kiselev, H. Hildebrandt and G.
Bachmanova contributed to the work presented here from our laboratory.
This work was in part supported by the Deutsche Forschungsgemeinschaft
(Sonderforschungsbereich 330) and by the Association for International
Cancer Research. H.G. was a recipient of a stipendship from the Alexander
von Humboldt Stiftung. Many more colleagues than mentioned here helped
through discussions and publications to refine our view of the membrane
topology of cytochrome P-450; the reference list is selective but not
eclectic. We thank Mrs. I. Fried for her desktop publishing activity.

References

1. R. Greinert, H. Staerk, A. Stier, and A. Weller, E-type
    delayed fluorescence depolarization, a technique to probe
    rotational motion in the microsecond range. J. Biochem.
    Biophys. Methods. 1: 77 (1979).
2. R. Greinert, S.A.E. Finch, and A. Stier, Cytochrome P-450
    rotamers control mixed-function oxygenation in
    reconstituted membranes. Rotational diffusion studied by
    delayed fluorescence depolarization. Xenobiotica 12: 717
    (1982a).
3. R. Greinert, S.A.E. Finch, and A. Stier, Conformation and
    rotational diffusion of cytochrome P-450 changed by
    substrate binding. Bioscience Report 2: 991 (1982b).
4. C. Richter, K.H. Winterhalter, and R.J. Cherry, Rotational
    diffusion of cytochrome P-450 in rat liver microsomes,
    FEBS Lett. 102: 151 (1979).
5. S. Kawato, J. Gut, R.J. Cherry, K.H. Winterhalter, and
    C. Richter, Rotation of cytochrome P-450. I. Invetigation
    of protein-protein interaction of cytochrome P-450 in
    phospholipid vesicles and liver microsomes, J. Biol.
    Chem. 257: 7023 (1982).
6. J. Gut, C. Richter, R.J. Cherry, K.H. Winterhalter, and
    S. Kawato, Rotation of cytochrome P-450. II. Specific
    interactions of cytochrome P-450 with NADPH-cytochrome
    P-450 reductase in phospholipid vesicles, J. Biol. Chem.
    257: 7030 (1982).
7. H. Garda, S.A.E. Finch, V. Krüger, J. Sidhu, and A. Stier,
    Microheterogeneity of rabbit liver cytochrome P-450 LM2:

biochemical and structural differences of two subforms
Eur. J. Biochem. submitted (1989).

8. S. Kawato and K. Kinosita, Time-dependent absorption
    anisotropy and rotational diffusion of proteins in
    membranes, <u>Biophys.J</u>. 36: 277 (1981).

9. A. Stier, S.A.E. Finch, R. Greinert, and H. Taniguchi,
    Membrane protein interactions in biotransformation, <u>in</u>:
    "Biochemistry, Biophysics and Induction", L. Vereczky and
    K. Magyar, eds., Akademiai Kiado, Budapest (1985).

10. T.L. Poulos, B.C. Finzel, and A.J. Howard, High-resolution
    crystal structure of cytochrome P-450 CAM, <u>J. Mol. Biol</u>.
    195: 687 (1987).

11. M. Sakaguchi, K. Mihara, and R. Sato, A short amino-
    terminal segment of microsomal cytochrome P-450 functions
    both as an insertion signal and as a stop-transfer
    sequence, <u>EMBO J</u>. 6: 2425 (1987).

12. O. Gotoh and Y. Fuji-Kuriyama, Evolution, structure, and
    gene regulation of cytochrome P-450, <u>in</u>: "Basis and
    Mechanisms of Regulation of Cytochrom P-450", Vol. 1,
    K. Ruckpaul and H. Rein, eds., Taylor & Francis, London
    (1989).

13. B.D. Hughes, B.A. Pailthorpe, L.T. White, and W.H. Sawyer,
    Extraction of membrane microvisosity from translational
    and rotational diffusion coefficients, <u>Biophys. J</u>. 37: 673
    (1982).

14. V.L. Tsuprun, K.N. Myasoedova, P. Berndt, O.N. Sograf
    E.V. Orlova, V.Ya. Chernyak, A.I. Archakov, and
    V.P. Skulachev, Quaternary structure of the liver
    microsomal cytochrome P-450, <u>FEBS Lett.</u> 205: 35 (1986).

15. R. Peters and R.J. Cherry, Lateral and rotational
    diffusion of bacteriorhodopsin in lipid bilayers.
    Experimental test of the Saffman-Delbrück equations,
    <u>Proc. Natl. Acad. Sci. USA</u> 79: 4317 (1982).

16. D.R. Nelson and H.W. Strobel, Secondary structure
    prediction of 52 membrane-bound cytochromes P-450 shows a
    strong structural similarity to P-450 CAM, <u>Biochemistry</u>
    28: 656 (1989).

17. A. M. Lesk and C. Chothia, How different amino acid
    sequences determine similar protein structure: the
    structure and evolutionary dynamics of the globins,
    <u>J. Mol. Biol.</u> 136: 225 (1980).

18. C. Chothia and A.M. Lesk, Helix movements and the
    reconstruction of the haem pocket during the evolution of
    the cytochrome c family, <u>J. Mol. Biol</u>. 182: 151 (1985).

19. D.E. Koshland Jr., G. Nemethy, and D. Filmer, Comparison
    of experimental binding date and theoretical models in
    proteins containing subunits, <u>Biochemistry</u> 5: 365 (1966).

20. I. Klotz, D.W. Darnall, and N.R. Langerman, Quaternary
    structure of proteins, <u>in</u>: "The Proteins", Vol. 1,
    H. Neurath and R.L. Hill, eds, Academic Press,
    New york (1975).

21. P.P. Tamburini, S. Mac Farquhar, J.B. Schenkman, Evidence
    of binary complex formations between cytochrome P-450,
    cytochrome b5, and NADPH-cytochrome P-450 reductase of
    hepatic microsomes, <u>Biochem. Biophys. Res. Commun.</u> 134:
    519 (1986).

22. R. Bernhardt, R. Kraft, A. Otto, and K. Ruckpaul,
    Electrostatic interactions between cytochrome P-450 LM2
    and NADPH-cytochrome P-450 reductase, <u>Biomed. Biochim.
    Acta</u> 47: 581 (1988a).

23. B. Bösterling and J.R. Trudell, Association of cytochrome
    b5 and cytochrome P-450 reductase with cytochrome P-450 in
    the membrane of reconstituted vesicles, J. Biol. Chem.
    257: 4783 (1982).
24. I.A. Jansson, P.M. Epstein, S. Bains, and J.B. Schenkman,
    Inverse relationship between cytochrome P-450 phos-
    phorylation and complexation with cytochrome b5,
    Arch. Biochem. Biophys. 259: 441 (1987).
25. R. Müller, W.E. Schmidt, and A. Stier, The site of cyclic
    AMP-dependent protein kinase catalyzed phosphorylation of
    cytochrome P-450 LM2. FEBS Lett. 187: 21 (1985).
26. S.D. Black and M.J. Coon, Studies on the identity of the
    heme-binding cysteinyl residue in rabbit liver microsomal
    cytochrome P-450 isozyme 2, Biochem. Biophys. Res. Commun.
    128: 82 (1985).
27. C. De Lemos-Chiarandini, A.B. Frey, D.D. Sabatini and
    G. Kreibich, Determination of the membrane topology of the
    phenobarbital-inducible rat liver cytochrome P-450
    isoenzyme PB-4 using site-specific antibodies, J. Cell
    Biol. 104: 209 (1987).
28. C.Chothia, Principles that determine the structure of
    proteins, Annu. Rev. Biochem. 53:, 537 (1984).
29. E. Schulz, A critical evaluation of methods for prediction
    of protein secondary structures, Annu. Rev. Biophys.
    Biophys. Chem. 17: 1 (1988).
30. R. Bernhardt, N.T. Ngoc Dao, H. Stiel, W. Schwarze,
    J. Friedrich, G.-R. Jänig, and K. Ruckpaul, Modification
    of cytochrome P-450 with fluorescein isothiocyanate,
    Biochim. Biophys. Acta 745: 140 (1983).
31. R. Bernhardt, R. Kraft, and K. Ruckpaul, A simple
    determination of the sidedness of the NH_2-terminus in the
    membrane bound cytochrome P-450 LM2, Biochem. Int. 17:
    1143 (1988b).
32. Matsuura, Y. Fujii-Kuriyama, and Y. Tashiro, Quantitative
    immunoelectron microscopic analyses of the distribution of
    cytochrome P-450 molecules on rat liver microsomes,
    J. Cell Sci.36: 413 (1979).
33. G. Vergères, K.H. Winterhalter, and C. Richter,
    Identification of the membrane anchor of microsomal rat
    liver cytochrome P-450, Biochemistry 28: 3650 (1989).
34. P.A. Kiselev, H. Garda, S.A.E. Finch, A. Stier,
    N.A. Gurinovich, S.I. Khatyleva, and A.A. Akhrem,
    Rotational diffusion of cytochrome P-450 in the model
    membrane of different phospholipid-composition,
    Doklady Akademii Nauk SSR 307: 473 (1989).
35. J. Seelig, P.M. Macdonald, and P.G. Scherer, Phospholipid
    head groups as sensors of electric charge in membranes,
    Biochemistry 26: 7535 (1987).
36. B.H. Honig, W.L. Hubbell, and R.F. Flewelling,
    Electrostatic interactions in membranes and proteins, Ann.
    Rev. Biophys. Biophys. Chem. 15: 163 (1986).
37. S. McLaughlin, The electrostatic properties of membranes,
    Annu. Rev. Biophys. Biophys. Chem. 18: 113 (1989).
38. T. Y. Tsong and R.D. Astumian, Electroconformational
    coupling and membrane protein function, Progr. Biophys.
    Mol. Biol. 50: 1 (1987).
39. A. Stier, S.A.E. Finch, R. Greinert, R. Müller, and
    H. Taniguchi, Structure of endoplasmic reticulum membrane
    and regulation of drug metabolism, in: "Pharmacological,

Morphological and Physiological Aspects of Liver Aging",
C.F.A. van Bezooijen, ed., EURAGE, Rijswijk (1984).

40. W.L. Dean and R.D. Gray, Relationship between state of
    aggregation and catalytic activity for cytochrome P-450
    LM2 and NADPH-cytochrome P-450 reductase, J. Biol. Chem.
    257: 14679 (1982).

41. S.L. Wagner, W.L. Dean, and R.D. Gray, Zwitterionic
    detergent mediated interaction of purified cytochrome
    P-450 LM4 from 5,6-benzoflavone-treated rabbits with
    NADPH-cytochrome P-450 reductase, Biochemistry 26: 2343
    (1987).

42. A. Stier, H.Garda, P. Kisselev, V. Krueger, S.A.E. Finch,
    and A.A. Akhrem, Membrane protein interaction in biotrans-
    formation, in: "Cytochrome P-450 Biochemistry and
    Biophysics", I. Schuster, ed., Taylor and Francis,
    London (1989).

43. I.P. Kanaeva, E.D. Skotselyas, I.F. Turkina,
    E.V. Petrochenko, D.R. Davydov, S.R. Kondrashin,
    Ch.S. Dzuzenova, G.I. Bachmanova, and A.I. Archakov,
    Reduction and catalytic properties of cytochrome P-450 in
    reconstituted system containing monomeric carriers,
    Biochem. Biophys. Res.Comm. 147: 1295 (1987).

44. P. Hildebrandt, H. Garda, A. Stier, G.I. Bachmanova,
    I.P. Kanaeva, and A.I. Archakov, Protein-protein
    interactions in microsomal P-450 isozyme LM2 and their
    effect on substrate binding, Eur. J. Biochem., in press.
    (1989).

45. J. Gut, C. Richter, R.J. Cherry, K.H. Winterhalter, and
    S. Kawato, Rotation of cytochrome P-450. Complex formation
    of cytochrome P-450 with NADPH-cytochrome P-450 reductase
    in liposomes demonstrated by combining protein rotation
    with antibody-induced crosslinking, J. Biol. Chem. 258:
    8588 (1983).

46. J. Monod, J. Wyman, and J.-P. Changeaux, On the nature of
    allosteric transitions: A plausible model, J. Mol. Biol.
    12: 88 (1965).

47. M. Eigen, New looks and outlooks on physical enzymology,
    Quart. Rev. Biophys. 1: 3 (1968).

48. J. Ricard, Organized polymeric enzyme systems: Catalytic
    properties, in: "Organized Multienzyme Systems: Catalytic
    Properties", R. Welch, ed., Academic Press, Orlando
    (1985).

49. A. Levitzki and D.E. Koshland Jr., The role of negative
    cooperativity and half-of-the-sites reactivity in enzyme
    regulation, Curr. Top. Cell Regul. 10: 1 (1976).

50. J.C. Gerhart and H.K. Schachman, Allosteric interactions
    in aspartate transcarbamylase. II. Evidence for different
    conformational states of the protein in the presence and
    absence of specific ligands, Biochemistry 7: 538 (1968).

51. K. Netter, Inhibition of oxidative drug metabolism in
    microsomes,, in: "Cytochrome P-450 Monooxygenase System",
    J.B. Schenkman and D. Kupfer, eds., Pergamon Press,
    Oxford, New York, Toronto, Paris, Frankfurt (1982).

52. D.L. Cinti, Agents activating the liver microsomal mixed
    function oxidase system, in: "Hepatic Cytochrome P-450
    Monooxygenase System", Schenckman, J.B. and D. Kupfer,
    eds., Pergamon Press, Oxford, New York, Toronto, Paris,
    Frankfurt (1982).

53. J.M. Lasker, M.-T. Huang, and A.H. Conney, In vitro and in vivo activation of oxidative drug metabolism by flavonoids, J. Pharmacol. Exp. Ther. 229: 162 (1984).

54. E.F. Johnson, G.E. Schwab, and L.E. Vickery, Positive effectors of the binding of an active site-directed amino steroid to rabbit cytochrome P-450 3c, J. Biol. Chem. 263: 17672 (1988).

55. G.E. Schwab, J.L. Raucy, and E.F. Johnson, Modulation of rabbit and human hepatic cytochrome P-450 catalyzed steroid hydroxylations by alpha-naphthoflavone, Mol. Pharmacol. 33: 493 (1988).

56. M. Ingelman-Sundberg, I. Johansson, and A. Hansson, Catalytic properties of the liver microsomal hydroxylase system in reconstituted phospholipid vesicles, Acta biol. med. germ. 38: 379 (1979)

57. W.M. Atkins and S.G. Sligar, Deuterium isotope effects in norcamphor metabolism by cytochrome P-450 CAM kinetic evidence for the two-electron reduction of a high-valent iron-oxo-intermediate, Biochemistry 27: 1610 (1988).

58. H. Sies and B. Brauser, Interaction of mixed function oxidase with its substrates and associated redox transitions of cytochrome P-450 and pyridine nucleotides in perfused rat liver, Eur. J. Biochem. 15: 531 (1970).

59. T. Iyanagi, T. Suzaki, and S. Kobayashi, Oxidation-reduction states of pyridine nucleotide and cytochrome P-450 during mixed-function oxidation in perfused rat liver, J. Biol. Chem. 256: 12933 (1981).

60. H.H. Ruf, Reduction kinetics of microsomal cytochrome P-450. A reexamination, in: "Biochemistry, Biophysics and Regulation of Cytochrome P-450", J.A. Gustafsson, H.J. Carlstedt-Duke, A. Mode and J. Rafter, eds., Elsevier/North Holland Biomedical Press, Amsterdam, New York, Oxford (1980).

61. W.L. Backes, S.G. Sligar, and J.B. Schenkman, Cytochrome P-450 reduction exhibits burst kinetics, Biochem. Biophys. Res. Comm. 97: 860 (1980).

62. J. Blanck, G. Smettan, O. Ristau, M. Ingelman-Sundberg, and K. Ruckpaul, Mechanism of rate control of the NADPH-dependent reduction of cytochrome P-450 by lipids in reconstituted phospholipids, Eur. J. Biochem. 144: 509 (1984).

63. H. Taniguchi, Y. Imai, T. Iyanagi, and R. Sato, Interaction between NADPH-cytochrome P-450 reductase and cytochrome P-450 in the membrane of phosphatidylcholine vesicles, Biochim. Biophys. Acta 550: 341 (1979).

64. J.A. Peterson, R.R. Ebel, D.H. O'Keefe, I. Matsubara, and R.W. Estabrook, Temperature dependence of cytochrome P-450 reduction, J. Biol. Chem. 251: 4010 (1976).

65. D.R. Davydov, A.V. Karyakin, B. Binas, B.I. Kurganov, and A.I. Archakov, Kinetic studies on reduction of cytochromes P-450 and b5 by dithionite, Eur. J. Biochem. 150: 155 (1985).

66. T.M. Lohman and W. Bujalowski, Negative cooperativity within individual tetramers of Escherichia coli single strand binding protein is responsible for the transition between the (SSB)35 and (SSB)56 DNA binding modes, Biochemistry 27: 2260 (1988).

67. M. Ingelman-Sundberg, Cytochrome P-450 organization and membrane interaction,, in: "Cytochrome P-450", P.R. Ortiz de Montellano, ed., Plenum Press, New York (1986).

68. C.-C. Wang, I.D. Goldfine, Y. Fujita-Yamaguchi,
    H.G. Gattmer. D. Brandenburg, and P. De Meyts, Negative
    and positive site-site interactions, and their modulation
    by pH, insulin analogs, and monoclonal antibodies, are
    preserved in the purified insulin receptor, Proc. Natl.
    Acad. Sci. USA 85: 8400 (1988).
69. J.-L. Gu, I.D. Goldfine, J.R. Forsayeth, and P. De Meyts,
    Reversal of insulin-induced negative cooperativity by
    monoclonal antibodies that stabilize the slowly
    dissociating ("K-super") state of the insulin receptor,
    Biochem. Biophys. Res. Comm. 150: 694 (1988).
70. S. Gammeltoft, Insulin receptors: Binding, kinetics and
    structure-function relationship of insulin, Physiol. Rev.
    64: 1321 (1984).

NADPH-Dependent Cytochrome P450 Reductase

Anthony Y. H. Lu

Merck Sharp & Dohme Research Laboratories
Rahway, NJ 07065 USA

INTRODUCTION

The pioneering studies of Horecker (1), Phillips and Langdon (2), Williams and Kamin (3) established that the microsomal NADPH-cytochrome c reductase is a flavoprotein capable of reducing various electron acceptors. Subsequent studies by other investigators (4-6) suggested an involvement of this reductase in microsomal hydroxylation since like cytochrome P450, it is inducible by phenobarbital. In addition, microsomal monooxygenase activity can be inhibited by cytochrome c (7) and antibodies against the reductase (8-10). A direct involvement of this enzyme in hydroxylation was established when the liver microsomal monooxygenase system was solubilized, resolved and reconstituted (11, 12). In these studies, the flavoprotein was shown to be an obligatory component in hydroxylation, transferring reducing equivalents from NADPH to cytochrome P450. Since then, the term NADPH-cytochrome P450 reductase has been used to reflect its physiological function.

Purification

The development of highly specific affinity ligands, 2',5'-ADP (13) and NADP (14), makes it possible to design simple and rapid methods for the purification of NADPH-cytochrome P450 reductase. As shown in Table 1, the reductase purified from detergent-solubilized microsomes differs in some very important aspects from the enzyme purified from protease or lipase-solubilized microsomes. These differences provide important information to define the structure and function of the reductase.

Characterization and Mechanism

A. Hydrophobic and hydrophilic domains - Treatment of the native reductase with various hydrolytic enzymes produces a large soluble peptide (M.W. ~71,000) and a small hydrophobic fragment (M.W. ~6,000). The primary

Table 1. Comparison of Purified NADPH-Cytochrome P450
Reductase Preparations Obtained from Detergent-
vs Hydrolytically-solubilized Liver Microsomes

Property	Detergent-solubilized Reductase	Hydrolytically-solubilized Reductase
Molecular weight:		
(a) by SDS-polyacrylamide gel electrophoresis	79,000	71,000
(b) in aqueous medium	aggregate	71,000
Prosthetic group	FAD and FMN (1:1)	FAD and FMN (1:1)
Interaction with antibody against protease-solubilized reductase	Yes	Yes
Interaction with biological membranes:		
(a) binding to liver microsomes	Yes	No
(b) binding to phospholipid vesicles	Yes	No
Catalytic activity:		
(a) reduction of cytochrome c	Yes	Yes
(b) reduction of ferricyanade	Yes	Yes
(c) reduction of cytochrome b5	Yes	Only in the presence of high salt
(d) reduction of cytochrome P450 in the presence of phospholipid	Yes	No
(e) function in P450-dependent hydroxylation	Yes	No

tryptic attack on the reductase is at the Lys 56-Ile 57 bond (Fig. 1), generating the hydrophilic and hydrophobic domains (15-17). With the rabbit reductase (15), trypsin attack also occurs at the Lys 46-Lys 47-Lys 48 sequence to yield three polypeptides, i.e., the hydrophobic domain (N-terminus to Lys 46 or Lys 47), the connecting region (Lys 47 or Lys 48 to Lys 56) and the hydrophilic domain (Ile 57 to the C-terminus, Ser 679). The hydrophilic peptide contains both FAD and FMN and retains the spectral characteristics of the native enzyme. Although it is functional in the reduction of cytochrome c, it can no longer reduce cytochrome P450 and thus not functional in the cytochrome P450-catalyzed hydroxylation reaction. While the hydrophilic peptide exists as a monomer in aqueous solution, the hydrophobic fragment is highly aggregated in aqueous medium. This fragment is hydrophobic judging from its amino acid composition (Table 2). It is essential for the proper interaction of the reductase with either cytochrome P450 or cytochrome b_5, and the anchoring of reductase to the biological membranes. For example, Gum and Strobel (20) have shown that homogeneous detergent-solubilized NADPH-cytochrome P450 reductase can be incorporated spontaneously into microsomes and synthetic phospholipid vesicles. This incorporated reductase is active in electron transfer in cytochrome P450-mediated demethylation of benzphetamine. In contrast, steapsin-solubilized reductase, devoid of the hydrophobic peptide, is not incorporated into the biological membranes.

B. Electron transport sequence – The presence of both FAD and FMN as prosthetic groups in the reductase was first reported by Iyanagi and Mason (21). Spectrophotometric and potentiometric titrations of the reductase with dithionite or NADPH under anaerobic conditions indicate that the two flavins have different redox potentials (22-24). The calculated midpoint potential of the high potential flavin is -0.190V, and that of the low potential flavin is -0.328V. Spectrophotometric experiments with the FMN-depleted enzyme establish that the low and high potential flavins of the reductase are FAD and FMN, respectively (23).

The flavins can be selectively removed from the reductase by treating the purified enzyme with 2M KBr at different pH, i.e., FAD at pH 6.5 and FMN at pH 8.5 (23, 25). The FAD-depleted reductase has negligible activity toward cytochrome c and other electron acceptors (25), whereas the FMN-depleted enzyme has lost most of its activity toward cytochrome c, menadione and 2,6-dichlorophenolindophenol but retained the ability to reduce ferricyanide (23). These results indicate that FMN is not a competent electron acceptor from NADPH and that the FAD site is the only point of electron uptake from NADPH. Thus, the available evidence supports a sequence of electron transport in the microsomal hydroxylation system as follows: NADPH $\rightarrow$ FAD $\rightarrow$ FMN $\rightarrow$ cytochrome P450.

Table 2. Properties of Hydrophobic Peptides from Microsomal
NADPH-Cytochrome P450 Reductase

Properties	Peptide Isolated from Rat by Gum and Strobel (18)	Peptide Isolated from Rabbit by Black et al (19)
M.W.	6,400	6,100
Total Amino Acid Residues	60	56
Hydrophobic AA	30%	34%
Effect on P450-Dependent Oxidation	None	Inhibition

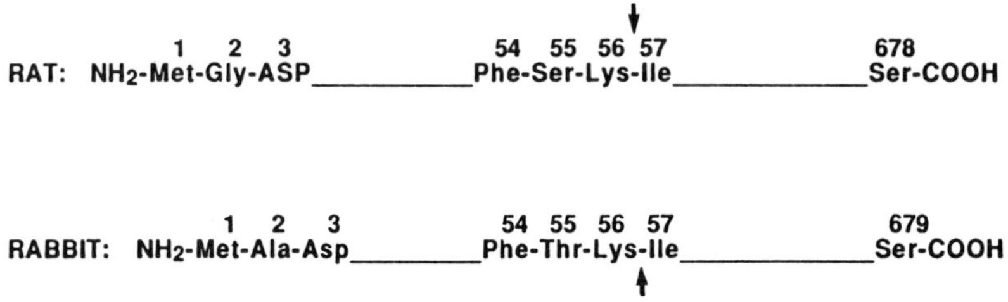

RAT: NH₂-Met-Gly-ASP___ 54 55 56 57 Phe-Ser-Lys-Ile___ 678 Ser-COOH

RABBIT: NH₂-Met-Ala-Asp___ 54 55 56 57 Phe-Thr-Lys-Ile___ 679 Ser-COOH

Figure 1. The primary tryptic attack site on microsomal NADPH-cytochrome P450-reductase. The amino acid sequence of the rat (16) and rabbit (17) enzymes were deduced from the base sequence of cDNA clones.

C. Active site - Since the three-dimensional structure of
    the reductase is not available, indirect approaches
    (such as the use of chemical agents to modify the struc-
    ture) have been used to study the topology of the active
    site of the enzyme. The results of various studies (26-
    35) suggest the involvement of thiol groups at or near
    the catalytic site, lysine residue near the binding site
    of 2'-phosphate moiety of the NADPH molecule, carboxyl
    groups at the cytochrome c binding site and tyrosine
    residue near the FMN-binding domain (Table 3).

    Perhaps because of the fact that the reductase functions
    as an electron carrier, very little effort is devoted to
    the development of an active site-directed reagent for
    the reductase. Ebel (36) found that the reductase is
    rapidly and irreversibly inactivated by diazotized 3-
    aminopyridine   adenine   dinucleotide   phosphate.
    Inactivation requires the presence of NADPH and one mole
    of the nucleotide is incorporated per mole of enzyme.
    Since the diazonium derivative of the 3- amino function
    behaves as a site-specific -SH reagent, the interaction
    between this compound and the reductase is probably
    through a sulfhydryl residue at the active site.

Interaction with Cytochrome P450

    The NADPH-dependent reductase must interact with
cytochrome P450 to allow efficient electron transfer from
flavin to heme during catalysis. Various studies have
indicated a 1:1 complex formation between the reductase and
cytochrome P450 (37-41). Whether this interaction involves
stable reductase-cytochrome P450 complexes or transient
complexes due to random collision in the membrane is still
not totally known. Interaction between cytochrome P450 and
NADPH-cytochrome P450 reductase may occur by a charge pairing
mechanism. This conclusion is supported by the following
observations. (a) In the reconstituted system, high ionic
strength inhibits benzphetamine N-demethylation (39). (b)
Chemical modification of lysine 384 on rabbit P450 LM2 by
isothiocyanate inhibits electron transfer from the reductase
to cytochrome P450 (42).   (c)   Modification of carboxyl
groups on NADPH-cytochrome P450 reductase by carbodiimide and
methylamine inhibits the interaction of reductase with
cytochrome P450 (33-35).

Membrane Topology

    NADPH-cytochrome P450 reductase is amphiphilic. The
hydrophilic catalytic domain contains the prosthetic groups
and the C-terminal whereas the hydrophobic domain is the N-
terminal peptide. The possible orientation of the reductase
in the microsomal membranes has been proposed by Black and
Coon (15). In this model, the hydrophilic domain is exposed
to the cytoplasm. The NH_2- and COOH-terminal regions are
either on opposite sides or on the same side of the membrane.
Cleavage of the Lys 56-Ile 57 bond by trypsin generates the
hydrophobic and hydrophilic domains. A 17 amino acid segment
(from Val 28 to Phe 44), which is 76% hydrophobic, may represent the actual membrane peptide. On either side of the

Table 3. Proposed Essential Amino Acid Residues Involved in the Catalytical Function of NADPH-Cytochrome P450 Reductase

Catalytic Function	Essential Amino Acid Residues	Reference
FAD-binding domain	cysteine	26, 27
FMN-binding domain	tyrosine, tryptophan	28, 29
NADPH-binding domain	lysine, cysteine	27, 30, 31
Cytochrome c-binding domain	aspartate, glutamate	32
Cytochrome P450-binding domain	Hydrophobic and charge pairing (aspartate and glutamate)	20, 33-35

hydrophobic region (i.e., the membrane-cytosol interface), charged residues are found in the hydrophilic segments.

Recent studies (43, 44) have shown that cytochrome P450 is bound to the endoplasmic reticulum membrane by only one or two transmembrane segments, located at the NH_2-terminal end of the molecule. The bulk of the protein, from residue 50 or 66 to the COOH terminus, contains the heme and the catalytic domain and is exposed to the cytosolic side of the membrane. Thus, as shown in Figure 2, the catalytic domains of the reductase and cytochrome P450 interact at the cytosolic side to affect electron transfer whereas the NH_2-terminal segments anchor these two proteins to the bilayer either by spanning the hydrophobic domain through the membrane (as shown) or by forming a hairpin structure (not shown).

Molecular Biology

Full-length cDNA's for rabbit and rat liver microsomal NADPH-cytochrome P450 reductase mRNA have been cloned by Katagiri *et al (17)*, and Kasper and coworkers (16). The rabbit clone contains 2,269 nucleotides. The single open reading frame of 2,037 nucleotides codes for a 679-amino acid polypeptide with a calculated molecular weight of 76,583. The rat clone contains 2,401 nucleotides. The single open reading frame of 2,034 nucleotides codes for a 678-amino acid polypeptide with a molecular weight of 76,962. Comparison of the amino acid sequence of the rat and rabbit reductase with that of flavoproteins of known three-dimensional structures suggests that the FMN-binding domain is near the amino-terminal portion of the protein while the FAD- and NADPH-binding domains are at the carboxyl-terminal half of the molecule (45). Site-directed mutagenesis studies show that replacement of Tyr-178 with aspartate abolishes FMN binding and cytochrome c reduction, indicating that Tyr-178 is required for FMN-binding. Replacement of Tyr-140 with aspartate has no effect on FMN binding but reduces cytochrome c reductase activity, suggesting that FMN may be bound in a less favorable conformation for efficient electron transfer.

The mechanism of reductase induction by phenobarbital has been investigated by Kasper and coworkers (46, 47). These investigators found a rapid increase in the rates of transcription and the appearance of elevated level of mRNA following phenobarbital treatment. These results indicate that phenobarbital elevates the level of NADPH-cytochrome P450 reductase by increasing the rate of transcription of the gene.

Role of Reductase in Reductive Metabolism and Metabolic Activation

In addition to its electron carrier role in cytochrome P450 mediated reactions, the reductase alone can catalyze the reduction of a variety of foreign chemicals (48-50), including many therapeutically important anti-tumor agents and antibiotics (Table 4). In these NADPH-dependent reductive reactions, reactive oxygen species and carbon-centered free radicals and reactive metabolites can be formed. For example, quinones are reduced by the reductase in a one-

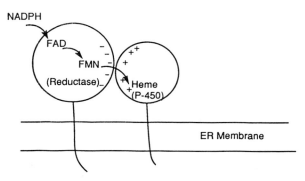

Figure 2.   Diagram showing the interaction between
             cytochrome P450 and NADPH-cytochrome P450
             reductase.   To simplify the drawing, the
             NH_2-terminus is shown on the opposite side
             of the hydrophilic domain with the
             hydrophobic segment spanning the bilayer.
             Alternatively, both the NH_2 and COOH
             termini can be on the cytoplasmic side of
             the membrane, with the apolar region
             embedded in a hairpin fashion.

Table 4. Compounds Known to be Reduced by NADPH-Cytochrome P450 Reductase

Class	Compounds
anticancer drugs and antibiotics	adriamycin mitomycin c daunorubicin bleomycin porfiromycin
quinones	1,4-naphthoquinone benzo(a)pyrene-3,6-quinone menadione resorufin
nitroimidazoles	metronidazole ronidazole misonidazole
aromatic nitro compounds	nitrofurantoin 4-nitroquinoline N-oxide
herbicide	paraquat
azo dyes	neoprontosil dimethylaminoazobenzene

electron step to semiquinone radicals which can covalently bind to nucleic acids or proteins, resulting in genotoxicity or cytotoxicity. Alternatively, semiquinone can transfer one electron to molecular oxygen to form superoxide anion radical which can undergo a variety of reactions generating hydrogen peroxide as well as hydroxyl radical. The hydroxyl radical is the most reactive oxygen species and is believed to be responsible for DNA-strand breaks, lipid peroxidation and enzyme inactivation during redox cycling of quinones. Thus, like cytochrome P450, the reductase also plays an important role in metabolic activation of some toxic foreign chemicals.

Acknowledgement

I wish to thank Terry Rafferty for her help in preparing this manuscript.

References

1.   B. L. Horecker, Triphosphopyridine nucleotide-cytochrome c reductase in liver, J. Biol. Chem. 183:593 (1950).
2.   A. H. Phillips and R. G. Langdon, Hepatic triphosphopyridine nucleotide-cytochrome c reductase: isolation, characterization and kinetic studies, J. Biol. Chem. 237:2652 (1962).
3.   C. H. Williams and H. Kamin, Microsomal triphosphopyridine nucleotide-cytochrome c reductase of liver, J. Biol. Chem. 237:587 (1962).
4.   L. Ernster and S. Orrenius, Substrate-induced synthesis of the hydroxylating enzyme system of liver microsomes, Fed. Proc. 24:1190 (1965).
5.   L. Shuster and H. Jick, The turnover of microsomal protein in the livers of phenobarbital-treated mice, J. Biol. Chem. 241:5361 (1966).
6.   Y. Kuriyama, T. Omura, P. Siekevitz and G. E. Palade, Effects of phenobarbital on the synthesis and degradation of the protein components of rat liver microsomal membranes, J. Biol. Chem. 244:2017 (1969).
7.   J. R. Gillette, B. B. Brodie and B. N. La Du, The oxidation of drugs by liver microsomes: on the role of TPNH and oxygen, J. Pharmacol. Exp. Ther. 119:532 (1957).
8.   F. Wada, H. Shibata, M. Goto and Y. Sakamoto, Participation of microsomal electron transport system involving cytochrome P450 in ω-oxidation of fatty acids, Biochim. Biophys. Acta 162:518 (1968).
9.   R. I. Glazer, J. B. Schenkman and A. C. Sartorrelli, Immunochemical studies on the role of reduced nicotinamide adenine dinucleotide phosphate-cytochrome c reductase in drug oxidation, Mol. Pharmacol. 7:683 (1971).
10. B. S. S. Masters, J. Baron, W. E. Taylor, E. L. Isaacson and L. Spalluto, Immunochemical studies on electron transport chains involving cytochrome P450. I. Effects of antibodies to pig liver microsomal reduced triphosphopyridine nucleotide-cytochrome c reductase and the non-heme iron protein from bovine adrenocorticol mitochondria, J. Biol. Chem. 246:4143 (1971).

11. A. Y. H. Lu and M. J. Coon, Role of hemoprotein P450 in fatty acid ω-hydroxylation in a soluble enzyme system from liver microsomes, J. Biol. Chem. 243:1331 (1968).

12. A. Y. H. Lu, K. W. Junk and M. J. Coon, Resolution of the cytochrome P450-containing ω-hydroxylation system of liver microsomes into three components, J. Biol. Chem. 244:3714 (1969).

13. Y. Yasukochi and B. S. S. Masters, Some properties of a detergent-solubilized NADPH-cytochrome c reductase purified by biospecific affinity chromatography, J. Biol. Chem. 251:5337 (1976).

14. J. D. Dignam and H. W. Strobel, NADPH-cytochrome P450 reductase from rat liver: purification by affinity chromatography and characterization, Biochemistry 16:1116 (1977).

15. S. D. Black and M. J. Coon, Structural feature of liver microsomal NADPH-cytochrome P450 reductase: hydrophobic domain, hydrophilic domain and connecting region, J. Biol. Chem. 257:5929 (1982).

16. T. D. Porter and C. B. Kasper, Coding nucleotide sequence of rat NADPH-cytochrome P-450 oxidoreductase cDNA and identification of flavin-binding domains, Proc. Natl. Acad. Sci. USA 82:973 (1985).

17. M. Katagiri, H. Murakami, Y. Yabusaki, T. Sugiyama, M. Okamoto, T. Yamano and H. Ohkawa, Molecular cloning and sequence analysis of full-length cDNA for rabbit liver NADPH-cytochrome P450 reductase mRNA, J. Biochem. 100:945 (1986).

18. J. R. Gum and H. W. Strobel, Isolation of the membrane-binding peptide of NADPH-cytochrome P450 reductase: characterization of the peptide and its role in the interaction of reductase with cytochrome P450, J. Biol. Chem. 256:7478 (1981).

19. S. D. Black, J. S. French, C. H. Williams and M. J. Coon, Role of a hydrophobic polypeptide in the N-terminal region of NADPH-cytochrome P450 reductase in complex formation with P450LM, Biochem. Biophys. Res. Commun. 91:1528 (1979).

20. J. R. Gum and H. W. Strobel, Purified NADPH-cytochrome P450 reductase: interaction with hepatic microsomes and phospholipid vesicles, J. Biol. Chem. 254:4177 (1979).

21. I. Iyanagi and H. S. Mason, Some properties of hepatic reduced nicotinamide adenine dinucleotide phosphate-cytochrome c reductase, Biochemistry 12:2297 (1973).

22. T. Iyanagi, H. Makino and H. S. Mason, Redox properties of the reduced nicotinamide adenine dinucleotide phosphate-cytochrome P450 and reduced nicotinamide adenine dinucleotide-cytochrome b5 reductases, Biochemistry 13:1701 (1974).

23. J. L. Vermillion and M. J. Coon, Identification of the high and low potential flavins of liver microsomal NADPH-cytochrome P450 reductase, J. Biol. Chem. 253:8812 (1978).

24. J. L. Vermillion and M. J. Coon, Purified liver microsomal NADPH-cytochrome P450 reductase: spectral characterization of oxidation-reduction states, J. Biol. Chem. 253:2694 (1978).

25. G. P. Kurzban and H. W. Strobel, Preparation and characterization of FAD-dependent NADPH-cytochrome P450 reductase, J. Biol. Chem. 261:7824 (1986).
26. Y. Nisimoto and Y. Shibata, Studies on FAD- and FMN-binding domains in NADPH-cytochrome P450 reductase from rabbit liver microsomes, J. Biol. Chem. 257:12532 (1982).
27. M. Haniu, T. Iyanagi, K. Legesse and J. E. Shively, Structural analysis of NADPH-cytochrome P450 reductase from porcine hepatic microsomes: sequences of proteolytic fragments, cysteine-containing peptides and a NADPH-protected cysteine peptide, J. Biol. Chem. 259:13703 (1984).
28. Y. Nisimoto, F. Hayashi, H. Akutsu, Y. Kyogoku and Y. Shibata, photochemically induced dynamic nuclear polarization study on microsomal NADPH-cytochrome P450 reductase, J. Biol. Chem. 259:2480 (1984).
29. A. L. Shen, T. D. Porter, T. E. Wilson and C. B. Kasper, Structural analysis of the FMN binding domain of NADPH-cytochrome P450 oxidoreductase by site-directed mutagenesis, J. Biol. Chem. 264:7584 (1989).
30. H. Inano and B. Tamaoki, Chemical modification of NADPH-cytochrome P450 reductase: presence of a lysine residue in the rat hepatic enzyme as the recognition site of 2-phosphate moiety of the cofactor, Eur. J. Biochem. 155:485 (1986).
31. I. A. Slepneva and L. M. Weiner, Affinity modification of NADPH-cytochrome P450 reductase, Biochem. Biophys. Res. Commun. 155:1026 (1988).
32. Y. Nisimoto, Localization of cytochrome c binding domain on NADPH-cytochrome P450 reductase, J. Biol. Chem. 261:14232 (1986).
33. S. G. Nadler and H. W. Strobel, Role of electrostatic interactions in the reaction of NADPH-cytochrome P450 reductase with cytochrome P450, Arch. Biochem. Biophys. 261:418 (1988).
34. P. P. Tamburini and J. B. Schenkman, Differences in the mechanism of functional interaction between NADPH-cytochrome P450 reductase and its redox partners, Mol. Pharmacol. 30:178 (1986).
35. R. Bernhardt, K. Pommerening and K. Ruckpaul, Modification of carboxyl groups on NADPH-cytochrome P450 reductase involved in binding of cytochrome c and P450 LM2, Biochem. Int. 14:823 (1987).
36. R. E. Ebel, Selective inactivation of NADPH-cytochrome P450 reductase by diazotized 3-aminopyridine adenine dinucleotide phosphate, Arch. Biochem. Biophys. 211:227 (1981).
37. G. T. Miwa, S. B. West, M. T. Huang and A. Y. H. Lu, Studies on the association of cytochrome P450 and NADPH- cytochrome c reductase during catalysis in a reconstituted hydroxylating system, J. Biol. Chem. 254:5695 (1979).
38. J. S. French, F. P. Guengerich and M. J. Coon, Interactions of cytochrome P450, NADPH-cytochrome P450 reductase, phospholipid and substrate in the reconstituted liver microsomal enzyme system, J. Biol. Chem. 255:4112 (1980).

39. B. Bosterling and J. R. Trudell, Association of cytochrome b5 and cytochrome P450 reductase with cytochrome P450 in the membrane of reconstituted vesicles, J. Biol. Chem. 257:4783 (1982).

40. Y. Nisimoto, K. Kinosita, A. Ikegami, N. Kawai, I. Ichihara and Y. Shibata, Possible association of NADPH-cytochrome P450 reductase and cytochrome P450 in reconstituted phospholipid vesicles, Biochemistry, 22:3586 (1983).

41. J. Gut, C. Richter, R. J. Cherry, K. H. Winterhalter and S. Kawato, Rotation of cytochrome P450. II. Specific interactions of cytochrome P450 with NADPH-cytochrome P450 reductase in phospholipid vesicles, J. Biol. Chem. 257:7030 (1982).

42. R. Bernhardt, A. Makower, G. R. Janig and K. Ruckpaul, Selective chemical modification of a functionally linked lysine in cytochrome P450 LM2, Biochim. Biophys. Acta 785:186 (1984).

43. D. R. Nelson and H. W. Strobel, On the membrane topology of vertebrate cytochrome P450 proteins, J. Biol. Chem. 263:6038 (1988).

44. C. A. Brown and S, D. Black, Membrane topology of mammalian cytochromes P450 from liver endoplasmic reticulum: determination by trypsinolysis of pheno-barbital-treated microsomes, J. Biol. Chem. 264:4442 (1989).

45. T. D. Porter and C. B. Kasper, NADPH-cytochrome P450 oxidoreductase: flavin mononucleotide domains evolved from different flavoproteins, Biochemistry 25:1682 (1986).

46. F. J. Gonzalez and C. B. Kasper, Phenobarbital induction of NADPH-cytochrome P450 oxidoreductase messenger ribonucleic acid, Biochemistry 19:1790 (1980).

47. J. P. Hardwick, F. J. Gonzalez and C. B. Kasper, Transcriptional regulation of rat liver epoxide hydratase, NADPH-cytochrome P450 oxidoreductase and cytochrome P450b genes by phenobarbital, J. Biol. Chem. 258:8081 (1983).

48. H. Kappus, Overview of enzyme systems involved in bioreduction of drugs and in redox cycling, Biochem. Pharmacol. 35:1 (1986).

49. G. Powis, Free radical formation by antitumor quinones, Free Radical Biol.& Med. 6:63 (1989).

50. B. K. Sinha, Free radicals in anticancer drug pharma-cology, Chem.-Biol. Interactions 69:293 (1989).

ESSENTIAL FEATURES OF NADH DEPENDENT CYTOCHROME b5 REDUCTASE

AND CYTOCHROME b5 OF LIVER AND LUNG MICROSOMES

Emel Arınç

Joint Graduate Program in Biochemistry
Department of Biology, Middle East Technical
University, 06531 Ankara, Turkey

INTRODUCTION

Endoplasmic reticulum of mammalian liver cells contains two major electron transport chains. One of these is NADH dependent where the reducing equivalents are transfered from NADH to cytochrome b5 through a flavoprotein, cytochrome b5 reductase. The second system is NADPH dependent and the reducing equivalents are transfered from NADPH to cytochrome P450 through cytochrome P450 reductase containing one mole each of FAD and FMN. NADH dependent cytochrome b5 system catalyzes several reactions of lipid metabolism such as $\Delta 9$ desaturation of stearyl-CoA to oleyl-CoA while the NADPH dependent cytochrome P450 system participates monooxygenation of xenobiotics and of some endogenous compounds (1-5). Recently, these two pathways have been found to be interrelated and transfer of electrons from one pathway to another has been implicated as shown in Fig. 1. If cytochrome b5 acts as a link between these two electron transport systems, it is possible that there is such a regulation depending on availability of NADH and NADPH.

Similar to liver, endoplasmic reticulum of lung tissue also contains cytochrome P450 and cytochrome b5 electron transport chains. Although the concentration of each compound is relatively low in comparison to liver, they may participate in the similar biochemical reactions in lung as they do in liver (6-12).

The following discussion will be limited to the cytochrome b5-involved electron transport system. The most of the information will be on the liver cytochrome b5 system. This system has been purified, crystallized, and well characterized since 1952. The purification and properties of lung cytochrome b5 system will also be included. Because of the physiological importance of hydrophobic peptide domains of these proteins, at the end of the present review, the focus will be on the topography of the hydrophobic peptide segments of cytochrome b5 and b5 reductase in membranes.

Molecular Aspects of Monooxygenases and Bioactivation of Toxic Compounds
Edited by E. Arınç *et al., Plenum Press, New York,*

It is an amphipathic integral membrane protein
participating in a number of oxidation-reduction reactions in
the endoplasmic reticulum. Cytochrome b5 can receive
electrons from both NADH and NADPH through cytochrome b5
reductase and cytochrome P450 reductase and donates electrons
to a number of electron acceptors (1-3, 13-25) including
several acyl-CoA desaturases as illustrated in Fig. 1.

Δ6-Desaturase is one of the desaturases which is
critical for the synthesis of essential fatty acids,
prostaglandins, cyclic AMP and it may have a regulatory role
in the control of cell division. The activity of this enzyme
appears to be lacking in many types of cancer cells and it
has been suggested that this may be a possible reason for the
uncontrolled growth of cancer cells (26). Cytochrome b5 can
also act as cytochrome P-450 reductase and reduce some
specific isozymes of cytochrome P450 (27-30). Except the
reduction of methemoglobin, these reactions can only take
place in the presence of intact amphipathic reductases or
cytochromes.

Cytochrome b5 was first was solubilized with the
treatment of mammalian liver microsomes with partially
purified lipase (31-33). Lipase solubilized and partially
purified cytochrome b5 had a Mr of about 11000, contained
heme group and had a standard redox potential of +0.02 V at
pH 7.0. Cytochrome b5 is reduced fully or partially by
chemical reagents such as sodium dithionite, cysteine,
potassium borohydrate and is oxidized by reagents having an
appropriate redox potential, such as potassium ferricyanide,
ferric chloride, cytochrome c, and menadione (31-33). It was
subsequently shown that treatment of microsomes with ionic
and nonionic detergents liberates the amphipathic (native)
form of cytochrome b5 having a Mr of 16500. Native
cytochrome b5 has been purified from liver microsomes of many
species including bovine, rat, rabbit, pig, human, horse,
and chicken (14, 34-43).

Primary structure of cytochrome b5 from bovine
microsomes is illustrated in Fig. 2. It is a single
polypeptide chain consisting of 133 amino acid residues.
N-terminal amino group is blocked with an acetyl group
(36-39). Treatment of this cytochrome b5 with limited
concentrations of trypsin primarily cleaves the bond between
Lys 90-Ile 91, resulting in two peptide chains differing
markedly in their polar and nonpolar amino acid compositions.
In further studies, proteolytic cleavage of Ser 98-Ile 99
bond of horse liver cytochrome b5 with limited concentrations
of chymotrypsin in aqeous solution (40-44) and structural
analysis of hydrophobic segment (Ile 91 to Asn 133) of bovine
liver cytochrome b5 with Chou Fasman and circular dichroism
measurements (38) revealed the presence of a third peptide
segment so-called connecting peptide. In bovine liver
microsomal cytochrome b5, hydrophilic peptide (N terminus to
Lys 90) is linked to hydrophobic (nonpolar) peptide (Ile 91
to C terminus Asn 133) by a connecting peptide consisting of
a short sequence of amino acids from Ile 91 to Ser 97.

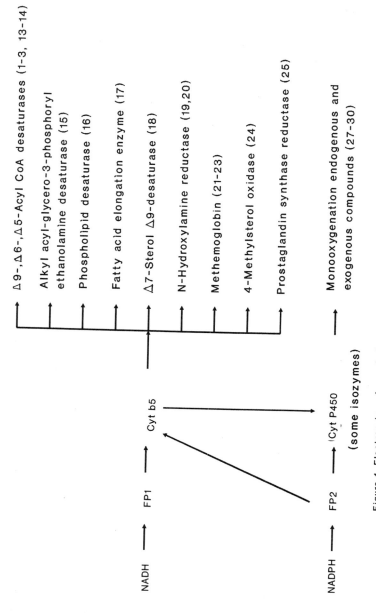

Figure 1. Electron transfer and physiological functions of electron transport chains of endoplasmic reticulum   FP1 : NADH-cytochrome b5 reductase
FP2 : NADPH-cytochrome P450 reductase

Δ9-, Δ6-, Δ5-Acyl CoA desaturases (1-3, 13-14)

Alkyl acyl-glycero-3-phosphoryl ethanolamine desaturase (15)

Phospholipid desaturase (16)

Fatty acid elongation enzyme (17)

Δ7-Sterol Δ9-desaturase (18)

N-Hydroxylamine reductase (19,20)

Methemoglobin (21-23)

4-Methylsterol oxidase (24)

Prostaglandin synthase reductase (25)

Monooxygenation endogenous and exogenous compounds (27-30)

NADH ⟶ FP1 ⟶ Cyt b5

NADPH ⟶ FP2 ⟶ Cyt P450 (some isozymes)

Figure 2 shows the primary structures of cytochrome b5 from liver microsomes of rat, rabbit, human, horse and pig in addition to that of bovine. These cytochromes b5 show an average of 88% identity when pairs of sequences are compared. Table 1 summarizes the general properties of native and hydrophilic peptide of purified cytochrome b5 obtained from detergent and hydrolytically solubilized liver microsomes, respectively. Hydrophilic peptide also called heme peptide has a Mr of approximately 11000, contains 82 to 97 amino acid residues depending somewhat on the species and solubilizing conditions, occupies NH_2 terminal of the protein and resides entirely in the aqueous medium. Although it catalyzes oxidation of cytochrome b5 reductase, it is not functional in the reduction of acyl CoA-desaturase, cytochrome P450 isozymes, and the other enzymes given in Fig. 1. In the absence of detergents, while hydrophilic peptide exists as a monomer, native cytochrome b5 forms octomer and the hydrophobic peptide is highly aggregated in aqeous solutions. Hydrophobic domain containing about 40 to 46 amino acid residues has a Mr of approximately 5000, occupies COOH terminal of the protein and is responsible for anchoring b5 tightly to the biological and synthetic membranes. The hydrophobic peptide sequences of cytochromes b5 given in Fig. 2 show an average sequence identity of 77% when any of the two sequences are compared.

Table 1. Properties of Purified Cytochrome Preparations Obtained from Detergent and Hydrolytically Solubilized Liver Microsomes

PROPERTIES	NATIVE CYTOCHROME b5	HYDROPHILIC PEPTIDE OF CYTOCHROME b5
Monomer Mr	16 500	11 000
Mr in Absence of Detergent	120 000	11 000
Absorption Maxima	413 nm Higher UV Abs.	413 nm
Oxidizes Oxidizes	b5 Reductase P450 Reductase	b5 Reductase P450 Reductase (in high salt conc. only)
Reduces Reduces Reduces	Acyl CoA Desaturase Some Cyt. P450 Other enzymes in in Fig. 1	Not able Not able Not able
Binds to Binds to	Endoplasmic Reticulum Phospholipid Vesicles	Not able Not able
Amino Acid Sequence in Six Species	Sequence Identity 88% (Hydrophobic Peptide 77%)	Identity 94%

```
                1                              10                                      20
BOVINE   Ac-Ala-Glu-Glu-Ser-Ser-Lys-Ala-Val-Lys-Tyr-Tyr-Thr-Leu-Glu-Glu-Ile-Gln-Lys-His-Asn
HORSE    Ac-Ala  -  Gln  -  Asp  -   -   -   -   -   -   -   -   -   -  Lys  -   -   -
RABBIT   Ac-Ala-Ala-Gln  -  Asp  -  Asp  -   -   -   -   -   -   -  Lys  -   -   -
PIG      H₂N-Ala  -  Gln  -  Asp  -   -   -   -   -   -   -   -   -   -   -   -
HUMAN    Ac-Ala  -   -   -  Asp-Glu  -   -   -   -   -   -   -   -   -   -   -   -
RAT      Ac-Ala  -  Gln  -  Asp  -  Asp  -   -   -   -   -   -   -   -   -   -  Lys

               21                              30                                      40
BOVINE   Asn-Ser-Lys-Ser-Thr-Trp-Leu-Ile-Leu-His-Tyr-Lys-Val-Tyr-Asp-Leu-Thr-Lys-Phe-Leu
HORSE    His  -   -   -   -   -   -   -   -   -  His  -   -   -   -   -   -   -   -
RABBIT   His  -   -   -   -   -   -   -   -   -  His  -   -   -   -   -   -   -   -
PIG       -   -   -   -   -   -   -   -   -   -  His  -   -   -   -   -   -   -   -
HUMAN    His  -   -   -   -   -   -   -   -   -  His  -   -   -   -   -   -   -   -
RAT      Asp  -   -   -   -  Val  -   -   -  His  -   -   -   -   -   -   -   -

               41                              50                                      60
BOVINE   Glu-Glu-His-Pro-Gly-Gly-Glu-Glu-Val-Leu-Arg-Glu-Gln-Ala-Gly-Gly-Asp-Ala-Thr-Glu
HORSE     -   -   -   -   -   -   -   -   -   -   -   -   -   -   -   -   -   -   -   -
RABBIT    -   -   -   -   -   -   -   -   -   -   -   -   -   -   -   -   -   -   -   -
PIG       -   -   -   -   -   -   -   -   -   -   -   -   -   -   -   -   -   -   -   -
HUMAN     -   -   -   -   -   -   -   -   -   -   -   -   -   -   -   -   -   -   -   -
RAT       -   -   -   -   -   -   -   -   -   -   -   -   -   -   -   -   -   -   -   -

               61                              70                                      80
BOVINE   Asn-Phe-Glu-Asp-Val-Gly-His-Ser-Thr-Asp-Ala-Arg-Glu-Leu-Ser-Lys-Thr-Phe-Ile-Ile
HORSE     -   -   -   -  Ile  -   -   -   -   -   -   -   -   -   -   -   -   -   -
RABBIT    -   -   -   -   -   -   -   -   -   -   -   -   -   -   -   -   -   -   -
PIG       -   -   -   -   -   -   -   -   -   -   -   -   -   -   -   -   -   -   -
HUMAN     -   -   -   -   -   -   -   -   -   -   -   -  Met  -   -   -   -   -   -
RAT       -   -   -   -   -   -   -   -   -   -   -   -   -   -   -   -   -   -   -

                                                  trypsin
               81                              90                                     100
BOVINE   Gly-Glu-Leu-His-Pro-Asp-Asp-Arg-Ser-Lys-Ile-Thr-Lys-Pro-Ser-Glu-Ser-Ile-Ile-Thr
HORSE     -   -   -   -   -   -   -   -   -   -   -  Ala  -   -  Val  -  Thr-Leu  -   -
RABBIT    -   -   -   -   -   -   -   -   -  Leu-Ser  -   -  Met  -  Thr-Leu  -   -
PIG       -   -   -   -   -   -   -   -   -   -  Ala  -   -   -   -  Thr-Leu  -   -
HUMAN     -   -   -   .   -   -   -   -  Pro  -  Leu-Ser  -   -  Pro  -  Thr-Leu  -   -
RAT       -   -   -   -   -   -   -   -   -   -  Ala  -   -   -   -  Thr-Leu  -   -

              101                             110                                     120
BOVINE   Thr-Ile-Asp-Ser-Asn-Pro-Ser-Trp-Trp-Thr-Asn-Trp-Leu-Ile-Pro-Ala-Ile-Ser-Ala-Leu
HORSE     -  Val  -   -   -  Ser  -   -   -   -   -  Val  -   -   -   -   -  Val
RABBIT    -  Val  -   -   -  Ser  -   -   -   -   -  Val  -   -   -   -   -   -
PIG       -  Val-Glu  -   -  Ser  -   -   -   -   -  Val  -   -   -   -   -   -
HUMAN     -   -   -   -  Ser-Ser  -   -   -   -   -  Val  -   -   -   -   -  Val
RAT       -  Val-Glu  -   -  Ser  -   -   -   -   -  Val  -   -   -   -   -   -

              121                             130    133
BOVINE   Phe-Val-Ala-Leu-Ile-Tyr-His-Leu-Tyr-Thr-Ser-Glu-Asn-COOH
HORSE    Val  -   -   -  Met  -  Arg-Ile  -   -  Ala  -  Asp
RABBIT   Ile  -   -   -  Met  -  Arg  -   -  Met-Ala-Asp-Asp
PIG      Val  -  Ser  -  Met  -   -  Phe  -   -   -   -   -
HUMAN    Val  -   -   -  Met  -  Arg  -   -  Met  -   -  Asp
RAT      Val  -   -   -  Met  -  Arg  -   -  Met-Ala  -  Asp
```

Figure 2. Amino acid sequences of mammalian liver microsomal cytochromes b5. Compiled from References 14, 34-43.

The absorption spectrum of native cytochrome b5 differs from hydrophilic peptide segment of b5 only in the near ultraviolet region. Three tryptophan and two tyrosine residues present in hydrophobic peptide gives higher absorption spectrum in that region.

Structural Features of Hydrophilic Heme Peptide

Hydrophilic heme peptide of cytochrome b5 of calf liver was crystallized and x-ray crystallographic analysis was carried out at 2.0 Å resolution (44). The diameter of the molecule is approximately 31 Å and the height is about 37 Å. Charged amino acids are located on the molecular surface while most of the nonpolar amino acid residues are found in the interior of the molecule. The heme binding pocket extends about half way down to the molecule and the crevice for heme binding is located on a side opposite to both N- and C-terminus of the hydrophilic peptide (35, 44, 45).

The propionyl side chains of heme edge orients at the surface of the molecule, while the heme itself is almost completely shielded from solvent (35, 44, 45). The heme group of cytochrome b5 is found to be quite unreactive compared with some other heme proteins such as hemoglobin. It does not react with oxygen or carbon monooxide neither in oxidized nor in reduced states.

The distribution of positive and negative charged amino acid residues over the surface of hydrophilic segment of b5 molecule is uneven. The top of the molecule is largely negative (net charge of -9). The lower half has a net charge of +1 with seven acidic and eight basic groups. The top part of the molecule contains exposed propionyl groups of the heme, lacks any positively charged amino acid residues, forms a nearly polar surface of anionic charge comprising Glu 47, 48, 52, 60 and Asp 64 that is perpendicular to the plane of the sequestered proteheme group (35).

Functional Importance of Selected Amino Acid Residues of b5

A number of chemical modification studies using group specific reagents are carried out to examine the reactivity of various amino acid residues and to identify the essential amino acids involved in heme binding and in the catalytic action of cytochrome b5. These studies have been carried out largely on calf cytochrome b5 with some parallel experiments on rabbit cytochrome b5.

It has been observed that when amino groups on all nine lysine residues are acylated with acetic and succinic anhydride in hydrophilic peptide segment of calf liver cytochrome b5, the spectra of the oxidized and reduced forms of cytochrome b5 are unaffected by this complete acylation. Thus, the lysyl residues are apparently not involved in heme binding (46). This conclusion was further supported by the observation that apocytochrome b5, which had been acetylated under the conditions that resulted in complete acylation of amino groups but no other amino acid residues, would combine completely with heme to yield unaltered cytochrome b5 spectrum.

By chemical modification studies, Strittmatter (47) has shown that at least one histidine residue is involved in heme binding which is also confirmed by X-ray analysis (44, 45). However, modification of tyrosyl groups by acetylation or ionidation does not alter the absorption or circular dichroic heme spectra (48). X-ray analysis of b5 also shows no close contact between heme group and any of the tyrosyl or trptophanyl residues.

The carboxyl groups of the hydrophilic peptide of cytochrome b5 have been modified with methylamine, glycine ethyl ester or taurine using a water-soluble carbodiimide (35, 49). The first two chemicals neutralize the negative charge of a carboxyl group, while the third produces a bulkier negatively charged side chain. After the characterization of methylamine derivatives, Glu 47, Glu 48, Glu 52, and a heme propionate of b5 have been found to be modified. Asp 64 has also been tentatively identified as the fifth modified anionic group. Kinetics of cytochrome b5 reduction by cytochrome b5 reductase have demonstrated that while V_{max} has not been altered, K_m has increased by increasing the number of the modified groups. When five carboxyl groups have been modified, K_m has increased from 9 μM to 91 μM.

Glycine ethyl ester derivatives of heme peptide of cytochrome b5 with 2-8 modified carboxyl groups have been obtained (49). This peptide with five or more glycine-modified carboxyl groups yields K_m of 250-350 μM. Again, the V_{max} value remains unaltered. In contrast, taurine modified heme peptide yields unchanged K_m, but V_{max} reduces by upto 98%. This is expected since taurine modified b5 retains all the negative charges but the position of the anionic group, now a sulfonate, is displaced approximately 4 Å away (49). Thus, carboxyl groups are identified as being the most important groups for binding of cytochrome b5 to cytochrome b5 reductase. The involvement of these groups in the catalytic action of cytochrome b5 and its reductase will be further discussed in the liver cytochrome b5 reductase section.

LUNG CYTOCHROME b5

Lung is an active organ in the synthesis of essential fatty acids and prostaglandins. Thus, cytochrome b5 dependent electron transport chain has a great importance to this tissue. One approach to better understand this system involves the purification and characterization of cytochrome b5 from lung tissue.

In our laboratory, cytochrome b5 has been partially purified from sheep lung microsomes and has been characterized (12, 50). Sheep lung, like the rabbit lung, contains high amounts of cytochrome P450LgM$_2$ isozyme (51, 52) which can accept the electrons from cytochrome b5. The purification method involved solubilization of sheep lung microsomes with cholate and Emulgen 913 in the presence of protease inhibitors, phenylmethylsulfonyl fluoride (PMSF) and ϵ-amino-n-caproic acid (ϵ-ACA), followed by the ion-exchange

Table 2. Molecular Weight and Spectral Properties of
Liver and Lung Microsomal Cytochrome b5

| TISSUE | Molecular Weight | Absolute Spectrum | |
		Oxidized nm	Reduced nm
LIVER[*]	16 500	413, 538	423, 526, 556
LUNG[**]	17 000	413, 540	423, 526, 555

[*]calf liver determined from amino acid sequence.
[**]sheep lung determined by SDS-PAGE.

chromatography of the solubilized microsomes on two
subsequent DEAE-cellulose columns, concentration of b5 by
Diaflo PM-30 membrane, and removal of excess non-ionic
detergent by Porapak Q column chromatography.

Purified lung cytochrome b5 was free of NADPH cytochrome
c reductase and NADH cytochrome b5 reductase activities and
cytochrome P-450. As shown in Table 2, its visible absorption
spectrum, like its liver counterpart, has a Soret band at 413
nm which shifts to 423 nm when reduced. The molecular weight
of lung cytochrome b5 is determined by sodium dodecyl sulfate
polyacrylamide gel electrophoresis (SDS-PAGE) and is found to
be approximately 17000.

LIVER NADH-CYTOCHROME b5 REDUCTASE

NADH-cytochrome b5 reductase (EC 1.6.2.2) is a
flavoprotein containing one mol of FAD per mol of the enzyme
and transfers the reducing equivalents from NADH to
cytochrome b5 in the endoplasmic reticulum. Cytochrome b5
reductase was originally isolated by treatment of calf liver
microsomes with crude snake venom (33). Takesue and Omura
(53) have shown that, as in the case of cytochrome b5,
reductase solubilized by the action of proteolytic enzymes
has only the FAD containing hydrophilic peptide segment of
the enzyme. The amphipathic reductase was purified by Spatz
and Strittmatter (54) and Mihara and Sato (55, 56) by the
solubilization of microsomes with detergents. In recent
years, introduction of 5'-ADP-agarose affinity column
chromatography (57) has resulted in the preparation of highly
purified liver b5 reductase enzyme in large quantities that
enables the structural analysis of b5 reductase.

Amino acid sequence of amphipathic steer liver
microsomal cytochrome b5 reductase has been recently
determined by Ozols et al. (58). Amphipatic cytochrome b5
reductase contains 300 amino acids consisting of a large
cytoplasmic, catalytically active, FAD-containing hydrophilic
domain (from C-terminus, Phe 300 to Ser 29), has a Mr of
approximately 28000, is responsible for electron transport

and is connected by a short, flexible linking region to a NH_2 terminal hydrophobic membrane binding segment (from Lys 24 to N-terminal myristylglycine).

After the elucidation of complete amino acid sequence of bovine liver cytochrome b5 reductase, its secondary structure has been predicted (58). Enzyme consists of about 25% α helix with 30% β sheet structures. The amino acid sequence of bovine liver microsomal b5 reductase is found to be 93% homologous to the 275-residue sequence of human erythrocyte cytochrome b5 reductase (58-59). Recently, cloning of cDNAs of human liver and placenta NADH-cytochrome b5 reductase has been reported (60). The amino acid sequences deduced from the nucleotide sequence of these cDNA clones overlapped each other and consisted of a sequence that completely coincides with that of human erythrocytes (59, 60) and a sequence of 19 amino acid residues extended at the amino terminal side. The latter sequence was found to be closely similar to that of hydrophobic membrane binding domain of bovine liver microsomes (58, 60).

In a recent study (61), both detergent and lysosome solubilized b5 reductases from pig liver were crystallized. Preliminary X-ray diffraction patterns of hydrophilic peptide of b5 reductase crystals indicated that the crystal is orthorhombic. The crystals of the detergent solubilized enzyme could not be grown upto an acceptable size for X-ray diffraction work (61).

NADH-Cytochrome b5 Reductase Complexes

When NADH is added to cytochrome b5 reductase in stoichiometric amounts under anaerobic conditions, the absorption spectrum of reductase changes. The flavin is reduced. Consequently, 390 nm and 460 nm flavin peaks disappear and a new peak at 317 nm appears (62). Upon exposure of the reaction mixture to air, the spectral changes are reversed and the initial oxidized spectrum of flavin is obtained. When the reductase is reduced anaerobically by methylviologen, the 390 nm peak disappears but the peak at 317 nm does not appear (63). However, subsequent addition of NAD causes the formation of 317 nm peak. These results imply that upon binding of NADH to the enzyme, a complex between the oxidized substrate (NAD) and the reduced enzyme containing $FADH_2$ is formed which shows a maximum absorbance at 317 nm.

The anaerobic reduction of cytochrome b5 reductase by NADPH is found to be quite different from that of NADH. When NADPH is added, it causes the reduction of the oxidized enzyme in an extremely slow reaction and the 390 nm peak disappears. However, the concomitant appearance of the 317 nm peak does not take place. This behavior is found to be similar that of methyl viologen and implies that formation of a stable complex between NADP and the reduced b5 reductase does not occur (63).

Essential Functional Amino Acids of Cytochrome b5 Reductase

Previous studies suggested that one specific and highly

reactive tyrosyl residue was involved in flavin binding to the aporeductase. Three classes of lysyl residues were identified to be important in the catalytic action of reductase (64). Only a single lysine residue was involved in binding to NADH. Another group of lysine residues was found to be essential for the interaction of the reductase with its natural electron acceptor cytochrome b5. The remaining lysyl residues were required for a stable haloenzyme structure (64). Further studies suggested that the interaction between the heme peptide of cytochrome b5 and b5 reductase is dominated by complementary charge pairing involving lysyl groups on the reductase with carboxyl groups on cytochrome b5 (35, 49). Recent studies of Hackett and Strittmatter (65) based on the formation of amide bonds between the carboxyl residues on liposome-bound cytochrome b5 and the lysyl residues on vesicle-bound b5 reductase by the catalytic action of a water soluble carbodiimide have provided compelling evidence for this complementary charge pairing (65). Recently, Hackett et al (66) identified a single cysteine residue, Cys 283, as an essential amino acid involved in NADH binding.

LUNG-NADH CYTOCHROME b5 REDUCTASE

Purification of cytochrome b5 reductase from lung has been found to be rather difficult. The purification procedures used for liver reductase could not be applied for the purification of the enzyme from the lung. Stability of the lung enzyme during the purification procedure has been a great problem that is to be overcomed (12).

Following the solubilization of sheep lung microsomes with Emulgen 913 and cholate in phosphate buffer pH 7.7 containing 20% glycerol, PMSF and ϵ-ACA, chromatography on DEAE-cellulose column resulted in the elution of cytochrome b5 reductase and cytochrome b5, separately. A second DEAE-cellulose column chromatography was introduced in order to concentrate b5 reductase in a mild manner without ultrafiltration to prevent FAD dissociation. In this step, further purification of b5 reductase was noted. The use of 5`-ADP agarose affinity column resulted in highly purified b5 reductase preparations.

As seen in Table 3, lung reductase has been found to be similar to liver counterpart in terms of monomer molecular weight, cleavage by mild trypsin, cofactor, and absolute spectrum. However, the apparent K_m values for the substrate cytochrome b5 obtained by sheep lung reductase and calf liver reductase are different, 1.67 μM (67) and 20.0 μM (33), respectively.

Kinetic studies are performed with lung reductase by varying NADH and cytochrome b5 concentrations at different fixed concentrations of cytochrome b5 or NADH. Lineweaver Burk plots of these substrates showed a series of parallel lines indicating a ping-pong type kinetic mechanism for interaction of NADH and cytochrome b5 with lung cytochrome b5 reductase (67). In this mechanism, NADH, in the absence of electron acceptor, reacts with oxidized flavo-reductase

158

Table 3. Some Properties of Liver and Lung Microsomal
Cytochrome b5 Reductase

Tissue	LIVER[*]	LUNG[**]
Turnover Number[+]	29000	27000
Mr of Monomer	34110[#]	34500[##]
Mr of Hydrophilic Peptide	28400	28000
Cofactor	FAD	FAD
Absolute Spectrum (nm)	390,461 Shoulder 485	390,461 Shoulder 490

[+]Moles of ferricyanide reduced/min/mole enzyme
[*]From steer liver; [**]From sheep lung
[#]From amino acid sequences of steer liver (58)
[##]Determined by SDS-PAGE (12)

(E.FAD) generating NAD^+ and reduced enzyme intermediate
(NAD^+-E.FADH$_2$). Subsequently, NAD^+ is released and the
reduced form of the enzyme (E.FADH$_2$) reacts with the
electron acceptor, cytochrome b5 (Fe^{3+}), and forms the
oxidized flavoprotein enzyme (E.FAD) and reduced cytochrome
b5 (Fe^{2+}). Although a kinetic study of this type has not
been carried out by liver b5 reductase, as discussed above, a
formation of a complex between the oxidized substrate
(NAD^+) and the reduced liver enzyme containing FADH$_2$
which shows a peak at 317 nm has been identified (62).

THE FUNCTIONAL IMPORTANCE AND TOPOLOGY OF HYDROPHOBIC PEPTIDE DOMAINS OF CYTOCHROME b5 AND CYTOCHROME b5 REDUCTASE

Hydrophic cytochrome b5 is reduced by hydrophilic
peptide of cytochrome b5 reductase and hydrophobic peptides
of both proteins are not required for this electron transfer.
However, for interaction with stearyl-CoA desaturase, with
some isozymes of cytochrome P450, and with the enzymes given
in Fig. 1, the entire amphipathic cytochrome b5 molecule
including the membrane-binding segment is required. Moreover,
the intact detergent solubilized forms of NADH cytochrome b5
reductase, NADPH cytochrome P-450 reductase, stearyl-CoA
desaturase and cytochrome P-450 are also found to be
necessary for interaction with amphipathic cytochrome b5 for
the reactions given in Fig. 1.

The membrane binding hyrophobic peptide domain of
NADH-cytochrome b5 reductase which occupies the amino
terminal myristylglycine through the first 28 amino acid
residues is isolated in cholate after a mild trypsin
treatment of cholate solubilized reductase, and after
exhaustive trypsin treatment of liposome bound reductase

(68). The detergent solubilized peptide shows high affinity
for phospholipid liposomes and can be constituted in
liposomes by the detergent dialysis method (68).

The hydrophobic peptide of reductase is necessary for
binding of b5 reductase to the microsomal membranes and
liposomes. Native cytochrome b5 reductase binds microsomes,
yielding microsomes enriched in b5 reductase in amounts upto
20% of the total protein as in the case of native cytochrome
b5 (69). The level of bound reductase can increase up to 100
fold, going from 0.07 nmole/mg microsomal protein in the
native microsomes to 7.2 nmoles/mg. The extra bound b5
reductase can not be removed from microsomes by washing in
0.5 M sodium chloride.

The hydrophobic peptide segment of liver b5 reductase is
strongly hydrophobic due to a sequence which contains a high
proportion of nonpolar residues and also due to acylation of
the amino terminal glycine with myristic acid (58, 68). The
fluorescence of Trp 16 of this peptide is found to be highly
sensitive to the polarity of the microenvironment. The
fluorescence quantum yield of this residue is 0.10 when the
hydropobic peptide is dispersed in 1% sodium cholate, but
0.46-0.52 when the peptide is reconstituted in dimyristoyl
phosphatidylcholine (DMPC) liposomes. The fluorescence energy
transfer analysis of this tryptophan residue in liposome
bound peptide shows that Trp 16 resides at a dept of 20-23 Å
in the lipid bilayer (68).

It is established that the hydrophobic amino-terminal
myristoyl group must contribute significantly to the
stabilization of the membrane binding domain within the
hydrocarbon region of the phosphalipid bilayer. However, the
precise location in the membrane remains to be solved in
future studies. This segment may be inserted to the membrane
in a trans-bilayer form or in a hair-pin structure (68).

In this area of research, much emphasis has been placed
on determining the topography and functional importance of
hydrophobic membrane binding domain of liver cytochrome b5.
There are several reasons for this: It is a small protein, is
easy to handle and is stable to changes of temperature and
pH.

The studies carried out in 1970s have shown that
hydrophobic segment of cytochrome b5 including the connecting
peptide, consisting of the C-terminal 43 amino acid residues
binds rapidly to phospholipid vesicles and fluorescence
emissions of Trp 109 increases about two folds upon insersion
of this peptide in liposomes. The location of this single
tryptophan residue in liposomes has been determined by
fluorescent energy transfer analysis method developed by
Koppel et al. (70). It is found that Trp 109 is located about
22 Å below the surface of the lipid bilayer (35, 71).

In the initial studies, the first question asked is
whether all 43 amino acid residues of hydrophobic peptide of
cytochrome b5 are required for binding of b5 to phospholipid
vesicles and for its catalytic interaction with stearyl CoA
desaturase enzyme (72). Amphipathic cytochrome b5 purified

from bovine microsomes is used in these experiments. Selectively shortened three derivatives of cytochrome b5 are obtained by carboxypeptidase A and/or Y digestion of native cytochrome b5. Derivative I, derivative II, and derivative III are found to be shortened by 6, 18, and 27 residues from C-terminus of b5, respectively. All three derivatives of b5 are incubated with DMPC vesicles at 30 ^{0}C for 18 hours. The removal of upto 18 residues from the C-terminus of b5 has no appreciable effect on binding of these derivatives (I and II) to phospholipid vesicles; But the removal of an additional nine residues results in the loss of the ability to bind the DMPC vesicles (72).

Then, the enzymatic activity of these three b5 derivatives to function as an electron carrier in the stearyl CoA desaturase system reconstituted in egg lecithin vesicles is studied. Derivative I and II are able to serve as electron donors for the desaturase. In contrast, derivative III derived from II by removal of the next nine COOH-terminal residues is ineffective in the desaturase system. Derivatives I and II exhibit the same flourescent enhancement as native cytochrome b5 in binding to phospholipid vesicles and fluorescent Trp 109 is also located about 20-22 Å below the outer surface of the unilameller vesicles (73). Furthermore, circular dichroism data indicate that both derivatives I and II but not derivative III essentially retain the predicted secondary structure of the whole hydrophobic peptide which contains about 50% helical and 25% β structure (72). The results demonstrates that the nine amino acid residues (107 - 115), seven of which are nonpolar, and their sequences are crucial for binding of b5 to vesicles in a physiologically active manner. In contrast, the following sequence of 11 residues (116-126) is not essential for this type of interaction (35).

While working with binding of native cytochrome b5 and its derivatives to artificial phospholipid membranes, the existence of two types of protein binding to the vesicles is noticed. In the first case, the transfer of cytochrome b5 between large and small vesicles and to microsomes takes place. In the second case, b5 is not transfered from phospholipid vesicles or from microsomes to either small or large vesicles (35, 74). These bindings are termed as "loose" and "tight" binding, respectively, to distinguish the more stable binding. Binding of native cytochrome b5 to DMPC or to microsomes is found to be "tight". The "loose" binding appeared to be a problem of a spontaneous insertion of b5 into the egg lecithin vesicles. The "tight" binding to egg lecithin vesicles is achieved by sonication or inclusion of detergents, peptides, or proteins in the vesicles.

In addition, when cytochrome b5 binds to liposomes in "loose" binding form, the carboxyl terminus of the hydrophobic peptide shows susceptibility to carboxypeptidase Y. In contrast, tightly bound cytochrome b5 is found to be resistant to carboxypeptidase Y digestion (74).

By using the intervesicle transfer of cytochrome b5 as a test system, Strittmatter et al. (35, 73) reexamined the nature of binding of derivatives I, II, and III of b5 to

phospholipid vesicles. They observed that all three
derivatives were bound to the phospholipid vesicles in
"loose" manner. Thus, even the removal of only six C-terminal
residues (derivative I) from Leu 128 to Asn 133, containing
five polar residues, has resulted in the loss of
characteristic "tight" binding of b5 either to phospholipid
vesicles or to microsomes (35, 73).

Moreover, the chemical modification of the four carboxyl
groups (Glu 96, Asp 103, Glu 132, Asn 133) of isolated
hydrophobic peptide segment with methylamine in the presence
of water soluble carbodiimide has produced a derivative
containing neutral methylamide groups in place of anionic
carboxyl groups and has also resulted in the loss of
characteristic "tight" binding (73). Thus, Strittmatter and
coworkers (35, 73) have concluded that the nine amino acid
residue sequence (107-115) including the fluorescent
tryptophanyl residue is sufficient for membrane interaction
in a physiologically functional manner, and that the short
polar C-terminal containing two carboxyl groups, Glu 132 and
Asn 133, of membrane binding domain is essential for the
lipid protein interaction which results in a normal "tight"
binding.

Even before this conclusion which establishes the
crucial functional importance of the negatively charged
C-terminal dipeptide Glu 132-Asn 133 in cytochrome b5
dependent reactions, several groups of laboratories have
found the presence of five polar amino acid sequence
including two negatively charged Glu 132 and Asn 133 residues
in the C-terminal of b5 rather intriguing, and concentrated
their work on the orientation of hydrophobic peptide of
cytochrome b5 in lipid bilayer. As a result, two main models
have been proposed (38, 71, 72, 75, 76). The first model
involves a transmembrane orientation in which COOH-terminal
of hydrophobic peptide spans the phospholipid bilayer and
resides in aqeous medium, on the side opposite to the NH_2-
terminal heme containing segment. The second model, cis
configuration, places COOH and NH_2 termini both on the same
external surface of the membrane.

In 1980, Tajima and Sato (76) prepared egg phosphatidylcholine liposome bound rabbit cytochrome b5 and allowed it to
react with iodine by lactoperoxidase. They found that both
tyrosine residues of nonpolar peptide, Tyr 126 and Tyr 129,
were iodinated. Thus, they suggested the "cis" configuration.
Subsequently, Daily and Strittmatter (77) isolated the
nonpolar peptide segment of b5 reconstituted into vesicles in
the tightly bound form. As in the case of amphipathic
cytochrome b5, the tightly bound nonpolar peptide did neither
transfer to other vesicles nor was released by exo- or
endo-peptidases (77). They found that C-terminal tyrosines,
Try 126 and Try 129, were rapidly ionized by the addition of
NaOH to external medium; Whereas, a pH-sensitive dye trapped
inside the vesicles ionized much more slowly. Furthermore,
Try 129 reacted with the membrane- impermeant probe
diazotized sulfanilic acid after it was added to the external
medium. They concluded that Try 129 and five residues from
C-terminus lie on or near the outer surface of phospholipid
bilayer. Thus, cis configuration model of b5 is favored.

However, in later years, the results of both papers have been questioned. It is understood that Tajima and Sato (76) prepared egg lecithin vesicles with loosely bound cytochrome b5 and Dailey and Strittmatter (77) used only nonpolar peptide segment and not the whole amphipathic cytochrome b5.

Using a procedure in which photoactivable phosphatidyl-choline derivatives were cross-linked to tightly bound amphipathic cytochrome b5, Takagaki et al (78) suggested that nonpolar peptide was inserted in "trans" configuration. However, the interpretation of the results reported by Takagaki et al (78) is complicated by very low yield of cross-linked protein and by relatively high degree of protein symmetry across the vesicles.

Although direct structural analysis by fluorescence spectroscopy (71), low resolution neutron diffraction (70), solution scattering (80), and X-ray diffraction (81) has established that the nonpolar peptide of tightly bound cytochrome b5 is deeply embedded in the membrane, the "cis" and "trans" configurations have not been distinguished. In addition, the X-ray diffraction analysis (81), carried out to a resolution of 15 Å, revealed a highly asymmetric electron density distribution across the membrane bilayer, consistent with a deep and asymmetric penetration of the bilayer by protein. This study did not also resolve the position of C-terminus in the phospholipid bilayer.

We (82) took a different experimental approach to identify the orientation of the carboxyl terminus of cytochrome b5 in phospholipid membranes by taking advantage of the presence of negatively charged amino acid residues, Glu 132 and Asn 133, at the carboxyl terminus of b5 and their possible interaction with a nucleopeptide in the presence of a catalyst, water soluble carbodiimide (EDC). During the study, amphipathic b5 was reconstituted asymmetrically into DMPC vesicles in the tight binding form such that the catalytic domain was expanded on the external surface. A membrane impermeant nucleophile, taurine, was incubated with DMPC- bound b5. If the carboxyl groups of C-terminal residues are exposed to aqueous medium, they would contact with taurine and would be bound covalently. The results indicate that this is the case.

In that study (82), the reactant concentrations and pH were optimized to ensure a relatively high taurine to b5 ratio. Reaction of cytochrome b5 bound to DMPC residues with a 335 fold molar excess of taurine in the presence of limited concentrations of EDC at pH 7.0 for three hours resulted in covalent modification of 6.7 moles of carboxyl group, on the average, per mole of b5. The modified b5 retained its negative charge and no denaturation of the protein was observed as judged by its UV/visible spectrum (82). The nonpolar peptide isolated after tryptic hydrolysis contained 1.8 mol of taurine per mol of nonpolar peptide. Subsequently, the C-terminal tetrapeptide containing residues Thr 130 and Asn 133 was generated by chymotryptic hydrolysis of radiolabeled hydrophobic peptide and was purified by gel filtration and ion exchange chromatography. Amino acid analysis of the C-terminal tetrapeptide showed that about 1.6

Table 4. Amino Acid Composition of Radiolabeled
C-terminal Tetrapeptide

Amino Acid	Residues/Peptide	
	Modified	Native
Aspartic Acid	0.99 (1)	0.94 (1)
Glutamic Acid	1.00 (1)	1.00 (1)
Serine	0.93 (1)	0.93 (1)
Threonine	0.90 (1)	0.98 (1)
Taurine	1.58	0.00

Amino acid composition of the chymotryptic peptide
isolated by chromatography on Sephadex G-25 and DEAE-
Sephacel was determined following the hydrolysis of the
sample in HCl for 24 hr. Results are the average of two
sets of separate experiments. Values in parentheses
indicate the number of residues predicted from the
amino acid sequence.
Reproduced with permission from Ref. 82

mol of taurine was cross-linked per mol of peptide (Table 4)
indicating that Glu 132 and and Asn 133 of the C-terminus
were primary sites of hydrophobic peptide modification. When
taurine was trapped inside vesicles and removed from the
external medium by chromatography, no cross linking to
cytochrome b5 was observed after addition of carbodiimide
even at relatively high taurine concentrations (82). Since
taurine did not leak out of the vesicles to a significant
degree, the absence of cross linking under the experimental
conditions suggests that covalent binding could not occur
inside the vesicles. Hence, the observed modification of the
C-terminal carboxyl residues of b5 with taurine could only
occur on the external surface of the vesicles. Thus, these
results in parallel with the results obtained with control
experiments indicated that hydrophobic peptide in membranes
is in the "cis" configuration, that is, it penetrates to
middle of the bilayer, loops back in a hair-pin configuration
to the outer surface to place both the NH_2 and COOH termini
on the same external surface of the phospholipid membrane.
The five COOH terminal polar residues of hydrophobic peptide
including Glu 132 and Asn 133 may contribute to stabilizing
both overall tertiary structure and ion pairing with the
phospholipid head groups.

The recent results of Ozols (39) with rabbit microsomal
cytochrome b5 confirmed the "cis" configuration of b5. In
that study (39), rabbit liver microsomes were treated with
limited concentrations of trypsin and subjected to gel

filtration and high-pressure liquid chromatography for purification of hydrophobic peptide of b5. (See Fig. 2 for the amino acid sequence of rabbit b5.) The hydrophobic peptide isolated from these microsomes lacked the COOH-terminal hexapeptide, indicating that when b5 is bound to intact microsomes, the carboxyl terminus is located on the cytosolic side of the membrane and does not extend to the lumen of the endoplasmic reticulum.

ACKNOWLEDGEMENT

This work was supported in part by a grant from Middle East Technical University AFP-89-01-08-04.

REFERENCES

1. N. Oshino, Y. Imai, and R. Sato, A function of cytochrome b5 in fatty acid desaturation by rat liver microsomes, J. Biochem. (Tokyo) 69:155-167 (1971).
2. N. Oshino and T. Omura, Immunochemical evidence for the participation of cytochrome b5 in microsomal stearyl CoA desaturase reaction, Arch. Biochem. Biophys. 157:395-404 (1973).
3. P. Strittmatter, L. Spatz, D. Corcoran, M. J. Rogers, B. Setlow, and R. Redline, Purification and properties of rat liver microsomal stearyl coenzyme A desaturase, Proc. Natl. Acad. Sci. USA 71:4565-4569 (1974).
4. A. Y. H. Lu and S. B. West, Multiplicity of mammalian microsomal cytochrome P-450, Pharmac. Rev. 31:277-295 (1980).
5. S. D. Black and M. J. Coon, Comparative structures of P-450 cytochromes, in: "Cytochrome P-450. Structure, Mechanism and Biochemistry", P. R. Ortiz de Montellano, ed., Plenum Press, New York (1986).
6. R. M. Philpot, E. Arınç, and J. R. Fouts, Reconstitution of the rabbit pulmonary microsomal mixed-function oxidase system from solubilized components, Drug Metab. Dispos. 3:118-126 (1975).
7. E. Arınç and R. M. Philpot, Preparation and properties of partially purified pulmonary cytochrome P-450 from rabbits, J. Biol. Chem. 251:3213-3220 (1976).
8. E. Arınç and M. Y. İşcan, Comparative studies of sheep liver and lung microsomal aniline 4-hydroxylase, Comp. Biochem. Physiol. 74C:151-158 (1983).
9. E. Arınç, Characterization of sheep liver and lung microsomal ethylmorphine N-demethylase, Comp. Biochem. Physiol. 80B:389-399 (1985).
10. Z. Parandoosh, V. S. Fujita, M. J. Coon, and R. M. Philpot, Cytochrome P-450 isozymes 2 and 5 in rabbit lung and liver : comparison of structure and inducibility, Drug Metab. Dispos. 15:59-67 (1987).
11. Y. Kikuta, E. Kusunose, S. Matsubara, Y. Funae, S. Imaoka, I. Kubota, and M. Kusunose, Purification and characterization of hepatic microsomal prostaglandin w-hydroxylase cytochrome P-450 from pregnant rabbits, J. Biochem. (Tokyo) 106:468-473 (1989).

12. T. Güray and E. Arınç, Purification of NADH-cytochrome
    b5 reductase from sheep lung and its electrophoretic
    spectral and some other properties, <u>Int. J. Biochem.</u>
    In Press.

13. T. C. Lee, R. C. Baker, and N. Stephens, Evidence for
    participation of cytochrome b_5 in microsomal $\Delta 6$-
    desaturation of fatty acids, <u>Fedn. Proc.</u> 36:672
    (1977).

14. N. Oshino, Cytochrome b_5 and its physiological
    significance, <u>in</u>: "Hepatic Cytochrome P-450
    Monooxygenase System", J. B. Shenkman and D. Kupfer,
    eds., Pergamon Press, New York (1980).

15. F. Paltauf, R. A. Prough, B. S. S. Masters, and J. M.
    Johnson, Evidence for the participation of
    cytochrome b_5 in plasmalogen sysnthesis, <u>J. Biol.
    Chem.</u> 249:2661-2662 (1974).

16. E. L. Pugh and M. Kates, Direct saturation of
    eicosatrienoyl lecithin to arachidonyl lecithin by
    rat liver microsomes, <u>J. Biol. Chem.</u> 252:68-73
    (1977).

17. S. R. Keyes, J. A. Alfano, I. Jansson, and D. L. Cinti,
    Rat liver microsomal elongation of fatty acids;
    possible involvement of cytochrome b_5, <u>J. Biol.
    Chem.</u> 254:7778-7784 (1979).

18. V. V. R. Reddy, D. Kupfer, and E. Kaspi, Mechanism of
    C-5 double bond introduction in the biosynthesis of
    cholesterol by rat liver microsomes. Evidence for
    the participation of microsomal cytochrome b_5, <u>J.
    Biol. Chem.</u> 252:2797-2801 (1977).

19. F. F. Kadlubar, E. M. McKee, and D. M. Ziegler, Reduced
    pyridine nucleotide-dependent N-hydroxylamine
    oxidase and reductase activities of hepatic
    microsomes, <u>Arch. Biochem. Biophys.</u> 156:46-57
    (1973).

20. F. F. Kadlubar and D. M. Ziegler, Properties of a
    NADH-dependent N-hydroxylamine reductase isolated
    from pig liver microsomes, <u>Arch. Biochem. Biophys.</u>
    169:83-92 (1974).

21. D. E. Hultquist and P. G. Passon, Catalysis of
    methemoglobin reduction by erythrocyte cytochrome
    b_5 and cytochrome b_5 reductase, <u>Nature</u> 229:252-
    254 (1971).

22. T. L. Poulos and A. G. Mauk, Models of complexes formed
    between cytochrome b_5 and subunits of
    methemoglobin, <u>J. Biol. Chem.</u> 258:7369-7373 (1983).

23. N. Borgese, G. Pietrini, and S. Gaetani, Concentration
    of NADH-cytochrome b_5 reductase in erythrocytes of
    normal and methemoglobinemic individuals measured
    with a quantitative radioimmunoblotting assay, <u>J.
    Clin. Inves.</u> 80:1296-1302 (1987).

24. H. Fukushima, G. F. Grinstead, and J. L. Gaylor, Total
    enzymic synthesis of cholesterol from lanosterol:
    cytachrome b5-dependence of 4-methyl sterol oxidase,
    <u>J. Biol. Chem.</u> 256:4822-4826 (1981).

25. P. Strittmatter, E. T. Machuga, and G. J. Roth, Reduced
    pyridine nucleotides and cytochrome b5 as electron
    donors for prostaglandin synthetase reconstituted in
    dimyristyl phosphatidylcholine vesicles, <u>J. Biol
    Chem.</u> 257:11883-11886 (1982).

26. N. Dippenaar, J. Booyens, D. Fabbri, and I. E. Katzeff, The reversibility of cancer: Evidence that malignancy in melanoma cells is gamma-linolenic acid deficiency-dependent, _S. Afr. Med. J._ 62:505-509 (1982).

27. Y. Imai, The roles of cytochrome b5 in reconstituted monooxygenase systems containing various forms of hepatic microsomal cytochrome P-450, _J. Biochem. (Tokyo)_ 89:351-362 (1981).

28. E. T. Morgan and M. J. Coon, Effects of cytochrome b5 on cytochrome P-450 catalyzed reactions, _Drug Metab. Dispos._ 12:358-364 (1984).

29. I. Jansson, P. P. Tamburini, L. V. Favreau, and J. B. Schenkman, The interaction of cytochrome b_5 with four cytochrome P-450 enzymes from the untreated rat, _Drug Metab. Dispos._ 13:453-458 (1985).

30. J. A. Peterson and R. A. Prough, Cytochrome P-450 reductase and cytochrome b_5 in cytochrome P-450 catalysis, _in_: "Cytochrome P-450. Structure, Mechanism and Biochemistry", P. R. Ortiz de Montellano, ed., Plenum Press, New York (1986).

31. P. Strittmatter and S. F. Velick, The isolation and properties of microsomal cytochrome, _J. Biol. Chem._ 221:253-264 (1956).

32. S. F. Velick and P. Strittmatter, The oxidation reduction stoichiometry and potential of microsomal cytochrome, _J. Biol. Chem._ 221:265-275 (1956).

33. P. Strittmatter and S. F. Velick, The purification and properties of microsomal cytochrome reductase, _J. Biol. Chem._ 228:785-799 (1957).

34. B. Hagihara, E. Furuyo, and T. Sugiyama, Chemical and Physical properties of cytochrome b_5, _in_: "Hepatic Cytochrome P-450 Monooxygenase System", J. B. Schenkman and D. Kupfer, eds., Pergamon Press, New York (1980).

35. P. Strittmatter and H. A. Dailey, Essential structural features and orientation of cytochrome b_5 in membranes, _in_: "Membranes and Transport Vol.1", A. N. Martonosi, ed., Plenum Press, New York (1982).

36. J. Ozols and P. Strittmatter, Correction of amino acid sequence of calf liver microsomal cytochrome b_5, _J. Biol. Chem._ 244:6617-6618 (1969).

37. L. Spatz and P. Strittmatter, A form of cytochrome b_5 that contains an additional hydrophobic sequence of 40 amino acid residues, _Proc. Nat. Acad. Sci. USA_ 65:1042-1046 (1971).

38. P. J. Fleming, H. A. Dailey, D. Corcoran, and P. Strittmatter, The primary structure of the nonpolar segment of bovine cytochrome b_5, _J. Biol. Chem._ 253:5369-5372 (1978).

39. J. Ozols, Structure of cytochrome b_5 and its topology in the microsomal membrane, _Biochim. Biophys. Acta_ 997:121-130 (1989).

40. J. Ozols, C. Gerard, and F. G. Nobrega, Proteolytic cleavage of horse liver cytochrome b_5, _J. Biol. Chem._ 251:6767-6774 (1976).

41. J. Ozols and C. Gerard, Covalent structure of the membranous segment of horse cytochrome b_5, _J. Biol. Chem._ 252:8549-8553 (1977).

42. A. Tsugita, M. Kobayashi, S. Tani, S. Kyo, M. A. Rashid, Y. Yoshida, T. Kajihara, and B. Hagihara, Comparative study of the primary structures of cytochrome b_5 from four species, Proc. Acad. Sci. USA 67:442-447 (1970).

43. K. Kondo, S. Tajima, R. Sato, and K. Narita, Primary structure of membrane binding segment of cytochrome b_5, J. Biochem. (Tokyo) 86:1119-1128 (1979).

44. F. S. Mathews, P. Argos, and M. Levine, The structure of cytochrome b_5 at 2.0 Å resolution, Cold Spring Harbor Symp. Quant. Biol. 36:387-395 (1971).

45. F. S. Mathews, E. W. Czerwinski, and P. Argos, The X-ray crystallographic structure of calf liver cytochrome b_5, in: "The Porphryrins Vol VII", D. Dolphin, ed., Academic Press, New York (1979).

46. J. Ozols and P. Strittmatter, The reactivity of lysyl residues of cytochrome b_5, J. Biol. Chem. 241: 4793-4797 (1966).

47. P. Strittmatter, The nature of heme binding in microsomal cytochrome b_5, J. Biol. Chem. 235:2492-2497 (1960).

48. T. E. Huntley and P. Strittmatter, The reactivity of the tyrosyl residues of cytochrome b_5, J. Biol. Chem. 247:4648-4653 (1972).

49. H. A. Dailey and P. Strittmatter, Modification and identification of cytochrome b_5 carboxyl groups involved in protein protein interactions with cytochrome b_5 reductase, J. Biol. Chem. 254:5388-5396 (1979).

50. M. Y. İşcan and E. Arınç, Kinetic and structural properties of biocatalytically active sheep lung microsomal NADPH-cytochrome c reductase, Int. J. Biochem. 18:731-741 (1986).

51. O. Adalı and E. Arınç, Characterization of sheep lung cytochrome P-450 isozymes 2, in: "Cytochrome P-450, Biochemistry and Biophysics", I. Schuster, ed., Taylor and Francis, London (1989).

52. O. Adalı and E. Arınç, Electrophoretic, spectral, catalytic and immunochemical properties of highly purified cytochrome P-450 from sheep lung, Int. J. Biochem. In Press.

53. S. Takesue and T. Omura, Solubilization of NADH cytochrome b_5 reductase from liver microsomes by digestion, J. Biochem. (Tokyo) 67:259-266 (1970).

54. L. Spatz and P. Strittmatter, A form of reduced nicotinamide adenine dinucleotide-cytochrome b_5 reductase containing both the catalytic site and additional hydrophobic membrane binding segment, J. Biol. Chem. 248:793-799 (1973).

55. K. Mihara and R. Sato, Partial purification of NADH-cytochrome b_5 reductase from rabbit liver microsomes with detergents and its properties, J. Biochem. (Tokyo) 71:725-735 (1972).

56. K. Mihara and R. Sato, Purification and properties of the intact form of NADH-cytochrome b_5 reductase from rabbit liver microsomes, J. Biochem. (Tokyo) 78:1057-1073 (1975).

57. D. A. Schafer and D. E. Hultquist, Purification of bovine liver microsoal NADH-cytochrome b_5 reductase using affinity chromatography, *Biochem. Biophys. Acta* 95:381-387 (1980).

58. J. Ozols, G. Korza, F. S. Heinemann, M. A. Hediger, and P. Strittmatter, Complete amino acid sequence of steer liver microsomal NADH-cytochrome b_5 reductase, *J. Biol. Chem.* 260:11953-11961 (1985).

59. T. Yubisui, T. Miyata, S. Iwanaga, M. Tamura, and M. Takeshita, Complete amino acid sequence of NADH-cytochrome b_5 reductase purified from human erythrocytes, *J. Biochem. (Tokyo)* 99:407-422 (1986).

60. T. Yubisui, Y. Naitoh, S. Zenno, M. Tamura, M. Takeshita and Y. Sakaki, Molecular cloning of cDNAs of human liver and placenta NADH-cytochrome b_5 reductase, *Proc. Natl. Acad. Sci. USA* 84:3609-3613 (1987).

61. K. Miki, S. Kaida, N. Kasai, T. Iyanagi, K. Kobayashi, and K. Hayashi, Crystallization and preliminary X-ray crystallographic study of NADH-cytochrome b_5 reductase from pig liver microsomes, *J. Biol. Chem.* 262:11801-11802 (1987).

62. P. Strittmatter, The properties of nucleotide complexes with microsomal cytochrome reductase, *J. Biol. Chem.* 234:2665-2669 (1959).

63. P. Strittmatter, The interaction of nucleotides with microsomal cytochrome reductase, *J. Biol. Chem.* 233: 748-753 (1958).

64. A. Loverde and P. Strittmatter, The role of lysyl residues in the structure and reactivity of cytochrome b_5 reductase, *J. Biol. Chem.* 243:5779-5787 (1968).

65. C. S. Hackett and P. Strittmatter, Covalent cross linking of the active sites of vesicle-bound cytochrome b_5 and NADH-cytochrome b_5 reductase, *J. Biol. Chem.* 259:3275-3282 (1984).

66. C. S. Hackett, W. B. Novoa, J. Ozols, and P. Strittmatter, Identification of the essential cysteine residue of NADH-cytochrome b_5 reductase, *J. Biol. Chem.* 261:9854-9857 (1986).

67. T. Güray, Purification and spectral and catalytical characterization of NADH-cytochrome b5 reductase from lung microsomes, Ph.D. Thesis, Middle East Technical University, Ankara (1989).

68. C. R. Kensil and P. Strittmatter, Binding and fluorescence properties of the membrane domain of NADH-cytochrome b_5 reductase, *J. Biol. Chem.* 261: 7316-7321 (1986).

69. M. J. Rogers and P. Strittmatter, The binding of reduced nicotine amide adenine dinucleotide-cytochrome b_5 reductase to hepatic microsomes, *J. Biol. Chem.* 249: 5565-5569 (1974).

70. D. E. Koppel, P. J. Fleming, and P. Strittmatter, Intramembrane positions of membrane bound chromophores determined by excitation energy transfer, *Biochemistry* 24:5450-5457 (1979).

71. P. J. Fleming, D. E. Koppel, A. L. Y. Lau, and P. Strittmatter, Intramembrane position of the fluorescent tryptophanyl resideu in membrane bound cytochromes, *Biochemistry* 24:5458-5464 (1979).

72. H. A. Dailey and P. Strittmatter, Structural and functional properties of the membrane binding segment of cytochrome b_5, J. Biol. Chem. 253:8203-8209 (1978).
73. H. A. Dailey and P. Strittmatter, The role of COOH terminal anionic residues in binding cytochrome b_5 to phospholipid vesicles and biological membranes, J. Biol. Chem. 256:1677-1680 (1981).
74. H. G. Enoch, P. J. Fleming, and P. Strittmatter, The binding of cytochrome b_5 to phospholipid vesicles and biological membranes: effect of orientation on intermembrane transfer and digestion by carboxypeptidase Y, J. Biol. Chem. 254:6483-6488 (1979).
75. C. Tanford, The Hydrophobic Effect, pp.205-211, John Wiley and Sons, New York (1980).
76. S. Tajima and R. Sato, Topological studies of the membrane binding segment of cytochrome b_5 embedded in phosphatidylcholine vesicles, J. Biochem. (Tokyo) 87:123-134 (1980).
77. H. A. Dailey and P. Strittmatter, Orientation of the carboxyl and NH_2 termini of the membrane binding segment of cytochrome b_5 on the same side of phospholipid bilayers, J. Biol. Chem. 256:3951-3955 (1981).
78. Y. Takagaki, R. Radhakrishnan, K. W. A. Wirtz, and H. G. Khorana, The membrane embedded segment of cytochrome b_5 as studied by cross linking with photoactive phospholipids. II. The nontransferable form, J. Biol. Chem. 258:9136-9142 (1983).
79. E. P. Gogol, D. M. Engelman, and G. Zaccai, Neutron diffrection analysis of cytochrome b5 reconstituted in deuterated lipid multilayers, Biophys. J. 43:285-292 (1983).
80. E. P. Gogol and D. M. Engelman, Neutron scattering shows that cyctochrome b5 penetrates deeply into the lipid bilayer, Biophys. J. 46:491-495 (1984).
81. L. M. Rzepecki, P. Strittmatter, and L. G. Herbette, X-ray diffraction analysis of cytochrome b5 reconstituted in egg phosphatidylcholine vesicles, Biophys. J. 49:829-838 (1986).
82. E. Arınç, L. M. Rzepecki, and P. Strittmatter, Topography of the C-terminus of cytochrome b5 tightly bound to dimyristoylphosphatidylcholine vesicles, J. Biol. Chem. 262:15563-15567 (1987).

ELECTRON TRANSFER FROM CYTOCHROME b5 TO CYTOCHROME P450

Claude Bonfils, Jean-Louis Saldana, Claude Balny
and Patrick Maurel

Service MPK, INSERM Unite 128
SANOFI RECHERCHE, Rue Blayac
34082 MONTPELLIER, FRANCE

INTRODUCTION

There are four electron carriers in the endoplasmic reticulum of liver cells, two flavoproteins: NADH cytochrome b5 reductase (FP1) and NADPH cytochrome P450 reductase (FP2) and two hemoproteins: cytochrome b5 and cytochrome P450. The two reductases were first evidenced in 1950 by Hogeboom and Schneider (1) by their strong cytochrome c reductase activity. They were isolated respectively by Spatz and Strittmatter (2) and Van der Hoeven and Coon (3). The cytochrome b5 was described by Strittmatter and Ball (4). It was isolated after tryptic digestion of microsomes (5) and then purified with full length protein chain by Spatz and Strittmatter (6). The last electron carrier discovered in microsomes was cytochrome P450, the CO-binding pigment evidenced by Klingenberg (7) and Garfinkel (8) in 1958.

These four components are all membrane proteins with amphipatic properties. Their protein chain presents a hydrophobic as well as a hydrophilic part (Table 1). The latter serves to anchor the protein in the membrane.

Table 1 Amphipatic nature of microsomal electron carriers

Enzymes	Molecular weight Native form	 Hydrophilic part
Cytochrome b5 (6)	16,700	11,000
NADPH cyt P450 reductase (9)	78,000	71,000
NADH cyt b5 reductase (10)	34,100	28,000

Molecular Aspects of Monooxygenases and Bioactivation of Toxic Compounds
Edited by E. Arınç *et al., Plenum Press, New York,*

In the case of cytochrome P450, there are several hydrophobic zones along the protein chain. This cytochrome is embedded in the reticulum membrane with polar segments protruding on both sides of the lipid bilayer (11).

In liver microsomes from untreated animals, the electron transferring enzymes are present at different concentrations, 0.5, 1.0 and 0.05 nmol/mg of protein for b5, P450 and the two reductases respectively (12). As electron transfer among the flavoproteins and the cytochromes requires the contact of catalytically active portions, their precise distribution within the membrane is not well known.

The microsomal electron carriers are composed of two distinct electron chains (Fig. 1). In the first chain, electrons provided by NADH are transferred to cytochrome b5 via the flavoprotein FP1. These electrons are further used by Stearyl CoA desaturase (13,14). The second electron pathway contains the flavoprotein FP2 which gives electrons from NADPH to cytochrome P450 as established by Omura and Sato (15). This last hemoprotein catalyses the hydroxylation of drugs (16,17).

NADH $\rightarrow$ FP1 $\rightarrow$ Cytochrome b5 $\rightarrow$ Stearyl CoA desaturase

NADPH $\rightarrow$ FP2 $\rightarrow$ Cytochrome P450 $\rightarrow$ Drug hydroxylation

Fig. 1. Electron chains of microsomes

Several experimental results have suggested that these two electron chains are not completely independent and that electrons from one chain may be transferred to acceptors of the second chain. First evidence was provided by observation of the synergism of NADH on hydroxylase activity as reported by Conney et al.(18). They showed that while NADH itself could only drive N-demethylation at 10-15% of the rate of NADPH, when added to the NADPH supported reaction, NADH elevated the rate more than 30%. This synergism was described on the hydroxylation pathway but did not exist on the path of fatty acyl CoA desaturase (19). This is indicative of probable electron flow from the NADH to the NADPH chain. The relationship between the two systems has been the subject of extensive work over the past two decades. The electron flow from the first electron way to cytochrome P450 via cytochrome b5 is now well documented.

ELECTRON FLOW FROM CYTOCHROME b5 TO P450

Previous experiments designed to study the relationship between the two microsomal electron chains were performed directly on microsomes. Several years later when the electron carriers were isolated and purified, these studies were completed by experiments on reconstituted systems of controled composition.

Studies on microsomes

To explain the synergism of NADH on hydroxylating reactions, Hildebrandt and Estabrook (20) attempted to show where the locus is for the effect of NADH on the overall reaction. They examined the spectrum change of microsomal pigments during the steady state of hydroxylating reaction in the presence of O_2, substrate, NADPH or NADH. Cytochrome b5 is partly reduced during the aerobic steady state after adding NADPH and NADH to the medium. Addition of substrate gives rise to a partial reoxidation of b5. This seems to indicate that during the steady state of hydroxylation, there is an acceptor for the electrons of reduced b5. In conclusion, Hildebrandt and Estabrook (20) suggest that cytochrome b5 participates directly in hydroxylations catalysed by P450 as a donor of the second electron. The first electron is given by FP2 and the second may be given either by FP2 or by b5.

The electron flow was also investigated by Noshiro and Omura (21) by measuring the inhibiting effect of antibodies on NADH supported hydroxylations. Anti FP1 antibodies have a weak effect, anti FP2 antibodies block the hydroxylation of the four substrates used, and anti cytochrome b5 antibodies have a partial effect on only two substrates. In another work, these authors (22) measured antibody action on NADH as well as NADPH supported reactions on microsomes from untreated and treated animals. They concluded that FP2 participate in NADH as well as NADPH supported hydroxylations, flavoprotein FP1 is mainly involved in NADH supported reactions and cytochrome b5 is an intermediate in both electron pathways, but this depends upon the type of substrate and the type of cytochrome P450 (Fig. 2).

Microsomes are complex mixtures of lipids and proteins. It is not easy to deduce from spectral measurements what the precise role of each component is. Furthermore, it cannot be assumed that antibodies have free acces to their target

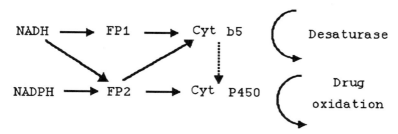

Fig. 2.   Interrelations between NADH and NADPH electron pathways

proteins.Other experiments have been conducted in many laboratories on reconstituted hydroxylating systems composed of a limited number of electron carriers.

Experiments on Reconstituted Systems

Drug hydroxylations. Hydroxylation of benzphetamine by a reconstituted system composed of P450 LM2, FP2 and micellar lipids was studied by Imai and Sato (23). Effect of b5 on NADH, NADPH and O_2 consumption as well as formaldehyde production was measured. The presence of b5 improves the coupling of NADPH oxidation to N-demethylation. The stoichiometry is 1/1 for the ratio NADPH/ formaldehyde and 1/2 for the ratio NADH/formaldehyde. The supply of at least one electron to cytochrome P450 from NADPH via FP2 is absolutely required for the demethylase activity. It can also be concluded that the second electron is introduced almost exclusively from NADH via FP1 and b5 under the conditions employed. However, since NADPH dependent demethylation can take place in a system in which no electron flow from NADH exists, it is certain that the second electron can also be supplied from NADPH via FP2. Cytochrome b5 is reducible by NADH and FP1 much faster than by NADPH and FP2.

In similar experiments, Ingelman-Sundberg and Johansson (24), using a reconstituted system composed of FP2, P450 LM2, and lipid vesicles , studied the hydroxylation of 7-ethoxycoumarin and p-nitroanisole. They observed that b5 enhanced the rate of hydroxylation about five fold. The increase in product formation was found to be correlated with the concomitant decrease in the production of H_2O_2 and O_2^- which are by-products of the hydroxylation reaction. As there is also no effect of b5 on the reduction of oxidized cytochrome P450 (first electron), they concluded that b5 is a more efficient donor of the second electron to P450 than is FP2.

The role of cytochrome b5 as electron carrier received confirmation from the work of Sugiyama et al.(25,26). They isolated from rabbit liver microsomes a form of cytochrome P450 with a high affinity for cytochrome b5. In a reconstituted system composed of this cytochrome, FP1, FP2 and detergent micelles, omission of cytochrome b5 completely abolishes p-nitroanisole demethylation. The maximal activity was attained when the system contained P450 and b5 at a molar ratio of 1/1. Trypsin digestion abolished the capacity of this cytochrome to reconstitute the demethylase activity. In this example, the second electron needed for demethylation is donated to cytochrome P450 only by means of intact cytochrome b5.

All the experiments in the reconstituted systems did not lead to such obvious results (Table 2). In some observations, cytochrome b5 has been shown without effect or with an inhibiting effect on hydroxylase reaction. Bösterling and Trudell (31) pointed out that the effect of b5 may be modulated both by lipids and reductase FP2. They carefully studied the effect of b5 on benzphetamine N-demethylation in reconstituted systems containing different ratio of FP2 versus P450. Moreover,they studied this effect in two lipidic

Table 2. Effect of b5 on drug hydroxylation by reconstituted systems

Authors	Isozymes	Lipids	Substrate	Effect
Lu et al. (27)	P450 phenobarbital treated rats	micelles	benzphetamine	inhibition
	P448 methylcholanthrene treated rats	micelles	benzpyrene	no effect
Lu and Levin (28)	P450 phenobarbital treated rats	micelles	chlorobenzene	increase
	P448 methylcholanthrene treated rats	micelles	chlorobenzene	no effect
Imai and Sato (23)	P450 phenobarbital treated rabbits	micelles	benzphetamine	weak incease
		micelles	N-Ndimethylaniline	increase
Sugiyama et al. (29)	P450 high affinity for b5 , rabbit liver	micelles	p-nitroanisole	obligatory
Ingelman-Sundberg and Johansson (24)	P450 LM2 rabbit liver	vesicles	7-ethoxycoumarin	increase
		vesicles	p-nitroanisole	increase
Morgan et al. (30)	P450 LM2	micelles	acetanilide	increase
	P450 LM3b	micelles	acetanilide	no effect
	P450 LM4	micelles	acetanilide	increase

media: micellar lipids and vesicular lipids. The effect of b5 is not the same in micelles and in vesicles . In micelles, b5 causes both activation and inhibition of N demethylation or has no effect depending upon the ratio b5/P450. Conversely, in vesicles which contain a higher proportion of phospholipid to protein, b5 increases the hydroxylase activity, progressively. The effect of b5 is largely dependent upon the phospholipid organization. The influence of reductase is also important. The stimulatory ability of b5 as well as the efficiency of the electron transfer are greatest at low reductase/P450 ratios (as found in microsomes) and are lowest at high reductase/P450 ratios. This dependence was observed both in vesicle and micelle reconstituted systems.

Care must be taken before drawing general conclusions from the preceding results, because reconstituted systems are not exact reproductions of the native microsomal membrane. However, it seems that, when the ratio of the different electron carriers approaches those found in microsomes, a stimulatory effect of b5 is observed. This role of b5 may be explained either by its ability to transfer electrons or by an effector role on the hydroxylase system. This last point has been investigated by Hlavica (32). The catalytic function of P450 was tested on cumene hydroperoxide sustained N-oxidation of 4-chloroaniline. When an artificial oxygen donor substitutes NADPH, FP2 and O_2, the electron transfer usually functional in the monooxygenation is not operative. Hence, any cytochrome b5 induced change can be attributed to an action of the hemoprotein , distinct from its role as an electron carrier. As reported by Hlavica, presence of b5 causes a significant decrease in apparent Km of P450 catalysed hydroxylation of 4-chloroaniline (from 33 to 18 mM) without changing the Vmax. Increased affinity of 4-chloroaniline for P450 in the presence of cytochrome b5 was also observed by measurement of the high spin/low spin equilibrium.

These results suggest that cytochrome b5 exerts a control over hydroxylase activity at more than one level. Apart from its well known function in microsomal electron transfer, this pigment might regulate substrate binding to P450 by inducing a conformational change of the oxidase through protein-protein interaction.

Kinetic measurements on reconstituted systems. Taniguchi et al. (33) incorporated FP1, FP2, b5 and P450 from phenobarbital treated rabbits into liposomes by the cholate dialysis method. They studied

Table 3   Electron transfer rates between microsomal electron carriers Data compiled from Taniguchi et al. (33)

Electron pathway	Reduction rate	
	NADH as electron source	NADPH as electron source
FP1 ——————> b5	30 s-1	n.d.
FP2 ——————> b5	0.15 s-1	0.47 s-1
FP2 ——————> P450	0.50 s-1	1.1 s-1
b5 ——————> P450	0.046 s-1	0.046 s-1

electron transfer from NADH and NADPH to the four microsomal carriers by the stopped flow method. From the values of the rate constants (Table 3), it may be deduced that electrons are mostly supplied from NADH to cytochrome b5 via FP1 and from NADPH to b5 via FP2. This latter flavoprotein can utilize not only NADPH but also NADH as electron donor. When b5 and P450 are present, there is a simultaneous reduction of the two cytochromes by FP2. Moreover, there is also electron transfer between reduced b5 and oxidized P450, but this flow

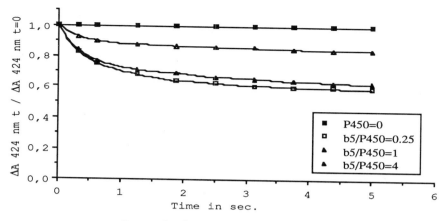

Reoxydation of reduced b5

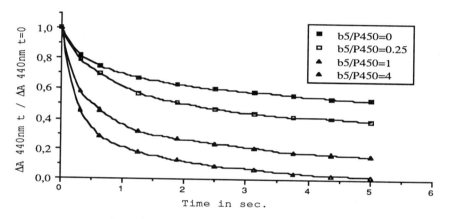

Reoxidation of P450 oxygenated complex

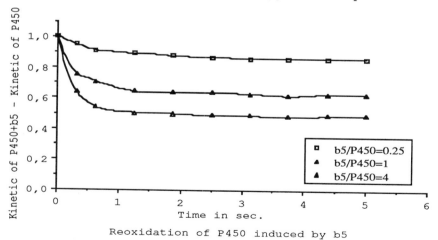

Reoxidation of P450 induced by b5

Fig. 3. Kinetics of reoxidation of P450 and b5 for various molar ratios

is very low compared to the reduction of cytochrome P450 by FP2. In conclusion, the first of the two electrons required for the monooxygenase reaction is supplied exclusively from FP2 with either NADPH or NADH as electron donor.

In our laboratory, we decided to follow the transfer of the second electron (34). We measured electron flow between reduced b5 and P450 2^+-O_2 (oxygenated complex) by the stopped flow method. The two cytochromes were added in vesicles of dilauroyllecithine prepared by cholate-gel filtration technique. The stopped flow apparatus we used has been built in our laboratory and allows us to reduce the two cytochromes directly in the first syringe. Cytochrome P450 and b5 are deprived of oxygen by bubbling argon then are photoreduced by the acrydine orange-methyl viologen system. The second syringe contains an oxygenated buffer. When the content of the two syringe is mixed, there is a rapid formation of P450 2^+-O_2, then reoxidation of b5 2^+ and P450 2^+-O_2. This reaction may be conducted in the presence or absence of substrate. It is possible to follow the kinetics of the two reactions by recording absorbance decrease at 424 nm for b5 and 440 nm for P450. Results are reported in Fig.3. Reoxidation of b5 alone is slow and monophasic. In the presence of P450, its reoxidation becomes biphasic with a rapid first phase. Conversely, reoxidation of the oxygenated complex of P450 alone is multiphasic. When b5 is added, the percentage of the first phase increases as b5 increases. It seems there are two phenomena : a spontaneous reoxidation of P450 oxygenated complex and a b5 induced reoxidation. By subtracting the two kinetics, we obtain the b5 induced kinetic which is monophasic. We observe in Table 4 that there is a close correlation in rate constants between the fast phase of reoxidation of b5 and the b5 induced reoxidation of P450 oxygenated complex.

Table 4 . Rate constant for b5-induced P450 reoxidation and b5 reoxidation at various molar ratios of b5/P450.

CONDITIONS	NO SUBSTRATE		1mM BENZPHETAMINE	
	P450	b5	P450	b5
b5/P450=0.25	k=2.1±0.2	kf=1.8±0.1	k=3.7±0.4	kf=3.7±0.2
b5/P450=1.0	k=2.4±0.2	kf=2.4±0.1	k=4.0±0.4	kf=4.5±0.3
b5/P450=4.0	k=2.5±0.2	kf=1.9±0.1	k=6.2±0.5	kf=7.7±0.4

At the end of this fast phase, the quantity of reoxidized b5 and the quantity of reoxidized P450 (Table 5) are closely related. Neither H_2O_2 nor O_2^- are able to produce the rapid reoxidation of b5. We may conclude that the rapid phase is due to a direct electron transfer from reduced b5 to the oxygenated complex of P450. As all the reducing equivalents present on b5 are not transferred to the oxygenated complex, we deduce that in the medium of the reaction, there is an equilibrium between free b5, free P450 and a b5-P450 1/1 complex.

EVIDENCE OF A 1/1 COMPLEX BETWEEN b5 AND P450

The presence of such a complex has been deduced from different experimental results. By using a two phase partition technique, Chiang (35) investigated interaction between b5 and cytochrome P450 (LM2 and LM4). In a dextran polyethyleneglycol two phase system, P450 is almost exclusively concentrated in the dextran rich bottom phase whereas b5 is distributed in the polyethyleneglycol top phase. After mixing P450 with b5, the partition coefficient change is indicative of a molecular interaction between the two cytochromes. The stoichiometry is about one, and only intact b5 is capable of interacting with P450.

Interaction between P450 and b5 has also been studied by the measure of spin equilibrium. Cytochrome b5 interacts with P450 by displacing this last cytochrome from low spin to high spin state (type I spectral change). Tamburini and Gibson (36) recorded spectral changes of rat P450 (PB-B) upon addition of b5. The plot of the magnitude of type I spectral change versus cytochrome b5 concentration results in anapparently hyperbolic relationship from which it is observed that these proteins interact to form a 1/1 complex.

Table 5 . Amount of P450 and b5 reoxidizing simultaneously at various molar ratios of b5 / P450.

b5/P450	P450 (μM)	b5 (μM)
NO SUBSTRATE		
1.32 : 5.18	0.72 ±0.12	0.62 ±0.09
5.36 : 5.45	2.01 ±0.30	1.72 ±0.26
10.25 : 2.61	1.32 ±0.20	1.53 ±0.22
1mM Benzphetamine		
1.30 : 5.13	0.92 ±0.14	0.85 ±0.13
2.36 : 2.20	0.92 ±0.14	1.02 ±0.15
10.37 : 2.01	1.30 ±0.20	1.76 ±0.30

We obtained the same results with rabbit P450 LM2 and b5 (34). These observations were completed by Jansson et al.(37) who reported the interaction of b5 with four P450 from untreated rats. Effect of b5 is of a different amplitude according to the type of cytochrome P450 isozyme. Cytochrome b5 causes a major low to high spin transition with RLM5 and RLM5a. In contrast, RLM2 and RLM3 are minimally influenced by b5. It is clear that interactions with b5 and formation of a 1/1 complex are largely dependent on the type of P450.

Another way to detect the presence of an intermolecular complex between proteins is the use of cross linking agent like carbodiimide. By this technique, Nisimoto (38-39) has evidenced a 1/1 complex in liposomes containing b5 and FP2. Tamburini and Shenkman (40), using a similar method, detected a complex in a mixture of P450 LM2 and cytochrome b5. They isolated and purified this complex. In acrylamide gel, it migrates as homogeneous protein band and exhibits a molecular weight of 68,000 Da. Consequently, this covalent complex is a heterodimer composed of one mole of P450 LM2 and one mole of b5. Moreover, several experimental results demonstrate that this complex fully retains its functional properties. The Kd of the complex with benzphetamine is identical to that observed with P450 LM2 and saturating amount of b5. The Km for FP2 in supporting either LM2 or LM2-b5 substrate hydroxylation is the same. These experiments provide a definitive proof of the presence of P450-b5 functional complex which has been previously hypothetized in kinetic experiments.

CONCLUSION

There are two electron chains in the endoplasmic reticulum of hepatocytes, one is implicated in fatty acid desaturation and the second in monooxygenation of xenobiotics. Reducing equivalents are not strictly funneled along each chain and electrons may be transferred from one pathway to the other.

Studies conducted on microsomes have suggested that, among the two electrons needed for monooxygenase reaction, one is given by flavoprotein FP2 and the second may be given either by cytochrome b5 or by FP2.

Electron transfer between cytochrome b5 and cytochrome P450 has been investigated in reconstituted systems which revealed that b5 may exert, in addition to transferring reducing equivalents to P450, an effector role in monooxygenation by inducing conformational change of P450. Kinetics measurements have also confirmed with at least one P450 isozyme that b5 may serve as donor for the second electron in an equimolecular complex of the two hemoproteins. This last hypothesis has received recent confirmation from purification and characterization of a covalent complex of P450 (LM2) and b5. However, this ability of b5 to form a functional complex with P450 is probably dependent upon the presence of a binding site on P450. This may mean that b5 as electron donor is not a general pathway with all the P450 isozymes present in hepatocytes.

REFERENCES

1. G.H. Hogeboom and W.C. Schneider,Cytochemical studies of mammalian tissues. III. Isocitric dehydrogenase and triphosphopyridine nucleotide-cytochrome c reductase of mouse liver, J.Biol.Chem., 186, 417 (1950).
2. L. Spatz and P. Strittmatter, A form of reduced nicotinamide adenine dinucleotide-cytochrome b5 reductase containing both the catalytic site and an additional hydrophobic membrane-binding segment, J. Biol.Chem., 248, 793 (1973).
3. T. A. Van der Hoeven and M. J. Coon, Preparation and properties of partially purified cytochrome P450 and reduced nicotinamide adenine dinucleotide phosphate-cytochrome P450 reductase from rabbit liver microsomes, J.Biol.Chem., 249:6302 (1974).
4. C.F. Strittmatter and E.G. Ball, A hemochromogen component of liver microsomes , Proc.Natl.Acad.Sci.USA, 38, 19 (1952)
5. P. Strittmatter and J. Ozols, The restricted tryptic cleavage of cytochrome b5, J.Biol.Chem., 241, 4787 (1966).
6. L. Spatz and P. Srittmatter, A form of cytochrome b5 that contains an additional hydrophobic sequence of 40 amino acid residues, Proc.Natl.Acad.Sci. USA, 68:1042 (1971).
7. M. Klingenberg, Pigments of rat liver microsomes, Arch.Biochem. Biophys., 75: 376 (1958).
8. D. Garfinkel, Studies on pig liver microsomes, I. Enzymatic and pigment composition of different microsomal fractions, Arch.Biochem.Biophys., 77: 493 (1958).
9. Y. Yassukochi and B.S.S. Masters, Some properties of a detergent solubilized NADPH-cytochrome c (cytochrome P450) reductase purified by biospecific affinity chromatography, J.Biol.Chem., 251:5337 (1976).
10. J. Ozols, G. Korza, F.S. Heinemann, M.A. Hediger and P. Strittmatter, Complete amino acid sequence of steer liver microsomal NADH cytochrome b5 reductase, J.Biol.Chem., 260:11953 (1985).
11. S.D. Black and M.J. Coon, P450 Sructure and Function, in:"Advances in Enzymology", A. Meister, ed., John Willey, N.Y., 60:35 (1987).
12. R. Sato and T. Omura, "Cytochrome P450", Kodansha, Tokio, Academic press, New-York, San Francisco, London (1978).
13. P.W. Holloway and J.T. Katz, A requirement for cytochrome b5 in microsomal stearyl coenzyme A desaturation, Biochemistry, 11:3689 (1972).
14. T. Shimakata, K. Mihara and R. Sato, Reconstitution of hepatic microsomal stearoyl coenzyme A desaturase system from solubilized components, J.Biochem., 72:1163 (1972).
15. T. Omura and R. Sato, A new cytochrome in liver microsomes, J.Biol.Chem., 237:1375 (1962).
16. D.Y. Cooper, S. Levin, S. Narasimhulu, O. Rosenthal and R.W. Estabrook, Photochemical action spectrum of the terminal oxidase of mixed function oxidase systems, Science, 147:400 (1965).

17. A.Y.H. Lu and M.J. Coon, Role of hemoprotein P450 in fatty acid w-hydroxylation in a soluble enzyme system from liver microsomes, J.Biol. Chem., 243:1331 (1968).
18. A.H. Conney, R.R. Brown, J.A. Miller and E.C. Miller, The metabolism of methylated aminoazodyes.VI. Intracellular distribution and properties of the demethylase system, Cancer Res., 17:628 (1957).
19. I. Jansson and J.B. Schenkman, Studies on three microsomal electron transfer enzyme systems. Specificity of electron flow pathways, Arch.Biochem.Biophys., 178:89 (1977).
20. A. Hildebrandt and R.W. Estabrook, Evidence for the participation of cytochrome b5 in hepatic microsomal mixed function oxidation reactions, Arch. Biochem. Biophys., 143:66 (1971).
21. M. Noshiro and T.Omura, Immunochemical study on the electron pathway from NADH to cytochrome P450 of liver microsomes, J.Biochem., 83:61 (1978).
22. M. Noshiro, N. Harada and T. Omura, Immunochemical study on the route of electron transfer from NADH and NADPH to cytochrome P450 of liver microsomes, J.Biochem., 88:1521 (1980).
23. Y. Imai and R. Sato, The roles of cytochrome b5 in a reconstituted N-demethylase system containing cytochrome P450, Biochem.Biophys.Res.Commun., 75:420 (1977).
24. M. Ingelman-Sundberg and I. Johansson, Cytochrome b5 as electron donor to rabbit liver cytochrome P450 LM2 in reconstituted phospholipid vesicles, Biochem.Biophys. Res.Commun., 97, 582 (1980).
25. T. Sugiyama, N. Miki and T. Yamano, NADH and NADPH dependent reconstituted p-nitroanisole O-demethylation system containing cytochrome P450 with high affinity for cytochrome b5, J.Biochem., 87, 1457 (1980).
26. T. Sugiyama, N. Miki, Y. Miyake and T. Yamano, Interaction and electron transfer between cytochrome b5 and cytochrome P450 in the reconstituted p-nitroanisole O-demethylase system, J.Biochem., 92, 1793 (1982).
27. A.Y.H. Lu, S.B. West, M. Vore, D. Ryan and W. Levin, Role of cytochrome b5 in hydroxylation by a reconstituted cytochrome P450-containing system, J.Biol. Chem., 249, 6701 (1974).
28. A.Y.H. Lu and W. Levin, Liver microsomal electron transport systems. III. The involvement of cytochrome b5 in the NADPH supported cytochrome P450 dependent hydroxylation of chlorobenzene, Biochem.Biophys. Res.Commun., 61, 1348 (1974).
29. T. Sugiyama, N. Miki and T. Yamano, The obligatory requirement of cytochrome b5 in the p-nitroanisole O-demethylation reaction catalysed by cytochrome P450 with a high affinity for cytochrome b5, Biochem.Biophys.Res. Commun., 90, 715 (1979).
30. E.T. Morgan, D.R. Koop and M.J. Coon, Heme requirement in the effects of cytochrome b5 on catalytic activity of rabbit liver cytochrome P450, Fed.Proc., 40, 697 (1981).
31. B. Bösterling, J.R. Trudell, A.J. Trevor and M. Bendix, Lipid-protein interactions as determinants of activation or inhibition by cytochrome b5 of cytochrome P450 mediated oxidations, J.Biol.Chem., 257, 4375 (1982).

32. P. Hlavica, On the function of cytochrome b5 in the cytochrome P450 dependent oxygenase system, Arch.Biochem.Biophys., 228, 600 (1983).

33. H. Taniguchi, Y. Imai and R.Sato, Role of the electron transfer system in microsomal drug monooxygenase reaction catalysed by cytochrome P450, Arch.Biochem.Biophys., 232, 585 (1984).

34. C. Bonfils, C. Balny and P. Maurel, Direct evidence for electron transfer from ferrous cytochrome b5 to the oxyferrous intermediate of liver microsomal cytochrome P450 LM2, J.Biol.Chem., 256, 9457 (1981).

35. J.Y.L. Chiang, Interaction of purified microsomal cytochrome P450 with cytochrome b5, Arch.Biochem.Biophys., 211, 662 (1981).

36. P.P. Tamburini and G.G. Gibson, Thermodynamic studies of the protein-protein interactions between cytochrome P450 and cytochrome b5, J.Biol.Chem., 258, 13444 (1983).

37. I. Jansson, P.P. Tamburini, L.V. Favreau and J.B. Schenkman, The interaction of cytochrome b5 with four cytochrome P450 enzymes from the untreated rat, Drug Metab.Dispos., 13, 453 (1985).

38. Y. Nisimoto and J.D. Lambeth, NADPH cytochrome P450 reductase-cytochrome b5 interactions: crosslinking of the phospholipid vesicle-associated proteins by a water-soluble carbodiimide, Arch.Biochem.Biophys., 241, 386 (1985).

39. Y. Nisimoto and H. Otsuka-Murakami, Cytochrome b5, cytochrome c, and cytochrome P450 interactions with NADPH-cytochrome P450 reductase in phospholipid vesicles, Biochemistry, 27, 5869 (1988).

40. P.P. Tamburini and J.B. Schenkman, Purification to homogeneity and enzymological characterization of a functional covalent complex composed of cytochrome P450 isozyme 2 and b5 from rabbit liver, Proc.Natl.Acad. Sci.USA, 84, 11 (1987).

FUNCTIONAL ASPECTS OF PROTEIN-PROTEIN INTERACTIONS OF CYTOCHROME P450

CYTOCHROME b5 AND CYTOCHROME P450 REDUCTASE

John B. Schenkman

Department of Pharmacology
University of Connecticut Health Center
Farmington, CT   06032, USA

INTRODUCTION

While cytochrome P450 has only been known for a relatively short time (1), and its role in xenobiotic metabolism has been known for an even shorter period (2), cytochrome b_5 has been known for a much longer time (see review, 3). Both of these hemoproteins are present in the endoplasmic reticulum of a number of tissue, particularly the liver. Only in the last two decades has cytochrome b_5 been found to play a role in liver microsomal monooxygenations (4-6). The original report of cytochrome b_5 involvement in P450 monooxygenations (7) was an attempt to explain NADH synergism of NADPH-supported drug oxidations (8). Subsequent studies including those discussed in this lecture, have resulted in implication of cytochrome b_5 in P450 monooxygenations, but do not explain the earlier reported NADH-synergism.

Cytochrome b5 in P450 monooxygenations

A sketch of the cytochrome P-450 cycle as it is currently believed to function is shown in Figure 1. To this we have added NADH cytochrome b_5 reductase (F_{PD}) and cytochrome b_5 (b_5) to indicate how cytochrome b_5 was initally suggested to function in the monooxygenase reaction. It was suggested to provide the second, rate-limiting, electron for the monoxygenase reaction (7). The second electron was suggested to be derived from either F_{PD} or from NADPH-cytochrome P450 reductase (F_{PT}). In fact, NADH-synergism did not occur in the reconstituted system consisting of F_{PT} plus P450 and cytochrome b_5 in the reconstituted system in dilauroyl phosphatidylcholine vesicles to which F_{PD} was added (I. Jansson and J.B. Schenkman, unpublished data). Nevertheless, cytochrome b_5 alone is capable of stimulating the NADPH-supported monooxygenase reaction, as discussed below.

The P450 monooxygenase cycle (Fig. 1) consists of the ferric P450 monooxygenase ($m°$), in an equilibrium between low spin (lS) and high spin (hS) configurations. Most (but not all) of the P450 forms are primarily in the low spin configuration. Substrate interaction frequently increases the proportion of high spin configuration ($m°s_{hS}$) due to the greater affinity of substrate for the high spin state (9). The high spin configuration has been shown to undergo more rapid reduction than the low spin configuration (10), due to a more favorable redox potential (9). In

Molecular Aspects of Monooxygenases and Bioactivation of Toxic Compounds
Edited by E. Arınç *et al., Plenum Press, New York,*

185

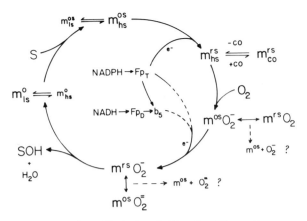

Figure 1.   Sketch of the P-450 cycle.

the cycle, oxygen is shown binding to the reduced hemoprotein
(m^{rs}) to yield the oxyferrous state (m^{rs} O_2), and with a subsequent
input of a second electron, the peroxy state is reached (m^{os} $O_{\bar{2}}$).
This form then disproportionates to water and a state of iron-bound
atomic oxygen, believed to be $Fe^V=O$, which allows the oxygen atom to
become incorporated into the substrate, thereby yielding the product.   As
reported earlier (11), some leakage of partially reduced oxygen occurs
(uncoupling), and this eventually appears as hydrogen peroxide, or as
water (11, 12).   Cytochrome b_5 has been shown to decrease the amount of
hydrogen peroxide formed while increasing the amount of product formed
(termed metabolic switching), when added to the reconstituted system (11,
12).

      In the reconstituted monooxygenase system, the effect of cytochrome
b_5 appears to depend upon both the form of cytochrome P450 used and
upon the substrate being oxidized (4-6, 11).   Not all forms of cytochrome
P450 are influenced by cytochrome b_5.   For example, some forms isolated
from rat liver microsomes by our laboratory, RLM3 and RLM2, are
unaffected by the presence of cytochrome b_5 with a number of
substrates, including drugs, chemicals and endogenous compounds such as
testosterone (6, 11).   In contrast, RLM5 is stimulated about 25% in
metabolism of certain drug substrates and about 35% in metabolism of
testosterone, and 3.5 fold in metabolism of p-nitroanisole by cytochrome
b_5.   Another form of cytochrome P450, RLM5a, showed a doubling of
activity toward the same substrates, and a 14-fold stimulation of
p-nitroanisole dealkylation by cytochrome b_5 (6).   The stimulatory
influence of cytochrome b_5 was shown to be due to an influence on the
V_{max} and not to the K_m for substrate (6).

Cytochrome b_5 interaction with cytochrome P450

      Attempts were made to determine the effect of cytochrome b_5 on
cytochrome P-450.   The addition of cytochrome b_5 to cytochrome P450
causes a shift in the spin equilibrium of the latter toward the high spin
configuration (Fig. 2) as reported earlier (13), and similar to the shift
seen on addition of certain substrates of the monooxygenase (14).   The
addition of substrate after cytochrome b_5 caused a further increase in

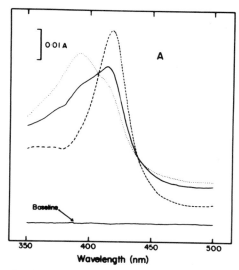

Figure 2. Spectral observation of spin equilibrium changes caused by addition of cytochrome b_5 (solid line) and subsequent addition of benzphetamine (dotted line) to P450 LM2 (dashes). From (13) with permission.

the proportion of high spin cytochrome P450 (Fig. 2). It doesn't matter which is added first; the net spin shift is the same when both perturbants are present with the cytochrome P450. The effect of cytochrome b_5 on the spin equilibrium of different cytochromes P450 varied with the forms examined (6); RLM5 underwent the greatest extent of spin shift while RLM2 barely showed a spectral shift indicative of spin equilibrium change. Further, the affinity of the different cytochromes P450 for cytochrome b_5 differed several fold.

Cytochrome b_5 is known to interact with its redox partners by a complementary charge pairing involving its highly conserved anionic amino acid residues, glutamate and aspartate. For example, its interaction with cytochrome c involves pairing with highly conserved cationic residues on that protein (15). Similarly, interaction between cytochrome b_5 and NADH-cytochrome b_5 reductase (F_{pD}) was shown to involve a charge pairing between cationic amino acids on the F_{pD} and anionic amino acids residues on the cytochrome b_5 (16). When the structure of cytochrome b_5 is viewed by molecular 3-dimensional imaging, it can be seen that around the exposed heme edge are 4 or 5 anionic amino acid residues. These form a mouth-like ring around the exposed heme with its tongue-like propionic acid residue protruding. Several of these conserved anionic residues, Glu-41, Glu-42, Glu-47, Glu-48, Glu-52, Glu-60, on cytochrome b_5 have been most frequently suggested to participate in complementary charge pairing reactions with redox partners (15, 17). In its interaction with cytochrome P-450 cytochrome b_5 also utilizes a complementary charge pairing mechanism (Table 1). As seen below, increases in ionic strength caused increases in the dissociation constant for the binding of cytochrome b_5 to RLM5 (18). The binding was monitored by the changes in spin equilibrium between 417nm (low spin) and 395nm (high spin), as shown in Figure 2. The increase in the dissociation constant is a clear indication of charge pairing, where the cations and anions of the salt insulate the protein charged residues from each other.

Table 1. The effect of ionic strength on spectral interactions between cytochrome b_5 and RLM5 at pH 7.5

Sodium phosphate (mM)	Conductivity (m , 25°C)	Dissociation constant (nM)
5.6	0.36	63
50.0	2.53	358
100.0	4.80	517
189.0	8.92	2322

In an attempt to gain an understanding of the interaction between cytochromes P450 and b_5 we resorted to chemical modification of anionic and cationic residues on the cytochromes, and examined the influences on the spectral binding constants, the dissociation constant (Kd) and maximal spectral change (A_{max}) from titrations (13). In Table 2 is shown the effect of selective neutralization of cytochrome b_5 charges on the ability to bind to rabbit LM2 and rat RLM5. Neutralization of cationic residues by acetylation (Acet b_5), if anything, lowered the dissociation constants for the interactions. In contrast, with both forms of P450, neutralization of carboxyl charges on cytochrome b_5 by methylamidation (MAb$_5$) resulted in almost complete loss of spectral changes such that Kd determinations were not possible (ND).

As shown in Table 2, cytochrome b_5 binding affinity to cytochromes P450 is intensified by prior addition of the substrate benzphetamine (BZP); the dissociation constant drops from 250nM to about 80nM with RLM5 and from 540nM to 80nM with LM2. The binding affinity became so tight in the presence of cytochrome b_5 that it was necessary to do iterative calculations correcting for bound cytochrome b_5 to obtain the apparant dissociation constants. Even as benzphetamine increased the affinity for cytochrome b_5, the presence of the latter hemoprotein when titrating P450 with benzphetamine decreased the dissociation constants for the substrate, from 3 to 7 fold (expt. 5 vs 6). Neutralization of cationic charges on the cytochrome b_5 were without influence on the K_d for benzphetamine with either P450 form, but neutralization of anionic charges on the cytochromes P450 resulted in loss of the increase in substrate affinity (Table 2). Loss of as few as a mean of 4.7 of the 35 mols of anionic charges per mol of cytochrome b_5 resulted in loss of the ability of b_5 to bind to P450 (13), as judged by the ability to perturb the spin equilibrium.

Table 2. Influence of chemical modification of cytochrome b_5 on binding to P450

Expt.	Titrant	Addition to	Spectral binding parameters RLM5		LM2	
			K_d	ΔA_{max}	K_d	ΔA_{max}
			μM	μM^{-1} P450	μM	μM^{-1} P450
1	b_5	P450	0.25	0.044	0.54	0.055
2	Acet b_5	P450	0.16	0.036	0.11	0.043
3	MAb$_5$	P450	ND	0.006	ND	<0.002
4	b_5	P450 + BZP	(0.08)	0.017	(0.08)	0.033
5	BZP	P450	66	0.033	139	0.055
6	BZP	P450 + b_5	20	0.014	20	0.034
7	BZP	P450 + Acetb$_5$	19	0.022	17	0.034
8	BZP	P450 + MAb$_5$	60	0.017	126	0.046

The effect of neutralization of anionic residues of cytochrome b_5 on substrate turnover is shown in Table 3. In the complete reconstituted monooxygenase system both LM2 and RLM5 metabolize p-nitroanisole at a fairly slow rate. The addition of equimolar cytochrome b_5 stimulates these turnovers by about one order of magnitude. Neutralization of cytochrome b_5 cationic residues did not appreciably influence product

Table 3. The effect of chemical modification of charged groups of cytochrome b_5 on the ability to stimulate P450 turnover of p-nitroanisole

Additions	mol COOH modified per mol b_5	P-nitroanisole turnover	
		RLM5	LM2
		(nmol/nmol P450/min)	
none		0.04	0.29
b_5	0	0.53	2.19
Acetb$_5$	0	0.46	1.69
MAb$_5$	3.9	0.13	0.83
MAb$_5$	8.8	0.02	0.15

formation. However neutralization of as few as 3.9 mols of carboxyl residues by methylamidation caused very appreciable losses in the stimulation afforded by cytochrome b_5. Further neutralization of carboxyl residues to a mean of 8.8 not only caused a complete loss of the stimulation, but even reduced activity to below that in the absence of cytochrome b_5. This latter affect is attributed (see below) to competition between cytochromes b_5 and P450 for the reductase in the reconstituted system.

Interaction between F_{PT} and P450

Cytochrome P450 interaction with NADPH-cytochrome P450 reductase (F_{PT}) also appears to be by a charge pairing mechanism. When oxidized P450 was titrated with oxidized F_{PT}, fairly large shifts in the spin equilibrium of RLM5 were seen, and considerably smaller shifts were seen with LM2 (Fig. 3); interestingly, at least two forms of P450, RLM2 and RLM3, did not undergo spin shifts on addition of F_{PT}. Half-maximal binding with RLM5 was about 8μM. Such studies, while suggesting functional interaction between cytochrome P450 and F_{PT} is similar to that between cytochromes P450 and b_5, do not take into account that functional interactions involves an oxidized cytochrome P450 and a reduced F_{PT}.

The interaction between cytochrome P50 and reduced cytochrome P450 reductase involves biphasic reduction kinetics (20). The fast or burst phase of reduction has been related to the amount of high spin cytochrome P450 (9, 10, 21). This fast phase also relates to the turnover of substrates. When anionic residues of P450 reductase were neutralized with methylamine, as few as a mean of 5mol of carboxyl groups neutralized per mol of F_{PT} were enough to decrease the rate constant for reduction by about half, and by 8 residues neutralized only 30% of the rate was obtained (19). Interestingly, the rate constant of the slow phase of P450 reduction was unchanged, indicating the slow phase of electron transfer is not controlled by charge pairing interaction. F_{PT} has 96 carboxylic amino acid residues per molecule (22), so the important neutralized residues are quite accessible to the aqueous medium as might be expected for charge pairing interactions.

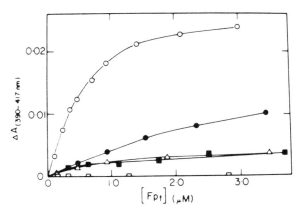

Figure 3. Effect of F_{PT} addition on the spin equilibrium of different forms of P450. O = RLM5; $\bullet$ = LM2; Δ = RLM3; $\blacksquare$ = RLM5A and $\square$ = RLM2. P450 = 0.56µM.

Examination of the affect of reagents specifically influencing charged amino acid residues on interaction of F_{PT} with redox partners revealed (19) substitution of the cationic amino group of lysine by amidination for a bulky amidine group did not impair reduction of either cytochromes c or b_5, or substrate metabolism by several forms of P450. In contrast, carboxyl group neutralization on the F_{PT} severely impaired substrate metabolism by three different forms of P450 (Fig. 4).

Testosterone metabolism by RLM3, benzphetamine metabolism by LM2 and 7-ethoxycoumarin dealkylation by LM2, RLM3 and RLM5 were all severely impaired by as little as 5 mols of carboxyl neutralized per mol of F_{PT} (Fig. 4).

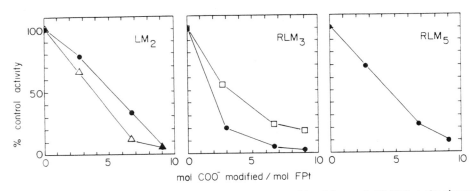

Figure 4. Influence of carboxyl group neutralization of NADPH-cytochrome P450 reductase on the ability of the reductase to support substrate turnover by different forms of cytochrome P450. Testosterone = –□–; benzphetamine = –Δ–; 7-ethoxycoumarin = –●–. From (19) with permission.

190

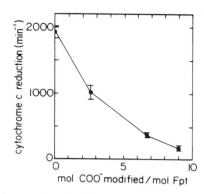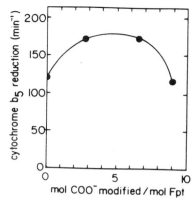

Figure 5. Influence of carboxyl group neutralization of F_{PT} on electron transfer to cytochrome c and b_5. From (19) with permission.

Interaction between F_{PT} and other redox partners

Attempts were made to learn the nature of interaction between NADPH-cytochrome P450 reductase and its redox partners. It interacts with two electron acceptors which charge pair with cytochrome b_5, P450 and cytochrome c. Cytochrome b_5 makes use of its anionic residues in such charge pairing interactions. When anionic charges on F_{PT} were neutralized by methylamidation, strong inhibition of cytochrome c reductase activity was obtained (50%) when as few as 3 mol of COOH were neutralized per mol of reductase (Fig. 5A). This was somewhat similar to the effect of anionic charge neutralization of P450 reduction (19), and was expected from the studies of Ng, et al. (15) showing cytochrome c to make use of conserved cationic residues in pairing with redox partners. However, such anionic charge neutralization resulted in a stimulation of electron transfer to cytochrome b_5 (Fig. 5B). If you recall, we indicated neutralization of carboxyl group charge on cytochrome b_5 impaired its ability to bind to cytochrome P450 (Table 2 and ref. 13). In contrast to these effects, such carboxyl charge neutralization on cytochrome b_5 greatly stimulated the reduction by F_{PT} (Fig. 6).

Whereas amidination of the cytochrome b_5 slightly slowed its rate of reduction, methylamidation resulted in a reduction so rapid as to be complete in the mixing time of the experiment (Fig. 6). This indicates that cytochrome b_5 and NADPH-cytochrome P450 reductase, while each utilizing carboxyl residues for interacting with redox partners, also interact with each other by complementary charge pairing. During redox pairing a charge repulsion exists between carboxyl groups on each of them, and neutralization of carboxyl residues on either partner results in an enhanced affinity for interaction between the two proteins. As suggested above, extensive anionic charge neutralization of cytochrome b_5 actually impaired substrate metabolism by cytochrome P450 (Table 3) below that in the absence of cytochrome b_5. This effect is now explainable in terms of the data of Figure 6; carboxyl-neutralized cytochrome b_5 no longer binds to cytochrome P450, but instead competes more effectively with it for interaction with F_{PT}.

Mechanism of cytochrome b₅ stimulation of P450 turnover

How does cytochrome b_5 elicit its stimulation of cytochrome P450 monooxygenation? Recent studies have shown P450 CAM, a soluble, crystallized form of P450, to have a high α-helical content (23). The heme of the cytochrome is tucked in between three α-helices. Mammalian forms of cytochrome P450 likewise have a high α-helical content (24).

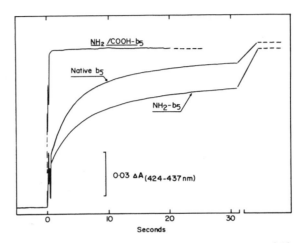

Figure 6. Reduction of native cytochrome b_5, amidinated b_5 (NH_2-b_5) and amidinated plus methylamidated b_5 $NH_2/COOH$-b_5 by NADPH-cytochrome P450 reductase. From (13) with permission.

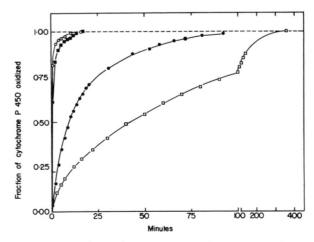

Figure 7. Circular dichroism spectra of 4μM cytochrome P450 IIB1 (PBRLM5) with and without cytochrome b_5.

Using circular dichroism we have found that the addition of cytochrome b_5 to cytochrome P450 causes a major change in the secondary structure of the P450, seen as a much greater than additive increase in the proportion of α-helical content (Fig. 7) of PBRLM5, the major phenobarbital-inducible form of P450 in the rat. Such an effect might affect electron transfer properties of the P450 since the heme moiety is located in the α-helical region of the apoprotein. The effect of cytochrome b_5 binding on electron transfer from P450 was examined (Fig. 8). In this study the cytochrome P450 was photochemically reduced under carbon monoxide, anaerobically, in a Thunburg cuvet. Ferric cytochrome b_5 was present in the sidearm. The reaction was started by tipping the ferric cytochrome b_5 into the ferrous cytochrome P450 solution and monitoring the change in absorbance (decline) at 450nm or change (increase) in the absorbance at 424nm. Within 0.8 min 80% of the cytochrome P450 was oxidized by native cytochrome b_5 (open circles) or by acetylated cytochrome b_5 (closed squares). In contrast, neutralization of 11 carboxyl groups per cytochrome b_5 (solid circles) decreased the rate of P450 oxidation by over 25 fold; a half-time of over 10 min was obtained. This impairment of the oxidation of P450 by carboxyl group neutralization is not due to the alteration of the redox potential of the cytochrome b_5, as the alteration reported is to a more positive value than the +5mV for native cytochrome b_5 (25). Further, cytochrome c, which has a redox potential of +255mV is an even poorer electron acceptor from ferrous P450 (Fig. 8; open squares). Since cytochrome c makes use of highly conserved lysine residues for interaction with redox partners (15), it would not be expected to interact favorably with cytochrome P450, which also makes use of cationic residues for such interactions. Because cytochrome b_5 can accept electrons from cytochrome P450, perhaps the stimulatory action of cytochrome b_5 on the monooxygenase reaction involves electron transfer between these hemoproteins. The data in Table 4 supports this hypothesis. The addition of cytochrome b_5 to the cytochrome P450 LM2-containing medium stimulated p-nitroanisole O-demethylation almost 9-fold (27). Removal of the heme and reintroduction of the hemin to apocytochrome b_5 (rb_5) did not impair the ability to stimulate the monooxygenase reaction (Table 4). Replacement of the heme with heme demethyl ester (Me_2b_5) significantly reduced the stimulation at a 1:1 stoichiometry with LM2 (1:1). This is due to the diminished anionic charge on the cytochrome b_5, since Me_2b_5 binding affinity to P450 is decreased; one can still obtain maximal spin shifts, but higher levels

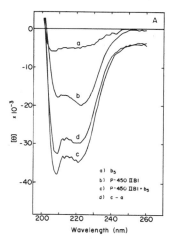

Figure 8. Reoxidation of LM2 by native b_5 (a) acetylated b_5 (b) amidated b_5 (c) and cytochrome c (d). From (13) with permission.

Table 4. Mechanism of cytochrome b_5 stimulation of P450
LM2 mediated p-nitroanisole demethylation

Additions to complete medium	p-nitrophenol formed (nmol/min/ml)
None	0.6
b_5 (1:1)	5.1
rb_5 (1:1)	5.4
Me_2b_5 (1:1)	1.8
Mnb_5 (1:1)	0.6

of Me_2b_5 are required (18). On the other hand, replacement of the
heme with manganous heme, where the iron of the heme is substituted by
manganese, resulted in loss of the stimulatory effect afforded by the
cytochrome b_5 on substrate oxidation. Manganous heme is poorly
reducible, and when present in the cytochrome b_5 it does not accept
electrons, although it binds to P450 and causes a shift in the spin
equilibrium toward the high spin configuration as does b_5. These
results indicate the influence of cytochrome b_5 resulting in enhanced
substrate metabolism by cytochrome P450 involves its ability to accept
electrons.

Cytochrome P450-cytochrome b_5 heterodimers

Using a water-soluble carbodiimide, EDC, a covalent complex was
prepared from cytochrome P450 LM2 and cytochrome b_5 (26). The EDC
catalyzes formation of an amide bond between closely alligned carboxyl
and amino groups. To ensure formation of specifically interacting
(charge pairing) rather than randomly colliding groups the carbodiimide
was only present for a short period in the presence of benzphetamine,
which increases the affinity to cytochrome b_5 (Table 2). The
complexing mixture was then passed through a DEAE-cellulose column to
recover uncomplexed cytochromes b_5 and LM2 as well as heterodimers
(26). The heterodimers were shown by migration on sodium lauroyl sulfate
polyacrylamide gel electrophoresis to have a molecular weight of 64,000,
the approximate sum of that LM2 (45,000) and cytochrome b_5 (18,000).
By spectroscopy the complexes were seen to contain equimolar amounts of
cytochromes P450 and b_5.

The orientation of the cytochrome b_5 in the complex was examined by
attempting its reduction with NADH-cytochrome b_5 reductase (F_{PT})
(Fig. 9). These two proteins interact for electron transfer by a
complimentary charge pairing involving lysyl residues on the reductase
and carboxyl residues on the b_5 (16, 27). On addition of NADH and
NADH-cytochrome b_5 reductase only about 20% of the cytochrome b_5
proved reducible (solid circles); none of the LM2 was reduced. Lack of
reduction of the major portion of cytochrome b_5 (80%) indicated the
orientation of cytochrome b_5 in complex with cytochrome P450 involved
the same carboxyl groups on the cytochrome b_5 which charge pair with
cytochrome b_5 reductase. In contrast, when NADPH-cytochrome P450
reductase and NADPH was added, about 80% of the cytochrome b_5 was
reduced (Fig. 9; open circles). The prior studies indicated that
cytochrome b_5 interacts with F_{PT} in a manner in which the carboxyl
groups of both hinder the interaction by repulsion. Both proteins also
use their carboxyl residues for interaction with cytochromes P450. Thus,
it is not unreasonable to conclude, since cytochrome b_5 can oxidize
cytochrome P450 (Fig. 8), that reduction of the complexed cytochrome b_5
by NADPH involves electron flow from NADPH-cytochrome P450 reductase.
The data also suggests that cytochrome b_5 binds to a site on cytochrome

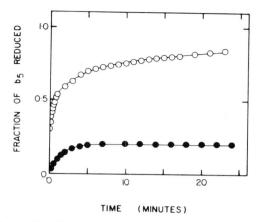

TIME (MINUTES)

Figure 9. The reduction of b_5 of LM2/b_5 covalent complex by F_{PD} and F_{PT}.

P450 that differs from the site to which the F_{PT} binds

A further indication that the cytochrome P450-cytochrome b_5 heterodimer is functionally identical to the natural complex formed on addition of cytochrome b_5 to the reconstituted monooxygenase system was seen by its interaction with substrates (26). As noted in Table 1, in the presence of cytochrome b_5 the Kd for benzphetamine binding to P450 LM2 was decreased one order of magnitude. Similarly, the Kd for benzphetamine with the LM2-cytochrome b_5 covalent complex was one order of magnitude lower than that with LM2 alone (26). This heterodimer is functionally identical to the complex formed by interaction of free cytochrome b_5 with P450 as seen by studying its interaction with F_{PT}

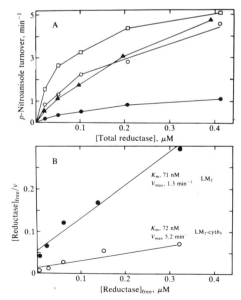

Figure 10. LM2 metabolism of p-nitroanisole with (open squares) and without (solid circles) cytochrome b_5, or covalently bound to b_5 (open circles and closed triangles).

195

(Fig. 10). The turnover of p-nitroanisole in the absence of cytochrome b_5 is low (Fig. 10A, solid circles); with cytochrome b_5 the activity is greatly stimulated (open squares). The activity of two different LM2-cytochrome b_5 covalent complexes at equal LM2 level produced activities similar to that of the free LM2 with cytochrome b_5 present (open circles and closed triangles). The Hanes-Woolf plot in Figure 10B shows the K_m for NADPH-cytochrome P450 reductase, calculated from iterative plots correcting for reductase bound to the LM2, is the same in the absence of cytochrome b_5 as with covalently bound cytochrome b_5. This data clearly indicates that the cytochrome b_5 binding site on P450 does not overlap or influence the binding site for Fp_T.

A sketch is shown in Figure 11 which summarizes our hypothesis as to the influence of cytochrome b_5 resulting in enhanced monooxygenase activity. The sketch shows ferric P450 (the Fe^{3+} indicates the heme iron charge) binding substrate then cytochrome b_5. It doesn't matter which binds first, the net effect is a higher affinity for both. Note that now we have a two heme-containing complex. The next step involves binding of reduced P450 reductase to the P450-b_5-substrate complex. The exact redox state of the reductase is not known as yet. Iyanagi et al. (28) have indicated that the redox potential of NADPH is more positive than and insufficient to produce four electron reduced reductase. However, Backes et al. (29) have produced data indicating the 4 electron reduced reductase is instrumental in substrate oxidation by cytochrome P450.

In the sketch transfer of a reducing equivalent to P450 is shown in brackets because as shown in this paper, the electron is rapidly taken from P450 by cytochrome b_5. Since this leaves the P450 in the ferric state, it can accept another electron from Fp_T before the latter dissociates from the P450-b_5 complex. Since the b_5 when reduced is unable to take another electron from the P450, oxygen may now bind forming an oxy-P450 complex. This has been estimated to have a redox potential of +50mV (30), or considerably more positive than the +5mV of b_5. This rapidly reoxidizes cytochrome b_5 and disproportionation of the dioxygen occurs to yield water and oxygenated substrate. The stimulation afforded by cytochrome b_5 in this hypothesis is the result of the b_5-P450 complex being a 2 electron acceptor. When cytochrome P450 alone binds to Fp_T it can only accept one electron. The reductase then dissociates, oxygen binds, and the reaction must await rebinding of

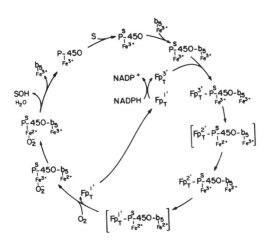

Figure 11. Sketch of P450 cycle with b_5 participation.

reductase for input of the second electron. Stimulation would be
afforded by two routes, a faster input of a rate limiting electron plus a
decrease in loss of 1-electron reduced oxygen before input of the second
electron to P450.

ACKNOWLEDGEMENTS

The studies reported were supported in part by grant GM26114 from the
United States Public Health Service.

REFERENCES

1. T. Omura and R. Sato, A new cytochrome in liver microsomes, J. Biol.
    Chem. 237:1375 (1962).
2. D.Y. Cooper, S. Levin, S. Narasimhulu, O. Rosenthal and R.W.
    Estabrook, Photochemical action spectrum of the terminal oxidase
    of mixed function oxidase systems, Science 147:400 (1965).
3. J.B. Schenkman, I. Jansson and K.M. Robie-Suh, The many roles of
    cytochrome b_5 in hepatic microsomes, Life Sci. 19:611 (1976).
4. S. Kuwahara and T. Omura, Different requirements for cytochrome b_5
    by two types of microsomal cytochrome P-450, Biochem. Biophys.
    Res. Comm. 96:1562 (1980).
5. E.T. Morgan and M.J. Coon, Effects of cytochrome b_5 on cytochrome
    P-450-catalyzed reactions, Drug Metab. Dispos. 12:358 (1984).
6. I. Jansson, P.P. Tamburini, L.V. Favreau and J.B. Schenkman, The
    interaction of cytochrome b_5 with four cytochrome P-450 enzymes
    from the untreated rat, Drug Metab. Dispos. 13:453 (1985).
7. A. Hildebrandt and R.W. Estabrook, Evidence for the participation of
    cytochrome b_5 in hepatic microsomal mixed function oxidation
    reactions Arch. Biochem. Biophys. 143:66 (1971).
8. A.H. Conney, R.R. Brown, J.A. Miller and E.C. Miller, The metabolism
    of methylated aminoazo dyes VI. Intracellular distribution and
    properties of demethylase system. Cancer Res. 17:628 (1957).
9. J.B. Schenkman, S.G. Sligar and D.L. Cinti, Substrate interaction with
    cytochrome P-450, p. 587 in "Hepatic Cytochrome P-450
    Monooxygenase System", J.B. Schenkman and D. Kupfer, eds.,
    Pergamon Press, New York (1982).
10. W.L. Backes, P.P. Tamburini, I. Jansson, G.G. Gibson, S.G. Sligar and
    J.B. Schenkman, Kinetics of cytochrome P-450 reduction: evidence
    for faster reduction of the high-spin ferric state, Biochemistry
    24:5130 (1985).
11. I. Jansson and J.B. Schenkman, Influence of cytochrome b_5 on the
    stoichiometry of the different oxidative reactions catalyzed by
    liver microsomal cytochrome P-450. Drug Metab. Dispos. 15:344
    (1987).
12. L.D. Gorsky and M.J. Coon, Effects of conditions for reconstitution
    with cytochrome b_5 on the formation of products in cytochrome
    P-450-catalyzed reactions, Drug Metab. Dispos. 14:89 (1986).
13. P.P. Tamburini, R.E. White and J.B. Schenkman, Chemical
    characterization of protein-protein interactions between
    cytochrome P-450 and cytochrome b_5, J. Biol. Chem. 260:4007
    (1985).
14. J.B. Schenkman, H. Remmer and R.W. Estabrook, Spectral studies of drug
    interaction with hepatic cytochrome, Mol. Pharmacol. 3:113 (1967).
15. S. Ng, M.B. Smith, H.T. Smith and F. Millet, Effect of modification of
    individual cytochrome c lysines on the reaction with cytochrome
    b_5, Biochemistry 16:4875 (1977).
16. A. Loverde and P. Strittmatter, The role of lysyl residues in the
    structure and reactivity of cytochrome b_5 reductase, J. Biol.
    Chem. 243:5779 (1968).
17. P. Strittmatter and H.A. Dailey, Essential features and orientation

of cytochrome b_5 in membranes, p. 71, in "Membranes and Transport", Vol. 1, A.N. Martonosi, ed., Plenum Publishing Corporation (1982).

18. P.P. Tamburini and J.B. Schenkman, Mechanism of interaction between cytochromes P-450 RLM5 and b_5: Evidence for an electrostatic mechanism involving cytochrome b_5 heme propionate groups, Arch. Biochem. Biophys. 245:512 (1986).

19. P.P. Tamburini and J.B. Schenkman, Differences in the mechanism of functional interaction between NADPH-cytochrome P-450 reductase and its redox partners, Mol. Pharmacol. 30:178 (1986).

20. P.L. Gigon, T.E. Gram and J.R. Gillette, Studies on the rate of reduction of hepatic microsomal cytochrome P-450 by reduced nicotinamide adenine dinucleotide phosphate: effect of drug substrates, Mol. Pharmacol. 5:109 (1969).

21. P.P. Tamburini, G.G. Gibson, W.L. Backes, S.G. Sligar and J.B. Schenkman, Reduction kinetics of purified rat liver cytochrome P-450. Evidence for a sequential reaction mechanism dependent on the hemoprotein spin state, Biochemistry 23:4526 (1984).

22. T.D. Porter and C.B. Kasper, Coding nucleotide sequence of rat NADPH-cytochrome P-450 oxidoreductase cDNA and identification of flavin-binding domains, Proc. Natl. Acad. Sci. USA 82:973 (1985).

23. T.L. Poulos, B.C. Finzel, I.C. Gunsalus, G.C. Wagner and J. Kraut, The 2.6A crystal structure of Pseudomonas putida cytochrome P-450, J. Biol. Chem. 260:122 (1985).

24. T. Shimizu, T. Nozawa, M. Hatano, H. Satake, Y. Imai, C. Hashimoto and R. Sato, Circular dichroism spectra of purified cytochromes P-450 from rabbit liver microsomes, Biochim. Biophys. Acta 579:122 (1979).

25. L.S. Reid, M.R. Mauk and A.G. Mauk, Role of heme propionate groups in cytochrome b_5 electron transfer, J. Amer. Chem. Soc. 106:2182 (1984).

26. P.P. Tamburini and J.B. Schenkman, Purification to homogeneity and enzymological characterization of a functional covalent complex composed of cytochromes P-450 isozyme and b_5 from rabbit liver, Proc. Natl. Acad. Sci. USA 84:11 (1987).

27. H.A. Dailey and P. Strittmatter, Modification and identification of cytochrome b_5 carboxyl groups involved in protein-protein interaction with cytochrome b_5 reductase, J. Biol. Chem. 254:5388 (1979).

28. T. Iyanagi, N. Makino and H.S. Mason, Redox properties of the reduced NADP-cytochrome P-450 and reduced NAD-cytochrome b_5 reductases, Biochemistry 13:1701 (1974).

29. W.L. Backes and C.E. Reker-Backes, The effect of NADPH concentration on the reduction of cytochrome P-450 LM2, J. Biol. Chem. 263:242 (1988).

30. F.P. Guengerich, Oxidation-reduction properties of rat liver cytochrome P-450 and NADPH-cytochrome P-450 reductase related to catalysis in reconstituted systems, Biochemistry 22:2811 (1983).

POSTTRANSLATIONAL MODIFICATION BY PHOSPHORYLATION: CONTROL OF CYTOCHROME P450 AND ASSOCIATED ENZYMES

Walter Pyerin[1] and Hisaaki Taniguchi

German Cancer Research Center
Institute of Experimental Pathology
D-6900 Heidelberg, Federal Republic of Germany

Regulation of proteins by posttranslational modification is of crucial importance for eucaryotic cells in particular because of the relatively long half-lives of their mRNAs and the resulting lack of rapid protein regulation at the nucleic acid level. The phosphorylation of proteins (and their dephosphorylation) has emerged as the most widespread and important posttranslational regulatory device. It is now known to occur in all types of cells and in virtually all cellular compartments and organelles (1,2).

Protein phosphorylation has a remarkable number of possibilities for exerting control on cytochrome P450 (P450) and associated enzymes, acting both at the level of their induction and their posttranslational tuning. In the course of the induction process, phosphorylation may influence the transmission of the inducing signal from outside the cell to its DNA, i.e., the receptors of the respective induction signal and the elements of the corresponding signal transmission chains, the transcriptional and posttranscriptional elements responsible for the production of mRNA species of these enzymes, the translational machinery for synthesizing the proteins, and the production of the diverse prosthetic groups. As fas as posttranslational tuning is concerned, phosphorylation may change the biological activity of enzymes, i.e., their interaction with substrates as well as with other enzymes, and it may be the determining factor for their half lives, i.e., for their degradation.

Our work and that of others has hitherto focused on the direct phosphorylation of enzyme protein. The result of the very first experiment (3) is shown in Fig. 1. A preparation of P450LM2, the major phenobarbital-inducible isoenzyme of rabbit liver microsomes (P450IIB4; 4), and a preparation of NADPH-dependent P450 reductase from the same source were investigated for phosphorylation. The transferred phosphoryl groups were ^{32}P-labelled and phosphorylation is therefore indicated by the

---

[1] Corresponding author

Molecular Aspects of Monooxygenases and Bioactivation of Toxic Compounds
Edited by E. Arınç *et al., Plenum Press, New York,*

incorporation of radioactivity. The autoradiogram shown reveals that P450 incorporated phosphoryl groups, while the reductase did so only slightly or not at all depending on its status, i.e., whether solubilized or part of a reconstituted monooxygenase complex, respectively. The incorporated phosphoryl groups came from ATP and the transfer was mediated by the catalytic subunit of cyclic AMP-dependent protein kinase (see below).

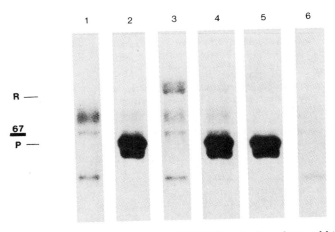

Figure 1. Phosphorylation of P450LM2 and NADPH-P450 reductase from rabbit liver microsomes by cyclic AMP-dependent protein kinase. P450 and P450 reductase preparations were incubated in the presence of phosphatidylcholine, catalytic subunit of cyclic AMP-dependent protein kinase and [γ-^{32}P]ATP followed by SDS-polyacrylamide gel electrophoresis and autoradiography. Shown is the autoradiogram. Lanes: 1, kinase alone; 2, P450/kinase; 3, reductase/kinase; 4, P450/reductase/kinase; 5, P450/reductase/phosphatidylcholine/kinase; 6, phosphatidylcholine/kinase. R, position of reductase; 67, position of marker protein (bovine serum albumin); P, P450LM2. For experimental details see Ref. 3.

The present contribution describes details of P450 phosphorylation and that of associated proteins, and, to provide the proper groundwork, reviews essential parts of the protein phosphorylation field and briefly mentions other forms of reversible covalent posttranslational protein modification.

PROTEIN PHOSPHORYLATION

The phosphorylation of proteins occurs exclusively posttranslationally (there is not any example available which would indicate that amino acids are phosphorylated prior to incorporation into a growing peptide chain) and is reversible (Fig. 2). The acceptor sites for the transferred phosphoryl groups are usually OH-group containing amino acid side chains, i.e., serine, threonine and tyrosine residues. Of the phosphoamino acids found in cell homogenates, usually over 90% is phosphoserine, up to 10% phosphothreonine, and far below 1% phosphotyrosine (1,2).

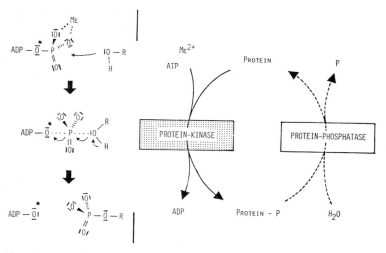

Figure 2. Phosphorylation of proteins: a reversible reaction. Me^{2+}, divalent metal ion; P, phosphate; protein-P, phosphorylated protein; R, protein body. For details see text.

The effect of phosphorylation on a protein is threefold. First, the replacement of a hydrogen atom by the comparably bulky phosphoryl group may, for spatial reasons, destort the backbone of the protein. Second, hydrogen bridges, which are important stabilizing elements of secondary and tertiary protein structures, are destroyed or their formation prevented. Third, the introduction of a phosphoryl group adds two negative charges and therefore may affect the intramolecular ionic bonding. Together these are enough to cause profound structural changes in a protein. One of the early convincing demonstrations of such changes used immunological means. Nairn et al. succeeded in producing specific antibodies against the phosphoform as well as the dephosphoform of a protein (5). Since antibodies are directed against epitopes, i.e., defined areas of the folded surface, a change in the structure of a protein changes its antigenicity. Accordingly, the binding of the antibody against the dephosphoform steadily decreased during the course of phosphorylation whereas that of the phosphoform-antibody increased with complementary kinetics, binding of phosphoform-antibody and phosphorylation degree of the protein nicely matching each other.

If the protein is an enzyme, phosphorylation usually changes its activity, its specificity, or its behaviour within enzyme complexes or the like. The first report of an enzyme regulation by phosphorylation/dephosphorylation appeared in 1955. Fischer and Krebs (6) and Sutherland and Wosilait (7) showed that glycogen phosphorylase, the enzyme catalyzing the phosphorylitic cleavage of glucose molecules from glycogen, exists in two interconvertible forms, one with and one without enzymatic activity. These forms were revealed as the phosphorylated and the dephosphorylated condition of one and the same enzyme. At present, well over a hundred enzymes are known to be regulated by phosphorylation/dephosphorylation (8). In this way, phosphorylation/dephosphorylation controls a broad array of cellular pathways such as the metabolism of carbohydrates, lipids, nucleic acids, amino acids, etc., and is involved in membrane signal transduction, DNA repair and transcription, viral oncogenesis, muscle cell contraction, interferon action, protein synthesis, etc. One of the most recent reports was a detailed mechanistic study by Bailey et al. (9) on tyrosine hydroxylase, a member of the monooxygenase family with tetrahydropterin as the prosthetic group. The phosphorylation results in an activity-determining change of the cofactor binding domain.

OTHER FORMS OF REVERSIBLE COVALENT PROTEIN MODIFICATION

Phosphorylation is the most widespread form of reversible covalent modification of proteins but not the only one. Other well-known modifications of proteins are (8):

ADP-ribosylation
The cosubstrate for this modification is NAD. Acceptor sites are arginine, glutamic acid, and lysine (terminal COOH) side chains. Examples include enzymes such as adenylate cyclase, DNA-ligase II, DNA-polymerases, DNA-topoisomerase I, RNase, and the large SV40 T antigen.

Nucleotidylylation
This means adenylylation and uridylylation, with ATP and UTP as the respective cosubstrates. Acceptor sites are tyrosine and serine side chains. Examples include glutamine synthetase (E. coli) and the large SV40 T antigen.

Methylation
Methylation proceeds with S-adenosyl-methionine as the cosubstrate. It occurs at aspartic acid, glutamic acid, lysine, histidine, and glutamine side chains. Examples include ACTH, calmodulin, ribonuclease, and cytochrome c.

Acetylation
The cosubstrate here is acetyl-CoA. Acceptor sites are lysine side chains. Examples are high mobility group proteins and histones.

Tyrosinolation
Here, tyrosine is fixed to the carboxy-terminus of a protein. Example is the tubulin α-subunit.

Sulfation
Sulfation occurs with 3-phosphoadenosine-5-phosphosulfate as the cosubstrate and tyrosine residues as the acceptor sites. Examples include fibrinogens, fibronectin, interleukin-2 receptor, gastrin II, complement C4, and immunoglobulins.

As far as P450 is concerned, there is a report by Lechner and Braz (10) pointing out that nuclear ADP-ribosyl transferase activity correlates with the phenobarbital induction of P450 monooxygenase activity in rat liver microsomes. A causal relationship, however, has not been established yet. Direct covalent modification of P450 has repeatedly been reported to occur through glycosylation. However, glycosylation belongs to the rather irreversible covalent modifications of proteins. It has not yet become clear what it may mean for P450, and moreover, the stochiometries determined show considerable deviations (reviewed in 11). Other forms of protein modification have not yet been reported with P450.

Another and more recently discovered type of covalent modification of proteins, probably reversible, is the attachment of lipid molecules such as phospholipid, diacylglycerol and various species of long chain fatty acids (12). The last one appears to exert effect also on endoplasmic P450 monooxygenase systems:

Acylation with long chain fatty acids

The most frequently reported acylations of proteins are those with the relatively rare myristic acid (tetradecanoic acid) and with plamitic, stearic- and oleic acid (hexa- and octadecanoic acids). While in all known examples myristoylation occurs at the amino-terminus resulting in an amide-linkage, the other modifications are of the ester-type linkage at serine, threonine, or cysteine side chains (12). NADH-cytochrome b5 reductase was reported to contain N-terminal myristic acid, i.e., at a glycine residue (13). The same modification is found in several other proteins (see below).

PROTEIN KINASES AND THEIR ACTION

The phosphorylation of proteins (see Fig. 2) is catalyzed by protein kinases (ATP: protein phosphotransferases, EC 2.7.1.37). Protein kinases represent a diverse and large family of enzymes. Their systematic naming is anything but easy. Substrates provide no proper basis since most kinases accept a number of different proteins and, on the other hand, different kinases may phosphorylate the same protein and even the same site in that protein. For example, glycogen synthase, the enzyme acting counter to glycogen phosphorylase, is phosphorylated in the N-terminal and the C-terminal region at no less then up to 12 different sites and by no less than 10 different protein kinases whereby synergistic phosphorylation may occur (14). In practice, therefore, protein kinases usually are classified as kinases transferring phosphoryl groups to protein alcohol groups, i.e., the serine and threonine kinases, and as kinases transferring phorphoryl groups to protein phenolic groups, i.e., the tyrosine kinases (1,2). Within each of these classes, individual groups of enzymes or subclasses exist which more often than not can be delineated on the basis of regulation of their activities by effector molecules. With serine/threonine kinases these are cyclic nucleotides, Ca^{2+}/calmodulin, diacylglycerol/Ca^{2+}/phospholipid, double-stranded RNA, hemin, etc. The respective kinases carry names such as cyclic AMP-dependent kinases, cyclic GMP-dependent kinases, phosphorylase kinase, myosin light chain kinase, protein kinase C, hemin-inhibited kinase, etc. A number of kinases are independent of this kind of regulatory compound. They comprise the so-called casein kinases type I and type II, several histone kinases, protease-activated kinases, etc. The other class, the tyrosine kinases, shows regulation principally by polypeptide hormones and growth factors. These kinases include the EGF receptor, the PDGF receptor or the insulin receptor. Tyrosine kinases with as yet unknown regulatory agents encompass the whole array of oncogene product kinases. The fact that these kinases are regulated by different stimuli and effectors and that multisite modifications occur not only expands the potential of the regulatory network of cells numerically, but creates the possibility for a competition between different converter enyzmes and, hence, between different regulatory systems within this network.

The mechanism of action of protein kinases resembles that of all the kinases producing O-phosphate esters (see Fig. 2): A purine nucleoside triphosphate - usually ATP - functions after complexation with divalent metal ions such as Mg^{2+} as the cosubstrate. The kinase transfers the terminal (the "γ" -) phosphoryl group of the triphosphate to the substrate protein resulting in a phosphoprotein and nucleoside diphosphate. The phosphorylation reaction starts by a nucleophilic attack of the oxygen of the amino acid's OH-group to the γ-phosphorus of the triphosphate. From the catalytic subunit of the cyclic AMP-dependent protein kinase (see below), which is the prototype of the whole kinase family and the kinase employed in the P450 phosphorylation experiment shown in Fig. 1, detailed information is already available. The general base in catalysis and the key elements of the phosphorylation appears to be an aspartate and a lysine residue in position 184 and 72, respectively (15). Although at quite distant positions in the amino acid chain, they closely face each other within the

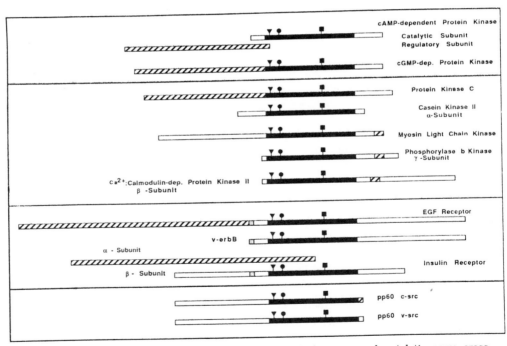

Figure 3. Sequence alignment of protein kinases. Black, conserved catalytic core; cross-hatched, known regulatory regions; stippled, putative membrane-spanning segment. Conserved residues correspond to the glycine-rich segment, Gly-50-Gly-55 (▼), Lys-72 (●), and Asp-184 (■). Reproduced with permission from Ref. 16.

protein. Lys-72 is in the center of the so called "nucleotide binding fold" and preferentially binds Mg-ATP. In doing so, it frees Asp-184 from its ionic bond. Asp-184, which belongs to the protein binding portion of the kinase, is now free to bind the protein's amino acid side chain to become phosphorylated. This binding causes the necessary increase in the nucleophilicity of the oxygen of the affected OH-group to start the nucleophilic attack.

Although strikingly divers, all the protein serine/threonine kinases and the tyrosine kinases seem to contain a homologous catalytic core (16). As shown in Fig. 3, this core contains the above-mentioned lysine residue, a conserved triad of glycines, and the above-mentioned aspartate residue. Protein kinases, therefore, obviously have evolved from a common origin and appear to have assembled modular structures by a mechanism that involved exon shuffling (16).

The modular structure of the kinases then show remarkable differences. While the cyclic AMP-dependent kinases have dissociable regulatory subunits, their closest relative, the cyclic GMP-dependent kinase, possesses a regulatory element which is part of a contiguous polypeptide chain constituting the N-terminal portion. Further, protein kinase C contains a binding region for Ca^{2+}, diacylglycerol and phosphatidyl serine N-terminal to the catalytic core, while myosine light chain kinase has a Ca^{2+} and calmodulin dependent binding domain C-terminal to the catalytic core. The casein kinase II (α-subunit) seems to contain a separate regulatory subunit (β-subunit). The epidermal growth factor (EGF) receptor and the insulin receptor both contain a hormone-binding domain N-terminal to a membrane-spanning domain and the catalytic core. In the cell, the hormone-binding region is located extracellularly, the catalytic domain intracellularly. However, while the EGF receptor is a single peptide chain, the insulin receptor consists of α and β subunits. Binding of the respective hormone outside the cell, therefore, activates the kinase inside. The product of the oncogene from Rous sarcoma virus, pp60[v-src], also belongs to the protein kinase family, as does its cellular counterpart pp60[c-src]. The latter possesses a regulatory segment C-terminal to the catalytic core which contains a phosphorylation site. Removal of this site suffices to convert the proto-oncoprotein into a transforming protein (17). Together, the differences in the non-catalytic regions (except for the oncogene proteins) are decisive for regulation, activation, and, of course, the subcellular localization of the respective kinases.

THE EMPLOYED PROTEIN KINASES

The kinases employed to date for phosphorylation of P-450 and associated enzymes were the cyclic AMP-dependent protein kinase, the Ca^{2+}/phospholipid/diacylglycerol-dependent kinase and the casein kinase type II. These kinases have the following characteristics:

The cyclic AMP-dependent protein kinase (PK A)

In mammalian cells, the cyclic AMP-dependent protein kinase is the major, if not the only, intracellular receptor for cyclic AMP. Hence, all of the diverse actions triggered by cyclic AMP are mediated by this kinase. The kinase consists of two types of subunit, the regulatory (R) and the catalytic (C) subunits. Together, they form the holoenzyme R_2C_2 which has no kinase activity. The cyclic nucleotide (four molecules) binds to the R-subunit dimer and causes a conformational change which lowers the affinity of R for C by a factor of roughly 10^4 leading, under physiological conditions, to a complete dissociation of R and C (Fig. 4). The released C-subunits (two of them per holoenzyme) can now act as protein kinases (1,2). The C-subunits from different sources appear to be essentially identical in terms of their physical and enzymatic properties, although slight structural differences are found at both the protein and the nucleic acid level (18-20). The R-subunits constitute two major forms, R I and R II, distinguishing two major types of holoenzymes, the type I and the type II holoenzyme (1,2). R I and R II can be distinguished by criteria such as autophosphorylation (specific for R II), high affinity for MgATP (specific for R I), amino acid sequence, binding properties for C-subunits and cyclic AMP-analogous, and tissue distribution.

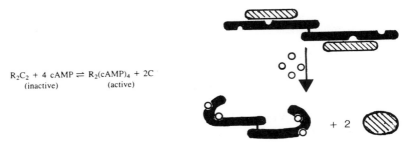

$$R_2C_2 + 4\ cAMP \rightleftharpoons R_2(cAMP)_4 + 2C$$
$$\text{(inactive)} \qquad\qquad \text{(active)}$$

Figure 4. Structure and activation of cyclic AMP-dependent protein kinase. Black, regulatory subunits; cross-hatched, catalytic subunits; circles, cyclic AMP molecules.

The monomeric C-subunit has a M_r of 40 kD and is a relatively symmetrical and globular protein (1,2,16). Interestingly, it has an unusual modification at its N-terminus. The N-terminus is acylated by myristic acid (21). The role of this covalent modification is not yet clear. Myristoylation, however, is also found in the oncogene product pp60[v-src] (reviewed in 12). In this case, myristoylation is the essential feature determining its subcellular localization (at the plasma membrane) and for its cell-transforming ability. Whether the myristoyl group of the C-subunit might play a similar role is unknown. Since the C-subunit is a soluble enzyme and is never found tightly associated with membranes, the group might rather serve as an intramolecular stabilization factor. Purified C-subunits contain endogenous phosphate at two positions, at Thr-197 and Ser-338 (16). This means that in in vitro-experimentation some autophosphorylation may always occur. The substrate specificity is broad with preference for "basic" substrates. Primary structural determinants appear to be dominant, the most preferred sequence being -Arg-Arg-X-Ser- (1,2). However, some synthetic peptides carrying this sequence are poor substrates compared to intact proteins suggesting that higher order determinants can be important. The preferred cosubstrate is MgATP; GTP, for example, is practically inactive.

When preparing the C-subunit, we make use of the fact that R-subunits and holoenzymes are rather similar in their isoelectric behaviour and differ from that of the

C-subunits. This allows a single step purification from tissue extracts simply by addition of cyclic AMP to the DEAE-column-bound, washed crude holoenzyme (22). The result is a highly pure C-subunit preparation (23).

The Ca^{2+}/phospholipid-dependent protein kinase (PK C)

This monomeric kinase was originally described as a proenzyme, requiring proteolysis for expression of its activity. Later it was shown that in the absence of proteolysis, it has activity that is dependent on Ca^{2+} and membrane phospholipids and enhanced by diacylglycerol (24). Accordingly, the current model (Fig. 5) defines a "hinge" region which connects a catalytically active fragment containing the substrate and the cosubstrate binding domains with a regulatory fragment containing the putative Ca^{2+}-binding domain and the phospholipid/diacylglycerol binding domain. The proteolytic cleavage occurs within this region generating a fragment that has kinase activity but is not dependent on Ca^{2+} and phospholipid.

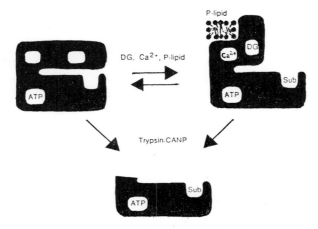

Figure 5. Structure and activation of Ca^{2+}/phospholipid-dependent protein kinase. DG, diacylglycerol; P-lipid, phospholipid; CANP, calpain; Sub, substrate binding site. Drawn according to Parker et al. (25a).

PK C occurs in a number of isoenzymes; a family of six unique genes gives rise to at least seven proteins (25). PK C purified from different sources has similar properties. The substrate specificity is broad with the "basic" histone H1 employed as model substrate when purifying and characterizing the kinase. Since the cyclic AMP-dependent kinase also accepts histones, PK C and PK A are often grouped as kinases with "basic" specificity. Although grouped together and phosphorylating a number of the same substrates - even the same site in a given substrate - PK A and PK C have clearly distinct specifities. PK C substrates often have basic residues flanking the phosphorylated serine or threonine residue on both the N-terminal and the C-terminal side; synthetic peptide studies suggest a requirement for basic residues on either the N-terminal side or the C-terminal side. The cosubstrate of PK C is ATP; GTP is not accepted.

The PK C preparations employed were the uncleaved kinase form with a M$_r$ of 77 kD and were obtained by DEAE-Sepharose followed by Phenyl-Sepharose resulting in reasonably pure preparations. Usually, however, further purification was achieved by either protamine agarose (26) or phosphatidylserine affinity chromatography (27).

The casein kinase type II (CK II)

This kinase is very unfortunately named, because it does not phosphorylate casein in vivo; casein was merely one of the model substrates employed in the early stages of discovery and biochemical characterization. CK II is a messenger-independent protein kinase with a number of putative key roles in cell regulation (28) which preferentially selects "acidic" proteins such as caseins or phosvitins in vitro. It therefore is an "acidic" counterpart to PK A and PK C. The enzyme phosphorylates exposed serine or threonine residues flanked C-terminally at one remove by one or several glutamic or aspartic acid residues, e.g., as in X-X-Ser-X-Glu-Glu (28). A similarly clearcut primary structural requirement as with PK A, however, does not exist. In contrast to PK A and PK C, CK II not only accepts ATP as a cosubstrate but also other purine triphosphates such as GTP or ITP (29). CK II's activity is influenced by diverse ions such as divalent and monovalent ions and, most typically, by anions such as fluoride and heparin.

The preparation of CK II needs several steps including DEAE-sepharose, phospho-cellulose and affinity chromatography on columns such as phosvitin-sepharose (29). CK II preparations yield an enzyme composed of α, α' and β subunits giving holoenzymes of the form $\alpha_2\beta_2$, $\alpha'\alpha\beta_2$, and $\alpha'_2\beta_2$ (29). Subunit β is autophosphorylated and able to control the holoenzyme's activity, while the α subunits contain the catalytic core, i.e., the cosubstrate and substrate binding domains. The M_rs are in the 40 kD range for subunits α and 26 kD range for subunit β (29). Both subunits of the human enzyme were cloned and sequenced recently (30,31) revealing extremely high sequence homologies between CK II's of evolutionary distant organisms.

PHOSPHORYLATION OF CYTOCHROME P450: HISTORICAL ASPECTS

The initial direct phosphorylation of P450LM2 (P450IIB4) by PK A shown above was carried out at a time when no reliable sequence information on P450s was available; it was a trial and error procedure. Earlier, although indirect, indications from animal studies were already in the literature. In 1962, Fouts described a dramatic depression of drug metabolism in liver cells upon repeated treatment of animals with catecholamines (32) which are known to act via an increase of the cellular cyclic AMP level. Similarly, Weiner et al. in 1972 (33,34) noted an increase in hexobarbital hypnosis in rats upon treatment with glucagon and theophyllin - again a treatment known to increase the cellular cyclic AMP level. Later, the direct involvement of cyclic AMP in the process was documented when Hutterer et al. in 1975 (35) applied a membrane-permeating derivative of cyclic AMP, namely, $N^6,O^{2'}$-dibutyryl cyclic AMP, to animals and noted a decrease of monooxygenase activity and of the concentration of cytochrome P450 and its induction.

The mechanism by which cyclic AMP exerts its effect, however, remained unknown. Our findings shown above suggested that the cellular target may be P450 itself.

IN VITRO-PHOSPHORYLATION OF CYTOCHROME P450 ISOENZYMES

Phosphorylation of P450 by PKA

The transfer of phosphoryl groups from ATP to P450LM2 by the catalytic subunit of cyclic AMP-dependent protein kinase (see Fig. 1) occurs regardless of the state of P450, i.e., as purified enzyme or in combination with lipids and cytochrome P450 reductase to give a complete reconstituted system (3,36). The phosphorylation reaction levels off at roughly 0.6 mol phosphate/mol P450; occasionally it exceeds 1 mol/mol (36,37). The apparent K_m value of this reaction is in the range of 2-8 µM P450 depending on the systems studied, i.e., P450 solubilized, in the presence of dilauroyl-phosphatidylcholine, and after incorporation into membranes of phosphatidylcholine/phosphatidylethanolamine/phosphatitic acid (molar ratio 2/1/0.6) as well as after reconstitution of the monooxygenase system with NADPH-P450 reductase at a molar ratio of P450/reductase/lipid of 1/1/600 (Table 1). Since this K_m value is similar to or even smaller than values measured in other phosphorylation reactions known to control various enzyme activities in vivo and, most important, since the cellular concentration of P450 in liver cells can be estimated roughly as 50 µM in phenobarbital-treated animals and since the cellular concentrations of both the kinase and the ATP are even higher than those employed in vitro, this phosphorylation reaction could well be of physiological relevance (36).

Table I. Kinetic constants of P450LM2 phosphorylation

Components	Apparent K_m[a]		Apparent V_{max}[b]	
	I	II	I	II
P-450	7.4	5.9	10.6	12.4
P-450 in DLPC	6.5	1.7	9.1	12.2
P-450 in membrane	6.0	7.6	2.9	3.1
P-450 in membrane plus reductase	6.0		3.1	

Values of two experimental series (I,II). P450 concentration range tested: 0.2-16 µM. For details see text and Ref. 36.
[a] [µM]
[b] [pmol phosphoryl groups/min]

The phosphorylation site (sites) was (were) obviously at a part of the P450 molecule protruding out of the membrane; it was accessible to the kinase not only in reconstituted systems but also in microsome-bound P450 (38). Microsomes prepared from phenobarbital-treated rabbits showed a strong incorporation of radioactivity into protein migrating upon electrophoresis with authentic P450 (Fig. 6A) when treated with the catalytic subunit of cyclic AMP-dependent kinase and [γ-^{32}P]ATP. Addition of a heat- and acid-stable inhibitor protein specific for the kinase inhibited this phosphorylation reaction, while addition of histones, the model substrate for the kinase, had no effect or even intensified phosphorylation of the microsomal protein, indicating a remarkable affinity between protein and kinase. It is not known, however, whether the myristoylation anchor of the kinase adds to this affinity by favouring the membrane environment of P450 or by interacting with hydrophobic regions within the microsomal protein. To test whether this microsomal protein was really P450 and not a protein migrating like P450 by chance, a comparative peptide mapping was carried out. For this purpose, the bands of authentic P450 and the corresponding microsomal protein were cut out from the gel and underwent partial proteolysis with V8-protease of *Staphyllococcus aureus* followed by SDS-gel electrophoresis and silver staining. The patterns obtained were highly similar throughout the lanes indicating identity of the digested proteins (Fig. 6B). The patterns were, moreover, clearly different from that of other P450 isoforms such as P450 LM4, the methylcholanthrene-inducible P450 species (P450IA; 4), or the NADPH-cytochrome P450 reductase.

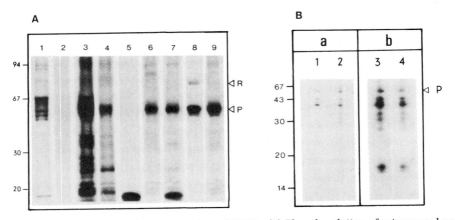

Figure 6. Phosphorylation of microsome-bound P450. (A) Phosphorylation of microsomal proteins by cyclic AMP-dependent protein kinase. Microsomal fraction from livers of phenobarbital-treated rabbits were incubated with catalytic subunit of cyclic AMP-dependent protein kinase and [γ-32P]ATP followed by SDS-polyacrylamide gel electrophoresis and autoradiography and tested for the influence by histone and a heat- and acid stable inhibitory protein (PKI) specific for the kinase. 1, protein pattern of microsomes; 2-9, autoradiograms: 2, phosphoprotein pattern in the absence of kinase; 3, phosphoprotein pattern in the presence of kinase; 4, as in 3 but in the presence of PKI; 5, histone phosphorylation by the kinase; 6, phosphoprotein pattern in the presence of kinase (different experimental set as shown in 2-4); 7, as in 6 but in the presence of histone; 8, phosphorylated authentic P450LM2 and NADPH-P450 reductase. R and P, position of reductase and P450 in the protein stained gel, respectively. (B) Comparative peptide mapping of P450LM2 and of phosphorylated microsomal protein migrating like P450LM2. The respective protein bands were cut out from gels, digested by V8-protease and subjected to electrophoresis followed by silver staining (a) and autoradiography of gel (b). 1, P450LM2; 2, microsomal protein comigrating with P450; 3, P450LM2; 4, microsomal protein comigrating with P450. P, position of undigested P450LM2. 67, 43, 30, 20, 14, positions of M_r-marker proteins (kD). For experimental details see Ref. 38.

In order to find out the nature of the phosphorylation site, radioactively phosphorylated solubilized P450 LM2 as well as microsomal P450 were cut out from the gel and underwent acid hydrolysis followed by separation of the hydrolysate by two-dimensional high voltage thin layer electrophoresis (38). In both cases, radioactivity was found exclusively at the place where phospho-serine was migrating (Fig. 7). P450LM2, therefore, is phosphorylated at serine residues in both its isolated and its microsome-bound form. That these were at the same sites in isolated and in microsome-bound P450 was shown by their comparable phosphopeptide patterns (see Fig. 6b), which showed not only identical numbers and positions of fragments but also identical relative intensitites of radioactivity independent of the phosphorylation time (38). Later, tryptic digestion of radioactively phosphorylated P450LM2 followed by sequencing of the main phosphorylated fragment revealed that there was a single phosphorylated serine residue, namely the serine residue in position 128 (39).

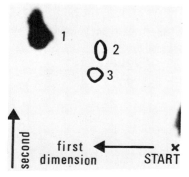

Figure 7. Phosphorylation site in microsome-bound P450 LM2. Microsomal protein phosphorylated and cut out from gel as in Fig. 6 followed by partial acid hydrolysis (6 N HCl) and separation by two-dimensional high voltage thin layer electrophoresis. Authentic phospho-amino acids localized by ninhydrin staining (1, phospho-serine; 2, phospho-threonine; 3, phospho-tyrosine), and radioactive phospho-amino acids by autoradiography.

The question then was whether this phosphorylation was a phenomenon observable with the rabbit P450 only. This led us to investigate the rat liver counterpart of LM2, P450 b (P450IIB1; 4) in detail (40).

P450b was isolated from liver microsomes of phenobarbital-treated rats and underwent the usual biochemical characterization (41). The highly pure preparation was phosphorylated by the catalytic subunit of the cyclic AMP-dependent protein kinase. The main results are summarized in Fig. 8. After only five minutes, incorporation of 0.5-0.6 mol phosphate/mol P450 was observed (A). Varying the kinase concentration within 0.1-1.0 μM - the cellular concentration range (1,2) - had little influence on these kinetics. The phosphorylation plateau was reached after 30-60 min at 0.7-1 mol phosphate/mol P450, depending on the kinase concentration employed. The phospho-amino acid analysis showed exclusively serine phosphate (B) . Tryptic

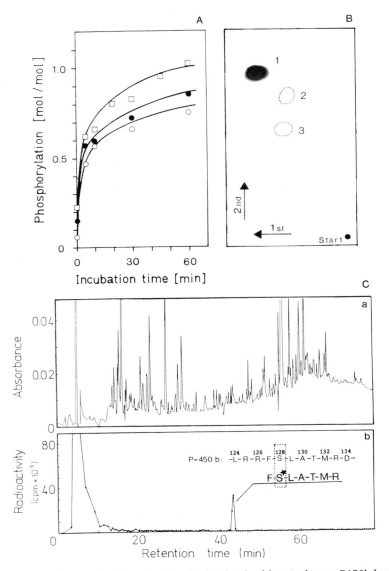

Figure 8. Phosphorylation of rat hepatic phenobarbital-inducible cytochrome P450b by catalytic subunit of cyclic AMP-dependent protein kinase. (A) Time course of phosphorylation. P450b was incubated for the indicated times in the presence of [γ-^{32}P]ATP and varying amounts of catalytic subunit of cyclic AMP-dependent protein kinase (◯ , 0.1 μM; ⬤ , 0.5 μM; ☐ , 1.0 μM) and the acid-precipitable radioactivity determined. (B) Phosphorylated amino acid side chain. Radioactively phosphorylated P450b was precipitated by trichloroacetic acid and subjected to analysis as described in Fig. 7 (1, phospho-serine; 2, phospho-threonine; 3, phospho-tyrosine). (C) Phosphorylation site within P450b. Radioactively phosphorylated P450b was digested with trypsin and the resulting tryptic peptides were separated by HPLC with a Vydac C18 reverse-phase column using a linear gradient of acetontrile. The column eluate was monitored at 210 nm (a), the eluted fractions were collected and tested for radioactivity (b). The radioactive peptide was sequenced. For experimental details see Ref. 40.

digestion of radioactively phosphorylated P450b followed by separation of the phospho-peptides via reversed phase high performance lipid chromatography resulted in only a single radioactive peak (C). This phosphopeptide was sequenced and found to have the composition F-S*-L-A-T-M-R. The sequence can unambiguously be assigned to the primary structure of P450b (42). Accordingly, the phosphorylated serine residue is Ser-128 which is flanked in the N-terminal direction by -Arg-Arg-Phe-. It therefore belongs to the sequence -Arg-Arg-Phe-Ser-, the kinase recognition sequence (1,2). The rat liver P450 isoenzyme therefore has the identical phosphorylation site as the rabbit liver enzyme. It seems, however, to become phosphorylated quicker and to a higher degree.

Is the phosphorylation observed with the major phenobarbital-inducible P450 isoenzymes a general phenomenon? The answer is clearly no. Of a series of 14 cytochrome P450 isoenzymes roughly half of them were phosphorylated by the cyclic AMP-dependent protein kinase (Table II) (43). A differential phosphorylation pattern was

Table II. Phosphorylation of various cytochrome P450 by cyclic AMP-dependent protein kinase and by Ca^{2+}/phospholipid-dependent protein kinase.

Cytochrome P450	Species	Relative extent of phosphorylation	
		Catalytic subunit of cyclic AMP-dependent kinase	Ca^{2+}/phospholipid-dependent kinase
LM 2	rabbit	+	-
LM 3c	rabbit	-	(+)
LM 4	rabbit	-	(+)
PB 1a	rat	-	+
PB 1b	rat	-	-
PB 2a	rat	-	+
PB 2d	rat	+	-
PB 2e	rat	(+)	+
PB 3a	rat	++	++
PB 3b	rat	+	-
MC 1a	rat	-	+
MC 1b	rat	-	-
UT 1	rat	-	-
Thromboxane-synthase	human	-	-

Degree of phosphorylation was estimated semi-quantitatively by densitometric scanning and expressed relative to either cytochrome P450 LM 2 (+; use of catalytic subunit of cyclic AMP-dependent protein kinase) or P450 PB2a (+; use of Ca^{2+}/phospholipid-dependent kinase). ++, strong phosphorylation; +, weak phosphorylation; -, no significant phosphorylation.
For experimental details and classification of P450 isotypes see Ref. 43.

also reported by Jansson et al. (44), who also used a random selection of P450 isoenzymes. Independent of any quantitative differences of the phosphorylation, it obviously had a strict isoenzyme specificity. In order to find the reason for this, we systematically analyzed the available amino acid sequences of P450 isotypes for the presence of an equivalent of Ser-128 (40).

P450 family	amino acid sequence	Form

	110 **P**	
I	GQSMTFNPDSGPLWAARRRLAQNALKSFSIASDPTLAS--SCYLEEHVSKE	rat c
IIA	GYGVAFS--SGERAKQLRR-------LSIATLRDFGVG-KRGVEERLLEE	rat a
IIB	EYGVIFA--NGERWKALRR-------FSLATMRDFGMG-KRSVEERIQEE	rat b
IIC	GLGIVFS--SGEKWKETRR-------FSLTVLRNLGMG-KKTIEERIQEE	rabbit 3b
IIC	GFGIVFS--NGNRWKEMRR-------FTIMNFRNLGIG-KRNIEDRVQEE	rat f
IID	GVILASY---GPEWREQRR-------FSVSTLRTFGMG-KKSLEEWVTKE	rat db1
IIE	NKGIIFN--NGPTWKDVRR-------FSLSILRDWGMG-KQGNEARIQRE	rat j
III	KKAVSISED---DWKRVRTLL-----------SPTFTSGKLKEMLPIIAQY	rabbit 3c
IV	GYGLLLL⸱⸱NGQPWFQHRRML-----------TPAFHYDILKPYVKNMADS	rat LAω
XI	PIGVLG-KKSG-TWKKDRVVL------NTEVMAPEAIKN-FIPLLNPVSQD	bovine SCC
XVII	QKGIAFA-DHGAHWQLHRKLALNA---FALFKDGNLK------LEKIINQE	bovine 17 α
XIX	EKGIIFN-NNTELWKTTRP-------FFMKALSGPGLVRMVTVCAESLKT	human arom
XXI	YPDLSLG-DYSLLWKAHKKLT---------RSALLLGI--RDSMEPVVEQL	human C21
LI	GKGVIYDCPNS-RLMEQKK--------FVKGALT---KEAFKSYVPLIAEE	yeast lan
CI	PTSM--DP---PEQRQFRALA------NQVV-----GMPVVDKLENRIQEL	P. putida cam

Figure 9. Comparison of amino acid sequences near the phosphorylation site of representatives of each family of the P450 gene superfamily. Underlined, kinase recognition sequence; P, phosphorylated serine (threonine) residue; boxed, the highly conserved arginine. P450 classification according to Nebert et al. (4). For details of sequence alignment and source of amino acid sequences see Ref. 40.

As shown in Fig. 9, Ser-128 equivalents flanked N-terminally by basic amino acid residues were found not to be statistically distributed over the P450 superfamily but to be restricted rather to isoenzymes of family II. The members of this family, which metabolize foreign compounds such as drugs (4), carry this serine residue or have a threonine residue at this site (e.g., rat f) which, however, accepts phosphoryl groups as well. Only the family II members possess the primary structure motif of basic amino acid residues close N-terminally to the Ser-128 equivalent. Other P450 isotypes such as family I members (e.g., rat c) do have the serine residue but the basic amino acid residues are separated by nine other residues instead of one thus destroying the kinase recognition sequence. Members of family III through XXI and the yeast P450lan as well as the *P. putida* P450cam do not show an equivalent of Ser-128.

This alignment resulted in a most suprising additional finding (40): Arg-125, which is part of the kinase recognition sequence, is present in P450 throughout all families (Fig. 9; boxed). No exception has been found, i.e., the Arg-125 equivalent exists down the evolutionary tree, even in microorganisms. In a few cases, the arginine residue is exchanged for a lysine residue, i.e., a conservative change had occurred. Such a highly conserved element should play an essential role for P-450. It is not known yet exactly what this role is, but it is tempting to speculate that it is an essential structural element which has a role in the binding of the heme. In P450cam, in fact, it definitely has been shown to play such a role (E. Johnson, this book).

What we have discovered about the phosphorylation of P450 isotypes through comparison of primary structures has been and will be a subject of experimental concern. As yet, prediction and observation perfectly match each other in a series of rabbit and rat isotypes of family II members and those of other families which we have isolated to homogeneity (41). The phosphorylated forms include aside others (37), isoenzymes such as P450j, the alcohol-inducible isoform from rat liver belonging to family II, which was also found by others (45), and P450 LgM2 from sheep lung. The latter was purified to homogeneity by E. Arinc and her group. The primary structure of P450 LgM2 is unknown but its properties are those of family II isoforms (46). (This is, by the way, a nice example of the application of the phosphorylation reaction as a classification tool.).

Phosphorylation of P450 by PK C and CK II

P450 isotypes may also become substrates of the Ca^{2+}/phospholipid-dependent protein kinase. As with PK A, PK C shows a pattern of selectivity for P450 isoenzymes (43). The patterns, however, differ but overlap (see Table II). Several of the P450-isoenzymes such as P450b were found to be substrates for both kinases; others such as P450PB1a or P450PB2a proved to be substrates only for PK C or not to be substrates for either kinase. This sort of a selective pattern of phosphorylation by PK C has, in principle, been confirmed recently by Epstein et al (47). The most promising report, however, came from an investigation by Vilgrain et al. (48). This group reported on the phosphorylation of P450scc, an enzyme of the inner mitochondrial membrane. It was shown to accept up to 4 mol phosphate/mol P450 within 15 min, and to have an apparent K_m value as low as 0.14 µM. Since this group also claims to have indications for the presence of PK C activity in inner mitochondrial membranes, their observation might possibly imply an involvement of PK C in the tuning of P450scc activity. Both a direct influence on P450scc steroidogenic activity and an indirect influence on the transfer of the cytosolic precursor of P450 into the mitochondrion were suspected.

Experiments with the "acidic" kinase CK II were less successful. None of the P450 isotypes probed so far for phosphorylation has shown any notable incorporation of phosphoryl groups, although some of them possess recognition sequences for CK II.

EFFECT OF PHOSPHORYLATION

The phosphorylation of family II P450 isoenzymes occurs at a serine residue which is close to a highly conserved arginine residue. Hence, phosphorylation introduces two negative charges near an essential basic element of the P450 molecule. It therefore should not be a "silent" phosphorylation but should have some effect on the P450 molecule. This is in fact the case.

Effect on the metabolic activity of P450

When rabbit P450LM2 was phosphorylated by PK A and the monooxygenase activity was measured in the reconstituted system, both the NADPH-dependent N-demethylase activity (benzphetamine) and O-deethylase activity (7-ethoxycoumarin) dramatically decreased (Fig. 10) (36). Neither incubation with ATP in absence of kinase nor incubation with kinase in the absence of ATP showed this effect. Phosphorylation of P450b by PK A decreased its monooxygenase activity within minutes (40). Koch and Waxman reported the same in a recent paper (49). Thus the introduction of a phosphoryl group on Ser-128 has an inactivating effect on P450.

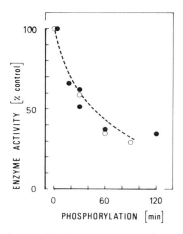

Figure 10. Effect of phosphorylation of P450LM2 on O-deethylase and N-demethylase activity. P450LM2 preincubated with dilauroylphosphatidylcholine than phosphorylated by PK A as described in Fig. 1, combined with NADPH-P450 reductase, and tested in the presence of NADPH for O-demethylase activity employing 7-ethoxycoumarin (●) or for N-demethylase activity employing benzphetamine (o). For experimental details see Ref. 36.

Effect on the conformation of P450

As pointed out above, phosphorylation might act via a structural change in a protein. A sensitive indicator of the conformational status of P450 is its light absorption spectrum. This was therefore analyzed more closely. Incubation of P450LM2 with PK A and ATP resulted in a slight red shift of the Soret band in the oxidized state thus indicating in fact some conformational change (Fig. 11A) (50). When the same preparation was reduced under saturating carbon monoxide, the observed change was rather dramatic. The major Soret peak shifted from 450 nm to 420 nm. The shift occurred in a time-dependent manner: the longer the phosphorylation, the stronger the shift. Measuring the phosphorylation degree in parallel it turned out that phosphorylation degree and conversion of the Soret peak from 450 nm to 420 nm nicely matched each other (37). The same effect has also been observed with the respective rat P450 (40) (Fig. 11B). Thus, P450 changes its conformation due to phosphorylation to give P420. P420 is known as the inactivated form of P450. Consistently, the effect was prevented by the addition of the heat- and acid-stable inhibitory protein specific for PK A (37,40). If incubated further in the presence of ATP and kinase, the total heme protein concentration decreased indicating heme detachment from P420 (50).

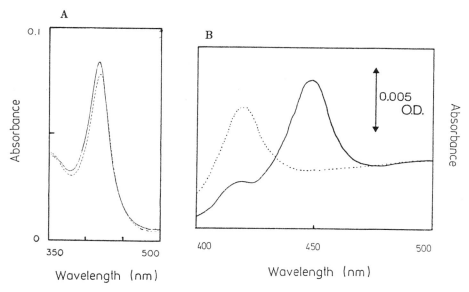

Figure 11. Effect of phosphorylation of P450LM2 on conformation. (A) Absolute spectrum in the oxidized state. Solid line, spectrum of P450LM2 before incubation with catalytic subunit of cyclic AMP-dependent protein kinase and ATP and that of P450LM2 incubated with kinase in the absence of ATP; broken line, spectrum of P450LM2 after incubation in the presence of kinase and ATP. For experimental details see Ref. 50. (B) CO-difference spectrum. Solid line, spectrum of P450b before incubation with kinase and ATP; broken line, spectrum of P450b after incubation in the presence of kinase and ATP. For experimental details see Ref. 40.

This actually is a strong hint that the highly conserved Arg-125 is involved in heme binding: introduction of a phosphoryl group near to it results in a conformational change creating a situation which, unlike that in unphosphorylated P450 and P450 isoforms lacking the phosphorylation site, releases heme easily to give the P450 apoprotein.

Further changes observed through phosphorylation

An intriguing aspect is the influence phosphorylation may have on the interaction of P450 with other component proteins of the monooxygenase system as well as with phase II enzymes. To this end, Schenckman and his group have carried out a series of experiments the results of which indicate a role of phosphorylation in the interaction of P450 and b5 (44,47,51; and this book).

All the phosphorylation experiments shown so far were carried out in cell-free systems with purified P450 isotypes or microsome-bound P450 and purified kinases thus bringing the in vivo relevance into question. This led us to investigate phosphorylation of the major phenobarbital-inducible P450 of rat liver cells. The hypothesis was that treatment of rats with phenobarbital should induce phosphorylatable P450 forms in their livers. Perfusion of the livers with collagenase-containing buffer should then separate the hepatocytes from each other and give a suspension culture, the so-called primary hepatocytes, consisting of cells which possess, in contrast to permanent cell lines, a high content of phosphorylatable P450. In order to activate the kinase, i.e., to dissociate the C-subunit of PK A from R-subunits (see above), the intracellular level of cyclic AMP had to be increased. This can be achieved directly or indirectly (Fig. 12). A direct way is the addition of membrane-permeating derivatives of cyclic AMP to the cell

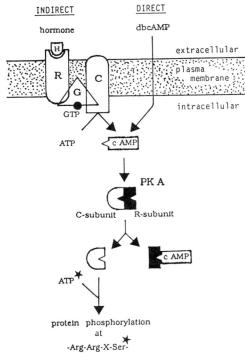

Figure 12. Activation of intracellular PK A activity by directly and indirectly increasing the cellular cyclic AMP level. H, hormone such as glucagon; R, hormone receptor; G, GTP-binding protein; C, adenylate cyclase; dbcAMP, dibutyryl cyclic AMP.

suspension, cyclic AMP itself being unable to traverse the plasma membrane. Such a derivative was N^6-$O^{2'}$-dibutyryl cyclic AMP. An indirect way is the stimulation of adenylate cyclase, an enzyme residing in the plasma membrane facing the cytoplasmic surface. The cyclase is activated via GTP-binding proteins which are, in turn, activated by receptor proteins, whose activity is dependent on binding of a suitable ligand such as the proteohormone glucagon. The phosphorylation of P450 should then occur, always presupposing the phosphorylation site, Ser-128, is accessible to the kinase in vivo and the released kinase molecules migrate to endoplasmic reticulum-bound P450. Finally, for the detection of phosphorylation, the intracellular ATP-pool had to be labelled since extracellularly added [γ-^{32}P]ATP cannot penetrate the plasma membrane. This can be achieved by the addition of ^{32}P-phosphate to the cell medium which labels the exchangeable phosphate pool of the cells including, among others, the γ-phosphate of ATP.

The results obtained fully verified the hypothesis (40). After SDS-polyacrylamide gel electrophoresis followed by autoradiography, the microsomal fraction of hepatocytes showed a dramatic increase in the phosphorylation of microsomal protein migrating with authentic P450b when the intracellular level of cyclic AMP was increased either directly by dibutyryl cyclic AMP or indirectly by glucagon (Fig. 13). Thus, a similar result was obtained as with isolated microsomes and isolated C-subunit of PK A. To make sure that the observed in vivo-phosphorylation was not an artifact produced by damaged cells or during preparation of the microsomes, a number of controls were carried out including addition of cyclic AMP to the culture medium and to all preparation buffers employed to provoke this artifact, and, on the other hand, addition of non-radioactive ATP to all preparation buffers at levels sufficiently high to suppress any radioactive phosphorylation. No noticeable phosphorylation of the microsomal protein was observed in the first case or any change of the phosphorylation pattern in the second. Moreover, hepatocytes prepared from rats not pretreated with phenobarbital and examined in parallel lacked cyclic AMP-stimulatable phosphorylation of the microsomal protein.

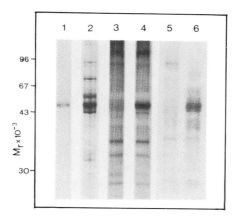

Figure 13. Phosphorylation of microsomal proteins in intact hepatocytes. Hepatocytes isolated from phenobarbital-pretreated rats were incubated in the presence of ^{32}P$_i$ with and without dbcAMP or glucagon. The microsomal fraction was analyzed by SDS-polyacrylamide gel electrophoresis followed by autoradiography. Shown are Coomassie blue stains (lanes 1 and 2) and autoradiograms (lanes 3 to 6). 1, isolated P450b; 2, microsomal protein pattern; 3 and 4, microsomal phosphoprotein pattern of hepatocytes incubated in the absence and in the presence of dbcAMP, respectively; 5 and 6, microsomal phosphoprotein pattern of hepatocytes incubated in the absence and in the presence of glucagon, respectively, with theophylline present in both.

Two different methods were employed to verify that the phosphorylated protein was really P450, namely, comparative peptide mapping and immunoprecipitation. For the comparative mapping, the phosphorylated protein bands were cut out from the gel, as described above for the in vitro-phosphorylation of microsomes, and partial proteolysis undertaken (see Fig. 6). The peptide fragments of the microsomal proteins suspected to be P450 were indeed highly similar to that of authentic P450b, as were the phosphopeptide patterns. Proteins migrating above and below P450 produced clearly different patterns (40).

For immunoprecipitation, antibodies against purified P450b were raised in rabbits and the serum made monospecific by affinity chromatography on P450b-bound Sepharose. This antibody preparation precipitated only one radioactive protein from the microsomal, detergent-solubilized fraction which migrated with authentic P450b in SDS-polyacrylamid gel electrophoresis (Fig. 14). As a control, the solubilized microsomes were treated with immunoglobulin obtained from the preimmune serum. In this case, very little, if any, radioactivity was precipitated (40).

1 2 3

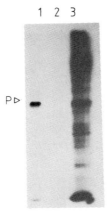

Figure 14. Immunoprecipitation of microsomal protein of hepatocytes co-migrating with P450b. Hepatocytes from phenobarbital-pretreated rats were incubated in the presence of $^{32}P_i$ and dbcAMP. The microsomal fraction was solubilized with detergents, incubated with monospecific antibody against P450b or with IgG obtained from preimmune serum, and precipitates subjected to SDS-polyacrylamide gel electrophoresis followed by autoradiography (1, precipitate with monospecific antibody; 2, precipitate with IgG from preimmune serum; 3, solubilized microsomes). P, position of purified P450b. For experimental details see Ref. 40.

These results clearly established the identity of the phosphorylated cellular protein as cytochrome P450.

The only amino acid phosphate ester detectable upon acidic hydrolysis of the P450 phosphorylated in intact cells was, as in the cell-free system, serine phosphate. The identity of the phosphopeptide maps of P450 phosphorylated in vivo and in vitro indicates that it is the same serine residue which becomes phosphorylated (40).

Does the phosphorylation reach a degree which would indicate physiological significance? To obtain an answer, the specific radioactivity of ATP was determined in cell extracts by measuring radioactivity and ATP concentration by a luciferin-luciferase ATP-assay combined with measuring ^{32}P-incorporation into phosphorylase b by phosphorylase kinase (52). Under the conditions employed, the degree of phosphorylation was in the range of 1-10%. This is a range which should be sufficiently high to be of some physiological significance. In fact, the P450-dependent metabolic activity of hepatocytes assessed with pentoxyresorufin as the substrate decreased faster in the presence of dibutyryl cyclic AMP than in control cells. However, this issue was not to be investigated further since, in our hands, primary hepatocytes are not suited for this kind of measurement. The P450 content of these cells is not stable; it decreases rather rapidly even in control cells (40).

Two reports from other laboratories appeared recently confirming our results. The one report came from Bartlomowicz et al. (51) and shows that under similar conditions as those we have used a significant decrease of pentoxyresorufin dealkylation as well as of testosterone metabolism occurs paralleled by a significant change of the relative amounts of diverse testosterone metabolites (Table III).

Table III. Isoenzyme-selective repression of monooxygenase activities in hepatocytes treated with cyclic AMP derivatives.

Treatment of hepatocytes	7-Pentoxy-resorufin dealkylation[a]	Testosterone metabolites[b] (nmol hydroxytestosterone/mg protein/min)						
		16β-OH	16α-OH	7α-OH	6β-OH	2β-OH	2α-OH	A
None (control)	2565 ± 81	0.81±0.10	2.27±0.20	0.49±0.07	1.10±0.21	~0.1[c]	0.51±0.05	2.45±0.25
cAMP derivatives[d]	1680 ± 52* (65)*	0.45±0.07* (56)*	1.56±0.14* (68)*	0.36±0.04* (73)	0.94±0.25 (85)	~0.1[c]	0.51±0.10 (100)	1.83±0.25 (74)

Microsomes were prepared from hepatocytes of phenobarbital-pretreated adult male Sprague-Dawley rats and incubated with 7-pentoxyresorufin or testosterone. The alkylation of 7-pentoxyresorufin was monitored by spectrophotofluorimetry. Testosterone metabolites were resolved and quantitated by HPLC. Values represent means ± S.E.M., numbers in brackets represent percent of control. [a] as (pmol resorufin/mg protein/min). [b] No further testosterone metabolite peaks unequivocally above background noise were noted in the HPLC chromatogram. The abbreviations denote the hydroxylated testosterone metabolites formed, "A" represents androst-4-ene-3,17-dione, the oxydation product of the 17β-hydroxy group of testosterone. [c] Clearly identifyable metabolite, but too close to background for accurate quantification. [d] 1 mM N6-dibutyryl-cAMP and 0.5 mM 8-thio-methyl-cAMP. * p < 0.05 compared with corresponding control.

Reproduced with permission from ref. 51.

The other report came from Koch and Waxmann (49). By following our procedures exactly, these authors not only obtained comparable results on the phosphorylation of P450 in hepatocytes but showed in addition and most importantly that the phosphorylation of phenobarbital inducible P450 can also occur in the livers of whole animals (49). Injection of dibutyryl cyclic AMP and theophylline to phenobarbital-induced rats after ^{32}P-phosphate application resulted in a microsomal phosphoprotein pattern which includes radioactive protein of the size of P450. Immunoprecipitation allowed then in fact to identify phosphorylated P450.

Together, the phosphorylation of P450 appears more and more to have become an established fact for the in vivo-situation as well. It therefore is tempting to assume that the regulation of P450 at the cellular level has to be extended to include the post-translational level. In addition to the well-known induction of P450, an isoenzyme-specific, postranslational modification by phosphorylation has to be considered as a regulatory device (Fig. 15). It is not known at the moment whether it is reversible (see below). Because the generated P420 loses its heme so easily, it probably is a mechanism of accelerated degradation, i.e., a mechanism to control the intracellular level of P450. This would be in line with the degradative process in other regulatory enzymes where phosphorylation acts as a device marking proteins for degradation (54).

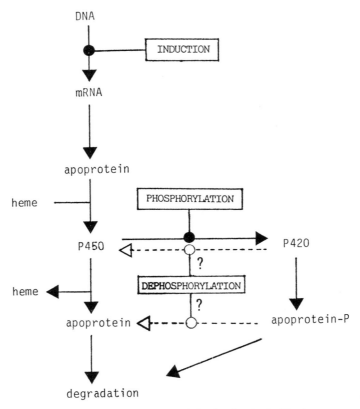

Figure 15. Regulation of hepatic P450 at the cellular level extended by isoenzyme-specific protein phosphorylation.

THE PHOSPHATASE EFFECT

Having established the phosphorylation of P450, one of the first questions was whether it might show reversibility. As pointed out above (see Fig. 2), the reverse reaction is catalyzed by phosphoprotein phosphatases, which hydrolytically remove phosphate groups from phosphoproteins. When protein phosphorylation/dephosphorylation became recognized as a widespread mechanism of regulation, the question arose

222

as to whether each protein was phosphorylated by a specific kinase and dephosphorylated by a corresponding phosphatase, or whether the interconverting enzymes
reacted with many different protein substrates. It appears now that a number of different phosphatases exist and that some of them are rather unspecific, i.e., they are
rather similar and show broad overlapping substrate specificites, while others are
clearly distinct and dephosphorylate a narrow range of substrates.

Phosphoprotein phosphatases

Phosphatases (phosphoprotein phosphohydrolases, EC 3.1.3.16) are a large
family of enzymes which presently are classified into three categories (55,56). Types 1
and 2 are the phosphoseryl/phosphothreonyl phosphatases, type 3 is the phosphotyr-
osyl phosphatases. There are a number of subclasses of types 1 and 2 which are
defined on the basis of their susceptibility to inhibition by heat-stable inhibitory proteins and of their activation by compounds such as basic proteins, Ca^{2+}/calmodulin,
divalent metal ions, etc., and of their preferences for certain phosphoproteins such as
the α and β subunit of phosphorylase kinase. The preferred substrate spectrum then
defines types 1, 2A, 2C, and 3 as phosphatases with broad specificity, type 2B with
narrow specificity. The catalytic subunits of the phosphatases are single proteins with
M_r in the range of roughly 40-60 kD.

The phosphatases employed were of the unspecific type, particularly phosphatases working optimally at pH values above 7. Such "alkaline phosphatases" are
present in almost all living organisms. Their physiological function, however, is still
largely obscure. In mammals several forms are distinguishable, namely, a so-called
liver/bone/kidney type, also known as tissue-unspecific type (in humans mapped to
chromosome 1), an intenstinal type (in humans mapped to chromosome 2), and additional types such those found in human term placenta. They all seem to be the product of a series of duplications of a single ancestral gene (55,56).

The rat liver alkaline phosphatase is an inducible enzyme, induced by ligating
the bile-duct combined with colchicine treatment, and further, expression is modulated by glucocorticoids, vitamin D, parathormone, etc. The rat liver alkaline phosphatase gene structure has recently been reported. It interestingly has two alternative
leader exons, i.e. a single gene with two promoters. As would be expected, cDNA-anal-
ysis has shown two types of mRNAs with divergent 5'-untranslated structures followed
by a common mRNA part (57).

The effect of phosphatase on microsomal monooxygenase

Incubation of liver microsomes from phenobarbital-treated rabbits, rats and
mice with alkaline phosphatase had a strong effect on monooxygenase activity: the
activity dramatically decreased (3,58,59). Thus the dephosphorylating enzyme had
the same effect on monooxygenase activity as the phosphorylating enzymes; there was
no simple reversion of the kinase effect. An artifact such as enzyme degradation by a
protease present in assays as an impurity of the phosphatase preparation was
unlikely since addition of protease inhibitor cocktails including phenylmethylsulfonyl-
fluoride (PMSF), Nα-p-tosyl-L-lysine-chloromethyl-ketone (TLCK), pepstatin A, and
trypsin inhibitor had no significant influence while addition of a phosphatase inhibitor
such as 5 mM EDTA was inhibitory (60). How can this apparent dichotomy between
kinase and phosphatase effect be explained?

As shown in Table IV, P450 has obviously little to do with the phosphatase effect. When investigating the effect of the phosphatase on microsomal P450 content and on conversion of P450 to P420, no significant change either in microsomes of rabbit liver nor in those of rat liver was observed (58,59). Likewise, reactions catalyzed by P450 alone, i.e. independent of any other protein component, were not affected either. For example, cumene hydroperoxide-dependent N-demethylase activity measured with N,N-dimethylaniline as substrate, did not change after the phosphatase treatment (60). However, when analyzing the influence of phosphatase on microsomal reductase

Table IV. Effect of alkaline phosphatase on microsomal monooxygenase activity, content of P450, and NADPH-P450 reductase activity. Microsomes from livers of phenobarbital-treated rabbits and rats were treated with alkaline phosphatase followed by determination of 7-ethoxycoumarin O-deethylation, CO-difference spectrum, and cytochrome c reduction. For experimental details see Ref. 59.

source of microsomes	Phosphatase treatment [min]	Monooxygenase activity[1] [%]	Cytochrome content[2] P-450	Cytochrome content[2] P-420	Reductase activity[3] [%]
rabbit liver	0	100	86	14	100
	60	10	83	17	8
rat liver	0	100	89	11	100
	60	30	86	14	23

[1] 7-ethoxycoumarin O-deethylation; absolute values (mean ± S.D.; n=5) were 2.2 ± .0.1 and 0.9 ± .0.1 nmol/min · mg protein of rabbit and rat microsomes, respectively

[2] taken from CO-difference spectra; quadruplicate samples; sum total of P-450 and P-420 set 100%

[3] cytochrome c reduction; absolute values (mean ± S.D.; n=4) were 0.22 ± 0.01 and 0.25 ± 0.01 μmol/min · mg protein of rabbit and rat microsomes, respectively.

activity, a pronounced effect was noted. Reductase activity measured as cytochrome c reduction dramatically decreased in both the rabbit and rat liver microsomes as well as in mouse liver microsomes (58,59). The decrease was of the same order of magnitude as that of the monooxygenase activity. Moreover, the kinetics of the decrease in monooxygenase activity and in reductase activity were comparable, as was their dependence on the phosphatase concentration (60). Obviously, the phosphatase effect on monooxygenase activity was brought about by an inactivation of the NADPH-P450 reductase.

In order to ensure that the phosphatase directly inactivates NADPH-P450 reductase, the enzyme was purified from the microsomal fraction (41,60) and incubated with phosphatases under the conditions employed with microsomes. This resulted in a time-dependent decrease of activity resembling the deactivation kinetics of the microsome-bound enzyme (Fig. 16). Thus the reductase is in fact the inactivated target of the phosphatase.

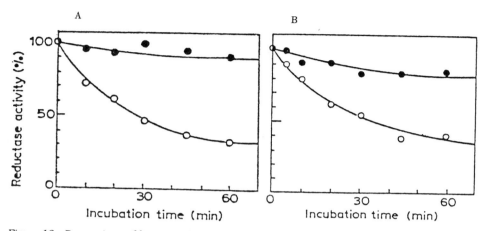

Figure 16. Comparison of kinetics of the alkaline phosphatase effect on microsome-bound and purified NADPH-P450 reductase. Microsomes from livers of phenobarbital-treated rabbits (A) and NADPH-P450 reductase purified from such microsomes (B) were incubated with alkaline phosphatase. At indicated times, aliquots were taken and tested for cytochrome c reductase activity (○). As control, microsomes and reductase were incubated in the absence of phosphatase (●). For details see Ref. 60.

Mechanism of the phosphatase effect

Dephosphorylation appears unlikely as the mechanism of inactivation because the reductase does not become significantly phosphorylated (see Fig. 1 and Table V), i.e., the phosphorylation degree obtained by any of the kinases employed remained below 0,05 mol phosphate/mol reductase.

The first hint as to the mechanism came from the determination of ferricyanide reduction in comparison to cytochrome c reduction (60). While electron acceptors such as cytochrome c - or P450 - need the presence of both prosthetic groups of the reductase, i.e., FAD and FMN, artificial electron acceptors such as ferricyanide can be reduced by the reductase even when it is depleted in FMN. Therefore, FMN depletion was achieved by dialysis against KBr, and found to show a much stronger effect on cytochrome c reduction than ferricyanide reduction (60). Since the phosphatase is known to accept FMN as a substrate but lacks pyrophosphatase activity and hence does not attack FAD, it was tempting to assume that the phosphatase effect on the reductase was caused by a loss in FMN. This hypothesis was tested.

The flavins of untreated and of phosphatase-treated reductase were extracted and analyzed chromatographically followed by fluorimetric determination of FAD, FMN and riboflavine (60). The ratio of FMN to FAD in untreated reductase preparations was approximately 0.8, i.e., both prosthetic groups were present in comparable amounts.

In contrast, phosphatase-treated reductase had unchanged FAD but lost roughly 80% of FMN. Riboflavine had increased at the expense of FMN (Fig. 17). This can easily be explained by the hydrolysis of the terminal phosphate group of FMN.

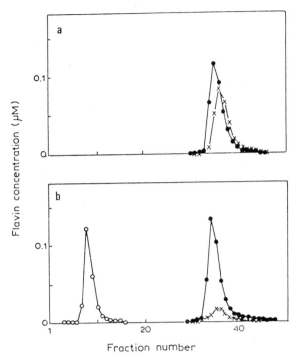

Figure 17. Flavin analysis of NADPH-P450 reductase treated by alkaline phosphatase. Purified reductase was incubated without (a) and with (b) alkaline phosphatase. Flavins were extracted and injected directly into a FPLC Mono Q Column. Concentrations of riboflavine (RF; o), FAD (●), and FMN (×) were determined in eluting fractions. For experimental details see Ref. 60.

The phosphatase therefore acts by an interruption of the electron transport chain from NADPH to cytochrome P450 by hydrolyzing the terminal phosphate ester of FMN and so inactivating NADPH-P450 reductase activity. As a consequence, the monooxygenase activity decreases. In contrast to the isoenzyme-specific phosphorylation of P450 and its selective effect on monooxygenase activity, the phosphatase effect concerns the overall monooxygenase activity.

PHOSPHORYLATION AND DEPHOSPHORYLATION OF FURTHER MONOOXYGENASE COMPONENTS

Phosphorylation

Significant phosphorylation of component proteins of the monooxygenase system other than P450 isoenzymes of family II was not observed (43). NADPH-P450 reductase was, as pointed out already, a poor substrate for the kinases employed, NADH-dependent cytochrome b5 reductase and cytochrome b5 were not phosphorylated either (Table V).

Table V. Phosphorylation of various enzymes involved in drug metabolism by cyclic AMP-dependent and by Ca^{2+}/phospholipid-dependent kinase.

Enzyme	Species	Relative extent of phosphorylation	
		Catalytic subunit of cyclic AMP-dependent kinase	Ca^{2+}/phospholipid-dependent kinase
Cytochrome P-450-reductase	rabbit	(+)	(+)
Cytochrome b5	rabbit	-	-
Cytochrome b5-reductase	rabbit	-	(+)
Epoxide hydrolase[a]	rabbit, rat	-	-
GSH-transferases[b]			
3-3	rat	-	-
1-2	rat	(+)	++
4-4	rat	(+)	+

For experimental details and estimation of phosphorylation degree see Ref. 43.

[a] Microsomal epoxide hydrolase (mEH$_b$) with broad substrate specificity for various xenobiotic epoxides, specifically including benzo(a)pyrene 4,5-oxide.

[b] Glutathione transferases A (Yb1 Yb1), B (Ya Yc), and X (Yb2 Yb2) have been termed 3-3, 1-2, and 4-4, respectively (61).

Phosphatase treatment

When the same phosphatase treatment of microsomes was carried out as described before and NADH-dependent ferricyanide reductase activity or NADH-dependent cytochrome c reductase activity was tested, no change was observed. Since NADH-dependent ferricyanide reduction is catalyzed by NADH-dependent cytochrome b5 reductase and NADH-dependent cytochrome c reduction by NADH-dependent cytochrome b5 reductase together with cytochrome b5, phosphatase has obviously no effect on cytochrome b5 and on its reductase. Further, the phosphatase has no effect on P450 (see above).

PHOSPHORYLATION OF PHASE II ENZYMES: THE GLUTATHIONE-S-TRANS-FERASES

Of the phase II enzymes screened for phosphorylation, a few proved to be substrates for one or the other of the three protein kinases employed. The only phosphorylation occurring to any mentionable degree and hence worth pursuing was that of certain isoenzyme types of the glutathione-S-transferase (GST) by the Ca^{2+}/phospholipid dependent PK C (43; see Table V). GST constitutes a family of isoenzymes catalyzing the conjugation of various compounds with glutathione on the sulfur atom of its cysteine residue thereby playing a major role in the detoxification of various compounds. In addition, GST binds with high affinity a variety of hydrophobic compounds such as heme, bilirubin, aryl hydrocarbons, and dexamethasone (61; and F. Oesch, this book). GST isoenzymes are homodimers or heterodimers comprising at least seven subunits (61). They constitute the transferases 1-1, 1-2, 2-2, etc. Six of these, namely 1-1, 1-2, 2-2, 2-3, 3-3, 3-4 were purified from rat liver cytosol and incubated with PK C in the presence and absence of Ca^{2+}/phospholipids. A differential phosphorylation was observed (27). Of the six GSTs, only isoenzymes 1-1, 1-2 and 2-2 were significantly phosphorylated in a strict Ca^{2+}/phospholipid-dependent manner, while the others incorporated phosphoryl groups to a considerably lower extent (Fig. 18). The composition of an isoenyzme has little effect on the subunit phosphorylation, i.e. the homodimers (1-1, 2-2) showed roughly twice as strong a phosphorylation than the respective subunits alone and the heterodimers (1-2), the sum of the phosphorylation of the two different subunits.

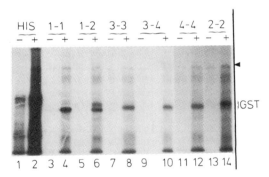

Figure 18. Phosphorylation of GST isoenzymes by Ca^{2+}/phospholipid-dependent kinase. GST isoenyzmes purified from rat liver cytosol were incubated with PK C in the absence (-) and in the presence (+) of Ca^{2+}/phospholipid followed by TCA-precipitation, SDS-polyacrylamide gel electrophoresis, and autoradiography. HIS, histone phosphorylation (control); 1-1, 1-2 etc, phosphorylation of respective GST isoenzyme; arrow head, autophosphorylated PK C; GST, position of GST isoenzymes. For experimental details see Ref. 27.

The phosphorylation of the best kinase substrate, the homodimer 1-1, reached after 60 min a plateau at a level corresponding to 1 mol phosphate/mol subunit. In the absence of Ca^{2+} and phospholipid, little phosphorylation was observed. The time course of the phosphorylation suggested that only one amino acid residue is specifically phosphorylated (27).

The homodimer 1-1 is known as "ligandin" (61). In addition to the binding capabilities mentioned above, this isoenzyme is involved also in lipid peroxidation. One of the main functions of "ligandin" is binding of bilirubin. This function was investigated in order to test for the significance of the phosphorylation. GST 1-1 was phosphorylated and bilirubin binding determined by measuring the quenching of the intrinsic tryptophan fluorescence through the bound bilirubin followed by fitting the data to a mass equation assuming one binding site per subunit (27). The effect observed was that GST 1-1 showed a decreased affinity towards bilirubin upon phosphorylation (Fig. 19). The dissociation constant was calculated to be 0.7 µM after

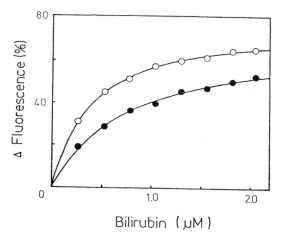

Figure 19. Effect of phosphorylation on the binding of bilirubin to GST 1-1. GST 1-1 incubated with kinase C in the presence (●) and in the absence (o) of Ca^{2+} and phospholipids followed by determination of the bilirubin connected quenching of intrinsic tryptophan fluorescence. For experimental details see Ref. 27.

phosphorylation while 0.32 µM was calculated for the control. The addition of Ca^{2+} and phospholipid in the absence of kinase C had little, if any, effect on bilirubin binding. Thus the incorporation of a phosphoryl group into GST 1-1 lowers the affinity towards bilirubin. Since ligandin is an abundant protein constituting about 5% of the total cytosolic protein of hepatocytes, the cellular concentration is much higher than the concentration adjusted in the phosphorylation reaction shown. Thus, the phosphorylation should also occur in vivo. In fact, there are reports by Satoh et al. (61) on the presence of "charge isomers" of GST in rat liver cytosol. Such heterogeneities could easily be explained by a posttranslational modification by phosphorylation.

References

1. Krebs, E.G., Beavo, J.A. (1979) Annu. Rev. Biochem. **48**, 923-959.
2. Edelmann, A.M., Blumenthal, D.K., Krebs, E.G. (1987) Annu. Rev. Biochem. **56**, 567-576
3. Pyerin, W., Wolf, C.R., Kinzel, V., Kübler, D., Oesch, F. (1983) Carcinogenesis **4**, 573-576
4. Nebert, D. et al. (1989) DNA **8**, 1-13.
5. Nairn, A.C., Detre, J.A., Casnellie, J.E., Greengard, P. (1982) Nature **299**, 734-736
6. Fischer, E.H., Krebs, E.G. (1955) J. Biol. Chem. **216**, 121-132
7. Sutherland and Wosilait (1955) Nature **175**, 169-170
8. Shacter, E., Chock, P.B., Rhee, S.G., Stadtman, E.R. (1986) In: The enzymes (Boyer, P.D., and Krebs, E.G., eds.), Vol. XVII, Academic Press, London, 21-42
9. Bailey, S.W., Dillard, S.B., Thomas, K.B., Ayling, J.E. (1989) Biochemistry **28**, 494-504
10. Lechner, M.C., Braz, J. (1985) Eur. J. Biochem. **151**, 621-624
11. Black, S.D., Coon, M.J. (1986) In: Cytochrom P-450 - Structure, Mechanism and Biochemistry. (Ortiz de Montellano, P.R., ed.), Plenum Press, New York/London, 161-216
12. Schmidt, M.F.G. (1989) Biochim. Biophys. Acta **988**, 411-426.
13. Ozols, J., Carr, S.A., Strittmatter, P. (1984) J. Biol. Chem. **259**, 13349-13354
14. Cohen, P. In: The enzymes (Boyer, P.D., and Krebs, E.G., eds.), Vol. XVII, Academic Press, London, 461-497
15. Büchler, J.A., Taylor, S.S. (1989) Biochemistry **28**, 2065-2070
16. Taylor, S.S. et al. (1988) Cold Spring Harbor Symp. Quant. Biol. **53**, 121-130
17. Cartwright, C.A., Eckhart, W., Simon, S., Kaplan, P.L. (1987) Cell **49**, 83-91
18. Kinzel, V., Hotz, A., König, N., Gagelmann, M., Pyerin, W., Reed, J., Kübler, D., Hofmann, F., Obst, C., Gensheimer, H.P., Goldblatt, D., Shaltiel, S. (1987) Arch. Biochem. Biophys. **253**, 341-349
19. Hotz, A., König, N., Taniguchi, H., Chrivia, J.C., Kinzel, V. (1989) Biochem. Biophys. Res. Commun. **160**, 596-601
20. Uhler, M., Chrivia, J., McKnight, G.S. (1986) J. Biol. Chem. **261**, 15360-15363
21. Carr, A.A., Biemann, K., Shoji, S., Parmelee, D.C., Titani, K. (1982) Proc. Natl. Acad. Sci. USA **79**, 6128-6131
22. Kinzel, V., Kübler, D. (1976) Biochem. Biophys. Res. Commun. **71**, 257-264
23. Pyerin, W., Gagelmann, M., Kübler, D., Kinzel, V. (1979) Z. Naturforsch. (Biosci.) **34**, 1186-1194
24. Kikkawa, V., Nishizuka, Y. (1986) In: The enzymes (Boyer, P.D., and Krebs, E.G., eds.), Vol. XVII, Academic Press, London, 461-497
25. Parker, P.J., Kour, G., Marais, R.M. Mitchell, F., Pears, C., Schaap, D., Stabel, S., Webster, C. (1989) Mol. Cell. Endocrinol. **65**, 1-11
25a. Parker, P.J., Coussens, L., Totty, N., Rhee, L., Young, S., Chen, E., Stabel, S., Waterfield, M.D., Ullrich, A. (1986) Science **233**, 853-859
26. Wooten, M.W., Vandenplas, M., Nel, A.E. (1987) Eur. J. Biochem. **164**, 461-467
27. Taniguchi, H., Pyerin, W. (1989) Biochem. Biophys. Res. Commun. **162**, 903-907
28. Tuazon, P.T., Traugh, J.A. (1990) In: Advances in Second Messenger and Phosphoprotein Research (Greengard, P., Robinson, G.A., eds.) Vol. **23**, in press
29. Pyerin, W., Burow, E., Michaely, K., Kübler, D., Kinzel, V. (1987) Biol. Chem. Hoppe-Seyler **368**, 215-227

30. Meisner, H., Heller-Harrison, R., Buxton, J., Czech, M.P. (1989) Biochemistry **28**, 4072-4076.

31. Jakobi, R., Voss, H., Pyerin, W. (1989) Eur. J. Biochem. **183**, 227-233

32. Fouts, J.R. (1962) Fed. Proc. **21**, 1107-1111

33. Weiner, M., Buterbaugh, G.G., Blake, D.A. (1972) Res. Commun. Chem. Pathol. Pharmacol. **3**, 249-263

34. Weiner, M., Buterbaugh, G.G., Blake, D.A. (1972) Res. Commun. Chem. Pathol. Pharmacol. **4**, 37-50

35. Hutterer, F., Dressler, K., Greim, H., Czygan, P., Schaffner, F., Popper, H (1975) Adv. Exptl. Med. Biol. **58**, 117-125

36. Pyerin, W., Taniguchi, H., Stier, A., Oesch, F., Wolf, C.R. (1984) Biochem. Biophys. Res. Commun. **122**, 620-626

37. Pyerin, W., Taniguchi, H. (1989) In: Biochemistry of Chemical Carcinogenesis (Hradec, J., Garner, R.C., eds.) IRL-Press, Oxford, in press

38. Pyerin, W., Marx, M., Taniguchi, H. (1986), Biochem. Biophys. Res. Commun. **134**, 461-468

39. Müller, R., Schmidt, W.E., Stier, A. (1985) FEBS Lett. **187**, 21-24

40. Pyerin, W., Taniguchi, H. (1989) EMBO J. **8**, 3003-3010

41. Taniguchi, H., Pyerin, W. (1989) J. Chromatogr. **476**, 299-308

42. Fujii-Kuriyama, Y., Mizukami, Y., Kawajiri, K., Sogawa, K., Muramatsu, M. (1982) Proc. Natl. Acad. Sci. USA **79**, 2793-2797

43. Pyerin, W., Taniguchi, H., Horn, F., Oesch, F., Amelizad, Z., Friedberg, T. Wolf, C.R. (1984) Biochem. Biophys. Res. Commun. **142**, 885-892

44. Jansson, I., Epstein, P.M., Satinder, B., Schenkman, J.B. (1987) Arch. Biochem. Biophys. **259**, 441-448

45. Johansson, I., Eliasson, E., Johansson, A., Hagbjörk, A.-L., Lindros, K., Ingelman-Sundberg, M. (1989) In: Cytochrome P450 - Biochemistry and Biophysics (Schuster, I., ed.) Taylor & Francis, London/New York/Philadelphia, 592-595

46. Adali, O., Arinc, E. (1989) NATO Advanced Study Institute on Molecular Aspects of Monooxygenases and Bioactivation of Toxic Compounds, Cesme/Izmir, Abstracts, p. 265

47. Epstein, P.M., Curti, M., Jansson, I., Huang, C.-K., Schenkman, J.B. (1989) Arch. Biochem. Biophys. **271**, 424-432

48. Vilgrain, I., Defaye, D., Chambaz, E.M. (1984) Biochem. Biophys. Res. Commun. **125**, 554-561

49. Koch, J.A., Waxman, D.J. (1989) Biochemistry **28**, 3145-3152

50. Taniguchi, H., Pyerin, W., Stier, A. (1985) Biochem. Pharmacol. **34**, 1835-1837

51. Jansson, I., Epstein, P.M., Curti, M., Schenkman, J.B. (1989) NATO Advanced Study Institute on Molecular Aspects of Monooxygenases and Bioactivation of Toxic Compounds, Cesme/Izmir, Abstracts, p. 267

52. England, P.J., Walsh, D.A. (1976) Anal. Biochem. **75**, 429-435

53. Bartlomowicz, B., Friedberg, T., Utesch, D., Molitor, E., Platt, K., Oesch, F. (1989) Biochem. Biophys. Res. Commun. **160**, 46-52

54. Rivett, A.J. (1986) Curr. Topics Cell. Regul. **28**, 291-337

55. Ballou, L.M., Fischer, E.H. (1986) In: The enzymes (Boyer, P.D., and Krebs, E.G., eds.), Vol. XVII, Academic Press, London, 311-361

56. Cohen, P. (1989) Annu. Rev. Biochem. **58**, 453-508

57. Toh, Y., Yamamoto, M., Endo, H., Misumi, Y., Ikehara, Y. (1989) Eur. J. Biochem. **182**, 231-237

58. Pyerin, W., Jochum, Ch., Taniguchi, H., Wolf, C.R. (1986) Res. Commun. Chem. Pathol. Pharmacol. **53,** 133-136
59. Pyerin, W., Horn, F., Taniguchi, H. (1987) J. Cancer Res. Clin. Oncol. **113,** 115-159
60. Taniguchi, H., Pyerin, W. (1987) Biochim. Biophys. Acta **912,** 295-302
61. Pickett, C.B., Lu, A.Y.H. (1989) Ann. Rev. Biochem. **58,** 743-764
62. Satoh, K., Kitahara, A., Sato, K. (1985) Arch. Biochem. Biophys. **242,** 104-111

PHYSIOLOGICAL AND PATHOPHYSIOLOGICAL ALTERATIONS IN RAT HEPATIC

CYTOCHROME P-450+

John B. Schenkman, Kenneth E. Thummel and Leonard V. Favreau

Department of Pharmacology
University of Connecticut Health Center
Farmington, CT   06032

INTRODUCTION

Cytochrome P-450 is a large family of related enzymes comprising a number of sub-families. The different cytochrome P-450 enzymes are present in just about every phylum examined for them. They are found in prokaryotes [1], unicellular eukaryotes [2], plants [3], fowl [4], insects [5], fish [6], invertebrates [7], and mammals [8]. While the prokaryotic enzyme, e.g., P-450CAM, is a soluble hemoprotein, that of higher organisms is membrane- bound. In mammals it is located in the mitochondrial inner membrane as well as in the endoplasmic reticulum membranes. In the mammal, different forms of cytochrome P-450 are involved in such diverse reactions as steroid hormone synthesis, at several steps, and the processing of compounds of exogenous origin as well as of endogenous origin for excretion from the body. The enzymes involved in the elimination of compounds are located in the endoplasmic reticulum of most organs and tissues in the body. The very large number of such forms found suggests roles for these other than just broad, overlapping spectra of chemical detoxication.

One of the important characteristics of cytochrome P-450 is its ability to respond to xenobiotic challenge. Administration to animals of pharmacological or toxicological doses of a diverse collection of compounds results in the selective elevation of certain forms of cytochrome P-450 in microsomal fractions of different tissues. For example, P-450c (IA_1) and P-450b (IIB_1) are elevated respectively by challenge of rats with 3-methylcholanthrene and phenobarbital [9]. In this paper we discuss a number of forms of cytochrome P-450 that respond to changes in the internal milieu. These P-450 forms are constitutive in the liver microsomes, some specific to the male and some to female rat, and some present in both sexes; their response to homeostatic changes in the animal indicate possible roles for these forms in homeostasis.

MATERIALS AND METHODS

Animals - Male or female rats, 8 weeks of age (Charles River, CD) were administered either 65mg/kg streptozotocin (Sigma) in 0.05M sodium citrate, pH 4.5, i.v. or the vehicle alone. The animals were allowed

+Reprinted with permission from Drug Metabolism Reviews, 20 (2-4) 557-584 (1989).

Molecular Aspects of Monooxygenases and Bioactivation of Toxic Compounds
Edited by E. Arınç et al., Plenum Press, New York

unlimited access to food (laboratory chow) and water. The extent of diabetes was monitored by urine glucose levels using Bili Lab Stix (Ames, Elkhart, IL). Diabetic rats given insulin received 10-15 IU/kg of Isophane insulin (Squibb) daily beginning the second week after streptozotocin. Rats given testosterone propionate received 1mg/kg/day, s.c. for 1 or 2 weeks. Spontaneous hypertensive rats (SHR) and their Wistar/Kyoto (WKY) controls were also obtained from Charles River Breeders. All animals were kept on a 12 hr light/dark cycle for at least one week before use.

Preparation of enzymes - RLM2 was purified by the method of Jansson, et al. (10), RLM2b by the method of Thummel, et al. (11), RLM3 and RLM5 by the method of Cheng and Schenkman (12) fRLM4 by the method of Schenkman, et al. (13), and RLM5b and RLM6 by the method of Favreau, et al. (14).

Preparation of antibodies - Antibodies were raised in New Zealand White rabbits or goat as described earlier (14, 15). The antibodies were made monospecific by absorption to cross-reacting antigens that were cross-linked to CNBr-activated Sepharose (16). For antibodies specific for the female P-450, fRLM4, male rat liver microsomes were cross-linked to the Sepharose. For antibodies to male-specific forms of cytochrome P-450, microsomes from female rats were cross-linked to the Sepharose, and the IgG fraction of the antiserum was purified by passage through a column of this antigen-bound Sepharose. Further purification of antibodies to male forms of cytochrome P-450 made use of Sepharose-coupled CMII fraction (contains RLM3, RLM5a and RLM5), CMI fraction (contains RLM2) and CMIII fraction (contains RLM6 and RLM2b). Monospecific antibodies to RLM5 were prepared as described (17). With the exception of anti-RLM2 and anti-RLM2b, all of the monospecific antibodies yielded single bands on Western blots of microsomes or single reactivity in 2-dimensional gel (isoelectric focussing/SDS-PAGE) Western blotting. Immuno-purified antibodies to RLM2 and RLM2b each recognized both RLM2 and RLM2b on Western blots of male rat liver microsomes and could not be made monospecific (11).

Analytical procedures - Western blots were performed by electrophoresis of microsomal samples or purified enzymes in SDS-PAGE, 9% acrylamide gels followed by transfer of the separated proteins to nitrocellulose sheets (0.2μM pore size, Schliecher and Schuell) as described [15, 18]. After electrophoretic transfer, protein bands were visualized by staining with ponceau S (Sigma). This dye is readily rinsed off with water or buffer prior to binding of the primary antibodies. Immunodetected proteins were quantitated by scanning densitometry, using alkaline phosphatase-conjugated anti-rabbit IgG (or anti-goat IgG for RLM5) as a second antibody and 5-bromo-4-chloro-3-indolyl phosphate/nitroblue tetrazolium for color development (Kirkegaard & Perry). In some instances quantitation was carried out using ^{125}I-labelled goat anti-rabbit IgG (New England Nuclear). Known amounts of the purified antigens were electrophoresed along side of the microsomal samples for preparation of standard curves.

Acetone was measured using a Varian model 3700 gas chromatograph equipped with a 6 ft. X 2mm glass column packed with 0.2% CW1500 on 80/100 mesh carbopack (Supelco). The injection temperature used was 160° with an oven temperature of 70°C. Collected blood samples were allowed to clot and then were centrifuged to obtain serum. Measurements of acetone and glucose were generally made within 24 hrs, but samples were stored at -80°C before use. A standard curve of acetone was run before each assay. Glucose was quantitated using the glucose oxidase kit (Sigma), monitoring absorbance at 450nm and a standard curve was run each time.

234

RESULTS

Constitutive forms of P-450 - A number of forms of cytochrome P-450 are constitutive, i.e., normally present, in liver microsomes of the untreated rat. These are shown in Table I. Of these, five forms are sex specific. Forms 2, 3, 5 and p are male-specific [10, 11, 13, 19] and f4 is female-specific [13]. The percentages of the different forms vary between the sexes. In the male rat, the major forms 3, 5 and 5a comprise some 83% of the spectrally determined total P-450. Form RLM5b is also detected at a level of about 15% of the P-450. The reported level of P-450p, 20% (form PB2a in ref. 19), appears higher than what is seen in our preparations of microsomes, although we have as yet not quantitated it. The total yield of P-450 proteins obtained by immunochemical means exceeds 100%, perhaps indicating the presence of apo-P-450 which would not be detected spectrophotometrically. In female rat liver microsomes, we find fRLM4 to be only about 20% of the total P-450 content, and the concentration of 5a to be about the same as in the male; because of the slightly lower levels of total P-450 in the female, 5a represents about 25% of the total. Imaoka, et al. [23], however, found f4 and 5a to each comprise about 45% of the total P-450 by immunochemical methods.

All of the forms in Table I have been tested for their ability to metabolize testosterone in reconstituted systems, and, with the exception of forms 6 [14] and UT8 [23], are capable of such metabolism. The constitutive P-450 enzymes exhibit broad, but overlapping, isomeric and epimeric specificities in their metabolite patterns. However, some of these forms have been shown to have major (>85%) responsibility for certain metabolites produced by liver microsomes [19, 24-26], despite the ability of other forms to produce these metabolites in the reconstituted system. This makes such catalytic activities excellent markers for specific enzymes when studying the effects of various conditions which perturb the levels of cytochromes P-450 in the microsomes.

Table I

Constitutive Forms of Rat Liver P-450

RLM	% P-450[f] M	F	Testosterone Metabolites	Gene	Inducer[e]
2	3.5	–	*15α,7β(?),6β,7α,16α	IIA2	–
2b	2.3	12	*7α,6α	IIA1	PCB, MC
3	23	–	6β,15α,16α,7α	IIC13	–
f4	–	27	*15α	IIC12	–
5	40	–	*2α, *16α, 17, 6β	IIC11	–
5a	20	g	2α, 16α, 17	IIC6	PB(?)
5b	15	g	16α	IIC7?	EtOH
6	9.8	14	–	IIE1	EtOH, Ac, isoniazid
p	20[a]	–	*6β	IIIA1	PCN, glucocorticoids
d	<5[b]		6β	IA2	MC, α-NF
e	<1[c]		16α, 16β	IIB2	PB
UT1	?[d]		2α, 16α		–
UT7	?[d]		6β		–
UT8	?[d]	–			–

Data calculated from reference: a = [19], b = [20], c = [21], d = [22],
      e: –, None; PB, phenobarbital; PCB, polychlorinated biphenyls; EtOH,
      ethanol; PCN, pregnenolone –16α-carbonitrile; MC,
      3-methylcholanthrene; α-NF, α-naphthoflavone.
f =   Immunochemically determined P-450 as a % of total spectrally determined
      P-450.
g =   not quantitated, but Western blot analysis indicates a similar content
      (nmole/mg) in male and female rats. * = Major microsomal metabolite by
      this P-450.

Although the forms of cytochrome P-450 shown in Table I are all constitutive in adult rat liver microsomes, a number of them have been reported to change in content on challenge of the animal with various xenobiotics. However, to date, some of the forms have not been seen to be induced by the various externally administered compounds; these are indicated in Table I by a dash. Elevation of several of the forms of P-450 by xenobiotics is often accompanied by a decline in levels of other forms. The fact that agents have not, as yet, been found which elevate levels of the sex-specific forms of P-450, RLM2, 3, f4, and 5 should not be taken as an indication that these forms cannot be induced. Levels of these forms can change, depending upon in vivo conditions, as will be shown below, and it is probably only a matter of time before some agent or other will be found as a perturbant of these forms.

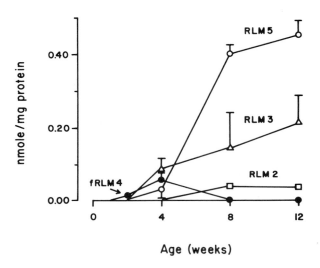

Age (weeks)

FIGURE 1.   Developmental changes in four sex-specific hepatic microsomal forms of cytochrome P-450 in the male rat. Western blots of microsomes were immunoquantitated with monospecific polyclonal antibodies as described under Materials and Methods.

Developmental changes - Several forms of cytochrome P-450 have been shown to be under developmental control, e.g., forms p [19], f4 [19, 27, 28], 5 [27, 29, 30], 3 [31], 2b [11, 19, 26], 2 [11, 32], and 6 [33, 34]. Both RLM3 and RLM5 (Figure 1), as well as RLM2 [11, 32] are not detected in liver microsomes from the immature rat of both sexes, but begin to appear in the male with the onset of puberty, reaching adult levels by 8 weeks of age. These enzymes never appear in liver microsomes of the female rat. fRLM4 is also absent in the newborn rat liver, and is expressed during sexual maturation (4 weeks, Figure 1). Expression is maintained only in the adult female rat and disappears with sexual development in the male rat.

RLM2b is not a sex-specific form of P-450. It does, however, undergo changes in content in liver microsomes of the developing rat [11, 19, 26]. It is present in the neonate of both sexes; levels rise to peak at 2 weeks of age in hepatic microsomes, but then decline in the male after sexual development.

In the adult rat, microsomal levels of RLM2b are approximately 4-fold higher in the female than in the male [11]. In contrast, RLM6 is present at low levels in microsomes of the neonate of both sexes, rising with age to peak at 2 weeks [34]. Thereafter, levels in both sexes decline, reaching a minimum at the onset of puberty and remaining at this level through adulthood in both sexes [33, 34]. There is essentially no sex difference in levels of this form in liver microsomes.

Perturbations by diabetes - In order to study the mechanism of physiological control over the various forms of cytochrome P-450 in rat liver microsomes, pathophysiological conditions were employed as a tool to perturb homeostatic conditions in the animal. It has long been known that diabetic rats exhibit altered rates of drug metabolism [35, 36]. The alteration has been suggested to be the result of a change in the specific content of cytochrome P-450 forms in the liver microsomes [36]. In agreement with those reports, we have observed that aminopyrine metabolism was greatly diminished and aniline hydroxylation was increased several fold (Table II). Similarly, differential effects were seen in production of isomeric metabolites of testosterone. Thus, compared with the untreated male rat, the rate of 16α-hydroxytes- tosterone production was diminished to only 12% in the diabetic rat. In contrast, testosterone 6β-hydroxylation and 7α-hydroxylation rates were considerably elevated. All observed changes could be reversed by the administration of insulin. From the known isomeric and epimeric specificities of testosterone metabolism by the different cytochromes P-450, it is evident that changes in the microsomal P-450 composition occur in diabetes. Further, it is possible to suggest from the data in Table II that RLM5, RLM2b and P-450p would be among the forms undergoing compositional changes.

Table II

Influence of Diabetes in Adult Male Rats[a]

	Control	Diabetic	Diabetic +Insulin
		% of control	
Blood sugar	141[b]	350	80
Blood acetone	0.07[c]	57,580	400
Cytochrome P-450	0.9[d]	150	100
Aminopyrine demethylase	6.8[e]	53	80
Aniline hydroxylase	0.5[e]	280	80
Testosterone[d]			
7α-OH	0.6	154	96
16α-OH	2.3	12	54
6β-OH	0.7	141	89

a) data in part from [17]. Rats were killed 4 weeks after a single I.V. dose (65mg/kg) of streptozotocin. Some rats received insulin (3-5 IU/day) 1 week before killing.
b) mg/dl
c) mM
d) nmol/mg protein
e) nmol/min/mg microsomal protein

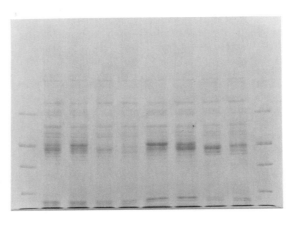

FIGURE 2. SDS-PAGE of liver microsomes from control and diabetic adult male and female rats. Lanes 2 and 3, and 6 and 7, were diabetic female and male, respectively. Lanes 4 and 5, and 8 and 9 were control female and male, respectively, four weeks after streptozotocin. Lanes 1 and 10 are molecular weight standards, 36K, 40K, 53K, 67K and 92K respectively, anode (bottom) to cathode.

In Figure 2 is shown the different proteins of liver microsomes of control and diabetic adult male and female rats, electrophoretically separated under denaturing conditions and stained with Coomassie Blue R. Cytochromes P-450 migrate in the 45-62 kDa region of the gel, and, as seen by the denser staining in this region, are major microsomal proteins. A comparison between the respective diabetic and control microsomes indicate clear differences exist; new dark bands present in the diabetic microsomes and some bands absent. Notable among the changing bands is one appearing at about 53 kDa in the diabetic microsomes. This band corresponds in molecular weight to RLM6, a form we isolated earlier [14] from the diabetic rat. In Figure 3 we see a comparison of levels of form 6 protein in male and female rat liver microsomes. The levels of form 6 protein is about the same in immature male and female microsomes as well as in adult microsomes. In the latter it is present at about 10% of the total P-450. In diabetes, the levels of form 6 rise about 4-fold (Figure 3), returning to control levels after two weeks of insulin therapy.

One other form of cytochrome P-450 also rises in diabetes, RLM5b. As reported earlier [15], three weeks after administration of streptozotocin levels of this enzyme had increased from 0.17nmol protein/mg microsomal protein to 0.28nmol protein/mg microsomal protein. Since the total microsomal cytochrome P-450 increased 50% in diabetes, this form remained about 17% of the microsomal P-450 content. Levels of both forms 6 and 5b

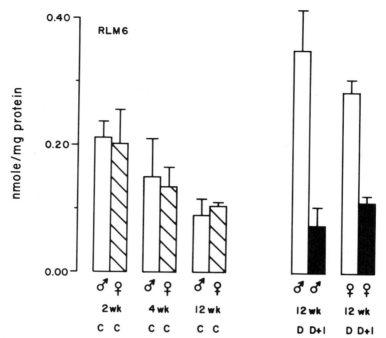

FIGURE 3. Developmental changes in RLM6 content in control male and female rate liver microsomes and the influence of diabetes and insulin on levels of this enzyme in the adult. Diabetic animals were given streptozotocin at 8 weeks of age and when administered, insulin therapy was initiated at 10 weeks of age. Enzyme quantitation was as in Figure 1.

were also elevated by 48 hr starvation [14], a condition metabolically similar to diabetes in that fat becomes a major source of total body energy needs. In both instances blood acetone levels are elevated. Chronic ethanol intake also elevated forms 6 and 5b [14], and, as noted earlier alcohol metabolism has several metabolic similarities to diabetes and starvation, with acetol and other metabolic products being generated [37]. As might be expected, acetone likewise was capable of rapidly elevating RLM6, when administered at a level of 1% in the drinking water [14]. RLM5b, however, was not elevated by acetone, and further, unlike form 6, was not induced by isoniazid [14]. LM3a, a form of rabbit cytochrome P-450 orthologous to RLM6, [38] and known to be inducible by acetone and ethanol [39] readily metabolizes acetone and acetol. Such activity was also found in rat liver microsomes [37] and shown to be catalyzed by RLM6 [14, 21].

Diabetes causes other changes in liver microsomal cytochrome P-450 concentration. As noted in Figure 2, some bands of lower molecular weight than form 6 were diminished in microsomes from the diabetic male rat. Two of these have been identified as RLM5 [17] and RLM3 [15]. A comparative time course of response of these two male-specific forms to diabetes is seen in Figure 4. Within 3 weeks of a single i.v. injection

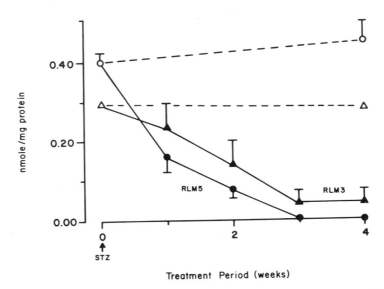

FIGURE 4.   Rapid decline in liver microsomal levels of RLM3 and RLM5 in
            adult male rats after onset of diabetes. Streptozotocin was
            administered to the rats at 8 weeks of age. Open symbols are
            untreated controls. Enzyme levels were quantitated as in
            Figure 1.

of streptozotocin (STZ), levels of form 5 had dropped below immunological
detection. Form 3 levels paralleled those of form 5, but did not
disappear completely. This precipitous drop in RLM5 explains the decline
in testosterone 16α-hydroxylase activity (Table II) in diabetes.
Levels of RLM5 are restored to normal by 2 weeks of insulin therapy, as
are levels of RLM3 (data not shown). Although form 3 levels declined,
and this major form of cytochrome P-450 primarily produces
6β-hydroxytestosterone from testosterone in the reconstituted system
[12], rates of 6β-hydroxytestosterone formation actually were increased
in diabetes, confirming a report that another form of P-450, P-450p, is
responsible for the observed microsomal 6β-hydroxylation of testosterone
[19].

    We have recently found [34] that in addition to RLM6 and 5b another
form of cytochrome P-450, RLM2b, is elevated after streptozotocin
treatment (Figure 5). A 5-fold increase is seen by 4 weeks, a time frame
during which minimal microsomal levels of forms 3 and 5 are found.
Levels of form 2b return to normal after 2 weeks of insulin therapy (not
shown). Form 2b is the major microsomal testosterone 7α-hydroxylase
[24-26], and the observed rise in this enzyme level corresponds to the
observed increase in catalytic activity in liver microsomes of the
diabetic rat (Table II).

    RLM2 is a male-specific form of cytochrome P-450, a testosterone
15α-hydroxylase with many structural and antigenic similarities to form
2b (10, 11). As a male specific form it might be expected to respond to
physiological perturbations in a manner similar to forms 3 and 5, two

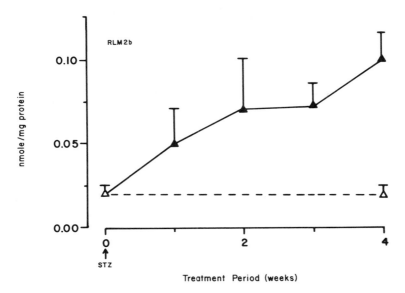

FIGURE 5.   Rise in RLM2b levels in liver microsomes of the adult male rat
after onset of diabetes.   Open symbols are untreated
controls.   Enzyme levels were quantitated as in Figure 1.
Treatments were as in Figure 4.

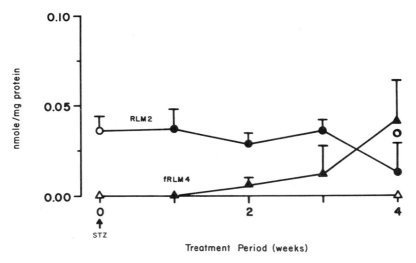

FIGURE 6.   Lack of change in levels of RLM2 and appearance of fRLM4 in
liver microsomes of adult male rats after development of
diabetes.   Treatments were as in Figure 4.   Enzyme levels were
quantitated as in Figure 1.

other male specific forms. However, unlike these two forms, RLM2 was largely unaffected by diabetes (Figure 6). No change in levels of form 2 were seen until 4 weeks after the administration of streptozotocin. In contrast, f4, a female specific form of P-450, not normally found in male rat liver microsomes, and also a testosterone 15α-hydroxylase, is just immunochemically detectable 2 weeks after the administration of streptozotocin (Figure 6). By the fourth week, readily quantitatable levels (0.040 nmol/mg microsomal protein) of fRLM4 were present. Other homeostatic changes must occur subsequently, as in chronic diabetic male rats (12 weeks) form f4 is not present in liver microsomes (data not shown).

Circulating testosterone levels have been reported to decline in the diabetic rat [40]. In our hands, using radioimmunoassay, plasma testosterone levels declined from a normal level of 3.8 ± 0.4 ng/ml in control male rats to 1.0 ± 0.4 ng/ml 4 weeks after streptozotocin administration. The administration of 1 mg/kg of testosterone propionate daily between weeks 2-4 of the study, resulted in plasma levels of 3.3 ± 1.0 ng/ml, and also resulted in a suppression of form f4 in the diabetic male rat (Figure 7). In contrast, form 2b, levels of which are much lower in adult male than adult female rats [11,26], remained elevated in diabetic rats treated with testosterone (Figure 7).

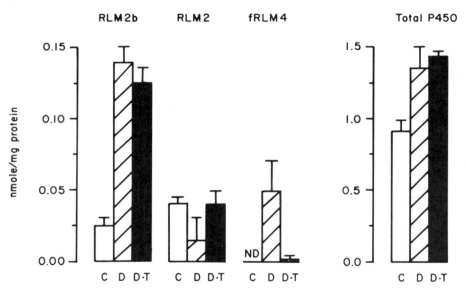

FIGURE 7. Influence of testosterone on liver microsomal levels of RLM2b, RLM2, fRLM4 and total cytochrome P-450 in the diabetic male rat. Animals were administered streptozotocin at 8 weeks of age, and where administered, daily testosterone was started at 10 weeks of age. All rats were killed at 12 weeks of age. Quantitation of enzymes were as in Figure 1. Total cytochrome P-450 levels were determined spectrophotometrically from the 450nm reduced carbon monoxide adduct.

Table III

Differences in adult male WKY and SHR rats. To determine these values, 10-12 week old rats were employed. Values are the average of three experiments.

	WKY	SHR	% of control
	100-125	170-200	
Blood Pressure (mm Hg)	72	99	(127%)
P-450 Reductase (units/mg)	0.56	0.71	(138%)
Specific Content (nmol P-450/mg)	7.6	12.5	(164%)
Aminopyrine N-demethylase (nmol/mg/min)	11.5	20.0	(174%)
Ethylmorphine N-demethylase (nmol/mg/min)	0.49	0.56	(114%)
Aniline hydroxylase (nmol/mg/min)	7.2	11.5	(160%)
Testosterone (nmol/mg/min)[a]	0.39	0.98	(251%)
6 β-hydroxylase	1.13	0.63	(56%)
7 α-hydroxylase	0.79	4.22	(235%)
16 α-hydroxylase			

[a] Total metabolites

243

Perturbations by hypertension – One other pathophysiological
condition known to have an influence on drug metabolism is hypertension.
Spontaneous hypertensive rats (SHR) have shorter hexobarbital sleeping
times and enhanced rates of microsomal ethylmorphine demethylase
activities [41, 42] than their Wistar/Kyoto (WKY) control. Since such
effects are suggestive of altered cytochrome P-450 populations, studies
were conducted using the SHR rat and its normotensive (WKY) control
(Table III). In the fully hypertensive adult male rat both total
microsomal cytochrome P-450 and NADPH-cytochrome P-450 reductase were
elevated to a small extent, compared to the control (WKY) rat. Rates of
both aminopyrine and ethylmorphine N-demethylation were increased
considerably more, 64 and 74%, respectively. In contrast, aniline
hydroxylase activity was essentially unchanged. Such data is indicative
of a change in the composition of microsomal forms of cytochrome P-450.
Examination of the products formed during testosterone metabolism by
hepatic microsomes from SHR and WKY age paired rats indicated a major
increase in the total catalytic activity in the hypertensive rat liver
microsomes (60%), and suggested that this change was accomplished by the
selective increase of certain forms of cytochrome P-450. In agreement,
the testosterone 6β-hydroxylase activity was elevated 2 1/2 fold, as was
the 16α-hydroxylase activity. In contrast, testosterone
7α-hydroxylase activity was half that of the age-matched WKY controls.
These changes are indicative of an increase in P-450p and RLM5, and a
suppression of RLM2b levels, respectively.

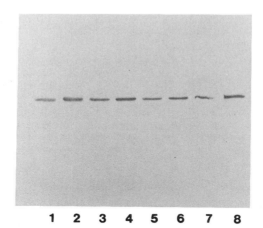

1   2   3   4   5   6   7   8

FIGURE 8a. Western blot analysis of levels of RLM5 in liver microsomes of
        age matched spontaneous hypertensive rats (SHR) and
        Wistar/Kyoto male rats (WKY). Odd numbered lanes were
        microsomes of SHR rats and even numbered lanes were microsomes
        of aged matched WKY rats at 10 weeks (lanes 1 and 2), 12 weeks
        (lanes 3 and 4), 14 weeks (lanes 5 and 6) and 15 weeks (lanes
        7 and 8). Blots were treated with monospecific polyclonal
        goat anti-RLM5 IgG and color was developed using alkaline
        phosphate conjugated rabbit anti-goat IgG.

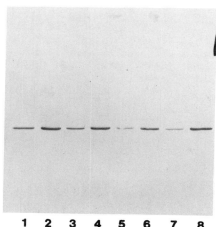

1 2 3 4 5 6 7 8

FIGURE 8b. Western blot analysis of levels of P-450p in liver microsomes
of age matched SHR rats and WKY male rats. Lanes 1, 3, 5, and
7 were microsomes from liver of 10 week, 12 week, 14 week and
16 week old male SHR rats, respectively. Lanes 2, 4, 6 and 8
were microsomes from age matched WKY rats, respectively. The
electrophoretic transfers were treated with mouse monoclonal
antibody to P-450 PCN (gift of Dr. Harry Gelboin) and color
was developed using alkaline phosphatase conjugated sheep
anti-mouse IgG.

To test whether these forms do, indeed, change in spontaneous hypertension, immunochemical detection means were employed. Figure 8 shows
Western blots of liver microsomes of SHR and age matched WKY control
rats. Levels of form 5 (A) were significantly higher in the SHR,
compared to WKY controls, as early as 10 weeks of age (track 2 vs 1).
This difference was maintained at least through 15 weeks of age (Figure
8A). Similarly, P-450p levels (Figure 8B) were several-fold higher in
the SHR rat liver microsomes (even tracks), compared with their
age-matched controls (odd tracks). As predicted, the liver microsomal
content of RLM2b was approximately 50% lower in SHR rats compared with
age-matched WKY control rats (Figure 8c).

Both RLM5 and P-450p are under developmental control, the former
appearing in male rat liver microsomes after the onset of puberty (Figure
1) [13, 19, 27, 29]. The latter enzyme is present in immature SHR rat
liver microsomes with levels declining slightly in males, but especially
suppressed in females at the onset of puberty [19]. Immature (3 week
old) WKY rats (Figure 9A, track 1, 2) do not have form 5. Similarly, the
enzyme is absent from immature SHR rat liver microsomes (track 3, 4).
The level of this enzyme in the 12 week old WKY rat is lower than in 12
week old SHR rat liver microsomes (tracks 5, 6 vs 7, 8). The immature
SHR rat liver microsomes had, however, considerably higher levels of
P-450p (compare Figure 9B, tracks 3, 4 vs tracks 1, 2) than the age
matched controls.

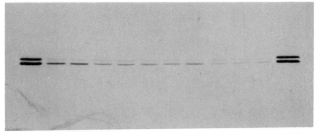

FIGURE 8c. Western blot analysis of RLM2b levels in liver microsomes of age matched SHR and WKY rats. Lanes 2, 3 and 6, 7, 8 were microsomes from 11 week and 15 week old WKY rats, respectively. Lanes 4, 5 and 9, 10, 11 were microsomes from 11 week and 15 week old SHR rats, respectively. Blots were treated with polyclonal rabbit anti-RLM2b IgG which recognized RLM2b (lower band) and RLM2 (upper band). Color was developed using alkaline phosphatase conjugated goat anti-rabbit IgG.

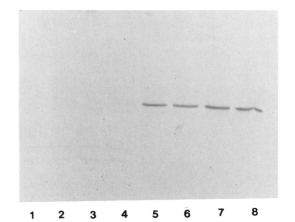

FIGURE 9a. Western blot analysis RLM5 levels in microsomes from immature and adult WKY and SHR male rats. Lanes 1, 2 and 5, 6 were microsomes from 3 week and 12 week old WKY rats, respectively. Lanes 3, 4 and 7, 8 were microsomes from 3 week and 12 week SHR rats, respectively. Blots were developed as in Figure 8a.

246

1 2 3 4 5 6 7 8

Figure 9b. Western blot analysis of P-450p levels in microsomes from
immature and adult WKY and SHR male rats. Lanes 1, 2 and 5, 6
were microsomes from 3 week and 12 week old WKY rats,
respectively. Lanes 3, 4 and 7, 8 were microsomes from 3 week
and 12 week old SHR rats, respectively. Blots were developed
as in Figure 8b.

Western blot analysis of SHR liver microsomes and their controls
(WKY) with anti-RLM5a IgG and anti-RLM6 IgG revealed (data not shown)
that there was no difference in levels of these enzymes. The latter is
in agreement with the absence of change in aniline hydroxylase activity
in the SHR rat (Table III).

DISCUSSION

The studies presented in this paper show clearly that the
constitutive forms of rat liver microsomal cytochromes P-450 are adaptive
enzymes, responding to changes in homeostatic conditions in the animal,
during both altered physiological and pathophysiological states. While
some of the constitutive forms of cytochrome P-450 also respond to
xenobiotic perturbation (RLM2b, 6, p), i.e., are induced to higher
levels, many of the forms are diminished in levels, and some forms do not
appear to change under xenobiotic challenge. However, almost all forms
of hepatic cytochrome P-450 vary in content in the microsomes under
certain conditions. Thus, some respond to development signals, some
respond to specific hormone levels, some respond to the absence of
specific hormones, some respond to xenobiotics and others respond to
changes in metabolic patterns. We suspect these changes to be an aid in
making life adaptive to changing environmental conditions.

Most of the constitutive cytochrome P-450 have been shown to be under
developmental control in the rat e.g., forms 2b [11, 19, 26], p [19], f4
[19, 27, 28], 5 [27, 29, 30], f [31], 3 [31], 6 [33, 34] and form 2 [11,
32].

The sex-specific forms, 2, 3, f4 and 5 are responsive to
physiological signals triggered during the period of sexual maturation.
In addition, a relationship has been established between unique patterns
of growth hormone secretion, which are neonatally imprinted [43-45] and
the appearance of form 5 in the male [29, 46, 47], and form f4 in the
female [28, 46-48]. Form 2 also begins to appear in the male rat liver
during puberty, however, its expression does not seem to be driven by the
male pulsatile secretion pattern of growth hormone, which appears to
regulate RLM5, and probably RLM3. Forms 2b, 6 and p are major components
of microsomal P-450 in the immature male and female rat liver, but
decline in relative content in one or both sexes as the rat continues to
develop. Growth hormone reportedly suppresses the microsomal
6β-hydroxylase, form p [49]. Forms 2b and p are suppressed in the male
[11, 26] and female [19], respectively, after the onset of puberty, while
the suppression of form 6 occurs after week 3, as the rat is weaned from
the maternal milk diet.

Many of the investigations that have examined the role of growth
hormone in the regulation of hepatic cytochromes P-450 in the rat have
relied heavily on use of the hypophysectomized rat model [28, 29, 32, 46,
49]. However, because the secretion of other trophic hormones in
addition to growth hormone, are also altered, as well as the secretion of
steroidal hormones from their target organs, the interpretation of
observed changes in hepatic P-450 may be overly simplified. We have
employed the use of a different animal model, the diabetic rat, as a tool
to further examine the regulation of the cytochrome P-450 enzymes.

In diabetes, a condition marked by altered energy metabolism, drug
[35, 50] and steroid [51] metabolism patterns are altered in liver
microsomes. It is now clear that these effects are the result of
changes in cytochrome P-450 isozyme composition in the microsomes.
Indeed, the level of form 2b increases some 5-fold, and forms 6 and 5b
increase 4 to 8-fold and 1.5-fold, respectively, in diabetic rat liver
microsomes [this paper,14]. Increases in levels of form 5b, 6, and b
(P-450b/e) are also observed in starvation [14, 21]. The decline in
testosterone 16α-hydroxylase activity in diabetes is the result of a
disappearance of form 5 from liver microsomes [15, 17]. RLM3, also a
male specific form of cytochrome P-450, likewise is severely diminished
in content in diabetic rat liver microsomes.

RLM5 has been shown to be under growth hormone control [29, 46],
requiring the male-specific pulsatile pattern of growth hormone (GH)
secretion [43] for complete expression. Since this pattern of secretion
of GH is lost within one day after administration of streptozotocin to
adult male rats [52], it is probable that herein lies the explanation for
the decline in levels of form 5 in diabetes. From the parallel decline
in levels of form 3, also a male-specific form of cytochrome P-450, it is
probable that this enzyme, too, is under control of growth hormone. In
contrast, levels of RLM2 which did not decline in diabetic male rats over
3 weeks and thus is probably not controlled by the episodic pattern of
growth hormone.

As already mentioned, growth hormone influences other forms of
cytochrome P-450 as well. P-450p, the microsomal 6β-hydroxylase, is
reportedly suppressed by growth hormone [49], perhaps explaining the rise
in testosterone 6β-hydroxylase activity [Table II] seen in the diabetic
rat. However, this enzyme is under developmental controls, rising after
birth to peak at about 4 weeks of age in the male and declining slightly

thereafter; in the female it reportedly declines more significantly after 2 weeks of age [19]. Presumably, the periods of very low blood GH concentration in the male rat [43, 44] allows for expression of form p, whereas in the female the moderate and more constant GH levels [43, 45] more strongly inhibits expression of P-450p [19].

The female testosterone 15α-hydroxylase, f4, also requires growth hormone for its expression [46-48]. It has been detected in liver microsomes of hypophysectomized [47] or intact [28] male rats receiving growth hormone infusions, or neonatally castrated male rats [28]. In this study, form f4 appeared in the 3rd and 4th week of diabetes, long after plasma growth hormone secretion has lost the episodic pattern with levels falling below trough concentrations found in the female, indicating some other factor(s) influenced its appearance. This factor may be plasma testosterone, the levels of which decline to less than 30% of normal by the fourth week after streptozotocin injection. Indeed, administration of testosterone, for two weeks prevented the appearance of f4 in the diabetic male liver microsomes. Form 2b, which rose immediately after streptozotocin administration, did not respond to testosterone treatment. This form is under developmental control [19, 26], rising to peak at one week of age and maintained at that level through adulthood in the female rat liver microsomes; in males the enzyme is suppressed after puberty [26]. If the male pattern of growth hormone secretion suppresses this enzyme, one would expect to see its levels rise in the diabetic male rat liver microsomes. Such an increase was observed, as noted in Figure 5, a response opposite to RLM5 (Figure 4). P-450b and P-450e, likewise, have been reported to be suppressed, both in male and female rats, by GH [53], in agreement with their elevation in starvation [21].

What is the significance of the changes in liver microsomal cytochrome P-450 composition? One possibility is that these forms are responsive to the homeostatic state of the animal and that changes in their contents in microsomes are attempts to adapt to perturbations from the norm. For example, one of the most marked cytochrome P-450 changes in diabetes is a 4 to 8-fold elevation of RLM6, a form present in both male and female rats. Microsomal levels of this enzyme do not appear to be controlled by growth hormone, which is high at birth, declines to a minimum by 13 days of age, and rises again by 30 days as the animals enter puberty [44]. In contrast, RLM6 is low at birth, rises to a maximum by two weeks of age, and declines somewhat between 2 and 4 weeks of age to levels seen in adult rats of both sexes [33, 34]. Although the decline in growth hormone after birth coincides with the rise in form 6, the subsequent decline in this enzyme preceeds the rise in growth hormone in puberty. Further, the different patterns of growth hormone secretion into the circulatory system of male and female rats is not reflected in differences in levels of form 6 between the sexes. On the other hand, starvation which influences form 6 as does diabetes [14], also influences the growth hormone secretion in a manner like diabetes in the male rat [54].

RLM6 levels may depend upon the metabolic state in the animal, suggesting a role for this enzyme. In newborn rats its levels rise as the animals begin nursing, and peaks just before the start of solid food ingestion after 2 weeks. Levels fall off sharply by 4 weeks [33, 34], after the animals have been fully weaned. Rat milk is 20% fat and fat provides 50% of the calories of the pups [55]. In untreated diabetes, conditions in which fat becomes a major source of energy for the body, due to an impairment in glucose uptake into cells, form 6 is elevated in the liver microsomes. Similarly, after 48 hours of starvation, when the body must depend upon fat as the major energy source, RLM6 is also elevated [14, 56]. Starvation and diabetic ketoacidosis, result in elevated levels of plasma ketone bodies, including acetone [57-60]. Acetone itself has been shown to both induce a specific type of P-450 mediated metabolism [61, 62] as well as P-450 RLM6 in rat [14, 63] and LM3a in rabbit [62].

As with diabetes and starvation, chronic excessive ethanol ingestion can cause elevated plasma acetone levels [57, 64]. Under these conditions RLM6 [14, 38] and LM3a [39] are both induced. Interestingly, RLM5b is also induced by chronic excessive ethanol ingestion, diabetes and starvation, but not by acetone or isoniazid [14]; RLM6 [14, 38] and isozyme 3a [62] are both induced by isoniazid, indicating control of hepatic RLM5b levels differs from that of RLM6.

Relationships between the many conditions that elevate RLM6-mediated metabolism [57], form 6 levels [14] and isozyme 3a levels [39] have been noted. Such conditions have in common the generation of elevated blood acetone levels. Argiles [65] has posed the question as to whether acetone plays a role in conversion of fat to carbohydrate, and concludes that acetone may, in certain metabolic conditions, be utilized by the body. In view of the observations by Casazza, et al. [37] of the existence of acetone and acetol monooxygenase activities in rats, hepatocytes and liver microsomes, the products of which initially yield acetol and methylglyoxal, respectively, the acetone monooxygenase pathways may serve as a salvage pathway under certain conditions, e.g., starvation to extract the most energy out of fat-derived metabolic products. Both activities have been shown to be catalyzed by the same enzyme in rabbits, isozyme 3a [39], and in rats, RLM6 [14, 21]. While Kosugi, et al. [66] have noted net incorporation of $2-^{14}C$ acetone into glucose and have concluded that this indicates some metabolism via lactate, major incorporation into carbon 3 and 4 of glucose (rather than $C_{1,2}$ and $C_{5,6}$) suggested pathways other than those proposed by Casazza, et al. [37]. Although more work is needed to determine the extent, in vivo, to which form 6 and its orthologs in other species contribute via salvage pathways to recapture acetone into intermediary metabolism, it is tempting to speculate that RLM6 and other constitutive forms of cytochrome P-450 have roles in maintaining homeostasis in vivo.

As shown in this study, other forms of cytochrome P-450, e.g., forms 5 and p, are altered in content in spontaneous hypertensive rats. Although the etiology of the hypertension developed by the SHR rat is still to be determined, it is noteworthy that both of the forms of cytochrome P-450 which are elevated have major contributions to microsomal metabolism of certain steroids in the rat. It will be of interest to determine the extent to which these forms contribute to steroid metabolism in the SHR rat, and perhaps other endogenous blood pressure regulating compounds, and the effect of inhibition of forms 5 and p.

In conclusion, a large number of forms of cytochrome P-450 exist in liver microsomes of the untreated rat. Since these have broad but overlapping substrate specificities for a large number of substrates, it is probable that their main function may relate to non-overlapping areas, e.g., the ability of RLM6 to metabolize acetone and acetol may be its role in vivo and forms p and 5 may be responsible for active steroid hormone elimination from the plasma. Alternatively, it is possible that these enzymes play a role in in vivo homeostasis by synthesizing biological messengers. The forms show isomeric and epimeric specificities in metabolism of endogenous compounds such as steroids [10, 67-69) fatty acids [70] and prostaglandins [71], hence it is possible such metabolies serve as effectors elsewhere in the body. Only time will tell as to which, if any, of the above suggestions are correct.

ACKNOWLEDGEMENT

We would like to acknowledge the expert help and advice of Dr. Ingela Jansson in preparing and finalizing this paper. This work is supported in part by USPHS grant GM26114.

REFERENCES

1. I.C. Gunsalus, Z. Physiol. Chem., 349, 1610 (1968).
2. A. Lindenmayer and L. Smith, Biochim. Biophys. Acta, 93, 445 (1964).
3. D.W. Russell, J. Biol. Chem., 246, 3870 (1971).
4. A. Poland and A. Kappas, Mol. Pharmacol., 7, 697 (1971)
5. J.W. Ray, Biochem Pharmacol., 16, 99 (1967).
6. J.J. Stegeman, R.L. Binder and A. Orren, Biochem Pharmacol., 28, 3431 (1979).
7. L.C. Quattrochi and R.F. Lee, Comp. Biochem. Physiol., 79C, 171 (1984).
8. T. Omura and R. Sato, J. Biol. Chem., 237, PC1375 (1962).
9. L.H. Botelho, D.E. Ryan and W. Levin, J. Biol. Chem., 254, 5635 (1979).
10. I. Jansson, J. Mole and J.B. Schenkman, J. Biol. Chem., 260, 7084 (1985).
11. K.E. Thummel, L.V. Favreau, J. Mole and J.B. Schenkman, Arch. Biochem. Biophys., 266, 319 (1988).
12. K.-C. Cheng and J.B. Schenkman, J. Biol. Chem., 257, 2378 (1982).
13. J.B. Schenkman, L.V. Favreau and I. Jansson in Biological Reactive Intermediates (J.J. Kocsis, D.J. Jollow, C.M. Witmer, J.O. Nelson and R. Snyder, eds.) Vol III, Plenum, N.Y. (1986) pp. 107-117.
14. L.V. Favreau, D.M. Malchoff, J.E. Mole and J.B. Schenkman, J. Biol. Chem., 262, 14319 (1987).
15. L.V. Favreau and J.B. Schenkman, Diabetes, 37, 577 (1988).
16. R. Axen, J. Porath and S. Ernback, Nature 214, 1302 (1967).
17. L.V. Favreau and J.B. Schenkman, Biochem. Biophys. Res. Commun., 142, 623 (1987).
18. H. Towbin, T. Staehelin and J. Gordon, Proc. Natl. Acad. Sci. USA, 76, 4350 (1979).
19. D.J. Waxman, G.A. Dannan and F.P. Guengerich, Biochemistry, 24, 4409 (1985).
20. P.E. Thomas, L.M. Reik, D.E. Ryan and W. Levin, J. Biol. Chem., 258, 4590 (1983).
21. I. Johansson, G. Ekström, B. Scholte, D. Puzycki, H. Jornvall and M. Ingelman-Sundberg, Biochemistry, 27, 1925 (1988).
22. Y. Funae, and S. Imaoka, Biochim. Biophys. Acta, 926, 349 (1987).
23. S. Imaoka, Y. Terano and Y. Funae, Biochim. Biophys. Acta, 916, 358 (1987).
24. D.J. Waxman, D.P. Lapenson, S.S. Park, C. Attisano and H.V. Gelboin, Mol. Pharmacol., 32, 615 (1987).
25. W. Levin, P.E. Thomas, D.E. Ryan and A.W. Wood, Arch. Biochem. Biophys., 258, 630 (1987).
26. K. Nagata, T. Matsunaga, J. Gillette, H.V. Gelboin and F.J. Gonzalez, J. Biol. Chem., 262, 2787 (1987).
27. K. Maeda, T. Kamataki, T. Nagai and R. Kato, Biochem. Pharmacol., 33, 509 (1984).
28. C. MacGeoch, E.T., Morgan, and J.A. Gustafsson, Endocrinol., 117, 2085 (1985).

29. E.T. Morgan, C. MacGeoch and J.A. Gustafsson, J. Biol. Chem., 260, 11895 (1985).
30. D.J. Waxman, J. Biol. Chem., 259, 15481 (1984).
31. S. Bandiera, D.E. Ryan, W. Levin and P.E. Thomas, Arch. Biochem. Biophys., 248, 658 (1986).
32. D.J. Waxman, G.A. LeBlanc, J.J. Morrissey, J. Staunton and D.P. Lapenson,J. Biol. Chem., 263, 11396 (1988).
33. P.E. Thomas, S. Bandiera, S.L. Maines, D.E. Ryan and W. Levin, Biochemistry, 26, 2280 (1987).
34. J.B. Schenkman, K.E. Thummel and L.V. Favreau, in Cytochrome P-450: Biochemistry and Biophysics pp. 851-854 (I. Schuster, ed), Taylor & Francis Pub., 1989.
35. R.C. Dixon, L.G. Hart and J.R. Fouts, J. Pharmacol. Exp. Ther., 133, 7 (1961).
36. M.R. Past and D.E. Cook, Biochem, Pharmacol., 31, 3329 (1982).
37. J.P. Casazza, M.E. Felver and R.L. Veech, J. Biol. Chem., 259, 231 (1984).
38. D.E. Ryan, D.R. Koop, P.E. Thomas, M.J. Coon and W. Levin, Arch. Biochem Biophys., 246, 633 (1986).
39. D.R. Koop and J.P. Casazza, J. Biol. Chem., 260, 13607 (1985).
40. F.T. Murray, J. Orth, G. Gunsalus, J. Weisz, J.B. Li., L.S. Jefferson, N.A. Musto and C.W. Bardin, Internat. J. Androl., 4, 265 (1981).
41. B.A. Merrick, M.H. Davies, D.E. Cook, T.L. Holcslaw and R.C. Schnell, Pharmacology, 30, 129, (1985).
42. P. Greenspan and J. Baron, Biochem, Pharmacol, 30, 678 (1981).
43. S. Eden, Endocrinology, 105, 555 (1979).
44. P. Walker, J.H. Dussault, G. Alvarado-Urbina and A. Dupont, Endocrinology, 101, 782 (1977).
45. S.R. Ojeda and H.E. Jameson, Endocrinology, 100, 881 (1977).
46. T. Kamataki, M. Shimada, K. Maeda and R. Kato, Biochem. Biophys. Res. Comm. 130, 1247 (1985).
47. R. Kato, Y. Yamazoe, M. Shimada, N. Murayama and T. Kamataki, J. Biochem., 100, 895 (1986).
48. C. MacGoech, E.T. Morgan, J. Halpert and J.-A. Gustafsson, J. Biol. Chem., 259, 15433 (1984).
49. Y. Yamazoe, M. Shimada, N. Murayama, S. Kawano and R. Kato, J. Biochem., 100, 1095 (1986).
50. R. Kato and J.R. Gillette, J. Pharmacol. Exp. Ther., 150, 285 (1965).
51. L.A. Reinke, H. Rosenberg and S.J. Stohs, Res. Comm. Chem. Pathol. Pharmacol., 19, 445 (1978).
52. G.S. Tannenbaum, Endocrinology, 108, 76 (1981).
53. Y. Yamazoe, M. Shimada, N. Murayama and R. Kato, J. Biol. Chem., 262, 7423 (1987).
54. G.S. Tannenbaum, O. Rostad and D. Brazeau, Endocrinology, 104, 1733 (1979).
55. P. Hahn and O. Koldovsky, in Utilization of Nutrients during post-natal development. Pergamon Press, Oxford, 1966.
56. J. Hong, J. Pan, F.J. Gonzalez, H.V. Gelboin and C.S. Yang, Biochem. Biophys. Res. Commun., 142, 1077 (1987).
57. K.W. Miller and C.S. Yang, Arch. Biochem. Biophys., 229, 483 (1984).
58. P.S. Schein, K.G.M.M. Alberti and D.H. Williamson, Endocrinology, 9 827 (1971).
59. K. Kosugi, V. Chandramouli, K. Kumaran, W.C. Schumann and B.R. Landau J. Biol. Chem., 261, 13179 (1986).
60. O.E. Owen, V.E. Trapp, C.L. Skutches, M.A. Mozzoli, R.D. Hoeldtke, G. Boden and G.A. Reichard, Jr., Diabetes, 31, 242 (1982).
61. Y.Y. Tu, R. Peng, Z.-F. Chang and C.S. Yang. Biol. Interact., 44, 247, (1983).

62. D.R. Koop, B.L. Crump, G.D. Nordblom and M.J. Coon, Proc. Natl. Acad. Sci. USA, 82, 4065 (1985).
63. C.J. Patten, S.M. Ning, A.Y.H. Lu and C.S. Yang, Arch. Biochem. Biophys., 251, 629 (1986).
64. A. Lefevre, H. Adler and C.S. Lieber, J. Clin, Invest., 49, 1775 (1970).
65. J.M. Argiles, Trends in Biological Sciences, 11, 61 (1986).
66. K. Kosugi, R.F. Scofield, V. Chandramouli, K. Kumaran, W.C. Schumann and B.R. Landau. J. Biol. Chem. 261, 3952 (1986)
67. Schenkman, J.B., Favreau, L.V., Jansson, I. and Mole, J.E. in Drug Metabolism from Molecules to Man, pp. 1-13, D. J. Benford, J.W. Bridges and G. G. Gibson, eds., Taylor and Francis, London, 1986.
68. D.J. Waxman, Biochem. Pharmacol., 37, 71 (1988).
69. A.W. Wood, D.E. Ryan, P.E. Thomas and W. Levin, J. Biol. Chem., 258 8839 (1983).
70. J. Capdevila, N. Chacos, J. Werringloer, R.A. Prough and R.W. Estabrook, Proc. Natl. Acad. Sci. USA, 78, 5362 (1985).
71. D. Kupfer, I. Jansson, L.V. Favreau, A.D. Theoharides and J.B. Schenkman, Arch. Biochem. Biophys., 261, 186 (1988).

ONTOGENESIS OF LIVER CYTOCHROMES P450

Claude Bonfils, Jean Combalbert, Thierry Pineau,
Christian Larroque, Reinhard Lange and Patrick
Maurel

Service MPK, INSERM Unité 128
SANOFI RECHERCHE, Rue Blayac
34082 MONTPELLIER, FRANCE

INTRODUCTION

The role of cytochromes P450 in the metabolism of
xenobiotics is now well documented. Multiple cytochromes P450
are present in the human liver as well as in laboratory
animals. The P450 isozymes are characterized by their
spectrum, molecular weight, immunologic properties as well as
their amino acid sequence. They are all monooxygenase
enzymes, but present overlapping substrate specificity. In
untreated animals, there is an equilibrium between the
different P450 forms which may be affected by treatment with
inducer drugs. This equilibrium may be also modified by
"natural" inducers like age, sex, and hormonal status. The
study of ontogenesis of cytochromes P450 is of both
biological and pharmacological interest. It is of
considerable importance in understanding hepatocytes
differenciation as well as drug metabolism during perinatal
life. We present here a survey of recent developments made in
several laboratories on rat and human P450 ontogenesis and we
report our own results on the rabbit liver.

PREVIOUS INVESTIGATIONS

Experimental data about ontogenic expression of
cytochromes P450 were obtained first from measurement of
hydroxylase activities of liver microsomes on different
substrates.

Laboratory Animals

Previous investigations(1,2) on animal species including
rat, mouse, rabbit, hamster, guinea pig and swine, indicated
that fetal liver possessed very low monooxygenase activity.
The increase to adult level was achieved within 3 to 8 weeks
post partum.

Human Monooxygenase

Development of hepatic monooxygenase activities in case of the human liver differs significantly from the schedule described above. Human fetal liver is able to metabolize different substrates (1,2) like aminopyrine, aniline, hexobarbital at significant rates (20 to 40 % of adult level) as early as the mid gestational period. At that time, the total cytochrome P450 content of liver is about one third of the adult level. Increase of liver P450 content was also followed in newborn infants(2) by measurement of serum half life of drugs like diphenylhydantoin or phenobarbital.

Table 1 . Nomenclature of Rat cytochrome P450, corresponding Human and Rabbit P450 forms. From Black and Coon (7) and Nebert et al.(8,9)

Rat				Rabbit	Human
Nebert	Waxman	Guengerich	Levin		
IA1		BNF-B	P450 c	P450 6	P450 P1
IA2		ISF-G	P450 d	P450 4	P450 P3
IIA2	P450 3	UT-F	P450 a		
IIB1	P450 PB4	PB-B	P450 b	(2=IIB4)	(P450LM2 IIB)
IIB2			P450 e		
IIC7			P450 f		(P450MP IIC)
IIC11	P450 2c	UT-A	P450 h	(3b=IIC3)	
IIC12	P450 2d	UT-I	P450 i	(1=IIC5)	
IID					(P450DB IID)
IIE1			P450 j	P450 3a	(P450j IIE)
IIIA2	P450PB2a	PCN-E		(3c=IIIA6)	(P450NF IIIA)
IVB					(LungP450 IVB)
	P450 PB1	PB-C			
					P450 9

Such studies have shown that drug metabolism is low the first days after birth, then it increases in few weeks to a higher level than in the adult. However, measurement of hydroxylase activities does not give precise indications of isozyme modifications because of the lack of specificity of cytochrome P450. With the recent purification of different P450 forms, it becomes possible to measure specifically several isozymes as a function of age by immunological methods. Results are now available concerning rat and human liver microsomes.

Qualitative measurements have been performed in several laboratories (3,4,5,6) by the use of antibodies directed against purified forms. Rat P450 nomenclature is quite confusing and Table 1 summarizes official and trivial names used in different laboratories.

Waxman et al.(6) measured eight forms in rat liver with aging. Five of them, PB-C, UT-F (IIA2), PB-B (IIB1), ISF-G (IA2) and BNF-B (IA1), develop soon after birth and remain constant in male as well as in female rats. Some forms, PB-C and UT-F, are highly concentrated (0.2 nmol / mg of protein) and the other three are less concentrated (0.05 nmol/mg of protein). Isozyme PB-B is not detectable in the fetus but develops soon after birth. Forms ISF-G and BNF-B do not develop before fifteen days of age.

The most interesting results involve sex specific isozymes, UT-A (IIC11), UT-I (IIC12) and PCN-E (IIIA2) P450

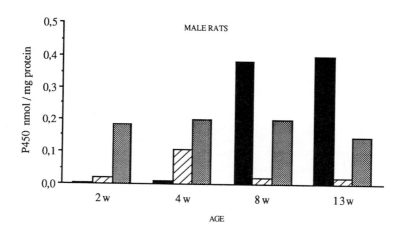

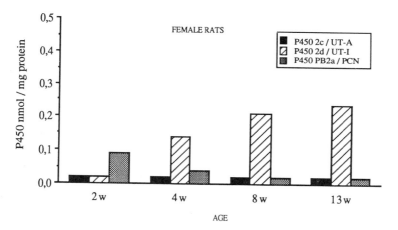

Fig. 1   Development of rat cytochrome P450. Data compiled from Waxman et al. (6).

UT-A is a male specific form and develops in one month old animals but not in females (Fig.1). UT-I develops in female rats and is repressed in males. PCN-E remains constant in male rats and is repressed in females. These cytochromes correspond to "imprinted" or programmed P450. Birth castration abolishes the expression of UT-A and PCN-E. This effect could be reversed by injection of testosterone during the first days of life. Estrogen exposure is not required for basal level expression of UT-I, but adult estrogen exposure stimulate the expression of this female specific P450. These investigations have led to the identification of sex specific P450 forms developing during the pubertal period. However, such isozymes have not yet been described in other species like man or rabbit.

ONTOGENESIS OF HUMAN CYTOCHROME P450

Immunoquantification of three isozymes, P450 5, 8, and 9 in fetal and adult human liver, was published by Cresteil et al.(10). Isozyme 5 is the major cytochrome present in fetal liver where it represents 85 % of total P450. P450 8 is not detectable in fetal liver and P450 9 is present at low level both in fetal and adult livers. It is not easy to compare these isozymes with orthologous forms described in the rat and the rabbit . A summary of human P450 nomenclature is given in Table 1. Since antibodies prepared against forms 5 and 8 recognize several human P450s, it seems that these forms are mixture of different P450s. However isozyme 5 probably contains P450 NF (IIIA) and isozyme 8 contains P450 MP (IIC).

From the human fetus, Kitada et al.(11,12,13) purified a cytochrome called P450 HFLa. Using specific antibodies, they observed that this cytochrome represented respectively 36 % and 5 % of fetal and adult liver P450. From its N- terminal aminoacid sequence, P450 HFLa belongs to the P450 NF gene family. This last cytochrome is inducible by dexamethazone, TAO, pregnenolone 16α-carbonitrile and is classified in the IIIA gene group. There are some substitutions in the beginning of the aminoacid sequence and it seems that fetal P450 HFLa and adult P450 NF are distinct gene products. Kitada's results concur with the previous work of Cresteil and confirm that a P450 of IIIA group develops during human fetal life.

ONTOGENESIS OF RABBIT CYTOCHROME P450

Hydroxylase Activities as a Function of Age

Such studies were performed ten years ago on rabbit liver microsomes by several authors (14,15).Numerous substrates were used such as aminopyrine, biphenyl, aniline, benzopyrene and ethoxycoumarin. All the xenobiotic metabolizing activities (Fig.2) were generally undetectable before 2 days of age and remained relatively low for the first 20 days of life. There after, a rapid 2 or 5 fold increase in all enzyme activities was noticed. Adult values were reached or exceeded by 30 days of age. There was one exception for aniline hydroxylase which was elevated soon after birth and remained

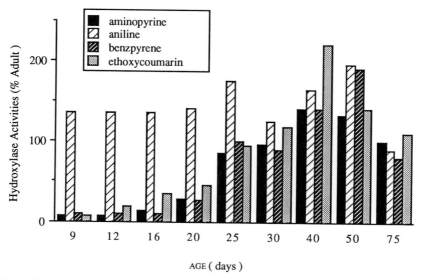

Fig. 2 Development of rabbit hydroxylating system. Data compiled from Tredger et al. (14).

at this level in older animals. Due to lack of specificity of P450 isozymes, it is difficult to deduce from these results the ontogenesisof one particular P450 form. However, aminopyrine is a preferencial substrate of 3b, aniline of 3a, benzopyrene of 6 and ethoxycoumarin of P450 2. The prevailing aniline hydroxylase activity during perinatal life favours the hypothesis that a 3a form is abundant soon after birth.

Isoenzymes Assay

These experiments were conducted in our laboratory. We prepared microsomes from fetus and young rabbits (from 1 week to 6 months post partum) and we measured total cytochrome P450 as well as isozymes 2, 3b, 3c, 4, and 6. P450 isozymes were specifically determined by immunologic methods (Immuno diffusion or Western blot assays). A detailed description of these results was presented in a recent paper by Bonfils et al. (16).
Development of total cytochrome P450 is in agreement with previous studies (14,15). A rapid increase is observed between 3 and 4 weeks of age which correspond to weaning in rabbit life. The development of P450 2, 3b, 3c, 4 and 6, as indicated in figure 3 and 4, roughly follows the total increase in cytochrome P450. However, the low concentration of P450 2 increases slightly during the second month after birth, P450 3c reaches a high level between one month and two months post partum, then decreases in the adult and cytochrome 3b develops in two steps. There is a first increase between 2 and 3 weeks and a second between 1 and 2 months.

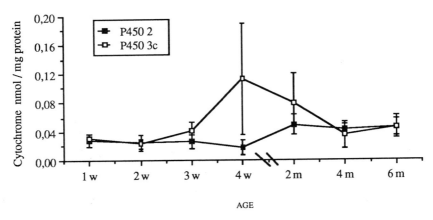

Fig.3   Ontogenesis of rabbit P540 2 and 3c.

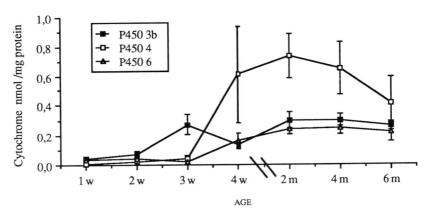

Fig. 4   Ontogenesis of rabbit P450 3b, 4 and 6.

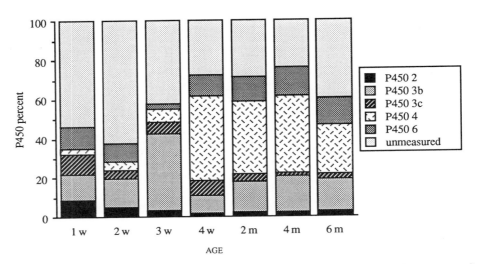

Fig. 5   Ontogenesis of five P450 forms expressed as total P450 content.

When the five isozymes we have detected immunologically are expressed as the per cent of total P450 content, it is possible to estimate the undetected isozymes as the function of age. From Fig.5, it is clear that one or more P450 isozymes specific of the newborn animals have not been detected by our specific antibodies since unmeasured P450 represent 60% in young animals and only 30% in the oldest.

The electrophoretic pattern of microsomal proteins as a function of age given in Fig.6 show that the P450 zone becomes increasingly complex with age. However, microsomes from young animals exhibit a protein band with an electrophoretic mobility intermediate between P450 3c and 2. This band contains a hemoprotein and react with the heme specific stain tetramethylbenzidine. It likely corresponds to the unmeasured P450 forms in microsomes from young animals. From 2 weeks old rabbit microsomes, we have isolated a cytochrome P450 which possesses the same electrophoretic mobility, we called it P450 2y.

Characterization of P450 2y

This cytochrome, homogeneous on SDS acrylamide gel exhibits an apparent molecular weight of 50,000 Da. Its absolute spectrum presents a wavelength maximum of 418 nm for oxidized form, 422 nm for reduced form and 452 for the CO-reduced form. This cytochrome presents an immunologic cross reactivity with anti P450 3a antibodies. P450 3a was first described by Koop et al.(17) and isolated from rabbit liver treated with ethanol. Its N-terminal aminoacid sequence is reported in Fig.7. There is a close similarity between P450 3a and 2y. However, the three first amino acids are lacking from about 50% of the cytochrome 2y we have purified and the 14th amino acid is Alanine instead of Valine in P450 3a.

These first results suggest that P450 2y is closely related with P450 3a prepared by Koop et al.(17). Some data clearly support the identity of the two cytochromes, they possess an immunologic cross reactivity, the same λ maximum of CO-reduced spectrum and close molecular weights. However, other results do not confirm this hypothesis, there is one difference in the first 20 amino acids, moreover, the spectrum of the oxidized form is low spin (418 nm) in the case of 2y and high spin (393 nm) in the case of 3a.

We further compared these two forms by two-dimensional gel electrophoresis which has a higher resolving power than SDS polyacrylamide gel. In spite of its wide use in membrane research, this technique has been rarely employed in microsomal analysis. We successfully employed it to resolve several P450 isozymes (20) and we applied this method to P450 3a and 2y. Electrophoretic pattern of microsomes from young animals revealed with anti 3a antibodies shows a single protein spot. When mixed with microsomes from adult, two major spots may be seen. This clearly indicates the non identity between adult and young 3a related forms.

In a recent paper, Khani et al.(21) isolated two highly similar genes coding for P450 3a family. They deduced from nucleotide sequence that gene 1 encodes for P450 3a and gene

261

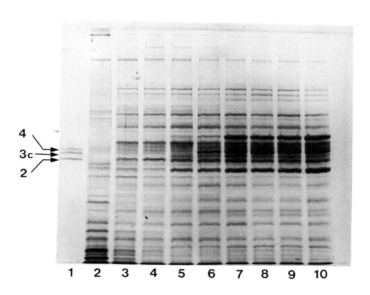

Fig. 6   SDS Polyacrylamide gel electrophoresis patterns of liver microsomal proteins as a function of age. Electrophoretic plate contains, lane 1:standard cytochrome P450 4, 3c and 2 (5 pmol), lane 2: microsomes from fetus, lane 3 to 6: microsomes from 1, 2, 3 and 4 week old rabbits respectively, lane 7 to 10: microsomes from 2, 4, 5 and 6 month old rabbits (12 μg of protein).

2 for an unknown protein. If we compare the N-terminal amino acid sequence given by gene 2 and that of P450 2y, they are exactly identical and the 14th amino acid is Alanine and not Valine. Furthermore, the comparison of total amino acid composition of P450 3a and protein encoded by gene 2 indicates that the latter possesses a lower isoelectric point than 3a. This is also in agreement with the results we obtained on two dimensional gels.

```
               5         10        15        20
P4502  M E F S L   L L L L A   F L A G L   L L L L F
   3b  M D L L I   I L G I C   L S C V V   L L S L W
   3c  M D L I F   S L E T W   V L L A A   S L V L L
   3a  A V L G I   T V A L L   G W M V I   L L F I S
   2y  A V L G I   T V A L L   G W M A I   L L F
```

Fig. 7 Comparison of N-terminal aminoacid sequence of P450 2y to other rabbit P450 forms. P450 2 sequence was compiled from Tarr et al.(18), P450 3b from Ozols et al.(19), P450 3c from Black and Coon (7) and P450 3a from Koop et al.(17). With P450 2y, the three first aminoacids were absent from about 50% of the purified protein.

A more exhaustive comparison is needed to assert that the gene isolated by Khani's group is the gene of P450 2y. Nevertheless, several experimental results are consistent with the hypothesis that 3a and 2y are distinct P450 forms. P450 3a is abundant in adult animals and 2y is characteristic of post natal life. These two cytochromes belong to the IIE gene family. Orthologous forms of P450 3a have been described in other animal species, P450 j in rats (22) and P450 HLj in human liver (23). But, so far, no evidence was obtained indicating the presence of several j subforms in these species.

CONCLUSION

The studies of cytochromes P450 ontogenesis have indicated quantitative as well as qualitative modifications of liver P450 content during perinatal life. These results are of considerable pharmacological interest. The ability of young liver to eliminate drugs must not be deduced from adult models.

According to the investigations performed in many laboratories, there is an important increase in total liver P450 content during the first post-natal weeks. The most common P450 isozymes develop during this period.

It has also been evidenced that some P450 forms appear at different developmental periods. These include the sex specific isozymes UT-A and UT-I from rat which develop during pubertal period, the human fetal form P450 HFLa and the rabbit post natal P450 2y.

If we compare fetal P450 HFLa and rabbit P450 2y, we observe that each form belongs to a gene family expressed in adulthood, IIIA and IIE, respectively. In both cases, the young form is closely related but not identical to the adult form. Young and adult forms are likely distinct gene products. The presence of two distinct isozymes in young and adult organism may be explained by a difference in substrate specificity. Young forms, which represent, at one developmental period, the predominant P450 form, are probably able to metabolize an important cell metabolite, but more investigations are necessary to determine what is the natural substrate of these P450 isozymes.

REFERENCES

1. J.R. Gillette and B. Stripp, Pre-and postnatal enzyme capacity for drug metabolite production, Federation Proc., 34:172 (1975).
2. A.H. Neims, M. Warner, P.M. Loughnan and J. V. Aranda, Developmental aspects of the hepatic cytochrome P450 monooxygenase system, An. Rev. Pharmacol. Toxicol., 16:427 (1976).
3. T. Cresteil, J. P. Flinois, A. Pfister and J. P. Leroux, Effect of microsomal preparations and induction on cytochrome P450-dependent monooxygenases in fetal and neonatal rat liver, Biochem.Pharmacol., 28:2057 (1979).
4. T. Cresteil, P. Beaune, C. Celier, J.P. Leroux and F.P. Guengerich, Cytochrome P450 isoenzyme content and monooxygenase activities in rat liver: effect of ontogenesis and pretreatment by phenobarbital and 3-methylcholanthrene, J. Pharmacol. Exp. Ther., 236:269 (1986).
5. E. T. Morgan, C. Mac Geoch and J. A. Gustafsson, Hormonal and developmental regulation of expression of the hepatic microsomal steroid 16α-hydroxylase cytochrome P450 apoprotein in the rat, J. Biol. Chem., 260:11895 (1985).
6. D. J. Waxman, G. A. Dannan, and F. P. Guengerich, Regulation of rat hepatic cytochrome P450: Age-dependent expression, hormonal imprinting, and xenobiotic inducibility of sex specific isoenzymes, Biochemistry, 24:4409 (1985).
7. S. D. Black and M. J. Coon, P450 structure and function, in:"Advances in Enzymology", A. Meister, ed., John Willey, N.Y., 60:35 (1987).
8. D. W. Nebert, M. Adesnik, M. J. Coon, R. W. Estabrook, F.J.Gonzalez F. P. Guengerich, I. C. Gunsalus, E. F. Johnson, B. Kemper, W. Levin, I. R. Phillips, R. Sato and M. R. Waterman, The P450 gene superfamily: recommended nomenclature, D N A, 6:1 (1987).

9. D. W. Nebert, D. R. Nelson, M. Adesnik, M. J. Coon, R. W. Estabrook, F. J. Gonzalez, F. P. Guengerich, I. C. Gunsalus, E.F. Johnson, B. Kemper, W. Levin, I. R. Phillips, R. Sato and M. R. Waterman, The P450 superfamily: updated listing of all genes and recommended nomenclature for the chromosomal loci, <u>D N A</u>,8:1 (1989).

10. T. Cresteil, P. Beaune, P. Kremers, C. Celier, F. P. Guengerich and J. P. Leroux, Immunoquantification of epoxide hydrolase and Cytochrome P450 isozymes in fetal and adult human liver microsomes, <u>Eur. J.Biochem.</u>,151:345 (1985).

11. M. Kitada, T. Kamataki, K. Itahashi, T. Rikihisa, R. Kato and Y. Kanakubo, Purification and properties of cytochrome P450 from homogenates of human fetal livers, <u>Arch. Biochem. Biophys.</u> 241:275 (1985).

12. M. Kitada, T. Kamataki, K. Itahashi, T. Rikihisa, R. Kato and Y. Kanakubo, Immunochemical examinations of cytochrome P450 in various tissues of human fetuses using antibodies to human fetal cytochrome P450, P450 HFLa, <u>Biochem. Biophys. Res. Commun.</u>,131:1154 (1985).

13. M. Kitada, N. Igoshi, T. Kamataki, K. Itahashi, T. Rikihisa and YKanakubo, The proteins immunochemically related to P450 HFLa, amajor form of cytochrome P450 in human fetal livers, are present in liver microsomes from various animal species, <u>Res. Commun Chem. Pathol. Pharmacol.</u>, 82:31 (1988).

14. J. M. Tredger, R. S. Chhabra and J. R. Fouts, Postnatal development of mixed-function oxidation as measured in microsomes from the small intestine and liver of rabbits, <u>Drug Metab.Dispos.</u>, 4:17 (1976).

15. S. A. Atlas, A. R. Boobis, J. S. Felton, S. S. Thorgeirsson and D.W. Nebert, Ontogenic expression of polycyclic aromatic compound-inducible monooxygenase activities and forms of cytochrome P450 in rabbit, <u>J. Biol. Chem.</u>, 252:4712 (1977).

16. C. Bonfils, J. Combalbert, T. Pineau, J. Angevin, C. Larroque, J.Derancourt, J.P. Capony and P. Maurel, Ontogenesis of rabbit liver cytochrome P450. Evidence for a P450 IIE (3a) related form prevailing during postnatal period, <u>Eur. J. Biochem.</u>, 188:187 (1990).

17. D. R. Koop, E. T. Morgan, G. E. Tarr and M. J. Coon, Purification and characterization of a unique isozyme of cytochrome P450 from liver microsomes of ethanol-treated rabbits, <u>J. Biol. Chem.</u>, 257:8472 (1982).

18. G. E. Tarr, S. D. Black, V. S. Fujita and M. J. Coon, Complete aminoacid sequence and predicted membrane topology of phenobarbital-induced cytochrome P450 (Isozyme 2) from rabbit liver microsomes, <u>Proc. Natl. Acad. Sci. USA</u>, 80:6552 (1983).

19. J. Ozols, F. S. Heinemann and E. F. Johnson, The complete aminoacid sequence of a constitutive form of liver microsomal cytochrome P450, <u>J. Biol. Chem.</u>, 260:5427 (1985).

20. C. Bonfils and J. Combalbert, Resolution of rabbit liver cytochrome P450 3b and 3c, P450 IIC3 and IIIA4, from microsomal membrane by two dimensional gel electrophoresis, <u>Electrophoresis</u>, 11:182 (1990).

21. S. C. Khani, T. D. Porter, V. S. Fujita and M. J. Coon,
    Organisation and differential expression of two highly
    similar genes in the rabbit alcohol-inducible cytochrome
    P450 subfamily, J. Biol.Chem., 263:7170 (1988).
22. D. E. Ryan., L. Ramanathan, S. Iida, P. E. Thomas, M.
    Haniu, J.E. Shively, C. S. Lieber and W. Levin,
    Characterization of a major form of rat hepatic
    microsomal cytochrome P450 induced by isoniazid J. Biol.
    Chem., 260:6385 (1985).
23. S. A. Wrighton, P. E. Thomas, D. T. Molowa, M. Haniu, J.
    E. Shively, S. L. Maines, P. B. Watkins, G. Parker, G.
    Mendez-Picon, W. Levin and P. S. Guzelian,
    Characterization of ethanol-inducible human liver
    N-nitrosodimethylamine demethylase, Biochemistry, 25:6731
    (1986).

MECHANISM OF STEROID HORMONE ACTION

Anders Berkenstam and Jan-Ake Gustafsson

Department of Medical Nutrition
Karolinska Institute
Huddinge University Hospital F60, NOVUM
S-141 86 Huddinge, Sweden

In general, specific genes are expressed in a temporal or tissue-specific manner or are regulated in response to extracellular signals (Mitchell and Tjian, 1989). Steroid hormones constitute one such group of extracellular signals that regulate the expression of certain genes by interacting with intracellular receptor proteins. These receptors belong to a family of ligand-activated enhancer-binding proteins i.e. the steroid/thyroid hormone receptor super-family, that also includes receptors for retinoic acid, thyroid hormone and some yet unidentified ligands (Evans, 1988).

The presence of steroid hormone receptors in a cell determines its physiological response to a given steroid. These intracellular, soluble receptor proteins make up approximately 0.001% of total cellular protein, corresponding to ~10,000 molecules per cell. The receptor proteins have a high affinity for their ligands (K_d ~10^{-9}) and are loosely associated with chromatin in the absence of hormone (Walters, 1985). In vitro, upon exposure of the hormone-bound receptor to e.g. heat treatment and/or increased ionic strength (Schmidt and Litwack, 1982), the steroid hormone-receptor complex acquires increased affinity for unspecific DNA (i.e. activation) possibly due to a conformational change and/or dissociation of the receptor associated heat shock protein 90 (Pratt et al., 1988; Denis et al., 1988). In vivo, the steroid receptor becomes tightly bound to the nucleus after binding of the hormone. Hormone-dependent interaction of ligand-bound receptor with specific chromosomal loci in vivo was first demonstrated by crosslinking of the insect steroid, ecdysterone, to polytene chromosomes of Drosophila melanogaster (Gronemeyer and Pongs, 1980). Later, Becker et al. (1986) demonstrated with the "genomic footprinting" technique that the glucocorticoid receptor (GR) interacts in vivo with its specific binding site (SRE) in the tyrosine aminotransferase gene and only in the presence of hormone. This binding site within the tyrosine aminotransferase gene has also been shown to interact with the purified GR in vitro (Jantzen et al., 1987).

The interaction of the hormone-occupied (i.e. activated) steroid receptor with the SRE is thought to be a prerequisite for the regulation of gene expression by steroid hormones, which occurs predominantly at the transcriptional level (Beato, 1989). However, both post-transcriptional mechanisms, e.g. effects on mRNA stability (Brock and Shapiro, 1983), and possibly also post-translational mechanisms e.g., effects on protein turnover (McIntyre and Samuels, 1985) have been implicated as possible targets for regulation of gene expression by steroid hormones.

Thus, the best characterized biological effects of steroid hormones, i.e. the regulation of specific gene products in response to changes in the extracellular environment, appear to be ultimately dependent on i) the ability of the steroid to interact with specific receptor proteins expressed in different target tissues; and on ii) the ability of the hormone-bound receptor to interact with the SRE in steroid regulated genes and thereby regulate the expression of specific genes by a largely unknown mechanism. The structural and functional requirements for these events to occur will be discussed in the following section.

STRUCTURAL AND FUNCTIONAL ORGANIZATION

Characterization of purified GR protein suggested a structural organization of the receptor in distinct functional domains (Carlstedt-Duke et al., 1987), separable by protease hypersensitive regions in the protein. Thus, the GR has been shown by limited proteolysis *in vitro* to be composed of three major functional domains; i) a C-terminal steroid binding domain; ii) a central DNA-binding domain; and iii) an N-terminal domain. The recent structural characterization by complementary DNA (cDNA) cloning for different steroid hormone receptors has confirmed this domain structure. Moreover, deletion mutant analysis of the expressed cDNAs has identified new discrete regions responsible for other functions of the receptors such as trans-activation, dimerization, nuclear translocation and interaction with the receptor-associated heat-shock protein 90 (Beato, 1989).

Comparison of the deduced primary amino acid sequences of the five classical steroid hormone receptors reveals six regions of varying homology, denoted A-F in Figure 1 (Green and Chambon, 1988a).

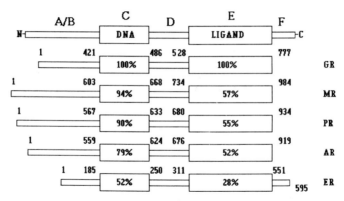

Figure 1. Schematic illustration indicating the structural and functional organization of the five classical human steroid hormone receptors and their sequence similarity. The upper part of the diagram shows the division of the receptor cDNAs into the six regions (A-F) based on maximum sequence similarity. The lower part of the diagram illustrates the amino acid numbers at the borders of the different regions and the sequence similarity (%) between the GR and the other steroid receptors.

Regions C and E are the most highly conserved structures among the different receptors within the family. This high sequence similarity indicates that these two domains are important for receptor functions common to all members of the steroid hormone receptor family, i.e. DNA- and ligand-binding, respectively. Similar sequence comparisons of the cDNAs for the different members of the steroid hormone receptor family indicate that they can be divided into two subfamilies. Thus, the GR, mineralocorticoid receptor (MR), progesterone receptor (PR), and androgen receptor (AR) appear to be more closely related to each other than to the estrogen receptor (ER). Moreover, both the structure of the natural ligands and the specific DNA sequences recognized by each hormone receptor (see below) are compatible with this apparent subdivision of the steroid hormone receptor family into two subgroups. The ER, on the other hand, appears to be more closely related to other members of the larger super-family of nuclear receptors, including the retinoic acid, thyroid hormone and vitamin D receptors as well as some receptor-like proteins with unknown ligands (Giguère et al., 1988; Miyajima et al., 1988; Wang et al., 1989).

The Ligand-binding Domain

Proteolytic fragments of steroid hormone receptors containing the steroid-binding site (Carlstedt-Duke et al., 1987) have been shown to correspond to approximately 250 amino acids of the protein, located in its C-terminal part. Thus, this domain (region E) is responsible for ligand-binding and is characterized by its large number of hydrophobic amino acids. The majority of these residues are probably involved in protein folding, but hydrophobic interactions are also known to be important for steroid binding (Hansen and Gorski, 1988).

The sequence similarity of the ligand-binding domain within the GR, MR, PR and AR subgroup is in the order of 50%, while these receptors share a similarity of only ~20-25% with the corresponding domain of the ER. Moreover, cross-linking studies by Carlstedt-Duke et al. (1988, 1989) have led to the identification of amino acid residues in contact with the steroid molecule in the GR and PR. One of the amino acids identified in the GR (Met-622) was shown to correspond exactly to one of the labelled amino acids in the PR (Met-759) and to be located within a hydrophobic segment present in both receptor proteins. The other amino acids identified did not correspond to each other but were all located within hydrophobic surfaces of the proteins.

The ligand-binding function of region E does not appear to be dependent on the other receptor domains. Thus, by deleting the A/B and/or C regions of the GR and ER, Rusconi and Yamamoto (1987) and Kumar et al., (1987) showed that the ligand-binding capacities of the proteins were not affected. However, the integrity of the whole domain *per se* seems to be important for ligand-binding since almost all deletions, point mutations or insertions within this domain destroy the ligand-binding function (Giguère et al., 1986; Kumar et al., 1987). Moreover, from experiments with PR expressed in E. coli it has been suggested that no post-translational modifications or other modifications unique to eukaryotic cells, are necessary for ligand-binding since the K_d for progesterone was reported to be in the same range as for the PR in eukaryotic cells (Eul et al., 1989).

In conclusion, the large number of hydrophobic amino acids and the integrity of the whole domain appear to be important for the ligand-binding function of the receptor protein. Furthermore, this region also seems to participate in nuclear transfer, transcription activation and dimerization functions (see below).

The DNA-binding Domain

In analogy to the steroid-binding domain, the region responsible for DNA-binding of the GR has been defined at the protein level (Carlstedt-Duke et al., 1987) by limited proteolysis *in vitro* of purified GR preparations. These studies confirmed the structure of the proposed DNA-binding domain obtained by deletion analysis of the expressed cDNA for the GR (Rusconi and Yamamoto, 1987). Thus, the DNA-binding domain of different steroid receptors has been defined to consist of a highly conserved ~65 amino acid long polypeptide region rich in cysteines and basic amino acids. The sequence similarity between GR, MR, PR and AR DNA-binding domains is 90% or more but only approximately 50% as compared to the corresponding ER domain.

Comparison of the deduced amino acid sequence of the DNA-binding domain of cloned steroid hormone receptors with those of other DNA-

binding proteins suggested that the steroid receptors belong to the group
of eukaryotic regulatory proteins that utilize the "zinc finger" motif
for DNA-binding (Berg, 1989). Thus, the steroid hormone receptors contain
a repetitive motif of cysteines in their DNA-binding domain (Figure 2).

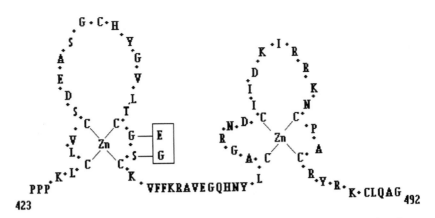

Figure 2. Putative structure of the DNA-binding domain drawn to
show the two zinc-coordinated fingers. Boxed amino acids indicate
residues of importance for template specificity (Danielsson et al.,
1989).

This motif is similar to the sequential and ordered occurrence of
histidines and cysteines found in *Xenopus* transcription factor IIIA,
which in turn originally was proposed by Miller et al. (1985) to contain
a "zinc-finger" motif. This motif has later been identified in a variety
of DNA-binding proteins (Klug and Rhodes, 1987). By expression in *E.coli*
and purification of the rat GR DNA-binding domain Freedman et al. (1988)
demonstrated using a number of techniques that region C of the GR does
indeed bind two zinc atoms and that each zinc atom is coordinated with
four cysteines. In addition, the specific DNA-binding activity *in vitro*
was shown to be dependent on metal-binding.

Studies on sequence-specific DNA-binding *in vitro* with purified
receptor preparations and gene transfer experiments (see below) suggested
that ER recognizes a different DNA sequence than GR. That region C is
responsible for specifying target gene activation was demonstrated by
experiments with hybrid receptors where the DNA-binding domain of the ER
and GR were switched, resulting in the predicted change in template
specificity and target gene activation (Green and Chambon, 1987).
Moreover, region C of the ER has been shown to be encoded by two
different exons (Ponglikitmongkol et al., 1988), each coding for one
"zinc-finger". The N-terminal exon contains hydrophobic amino acids while
the C-terminal exon is rich in basic amino acids in addition to the four
conserved zinc coordinating cysteines. Thus, it was proposed that the N-
terminal finger could specify DNA-binding and the C-terminal finger with
its more basic amino acid composition could be important for non-specific
anchorage to the acidic sugar-phosphate backbone on DNA. Later, Green et
al., (1988b) showed by experiments with ER and GR chimaeric receptors
that the N-terminal "zinc finger" appears to determine target gene
specificity. However, a more detailed mutational analysis of the ER
(Mader et al., 1989) showed that the amino acids responsible for template
specificity were not within the first zinc-finger but were located at the
C-terminal side of the first finger. Similar data, regarding structural
determinants required for template specificity, were presented by
Danielsen et al. (1989) and Umesono and Evans (1989) who also
demonstrated that the remainder of the DNA-binding domain normally

confers structural information required for preventing recognition of more than one SRE at the same time.

Trans-activating Region(s)

Compared to the relatively well defined, and by conventional biochemical and genetic methods separable steroid- and DNA-binding domains, the structure(s) responsible for transactivation appears to consist of multiple regions (Green and Chambon, 1988a). Also, in contrast to the steroid- and DNA-binding domains which appear to function relatively independently of the other parts of the receptor protein, the trans-activating region(s) seem to be dependent on, or interdigitated with other domains of the receptor protein. Thus, both the A/B region, the DNA-binding domain and the steroid-binding domain have been described to be of importance for trans-activation, depending on the particular receptor, target gene and host cell line under study.

In the case of the GR it has been demonstrated that regions outside the DNA-binding domain influence trans-activation. These trans-activating regions denoted tau_1 and tau_2, were shown to be localized in the A/B region and steroid-binding domain of the GR, respectively (Hollenberg and Evans, 1988). In addition, it was reported that these domains function in a cooperative and position-independent fashion. Although these domains are unrelated with regard to amino acid sequence, both the tau_1 and the tau_2 regions are acidic in character. Thus, it has been suggested that the GR may activate gene transcription through mechanisms similar to those proposed for the yeast transcription factors GAL4 and GCN4 (Ptashne, 1988). In this context, it is interesting to note that both the expressed ER (Metzger et al., 1988) and GR (Schena and Yamamoto, 1988) appear to function also in yeast. Moreover, in a recent report (Freedman et al., 1989), it was shown that a discrete segment of the GR, encompassing the DNA-binding domain and a trans-activating region expressed in *E. coli,* when added to nuclear extracts from *Drosophila* embryos enhanced transcription *in vitro* of several glucocorticoid responsive genes that were tested. Accordingly, at least some transcription factors from yeast, *Drosophila* and humans appear to utilize the same basic mechanisms for trans-activation.

A region similar to the tau_2 region of the GR has been identified within the steroid-binding domain of the ER. However, by using GAL4-ER fusion proteins, Webster et al. (1989) demonstrated that this region did not stimulate transcription significantly by itself. In these experiments, the whole steroid-binding domain of ER was suggested to be necessary (and sufficient) for maximal trans-activation.

Attempts to map trans-activating region(s) of the PR have been hampered by the fact that different groups have obtained different results under various experimental conditions. Thus, Gronemeyer et al., (1987) reported that expressed forms of the chick PR truncated in regions E or A/B, activated gene transcription poorly in HeLa cells, whereas similarly truncated PR forms were reported by others to exhibit almost wild-type activity (Carson et al., 1987) in CV-1 cells. Although still a controversial and largely unresolved issue, PR appears to be present in cytosolic extracts as two forms, denoted A and B (Schrader and O'Malley, 1972; Horwitz and Alexander, 1983). The B-form has a higher molecular weight than the A-form and it has been suggested that the latter form represents an N-terminally truncated form of B (Gronemeyer et al., 1987). A functional difference between these two forms of the PR in transcription activation of two target genes was reported by Tora et al., (1988), indicating a role of the N-terminal domain in steroid-specific transactivation.

Dimerization Region(s)

Comparison of several functional SREs in various hormonally regulated genes, showed that these elements generally appear in a partially palindromic configuration (Jantzen et al., 1987). An obvious implication of this configuration of SREs was that steroid hormone receptors might bind as dimers to their responsive elements.

Using a gel retardation assay and different forms of the bacterially expressed DNA-binding domain of the GR, Tsai et al. (1988) demonstrated a concentration-dependent binding *in vitro* of the recombinant GR to one or both halves of an imperfect palindromic sequence derived from the SRE of the tyrosine aminotransferase gene. These experiments suggested that binding to one half-site of the palindrome might facilitate binding to the other half-site. In contrast, this apparent cooperative binding could not be demonstrated with activated GR purified from rat liver. Consequently, since the employed purification protocol appears to select for a monomeric form of the GR, the binding of the native GR *in vitro* to the two half-sites may thus occur too rapidly to be differentiated into two phases. However, Wrange et al. (1989) reported that under certain conditions the purified transformed GR exists as a homodimer already before it has bound to DNA, as well as after it has bound to DNA.

Results of Sabbah et al. (1989) likewise suggested that the ER in solution is comprised of two hormone binding units, regardless of whether it exists in its non-activated (8-9S) or activated (5S) DNA-binding form. Moreover, it was found by trypsin cleavage of ER that the steroid-binding domain of the receptor protein was sufficient for dimerization (Sabbah et al., 1989). Kumar and Chambon (1988) identified two dimerization domains of the ER by expression of various truncated forms of the human ER cDNA (Green et al., 1986) in HeLa cells. One dimerization-domain was shown to be "hormone-dependent" and localized within the entire steroid-binding part of the ER. A second, weaker and hormone-independent dimerization structure was localized to the DNA-binding domain. However, these results are somewhat difficult to interpret in light of a recent report (Tora et al., 1989) demonstrating a reverse transcriptase induced cloning artefact in the particular ER cDNA clone used (OR8). This "point-mutation" has been reported to decrease the apparent affinity of ER for estradiol at 25 $^{\circ}$C, at which temperature the hormone-dependent dimerization was demonstrated (Kumar and Chambon, 1988).

In a recent report, Guiochon-Mantel et al. (1989) used a series of deletion mutants of the PR cDNA and a panel of monoclonal antibodies against the PR to show that the dimerization process demonstrated *in vitro* may be represented by a hormone-dependent interaction between receptor monomers *in vivo*. This finding, may have important functional implications for the hormone-dependent association of steroid hormone receptors with specific chromosomal binding-sites observed *in vivo* (Becker et al., 1986).

DNA-REGULATORY ELEMENTS

The function of various cis-acting elements, i.e. short stretches of DNA, has been shown to be dependent on direct or indirect interaction with specific trans-acting factors, thereby altering the transcription of specific target genes (Ptashne, 1988). These cis-acting elements can be classified as basic promoter elements, upstream promoter elements (UPE), enhancers or silencers (Maniatis et al., 1987). Promoters transcribed by RNA polymerase II can be classified as those that contain a TATA element and those that do not contain an obvious TATA box. The latter group of promoters can be divided into two classes; GC-rich promoters and not GC-rich promoters. The former group of promoters are frequently found in housekeeping genes (Sehgal et al., 1988) whereas the latter group of genes has been identified in many genes that are regulated during development or differentiation (Robertson, 1988).

The AT-rich region, i.e. the TATA element, is located approximately 30 bp upstream of the mRNA start-site ("cap site") and has been demonstrated to interact with general trans-acting factors (Buratowski et al., 1988). The TATA box has been shown to be important for correct initiation of transcription and/or initiation frequency. The upstream promoter elements are typically located between 40-100 bp upstream from the cap site. These elements quantitatively modulate the expression of a selective class of promoters. Several upstream promoter elements have been identified in both viral and cellular genes and their corresponding trans-acting factors have been purified and characterized (Mitchell and Tjian, 1989). In addition to the TATA box and the upstream promoter elements, cis-elements called enhancers or silencers have been identified at a larger distance from the transcription initiation site and shown to enhance or repress the activity of a particular promoter (Voss et al., 1986). Comparison of different UPEs and enhancers has shown that these elements have many overlapping criteria, but UPEs appear to be relatively distance-dependent in contrast to enhancers. The promoters and enhancers that control the activity of RNA polymerse II are composed of a multitude of cis-elements, or modules. These modules have in several cases been shown to interact with transcription factors both *in vitro* and *in vivo*. However, a module alone has only a weak or negligible effect on transcription and requires either a multimerization or combination with a number of other modules to elicit its function (Ondek et al., 1988).

Enhancers are typically found in genes coding for products which have to be made in large numbers or rapidly in response to extracellular signals. Examples are the immunoglobulin genes and the lysozyme gene . The immunoglobulin enhancers are located downstream of the promoter i.e., in the intron between the variable and constant regions. In contrast, the estradiol-responsive enhancer of the prolactin gene is located approximately two thousand basepairs upstream of the promoter (Maurer and Notides, 1987). Enhancers have also been localized in close proximity to the promoter region and experimentally manipulated enhancers have also been found to be functional in an inverted position or in front of a heterologous promoter (Chandler et al., 1983). Thus, the characteristics of enhancer sequences are that they function in a position-, orientation- and promoter-independent manner. According to their function, enhancers can be classified as constitutive, cell-type specific, or inducible. Moreover, certain genes such as the lysozyme gene contain many different enhancer elements with both constitutive, cell-type specific and inducible characteristics (Sippel et al., 1988). In conclusion, the combination and total number of the different regulatory cis-acting elements (TATA box, UPEs and enhancers/silencers) appear to determine cell-type specificity and strength, respectively, of a specific target promoter.

The mechanism(s) for enhancer-dependent activation of transcription is at present not known in detail. However, steroid hormone receptors have been shown to regulate gene transcription by interaction with SREs which share all the features of enhancers. Thus, mechanisms of steroid hormone regulated gene expression have been used as a model for enhancer-regulated eukaryotic gene expression.

Steroid Responsive Elements

Steroid responsive elements have been identified *in vivo* by gene transfer techniques and *in vitro* by DNA-binding studies with purified receptor preparations. The SREs for the different steroid hormone receptors appear to be closely related palindromic sequences. For example, it has been shown that the SRE of the MMTV-LTR (Cato et al., 1986), lysozyme (Renkawitz et al., 1984) and metallothionein-IIA (Slater et al., 1988) genes can be regulated both by progestins and glucocorticoids while the SRE of the vitellogenin gene appears to be regulated only by estrogens (Brock and Shapiro, 1983). Based on functional studies a 15 bp consensus sequences for GR- and ER-specific SREs have been formulated (Klock et al., 1987). From these studies it has been shown that 15 bp GR-specific SRE can be converted to an SRE inducible by estrogens by only a two base pair substitution. Thus, SRE consensus sequences consist of two 6 bp partially palindromic sequences separated by a 3 bp variable "spacer" sequence. However, no clear consensus recognition sequence has been identified for the PR, AR or MR but the 15 bp SRE for the GR is able to mediate hormone induction both by progestins and androgens (Strähle et al., 1987). The similarity of the SREs suggests that these regulatory cis-acting elements belong to one family of related sequences that possibly could have coevolved with the steroid hormone receptor family.

Binding studies *in vitro* of the interaction of GR and PR with the SREs of the MMTV-LTR and lysozyme gene have shown a very similar binding pattern for the two receptors (von der Ahe et al., 1985; Tsai et al., 1988) but also clear differences (von der Ahe et al., 1986; Chalepakis et al., 1988; Berkenstam et al., 1988). Thus, by DNase I protection experiments we have shown (Berkenstam et al., 1988), that the purified human PR protect GR binding sites 1.3 (Payvar et al., 1983) in a fragment of the MMTV-LTR (nucleotides -264 to -109) containing GR binding sites 1.3, 1.4 and 1.5´. However, no clear-cut binding occurred to the two other GR binding sites (1.4 and 1.5´). This difference in binding of PR and GR to the MMTV-LTR could possibly reflect differences in relative affinities of each receptor to the individual binding sites. Alternatively, sequences downstream of GR binding site 1.5´ (nucleotides downstream of -109) might be needed for binding of PR but not for GR. When a fragment from the MMTV-LTR containing GR binding sites 1.1, 1.2, 1.3, 1.4 and 1.5 (Payvar et al., 1983) was used, distinct footprints were obtained at binding sites 1.3 and 1.5 with the human PR (Hecht et al., 1988). Thus, nucleotides downstream of -109 in the MMTV-LTR are apparently necessary for binding of the human PR but not for the rat GR. This observation together with results of Cato et al. (1986) suggests that sequences downstream of -109 are required both for PR binding *in vitro* to binding site 1.5 and for induction of MMTV-LTR by progestins. Moreover, the study by Chalepakis et al. (1988), showed that one particular MMTV-LTR mutant demonstrated a much more pronounced decrease in responsiveness to glucocorticoids than to progestins. This difference in hormone sensitivity correlated with differences in relative binding-affinities towards the two receptors observed *in vitro*. Accordingly, subtle differences in DNA-sequence preference between GR and PR may constitute one mechanism underlying steroid-specific gene expression.

REGULATION OF STEROID HORMONE RESPONSE

In general, the biological response to a hormone is dependent on the concentration of free hormone (i.e. the amount of hormone available for receptor-binding), the concentration of receptors and the affinity (K_d) of the receptor for its cognate hormone. In line with this, a direct relationship between receptor concentration and degree of response has also been demonstrated for steroid hormones (Gehring et al., 1984). However, in other experimental systems (Darbre and King, 1987) and clinical situations (Thorpe and Rose, 1986) such a direct relationship between steroid hormone receptor concentration and biological effect is not as evident.

Regulation of Steroid Hormone Receptor Concentrations

Since the concentration of steroid hormone receptors in a target cell is one determinant of the cellular response to steroid hormones, the regulation of hormone receptors is an important aspect of physiological control of steroid regulated gene expression. Steroid hormone receptor levels have been shown to be regulated both by their own ligands (autologous regulation) and by other hormones (heterologous regulation).

Autologous receptor down-regulation appears to be a common property of most steroid hormones. Ligand-binding and dense amino acid incorporation studies suggested that the ER, PR and GR proteins are down-regulated by their own ligands (Svec and Rudis, 1981; Eckert et al., 1984; Mullick and Katzenellenbogen, 1986). Cloning of the cDNAs for these receptors enabled studies on the mechanisms of autologous receptor down-regulation and showed that the decrease in receptor protein was associated with a reduction in receptor mRNA levels. For the GR (Okret et al., 1986) and PR (Wei et al., 1988; Berkenstam et al., 1989) it has been suggested that this decrease of receptor mRNA mainly occurs through inhibition of transcription of the respective receptor genes and not through effects on mRNA stability. However, for the ER the predominant mechanism regulating ER mRNA levels appear to be a post-transcriptional event. In addition, the inhibition of ER gene transcription by estradiol has been shown to be only a transient and reversible effect (Saceda et al., 1988). The mechanism(s) for the post-transcriptional effect of estradiol on ER mRNA levels are at present not known but estradiol has been shown to specifically influence the stability of other mRNAs (Brock and Shapiro, 1983).

The heterologous regulation of steroid hormone receptors has not been studied in such detail as the autologous regulation. The induction of PR by estradiol in estrogen/progesterone target tissues (Horwitz et al., 1983) is probably the best studied heterologous regulation of steroid hormone receptor concentrations and has been shown to be associated with an increased steady-state level of PR mRNA molecules (Wei et al., 1988; Berkenstam et al., 1989). However, it is not known whether this regulation by estradiol occurs at the transcriptional or at the post-transcriptional level.

Most studies concerning the regulation of cellular steroid hormone receptor concentrations have so far focused mainly on the mechanisms underlying this regulation, and few studies have attempted to assess the possible biological significance of down-regulated receptor levels. In a recent study Yi-Li et al. (1989) showed that after an initial rapid increase of tyrosine aminotransferase activity in rat liver caused by glucocorticoids, the enzyme activity was decreased in parallel with a decrease in GR levels (as measured with a ligand exchange assay). However, based on different experimental models, it has been shown that the threshold-level of receptor concentration for one response to occur is quite different from the concentration required for another response to occur (Rabindran et al., 1987; Cook et al., 1988), indicating that other factors in addition to the receptor appear to determine the biological effect of steroid hormones.

Interdependence of Steroid Hormone Receptors and Transcription Factors

Steroid responsive elements in several hormonally regulated genes are often found adjacent to binding-sites for other transcription factors. These findings suggest that transcriptional control of steroid regulated genes might depend on the combinatorial effects of these cis-acting elements through a direct interaction of steroid hormone receptors with other transcription factors. Thus, it has been demonstrated that transcription activation of the MMTV-LTR by glucocorticoids requires binding of nuclear factor one (NFI) to its binding site (Cordingley et al., 1987), which in turn is a glucocorticoid dependent event.

Moreover, it has been shown that a single SRE immediately upstream of a TATA element is sufficient for steroid-dependent activation of transcription (Strähle et al., 1988), but not sufficient when placed further upstream of the TATA element. However, two copies of an SRE or a combination of one single SRE with a binding site for another transcription factor is able to confer hormone inducibility even from distance. In this way, various cis-elements have been shown to act cooperatively with different SREs (Schüle et al., 1988). In addition to this apparent interdependence between steroid receptors and transcription factors in the synergistic activation of transcription, it has been hypothesized that the inhibition of transcription by glucocorticoids occurs through interference with the activity, or binding of other important transcription factors (Akerblom et al., 1988).

The molecular basis for the apparent interdependence of steroid hormone receptors is at present not known, but recent reports (Cleat and Hay, 1989; Poellinger et al., 1989) have shown that one DNA-binding protein can increase the affinity of another DNA-binding protein for its binding-site on DNA. More importantly, the cooperative binding observed in vitro was correlated to a functional cooperativity in activation of transcription (Poellinger et al., 1989). Cooperative binding of two proteins to DNA has also been demonstrated for the GR to two adjacent SREs and has been proposed to explain the synergism between two SREs observed at the level of transcription (Schmid et al., 1989).

Variable Responsiveness of Hormonally Regulated Genes

Interaction between the steroid hormone-receptor complex and SREs on DNA appears to be a prerequisite for hormonal regulation of transcription. However, isolation of hormone-resistant cell lines has indicated that this may not be sufficient for a hormone response. From the large number of glucocorticoid-resistant cell variants isolated

(Yamamoto et al., 1976) and characterized, hormone resistance has been
shown in almost all cell lines to be due to either reduced levels or
altered forms of the GR. However, resistant cell lines have been isolated
containing GRs that allow only a selective expression of certain
glucocorticoid-mediated responses (Rabindran et al., 1987; Cook et al.,
1988). If the GR is structurally defective or if such cells require
higher GR concentrations for a particular response is at present not
known.

Darbre and King (1987) showed that in an unresponsive mouse mammary
cell line, where all endogenous hormone inducible parameters were lost, a
functional steroid-responsive machinery was detected when a steroid-
responsive recombinant gene was transfected. Moreover, by analyzing the
inducibility of a steroid-responsive promoter in different cell lines, Wu
and Pfahl (1988) in analogy to Dabre and King (1987) showed that the mere
presence of receptor in a cell is not always sufficient to allow the
hormone response to occur. More importantly, it was also shown that in
some cell lines, other factors are present that can substitute for the
steroid hormone-receptor complex.

We have studied the hormonal regulation of ER and PR mRNA
concentrations in MCF-7 and T47D$_{CO}$ breast cancer cells. Using a solution
hybridization assay for quantification of ER and PR mRNA it was shown
that estradiol down-regulates ER mRNA and up-regulates PR mRNA in MCF-7
cells, as also demonstrated by Wei et al., (1988). However, neither of
these estrogenic effects were seen in T47D$_{CO}$ cells which contained
approximately eight times as high PR mRNA levels as MCF-7 cells,
independent of estradiol. We then measured the ER content to evaluate if
this apparent estrogen insensitivity was due to lack of ER. By
conventional hormone-binding assays we could not detect any specific
estradiol binding. On the other hand with, an ER-ELISA assay
approximately 10% of the ER levels found in extracts from MCF-7 cells was
detected in extracts from T47D$_{CO}$ cells. By immunoblotting we could show
that this immunologically reactive ER in T47D$_{CO}$ cells was of the same
molecular weight as the ER in MCF-7 cells (Berkenstam et al., 1989). Next
we investigated possible mechanisms responsible for this apparent
estrogen resistance of T47D$_{CO}$ cells.

The primary structure of the ligand binding domain of the estrogen
receptors in T47D$_{CO}$ and MCF-7 cells, respectively was determined using
the polymerase chain reaction (PCR) technique. The two sequences were
found to be identical and coded for a glycine residue at amino acid
position 400, as compared to a previously reported valine residue at the
same position in the cloned estrogen receptor from MCF-7 cells. This
particular amino acid was identified as glycine also in the estrogen
receptor present in normal human breast glandular tissue. Since a
mutation within the ligand-binding domain of the estrogen receptor in
T47D$_{CO}$ cells could not be detected, attempts were made to assess at which
level abberation(s) in estrogen mechanism of action in T47D$_{CO}$ cells
occurred by using cellular transfection assays. Plasmids with or without
an estrogen responsive element cloned in front of the thymidine kinase
promoter and the CAT gene were introduced into cells grown with or
without estradiol. The presence of the responsive element resulted in an
~20-fold and an ~50-fold induction of CAT activity in T47D$_{CO}$ cells grown
in the absence or presence of estradiol, respectively. In contrast, high
basal activity or estradiol inducibility were not detected after
transfection of constructs where the estrogen responsive element was
replaced by a point-mutated estrogen responsive element or a
glucocorticoid/progesterone responsive element. These results demonstrate
that T47D$_{CO}$ cells do contain a functional estrogen receptor despite their
apparent estrogen resistance with regard to regulation of endogenous
genes. Moreover, gel mobility shift assays with nuclear extracts from
T47D$_{CO}$ cells indicated that proteins distinct from the estrogen receptor
interact in a hormone-independent manner with the estrogen responsive
element used in the transfection assays (Berkenstam, 1989). Thus, in
analogy to the results of Wu and Pfahl (1988) it is suggested that
proteins other than the ER may substitute for the ER in T47D$_{CO}$ cells and
lead to a hormone independent constitutive expression of estrogen target
genes.

Moreover, it has been shown that the long terminal repeats of Moloney murine leukemia virus (MoMuLV) and sarcoma virus (MoMuSV) both bind GR *in vitro* (Speck and Baltimore, 1987), yet only the enhancer of MoMuLV mediates glucocorticoid response *in vivo* (Overhauser and Fan, 1985). Thus, it has been suggested that proteins other than the receptor may modulate the responsiveness to steroid hormones, perhaps by influencing the availability of the SRE for hormone receptor binding (Speck and Baltimore, 1987). The synergistic action observed between steroid hormone receptors and transcription factors thus shows a pronounced cell-type dependence (Strähle et al., 1988; Schüle et al., 1988) both with regard to basal (i.e. hormone-independent or constitutive) and hormone-dependent transcription. This probably reflects the presence and/or the abundance of particular factors in different cell lines. Furthermore, it has recently been described that steroid hormone receptors may compete for factors that mediate their enhancer function and that such competition can lead to transcription interference between two steroid hormone receptors (Meyer et al., 1989).

References

Akerblom, I.E., Slater, E.P., Beato, M., Baxter, J.D., and Mellon P.L.,1988, Negative regulation by glucocorticoids through interference with a cAMP responsive enhancer. Science, 241:350.

Beato, M., 1989, Gene regulation by steroid hormones. Cell, 56:335.

Becker, P.B., Gloss, B., Schmid, W., Strähle, U., and Schütz, G., 1986, In vivo protein-DNA interactions in a glucocorticoid response element require the presence of the hormone. Nature, 324:686.

Berg, J.M., 1989, DNA binding specificity of steroid receptors. Cell, 57:1065.

Berkenstam, A., Glaumann, H., and Gustafsson, J-Å., 1988, Unspecific and sequence-specific deoxyribonucleic acid binding of the partially purified human progesterone receptor. Mol. Endocrinol., 2:571.

Berkenstam, A., Glaumann, H., Martin, M., Gustafsson, J-Å., and Norstedt, G., 1989, Hormonal regulation estrogen receptor messenger ribonucleic acid in T47D$_{CO}$ and MCF-7 breast cancer cells. Mol. Endocrinol., 3:22.

Berkenstam, A., 1989, PhD thesis, Karolinska Institute, Stockholm, Sweden.

Brock, M.L., and Shapiro, D.J., 1983, Estrogen stabilizes vitellogenin mRNA against cytoplasmic degradation. Cell, 34:207.

Buratowski, S., Hahn, S., Sharp, P.A., and Guarente, L., 1988, Function of a yeast TATA element-binding protein in a mammalian transcription system. Nature, 334:2148.

Carlstedt-Duke, J., Strömstedt, P-E., Wrange, Ö., Bergman, T., Gustafsson, J-Å., and Jörnvall, H., 1987, Domain structure of the glucocorticoid receptor protein. Proc. Natl. Acad. Sci. USA, 84:4437.

Carlstedt-Duke, J., Strömstedt, P-E., Persson, B., Cederlund, E., Gustafsson, J-Å., and Jörnvall, H., 1988, Identification of hormone-interacting amino acid residues within the steroid-binding domain of the glucocorticoid receptor in relation to other steroid hormone receptors. J. Biol. Chem., 263:6842.

Carlstedt-Duke, J., Strömstedt, P-E., Dahlman, K., Rae, C., Berkenstam, A., Hapgood, J., Jörnvall, H., and Gustafsson, J-Å., 1989, Structural analysis of the glucocorticoid receptor protein, in: "The steroid/thyroid hormone receptor family and gene regulation", Carlstedt-Duke, J., Eriksson, H., and Gustafsson, J-Å., ed., Birkhauser Verlag, Basel.

Carson, M.A., Tsai, M-J., Conneely, O.M., Maxwell, B.L., Clark, J.H., Dobson, A.D.W., Elbrecht, A., Toft, D.O., Schrader, W.T., and O'Malley, B.W., 1987, Structure- function properties of the chicken progesterone receptor A synthesized from complementary deoxyribonucleic acid. Mol. Endocrinol., 1:791.

Cato, A.C.B., Henderson, D., and Ponta, H., 1986, The hormone response element of the mouse mammary tumor virus DNA mediates the progestin and androgen induction of transcription in the proviral long terminal repeat region. EMBO J., 6:363.

Chalepakis, G., Arnemann, J., Slater, E., Brüller, H-J., Gross, B., and Beato, M., 1988, Differential gene activation by glucocorticoids and progestin through the hormone regulatory element of mouse mammary tumor virus. Cell, 53:371.

Chandler, V.L., Maler, B.A., and Yamamoto, K.R., 1983, DNA sequences bound specifically by glucocorticoid receptor in vitro render a heterologous promoter hormone responsive in vivo. Cell, 33:489.

Cleat, P.H., and Hay, R.T., 1989, Co-operative interactions between NFI and the adenovirus DNA binding protein at the adenovirus origin of replication. EMBO J., 8:1841.

Cook, P.W., Swanson, K.T., Edwards, C.P., and Firestone, G.L., 1988, Glucocorticoid receptor-dependent inhibition of cellular proliferation in dexamethasone-resistant and hypersensitive rat hepatoma cell variants. Mol. Cell. Biol.,8:1449.

Cordingley, M.G., Tate Riegel, A., and Hager, G.L., 1987, Steroid-dependent interaction of transcription factors with the inducible promoter of mouse mammary tumor virus in vivo. Cell, 48:261.

Danielsen, M., Hinck, L., and Ringold, G.M., 1989, Two amino acids within the knuckle of the first zinc finger specify DNA response element activation by the glucocorticoid receptor. Cell, 57:1131.

Darbre, P.D., and King, R.J.B., 1987, Progression to steroid insensitivity can occur irrespective of the presence of functional steroid receptors. Cell, 51:521.

Denis, M., Poellinger, L., Wikström, A-C., and Gustafsson, J-Å., 1988, Requirement of hormone for thermal conversion of the glucocorticoid receptor to a DNA-binding state. Nature, 333:686.

Eckert, R.L., Mullick, A., Rorke, E.A., and Katzenellenbogen, B.S., 1984, Estrogen receptor synthesis and turnover in MCF-7 breast cancer cells measured by a density shift technique. Endocrinology, 114:629.

Eul, J., Meyer, M.E., Tora, L., Bocquel, M.T., Quirin-Stricker, C., Chambon, P., and Gronemeyer, H., 1989, Expression of active hormone and DNA-binding domains of the chicken progesterone receptor in E.coli. EMBO J., 8:83.

Evans, R.M., 1988, The steroid and thyroid hormone receptor superfamily. Science, 240:889.

Freedman, L.P., Luisi, B.F., Korzun, Z.R., Basavappa, R., Sigler, P.B., and Yamamoto, K.R., 1988, The function and structure of the metal coordination sites within the glucocorticoid receptor DNA binding domain. Nature, 334:543.

Freedman, L.P., Yoshinaga, S.K., Vanderbilt, J.N., and Yamamoto, K.R., 1989, In vitro transcription enhancement by purified derivatives of the glucocorticoid receptor. Science, 245:298.

Gehring, U., Mugele, K., and Ulrich, J., 1984, Cellular receptor levels and glucocorticoid responsiveness of lymphoma cells. Mol. Cell. Endocrinol., 36:107.

Giguère, V., Hollenberg, S.M., Rosenfeld, M.G., and Evans, R.M., 1986, Functional domains of the human glucocorticoid receptor. Cell, 46:645.

Giguère, V., Yang, N., Segui, P., and Evans, R.M., 1988, Identification of a new class of steroid hormone receptors. Nature, 331:91.

Green, S., Walter, P., Kumar, V., Krust, A., Bornert, J-M., Argos, P., and Chambon, P., 1986, Human oestrogen receptor cDNA: sequence, expression and homology to v-erb-A. Nature, 320:134.

Green, S. and Chambon, P., 1987, Oestradiol induction of a glucocorticoid-responsive gene by chimaeric receptor. Nature, 325:75.

Green, S. and Chambon, P., 1988a, Nuclear receptors enhance our understanding of transcription regulation. Trends in Genet., 4:309.

Green, S., Kumar, V., Theulaz, I., Wahli, W., and Chambon, P., 1988b, The N-terminal DNA-binding "zink finger" of the oestrogen and glucocorticoid receptors determines target gene specificity. EMBO J., 7:3037.

Gronemeyer, H., and Pongs, O., 1980, Localization of ecdysterone on polytene chromosomes of Drosophila melanogaster. Proc. Natl. Acad. Sci. USA, 77:2108.

Gronemeyer, H., Turcotte, B., Quirin-Stricker, C., Bocquel, M.T., Meyer, M.E., Krozowski, Z., Jeltsch, J.M., Lerouge, T., Garnier, J.M., and Chambon, P., 1987, The chicken progesterone receptor: sequence, expression and functional analysis. EMBO J., 6:3985.

Guiochon-Mantel, A., Loosfelt, H., Lescop, P., Sar, S., Atger, M., Perrot-Applanat, M., and Milgrom, E., 1989, Mechanisms of nuclear localization of the progesterone receptor: evidence for interaction between monomers. Cell, 57:1147.

Hansen, J.C., and Gorski, J., 1988, Hormone-dependent and hormone-independent conformational transitions of the estrogen receptor, in: "Steroid receptors in health and disease", V.K., Moudgil, ed., Plenum Press, NY.

Hecht, A., Berkenstam, A., Strömstedt, P-E., Gustafsson, J-Å., Sippel, A.E., 1988, A progesterone responsive element maps to the far upstream steroid dependent DNase hypersensitive site of chicken lysozyme chromatin. EMBO J., 7:2063.

Hollenberg, S.M., and Evans, R.M., 1988, Multiple and cooperative trans-activation domains of the human glucocorticoid receptor. Cell, 55:899.

Horwitz, K.B., Mockus, M.B., Pike, A.W., Fennessey, P.V., and Sheridan, R.L., 1983, Progesterone receptor replenishment in T47D human breast cancer cells. J. Biol.Chem., 258:7603.

Horwitz, K.B. and Alexander, P.S., 1983, In situ photolinked nuclear progesterone receptors of human breast cancer cells: subunit molecular weight after transformation and translocation. Endocrinology, 113:2195.

Jantzen, H-M., Strähle, U., Gloss, B., Stewart, F., Schmid, W., Boshart, M., Miksicek, R., and Schütz, G., 1987, Cooperativity of glucocorticoid response elements located far upstream of the tyrosine aminotransferase gene. Cell, 49:29.

Klock, G., Strähle, U., and Schütz, G., 1987, Oestrogen and glucocorticoid responsive elements are closely related but distinct. Nature, 329, 734.

Klug, A., and Rhodes, D., 1987, "Zinc fingers": a novel protein motif for nucleic acid recognition. Trends Biochem. Sci., 12:464.

Kumar, V., Green, S., Stack, G., Berry, M., Jin, J-R., and Chambom, P., 1987, Functional domains of the huamn estrogen receptor. Cell, 51:941.

Kumar, V., and Chambon, P., 1988, The estrogen receptor binds tightly to its responsive element as a ligand-induced homodimer. Cell, 55:145.

Mader, S., Kumar, V., de Verneuil, H., and Chambon, P., 1989, Three amino acids of the oestrogen receptor are essential to its ability to distinguish an oestrogen from a glucocorticoid-responsive element. Nature, 338:271.

Maniatis,T., Goodbourn, S., and Fischer, J.A., 1987, Regulation of inducible and tissue- specific gene expression. Science, 236:1237.

Maurer, R.A., and Notides, A.C., 1987, Identification of an estrogen-responsive element from the 5′-flanking region of the rat prolactin gene. Mol. Cell. Biol., 7:4247.

McIntyre, W.R. and Samuels H.H., 1985, Triamcinolone acetonide regulates glucocorticoid-receptor levels by decreasing the half-life of the activated nuclear-receptor form. J. Biol. Chem., 260:418.

Metzger, D., White, J.H., and Chambon, P., 1988, The human estrogen receptor functions in yeast. Nature, 334:31.

Meyer, M-E., Gronemeyer, H., Turcotte, B., Bocquel, M-T., Tasset, D., and Chambon, P., 1989, Steroid hormone receptors compete for factors that mediate their enhancer function. Cell, 57:433.

Miller, J., McLachlan, A.D., and Klug, A., 1985, Repetitive zinc-binding domains in the protein transcription factor IIIA from Xenopus oocytes. EMBO J., 4, 1609.

Mitchell, P.J., and Tjian, R., 1989, Transcriptional regulation in mammalian cells by sequence-specific DNA binding proteins. Science, 245:371.

Miyajima, N., Kadowaki, Y., Fukushige, S, Shimizu, S., Semba, K., Yamanishi, Y., Matsubara, K, Toyoshima, K., Yamamoto, T., 1988, Identification of two novel members of the erbA superfamily by molecular cloning: the gene products of the two are highly related to each other. Nucleic Acids Res., 16:11057.

Mullick, A., and Katzenellenbogen, B.S., 1986, Progesterone receptor synthesis and degradation in MCF-7 breast cancer cells as studied by dense amino acid incorporation. J. Biol. Chem., 261:13236.

Okret, S., Poellinger, L., Dong, Y., and Gustafsson, J-Å., 1986, Down-regulation of glucocorticoid receptor mRNA by glucocorticoid hormones and recognition by the receptor of a specific binding sequence within a receptor cDNA clone. Proc. Natl. Acad. Sci. USA, 83:5859.

Ondek, B., Gloss, L., and Herr, W., 1988, The SV40 enhancer contains two distinct levels of organization. Nature, 333:40.

Overhauser, J., Fan, H., 1985, Generation of glucocorticoid-responsive moloney murine leukemia virus by insertion of regulatory sequences from murine mammary tumor virus into the long terminal repeat. Mol. Cell. Biol., 54:133.

Payvar, F., DeFranco, D., Firestone, G.L., Edgar, B., Wrange, Ö., Okret, S., Gustafsson, J-Å., and Yamamoto, K., 1983, Sequence-specific binding of glucocorticoid receptor to MTV DNA at sites within and upstream of the transcribed region. Cell, 35:381.

Poellinger, L., Yoza, B.K., and Roeder, R.G., 1989, Functional cooperativity between protein molecules bound at two distinct sequence elements of the immunoglobulin heavy-chain promoter. Nature, 337:573.

Ponglikitmongkol, M., Green, S., and Chambon, P., 1988, Genomic organization of the human oestrogen receptor gene. EMBO J., 7:3385.

Pratt, W.B., Jolly, D.J., Pratt, D.V., Hollenberg, S.M., Giguere, V., Capedon, F.M., Schweizer-Groyer, G., Catelli, M.G., Evans, R.M., and Baulieu, E.E., 1988, A region in the steroid binding domain determines formation of the non-DNA-binding, 9S glucocorticoid receptor complex. J. Biol. Chem., 263:267.

Ptashne, M., 1988, How eukaryotic transcriptional activators work. Nature, 335:683.

Rabindran, S.K., Danielsen, M., and Stallcup, M.R., 1987, Glucocorticoid-resistant lymphoma cell variants that contain functional glucocorticoid receptors. Mol. Cell. Biol., 7:4211.

Renkawitz, R., Schütz, G., von der Ahe, D., and Beato, M., 1984, Sequences in the promoter region of the chicken lysozyme gene required for steroid regulation and receptor binding. Cell, 37:503.

Robertson, M., 1988, Homeo boxes, pou proteins and the limits to promiscuity. Nature, 336:522.

Rusconi, S., amd Yamamoto, K., 1987, Functional dissektion of the hormone and DNA binding activities of the glucocorticoid receptor. EMBO J., 6:1309.

Sabbah, M., Redeuilh, G., and Baulieu, E., 1989, Subunit composition of the estrogen receptor. J. Biol. Chem., 264:2397.

Saceda, M., Lippman, M.E., Chambon, P., Lindsey; R.L., Ponglikitmongkol, M., Puente, M., and Martin, M.B., 1988, Regulation of estrogen receptor in MCF-7 cells by estradiol. Mol. Endocrinol., 2:1157.

Schena, M., and Yamamoto, K.R., 1988, Mammalian glucocorticoid receptor derivatives enhance transcription in yeast. Science, 241:965.

Schmid, W., Strähle, U., Schütz, G., Schmitt, J., and Stunnenberg, H., 1989, Glucocorticoid receptor binds cooperatively to adjacent recognition sites. EMBO J., 8:2257.

Schmidt, T.J. and Litwack G., 1982, Activation of the glucocorticoid-receptor complex. Physiological Rev., 62:1131.

Schrader, W.T., and O'Malley, B., 1972, Progesterone-binding components of chick oviduct. J. Biol. Chem., 247:51.

Schüle, R., Muller, M., Kaltschmidt, C., and Renkawitz, R., 1988, Many transcriptional factors interact synergistically with steroid receptors. Science, 242:1418.

Sehgal, A., Patil, N., and Chao, M., 1988, A constitutive promoter directs expression of the nerve growth factor receptor gene. Mol. Cell. Biol., 8:3160.

Sippel, A.E., Theisen, M., Borgmeyer, U., Strech-Jurk, U., Rupp, R.A.W. Püschel, A.W. Müller, A., Hecht, A., Stief, A., and Grussenmeyer, T., 1988, Regulatory function and molecular structure of DNaseI-hypersensitive elements in the chromatin domain of a gene. in: "Architecture of eukaryotic genes", Kahl, G., ed., VCH verlagsgesellschaft, Weinheim.

Slater, E.P., Cato, A.C.B., Karin, M., Baxter, J.D., and Beato, M., 1988, Progesterone induction of metallothionein-IIA gene expression. Mol. Endocrinol., 2:485.

Speck, N.A. and Baltimore, D., 1987, Six distinct factors interact with the 75-base-pair repeat of the moloney murine leukemia virus enhancer. Mol. Cell. Biol., 7:110.

Strähle, U., Klock, G., and Schütz, G., 1987, A DNA sequence of 15 base pairs is sufficient to mediate both glucocorticoid and progesterone induction of gene expression. Proc. Natl. Acad. Sci.USA, 84:7871.

Strähle, U., Schmid, W., and Schütz, G., 1988, Synergistic action of the glucocorticoid receptor with transcription factors. EMBO J., 7:3389.

Svec, F. and Rudis, M., 1981, Glucocorticoids regulate the glucocorticoid receptor in the AtT-20 cell. J. Biol. Chem., 256:5984.

Thorpe, S.M., and Rose, C., 1986, Oestrogen and progesterone receptor determinations in breast cancer: technology and biology. Cancer Surveys, 5:505.

Tora, L., Gronemeyer, H., Turcotte, B., Gaub, M-P., and Chambon, P., 1988, The N-terminal region of the chicken progesterone receptor specifies target gene activation. Nature, 333:185.

Tora, L., Mullick, A., Metzger, D., Ponglikitmongkol, M., Park, I., and Chambon, P., 1989, The cloned human oestrogen receptor contains a mutation which alters its hormone binding properties. EMBO J., 8:1981.

Tsai, S.Y., Carlstedt-Duke, J., Weigel, N.L., Dahlman, K., Gustafsson, J-Å., Tsai, M-J., and O'Malley, B.W., 1988, Molecular interactions of steroid hormone receptor with its enhancer element: evidence for receptor dimer formation. Cell, 57:361.

Umesono, K. and Evans, R., 1989, Determinants of target gene specificity for steroid/thyroid hormone receptors. Cell, 57:1139.

von der Ahe, D., Janich, S., Scheidereit, C., Renkawitz, R., Schütz, G., and Beato, M., 1985, Glucocorticoid and progesterone receptors bind to the same sites in two hormonally regulated promoters. Nature, 313:706.

von der Ahe, D., Renoir, J-M., Buchou, T., Baulieu, E-E., and Beato, M., 1986, Receptors for glucocorticosteroid and progesterone recognize distinct features of a DNA regulatory element. Proc. Natl. Acad. Sci. USA, 83:2817.

Voss, S.D., Schlokat, U. and Gruss, P., 1986, Trends Biochem. Sci., 11, 287.

Walters, M.R., 1985, Steroid hormone receptors and the nucleus. Endocrine Rev., 6:512.

Wang, L-H., Tsai, S.Y., Cook, R.G., Beatti, W.G.B., Tsai, M-J., and O'Malley, B., 1989, COUP transcription factor is a member of the steroid receptor superfamily. Nature, 349:163.

Webster, N.J.G., Green, S., Taset, D., Ponglikitmongkol, M., and Chambon, P., 1989, The transcriptional activation function located in the hormone-binding domain of the human oestrogen receptor is not encoded in a single exon. EMBO J., 8:1441.

Wei, L.L., Krett, N.L., Francis, M.D., Gordon, D.F., Wood, W.M., O'Malley, B.W., and Horwitz, K.B., 1988, Multiple human progesterone receptor messenger ribonucleic acids and their autoregulation by progestin agonists and antagonists in breast cancer cells. Mol. Endocrinol., 2:62.

Wrange, Ö., Eriksson, P., and Perlman, T., 1989, The purified activated glucocorticoid receptor is a homodimer. J. Biol. Chem., 264:5253.

Wu, K.C. and Pfahl, M., 1988, Variable responsiveness of hormone-inducible hybrid genes in different cell lines. Mol. Endocrinol., 2:1294.

Yamamoto, K.R., Gehring, U., Stampfer, M.R., Sibley, C.H., 1976, Genetic approaches to steroid hormone action. Rec. Prog. Hormone Res. 32:3.

Yi-Li, Y., Jin-Xing, T., and Ren-Bao, X., 1989, Down-regulation of glucocorticoid receptor and its relationship to the induction of rat liver tyrosine aminotransferase. J. Steroid Biochem., 32:99.

XENOBIOTIC REGULATION OF CYTOCHROME P-450 GENE EXPRESSION

Mikael Gillner, Jan Bergman[*], and Jan-Åke Gustafsson

Department of Medical Nutrition, Karolinska Institute,
Huddinge University Hospital, F60, NOVUM, S-141 86 HUDDINGE
and *Department of Organic Chemistry, Royal Institute of
Technology, S-100 44 STOCKHOLM, SWEDEN

INTRODUCTION

The relative amounts of some of the cytochrome P-450 enzymes are
known to be selectively increased by chemicals (Conney, 1982, Whitlock,
1986). Especially interesting is the induction of certain forms of
cytochrome P-450 by 5,6-benzoflavone (BNF), polycyclic aromatic hydrocarbons (PAH) such as 3-methylcholanthrene, and chloroaromatics like
2,3,7,8-tetrachlorodibenzo-p-dioxin (TCDD), since this induction profoun-
dly influences the metabolism of PAH to ultimate carcinogens that are
able to form specific adducts with DNA (Conney, 1982). TCDD is the most
potent inducer of these particular forms of cytochrome P-450 (c. f.
Poland and Knutson, 1982).

The major cytochrome P-450 inducible by TCDD, 3-methylcholanthrene
and BNF in the rat is called cytochrome P-450 IA1, whereas the major
isosafrole-inducible cytochrome P-450 in the rat is called cytochrome
P-450 IA2. BNF, PAH and TCDD are also capable of increasing the levels of
other enzymes involved in the metabolism of xenobiotics, among others
cytochrome P-450 IA2, DT-diaphorase, and glutathione S-transferase (cf.
Poland and Knutson, 1982).

TCDD and structurally related chlorinated aromatics also induce a
broad spectrum of toxic effects in epithelial tissues (Poland and
Knutson, 1982) including thymic involution in most species and changes in
differentiation in epidermis in humans (manifesting as chloracne). A
possible role for the thyroid gland in the mediation of toxic effects of
TCDD has been discussed, since certain signs of TCDD toxicity (Poland and
Knutson, 1982) resemble those of thyrotoxicosis (eg. loss of weight) or
hypothyroidism eg. bradycardia and increased levels of triglycerides
and/or cholesterol).

The induction of TCDD-inducible xenobiotic metabolizing enzymes (and
probably also the toxic effects of TCDD) is mediated by an intracellular
soluble receptor protein which binds TCDD, related chlorinated aromatic
hydrocarbons and certain PAH as well as BNF with high affinity and
specificity (Poland and Knutson, 1982). The association of one of these
inducers with the TCDD receptor increases its affinity for DNA (Hannah et
al., 1976), leading to an increased rate of transcription of the cytochromes P-450 IAI and IA2 genes (Tukey et al., 1982). This mechanism of

action appears analogous with the existing model for steroid hormone action (Gorski and Gannon, 1976; Yamamoto and Alberts, 1976). This model entails: 1) translocation of the ligand-receptor complex into the cell nucleus (although this first step is subject of controversy (Whitlock and Galeazzi, 1984)); 2) interaction of the complex with nuclear target elements resulting in an enhanced transcription of target genes.

The physico-chemical properties of the TCDD receptor are also in agreement with the corresponding properties of the steroid hormone receptors, and in particular, those of the glucocorticoid receptor (cf. Gustafsson, et al., 1987). Both the glucocorticoid and the TCDD receptor are assymetric proteins with a relative molecular mass of 90,000-100,000 (Poellinger et al., 1983; 1985). In view of these similarities, it is tempting to speculate that the TCDD receptor belongs to the same multigene family as the steroid and thyroid hormone receptors.

Steroid receptors are intracellular soluble proteins that mediate the complex effects of their ligands ie. steroid hormones, on development, growth and physiological homeostasis by selective modulation of gene expression, predominantly at the level of DNA transcription. Purification and recent cloning of several steroid hormone receptors, including the glucocorticoid, estrogen, progesterone and vitamin D receptors, have allowed a detailed biochemical characterization and shown that these proteins contain discrete DNA-and ligand-binding functional domains (reviewed in Gustafsson et al., 1987). The cystein-rich DNA-binding domain of the estrogen and glucocorticoid receptors is evolutionarily highly conserved and is common to all the other identified members of the steroid receptor gene family exhibiting in total a 50-60 % identity over a stretch of 66 or 68 amino acids (Petkovich et al., 1987 and references therein). In fact, nucleotide sequences encoding the DNA-binding domain of steroid hormone receptors have successfully been used as hybridization probes to screen for related but unrecognized or novel receptor structures under reduced stringency conditions. In such a manner, the mineralocorticoid (Arizza et al, 1987) and the androgen (Chang et al., 1988; Lubahn et al., 1987) receptors have recently been cloned. The v-erbA oncogene of avian erythroblastosis virus has also been shown to share sequence homology with the members of the steroid hormone receptor gene family. The cellular homologue of the v-erbA oncogene of avian erythroblastosis virus is the intranuclear thyroid (triiodothyronine) receptor (Sap et al., 1986; Weinberger et al., 1986; Thompson et al., 1987).

Moreover, a receptor for retinoic acid has been identified as a previously unknown member of this multigene family of transcriptional activators with ability to bind and become activated by lipophilic low molecular ligands (Petkovich et al., 1987; Giguère et al., 1987 and 1988; Brand et al., 1988). Retinoic acid has been implicated to function as a morphogen (ie. to initiate changes in levels of gene expression and to determine the further development of target cells) in the development of the chicken (Thaller and Eichele, 1987), and shown to block differentiation of several cell types (cf. Giguère et al, 1987). Retinoids inhibit tumour progression in animals and blocks the actions of tumour promoters in vitro (cf. Giguère et al., 1987). Thus, the retinoic acid receptor may possibly function as a modulator of oncogenesis.

Thyroid hormone also exerts profound effects on development and differentiation. It is thus required for metamorphosis in amphibians (Eberhardt et al., 1980). The thyroid hormones (3,5,3',5'-tetraiodo-L-thyronine (L-T4) and 3,5,3-triido-L-thyronine(L-T3) affect the growth and metabolic function of vertebrate tissues (cf. Samuels et al., 1988).

It is remarkable that the genes for steroid hormone, thyroid hormone and retinoic acid receptors exhibit sequence homology despite the fact that these receptors bind ligands with relatively differing chemical structures. It implies that chemically distinct molecules can exert their gene regulatory effects via the same general mechanism.

Therefore, we try to, by purification and cloning, as well as by ligand binding studies, map the relationship of the TCDD receptor to the receptors for steroid and thyroid hormones and for retinoic acid.

The nature of a putative endogenous ligand for the TCDD receptor is, however, not known, and the TCDD receptor does not bind steroid or thyroid hormones (Poland and Knutson, 1982). In view of the multiple effects of TCDD on cellular proliferative responses and differentiation processes, it is possible that the endogenous ligand for the TCDD receptor is a morphogen and/or a modulator of carcinogenesis. It can, however, not be excluded that the natural ligand for the TCDD receptor is of exogenous nature. TCDD and the related chlorinated aromatics are probably not the natural ligands for the TCDD receptor since they are environmental contaminants of industrial origin (Czuczwa and Hites, 1984). A third possibility is that the natural ligand (or precursors to it) for the TCDD receptor occurs in the diet. It has thus recently been suggested that UV-products of tryptophan constitute candidates for an endogenous TCDD receptor ligand (Rannug et al., 1987). It is also well-known that certain dietary indoles induce cytochrome P-450 IA 1 and 2 dependent enzymatic microsomal enzyme activities in rats (Loub et al., 1975). If these dietary compounds induce cytochrome P-450 IA 1 or 2 by the same mechanism as TCDD they ought to bind to the TCDD receptor. Therefore, we have used both a TCDD receptor binding assay and computer-supported molecular structure studies to identify critical properties necessary for TCDD receptor binding of various indoles (Gillner et al., 1985).

MATERIALS AND METHODS

Materials

Chemicals. Dibenz(a,h)anthracene, benz(a)anthracene and 7,12-dimethylbenz(a)anthracene were obtained from Sigma Chemical Co. (St. Louis, MO). Benzo(1,2-b:4,5-b')bis(1)benzothiophene was a kind gift from Dr. P. Kirby, Shell Co. (Sittingbourne Research Centre, Kent, UK). Isoquino(3,4-b)phenanthridine was generously provided by Dr LeRoy H. Klemm (University of Oregon, OR). These compounds were used without further purification.

Benzo(1,2-b:4,5-b')bisbenzofuran was prepared according to Bergman et al., (1979). 5,11-Dimethylindolo(3,2-b)carbazole was synthesized as described by Hünig and Steinmetzer (1976). 2,8-Dimethylindolo(3,2-b)carbazole, and 4,10-dimethylindolo(3,2-b)carbazole were prepared according to the procedure of Robinson (1963), with the corresponding phenylhydrazones as starting materials. 5,11-Diacetylindolo(3,2-b)carbazole was synthesized as described by Robinson (1963).

5,11-Diethylindolo(3,2-b)carbazole, and 5,11-dibutylindolo(3,2-b)-carbazole were prepared by alkylation of indolo(3,2-b)carbazole according to Bergman and Sand (1984). 6,12-Dihydroxyindolo(3,2-b)carbazole was prepared by reduction of the corresponding quinone with Zn in acetic acid; the indolo(3,2-b)carbazole-6,12-quinone was prepared as described by Osman et al. (1977). 5,11-Di(**N,N**-dimethylaminoethyl)indolo(3,2-b)carbazole was prepared according to Bergman. All indolo(3,2-b)carbazoles

were purified by sublimation under reduced pressure. The compounds
synthesized were at least 95 % pure as judged by mass-spectrometry.
Infrared spectra of 5,11-dimethylindolo(3,2-b)carbazole were recorded,
and no N-H signals were observed, indicating that these preparations
contained less than 0.5 % indolo(3,2-b)carbazole. Other chemicals were
obtained from the same sources as described earlier (Gillner et al.,
1985).

Animals. Sprague-Dawley rats (weighing 200 µg) were obtained from
the same source and treated as described earlier (Gillner et al., 1987).
5,6-Benzoflavone injections were given as described earlier. Implants
were made from 4.5 mm long silastic tubing (No. 602-285, Dow Corning,
Midland, MI) in which crystals of compunds (80-100 mg) were packed. The
implants were sealed with .5 mm wooden plugs and silicon glue (Type A,
Dow Corning) at the ends. The implants were inserted intraperitoneally
under ether anaesthesia. The animals were sacrificed after three days of
implantation, or 40 h after injection, by cervical dislocation.

Methods

Receptor binding studies. As a TCDD receptor source, rat liver
cytosol was used. Compounds insoluble in dimethylsulphoxide at room
temperature were dissolved in **N,N**-dimethylsulphoxide. The same
volume of **N,N**-dimethylacetamide was added to the incubations without
competitor. Binding assays were performed by electrofocusing in polyacrylamide gel, and determination of IC_{50}-values by logit-log plots were
carried out described earlier (Gillner et al., 1985).

7-Ethoxyresorufin **O**-deethylase activity assay. Preparation of liver
microsomal fractions and 7-ethoxyresorufin **O**-deethylase activity assay
were carried out as described earlier (Gillner et al., 1987).

RNA isolation and blot hybridization. RNA isolation and blot
hybridization was carried out as described earlier (Gillner et al.,
1987).

RESULTS AND DISCUSSION

We have used both a TCDD receptor binding assay and computer-supported molecular structure studies to identify critical properties necessary
for TCDD receptor binding of various indoles (Gillner et al., 1985). We
found that if the van der Waals radii of all atoms are considered, most
of the molecules binding with a high affinity to the TCDD receptor may be
included into a 6.8 x 13.7 Å rectangle. In contrast to the 3 x 10 Å
rectangle with halogen centers in its corners proposed by Poland and
Knutson (1982) to account for the TCDD receptor binding of halogenated
ligands, such a model accounts for the binding of both BNF and TCDD to
the TCDD receptor.

In the same study (Gillner et al., 1985), we also found a new type
of ligand binding to the TCDD receptor with high affinity. This is a
heteroaromatic polycyclic hydrocarbon that may be chemically derived
(Bergman, 1970) from some of the indoles earlier reported to induce aryl
hydrocarbon hydroxylase (AHH) (Wattenberg, 1980), a cytochrome P-450IA 1
and/or 2 dependent microsomal enzyme activity. Since this compound,
indolo(3,2-b)carbazole, binds with high affinity to the TCDD receptor
despite the fact that it is unhalogenated, it was of interest to investigate the influence on receptor binding affinity of different substituents
on this parent structure. We have therefore synthesized or acquired some

derivatives and analogues of this parent compound, examined them by means of receptor binding assays, and compared them to known TCDD receptor ligands.

Some poly- and heterocyclic aromatic hydrocarbons related to indolo(3,2-b)carbazole were also studied. Dibenz(a,h)anthracene may be viewed as an analogue of indolo(3,2-b)carbazole, where each nitrogen is replaced by two carbon atoms. The affinity of dibenz(a,h)anthracene for the TCDD receptor was higher (IC_{50} 2.5 nM) than that of indolo(3,2-b)carbazole (IC_{50} 3.6 nM). Replacement of the nitrogens of indolo(3,2-b)carbazole with sulfur, leading to benzo(1,2-b:4,5-b)bis(1)benzothiophene, increased the affinity (IC_{50} 3.3 nM), whereas replacement with oxygen, yielding benzo(1,2-b:4,5-b')bis(1)benzofuran, decreased the affinity (IC_{50} 29 nM). The dibenz(a,h)anthracene relatives benz(a)anthracene, isoquino(3,4-b)phenanthridine, and 7,12-dimethylbenz(a)anthracene had intermediate TCDD receptor affinities (IC_{50} 48 nM, 137 nM and 139 nM; respectively).

Two rutaecarpine alkaloids were also investigated (Gillner et al., 1989). The IC_{50} value for TCDD receptor binding of 7,8-dehydrorutaecarpine was estimated to 7 nM indicating a high-affinity interaction, whereas rutaecarpine appeared less active (IC_{50} 110 nM). These findings are of interest in view of the fact that analogues of these compounds may be formed following UV-irradiation of tryptophan and that such photo-products have been suggested to constitute (the) endogenous ligand(s) for the TCDD receptor. (Rannug et al., 1987). As further support of this notion, the rutaecarpine alkaloids investigated could be fitted into the 6.8 x 13.7 Å rectangle. In view of their structural similarity to dehydrorutaecarpine and the agreement of their mol. wt with that of the photoproduct with the highest affinity for the TCDD receptor, we suggested deaza-analogues of dehydrorutaecarpine to represent possible candidates for the endogenous receptor ligand.

Methylation in a critical position is known to affect the affinity of certain steroids for their receptors (c. f. Poellinger et al., 1985). Hence the effect of methylation of indolo(3,2-b)carbazole was investigated. **N**-methylation of indolo(3,2-b)carbazole, leading to 5,11-dimethylindolo(3,2-b)carbazole, increased the TCDD receptor affinity about three-fold (IC_{50} 1.2 nM), as compared to the parent compound (IC_{50} 3.6 nM). 4,10-Dimethylindolo(3,2-b)carbazole and 2,8-dimethylindolo(3,2-b)carbazole, had lower affinities, (IC_{50} 19 nM, and IC_{50} > 150 nM, respectively), indicating that the affinity decreased with the distance of the methyl group from the nitrogen.

Since **N**-methylation increased the TCDD receptor affinity of indolo-(3,2-b)carbazole, some other **N**-substituted indolo(3,2-b)carbazoles were studied. **N**-ethylation of indolo(3,2-b)carbazole leading to 5,11-diethyl-indolo(3,2-b)carbazole, decreased the binding affinity by a factor of three (IC_{50} 8.9 nM). Acetylation of indolo(3,2-b)carbazole, giving diacetylindolo(3,2-b)carbazole also resulted in lower affinity (IC_{50} 11.2 nM). Substitution on the nitrogens of indolo(3,2-b)carbazole with more bulky substituents like in 5,11-dibutylindolo(3,2-b)carbazole, and 5,11-di(**N,N**-dimethylaminoethyl)indolo(3,2-b)carbazole further decreased the affinity (IC_{50} > 150 nM and IC_{50} > 1500 nM, respectively). Introduction of oxygen in, or hydroxylation of, the middle ring of indolo(3,2-b)carbazole, giving indolo(3,2-b)carbazole-6,12-quinone, or 6,12-dihydro-xyindolo(3,2-b)carbazole also lowered the TCDD receptor affinity (IC_{50} > 1500 nM).

Thus, for the substituted indolo(3,2-b)carbazoles, steric factors (such as the size and position of the substituent) appeared to be correlated to TCDD receptor affinity. Furthermore, in a more extensive QSAR study of ten PAH binding to the TCDD receptor using 40 different steric, quantum chemical and NMR parameters, length and distance criteria were found to be the strongest contributors to the models (Johnels et al., 1989). Taken together, these results seem to stress the importance of steric factors for ligand binding to the TCDD receptor.

The indolocarbazoles, as well as some other isosteric heterocycles studied in this work, represent a new class of TCDD receptor ligands distinct from the previously known ligands (halogenated aromatic hydrocarbons and PAH). The TCDD receptor binding estimates of these heterocycles may be of value to the development of more general structure-affinity concepts than those presently available for binding of ligands to the TCDD receptor.

The molecules found to bind with high affinity to the TCDD receptor may be fitted into the 6.8 x 13.7 Å rectangle when their atomic van der Waals radii are included. However, examples may be found of compounds, such as 2,3,7,8-tetramethyldibenzo-p-dioxin, that can be fitted into the 6.8 x 13.7 Å pattern (ref), but have virtually no TCDD receptor affinity. Possibly, other factors such as charge distribution, or π-electron distribution within the ligand and/or polarizibility of the ligand, are also important determinants for dioxin-receptor binding. In order to correlate them with TCDD-receptor affinity, such properties of ligands are presently studied using computational chemistry.

Indolo(3,2-b)carbazole is virtually insoluble in customary organic solvents (Grotta et al., 1961). Due to this insolubility, it was not possible to administer the compound in corn oil or albumin solution to rats in order to investigate its biological activity. If a solution of indolo(3,2-b)carbazole (or its 5,11-dimethyl derivative) in dimethylsulfoxide was injected intraperitoneally, the compound could be found as a crystalline deposit at the administration site. In order to investigate if indolo(3,2-b)carbazoles induce a microsomal enzyme activity regulated by the TCDD receptor we have instead used intraperitoneal silastic implants containing crystalline preparations of the compounds.

Silicone rubber (dimethylpolysiloxane; Silastic) has been used in chronic implants because it does not cause host vs. grapht reactions even after long periods of time (Dziuk and Cook 1966). Steroids diffuse more readily through silicone rubber than other synthetic polymers (Kincl et al., 1968). The permeation through silicone rubber is favoured for more hydrophobic steroids (which have fewer hydroxyl groups) (Kincl and Rudel, 1971). The release of compounds is generally higher from silastic implants containing dry crystalline powder as compared to implants in which the compound is in solution (Kincl and Rudel, 1971). The release of steroids from silicone rubber implants has been shown in vitro to be proportional to the membrane surface area (Kincl and Rudel, 1971), and to give even plasma levels of steroid for prolonged periods of time in the rat (Kincl and Rudel, 1971). Intramuscular, adipose tissue, and subcutaneous implants have been used, with the efficiency for release of steroids decreasing in the order mentioned (Kincl and Rudel, 1971).

As shown in Table 1, implants containing BNF and 5,11-dimethylindolo-(3,2-b)carbazole produced significant induction of microsomal 7-ethoxy-resorufin O-deethylation in rat liver, whereas indolo(3,2-b)carbazole implants failed to do so. Furthermore, BNF implants as well as 5,11-dimethylindolo(3,2-b)carbazole implants induced cytochrome P-450 IA 1 and 2 mRNA in rat liver (not shown).

Table 1. Effect of indolo(3,2-b)carbazoles on 7-ethoxyresorufin O-deethylase activity of rat liver.

Injections of 5,6-benzoflavone (80 mg/kg) or corn oil were given i.p. 40 h before sacrifice. Silastic implants (4.5 mm) were inserted i.p. three days before sacrifice. Liver microsomes were prepared from the four rats of each treatment group, and 7-ethoxyresorufin O-deethylase activity determined as described in Materials and Methods.

Experiment	Pretreatment	Compound	7-ethoxyresorufin O-deethylase activity pmol/mg protein/min
1	2 implants	none	120 ± 39
1	2 implants	5,6-benzoflavone[a]	2101 ± 501
1	2 implants	indolo(3,2-b)carbazole	131 ± 33
2	injection	corn oil	172 ± 92
2	injection	5,6-benzoflavone	5208 ± 1500
2	1 implant	5,11-dimethylindolo-(3,2-b)carbazole[a]	546 ± 69

[a] Significantly different (p < 0.05) from control (Student's t-test).

The induction of microsomal 7-ethoxyresorufin O-deethylation in rat liver by 5,11-indolo(3,2-b)carbazole-containing intraperitoneal implants observed in the present study indicates that this compound is an agonist or a partial agonist with respect to a TCDD receptor mediated biochemical response. The induction of cytochrome P-450 IA 1 and 2 mRNA by 5,11-dimethylindolo(3,2-b)carbazole obtained here is also in line with this notion. The failure of indolo(3,2-b)carbazole-containing implants to induce 7-ethoxyresorufin O-deethylation despite the relatively high receptor affinity of indolo(3,2-b)carbazole may possibly be due to a rapid metabolism or a low permeation through silicone rubber of this compound as compared to 5,11-dimethylindolo(3,2-b)carbazole, which is conceivably more hydrophobic and resistant to metabolism. Another possible explanation for the low induction observed would be partial antagonism. Certain other TCDD receptor ligands have been demonstrated to be partial antagonists (Bannister, 1987 and references therein).

The characteristic spectrum of toxic effects elicited by halogenated aromatic hydrocarbons is not observed with PAH, although both classes of compounds are able to bind to the TCDD receptor and to induce AHH. To account for this discrepancy it has been suggested, since PAH is much more rapidly metabolized that chlorinated aromatic hydrocarbons, that enzyme induction is an early event, but that toxicity (especially epithelial cell proliferation and metaplasia) is a much later event which requires persistent receptor occupation and gene expression (Poland and Knutson, 1982). Alternatively, the stimulus of gene expression must be present at certain time-points to elicit full response, as has been shown with steroidal regulation of gene expression in some systems (Harris and Gorski, 1978). Since induction of a TCDD receptor regulated enzyme activity may be obtained by silastic implants containing PAH, these may probably be used to conveniently test if any of the two suggested explanations for the lower toxicity of PAH as compared to chlorinated receptor ligands is valid.

ACKNOWLEDGEMENTS

The skilful technical assistance of Ms Monica Alexandersson and Birgitta Fernström is gratefully acknowledged. This study was supported by grants from the Swedish Cancer Society and from the National Institute of Health (ESO3904).

REFERENCES

Arizza, J.L., Weinberger, C., Serelli, G., Glazer, T., M. Handeli, G., Housman, D. and Evans, R.M., 1987, Cloning of human mineralocorticoid receptor complementary DNA: structure and kinship with glucocorticoid receptor, Science, 237: 268-275.

Bannister, R., Davis, D., Zacharewski, T., Tizard, I., and Safe, S., 1987, Aroclor 1254 as a 2,3,7,8-tetrachlorodibenzo-p-dioxin antagonist: effects on enzyme induction and immunotoxicity, Toxicology, 46: 29-42.

Bergman. J. Manuscript in preparation.

Bergman, J., 1970, Condensation of indole and formaldehyde in the presence of air and sensitizers: a facile synthesis of indolo(3,2-b)carbazole, Tetrahedron, 26: 3353-3355.

Bergman, J., and Sand, P., 1984, A new simple procedure for alkylation of nitrogen heterocycles using dialkyl oxalates and alkoxides, Tetrahedron Lett., 1957-1961.

Bergman, J., Egestad, B. and Rajapaksa, D., 1979, An X-ray crystal and molecular structure determination of a benzo(1,2-b:4,5-b')bisbenzofuran derivative formed by base-induced condensation of 3(2H)-benzofuranone, Acta Chem. Scand. Ser. B Org. Chem. Biochem., 33: 405-409.

Brand, N., Petkovich, M., Krust, J., Chambon, P., Thé, H.H., Marchioo, A., Tiollais, P., Dejean, A., 1988. Identification of a second human retinoic acid receptor, Nature, 332: 850-853.

Chang, C., Kokontis, J., and Liao, S., 1988, Molecular cloning of human and rat complementary DNA encoding androgen receptors, Science, 240:324-326.

Conney, A.H., 1982, Induction of microsomal enzymes by foreign chemicals and carcinogenesis by polycyclic aromatic compounds, Cancer Res., 42: 4875-4917.

Czuczwa, J.M. and Hites, R.A., 1984, Environmental fate of combustion-generated polychlorinated dioxins and furans, Environ. Sci. Technol., 18: 444-450.

Dziuk, P.J., and Cook, B., 1966, Passage of steroids through silicone rubber, Endocrinology, 78: 208-211.

Eberhardt, N.L., Apriletti, J.W. and Baxter, J.D., 1980, pp 311-394 in: "Biochemical actions of hormones" Vol III, G. Litwack, ed. Academic Press, New York.

Evans, R.M., 1988, The steroid and thyroid hormone receptor superfamily, Science, 240: 889-895.

Giguère, V., Ong, E.S., Segui, P. and Evans, R.M., 1987, Identification of a receptor for the morphogen retinoic acid, Nature, 330: 624-629.

Giguère, V., Yang, N., Segui, P. and Evans, R.M., 1988, Identification of a new class of steroid hormone receptors, Nature, 331: 91-94.

Gillner, M., Bergman, J., Cambillau, C., Fernström, B. and Gustafsson, J.-Å., 1985, Interactions of indoles with specific binding sites for TCDD in rat liver, Mol. Pharmacol., 28: 357-363.

Gillner, M., Brittebo, E.B., Brandt, I., Söderkvist, P., Appelgren, L.E.
and Gustafsson, J.-Å., 1987, Uptake and specific binding of
2,3,7,8-tetrachlorodibenzo-p-dioxin in the olfactory mucosa of
mice and rats, Cancer Res., 47: 4150-4159.

Gillner, M., Bergman, J., Cambillau, C. and Gustafsson, J.-Å., 1989,
Interactions of rutaecarpine alkaloids with specific binding sites
for TCDD in rat liver, Carcinogenesis, 10: 651-654.

Gorski, J. and Gannon, F., 1976, Current models of steroid hormone
action: a critique, Ann. Rev. Physiol., 38: 425-450.

Gustafsson, J.-Å., Carlstedt-Duke, J., Poellinger, L., Okret, S.,
Wikström, A.-C., Brönnegård, M., Gillner, M., Dong, Y., Fuxe, K.,
Cintra, A., Härfstrand, A. and Agnati, L., 1987, Biochemistry,
molecular biology and physiology of the glucocorticoid receptor,
Endocrine Reviews, 8: 185-234.

Hannah, R., Lund, J., Poellinger, L., Gillner, M. and Gustafsson, J.-Å.,
1986, Characterizations of the DNA-binding properties of the
receptor for 2,3,7,8-tetrachlorodibenzo-p-dioxin, Eur. J. Biochem., 156: 237-242.

Harris, J. and Gorski, J., 1978, Evidence for a discontinuous requirement
for estrogen in stimulation of deoxyribonucleic acid synthesis in
the immature rat uterus, Endocrinology, 103: 240-245.

Hünig, S. and Steinmetzer, H.-C., 1976, Kondensierte Stickstoffheterocyc-
len, Liebigs Ann. Chem. 1090-102.

Johnels, D., Gillner, M., Nordén, B., Toftgård, R. Gustafsson, J.-Å.,
1989, Quantitative structure-activity relationship (QSAR) analysis
using the partial least squares (PLS) method: the binding of
polycyclic aromatic hydrocarbons (PAH) to the rat liver 2,3,7,8-
tetrachlorodibenzo-p-dioxin (TCDD) receptor. Quant. Struct.-Act.
Relat., 8, 83-89.

Kincl, F.A., Benagiano, G. and Angee, H.W., 1968, Sustained release
hormonal preparations. 1. Diffusion of various steroids through
polymer membranes, Steroids, 11: 673-680.

Kincl, A.F. and Rudel, H.W., 1971, Sustained release hormonal preparations, Acta Endocrinol. Suppl., 151: 5-30.

Loub, W.D:, Wattenberg, L.B. and Davis, D.W., 1975, Aryl hydrocarbon
hydroxylase induction in rat tissues by naturally occurring
indoles of cruciferous plants, J. Natl. Cancer. Inst., 54:
985-988.

Lubahn, D.B., Joseph, D.R., Sullivan, P.M., Willard, H.F., French, F. and
Wilson, E., 1988, Cloning of androgen receptor cDNA and localisa-
tion to X-chromosome, Science, 240: 327-330.

Osman, A.M., Hamman, A.S. and Salah, A.Th., 1977, New diindoloquinones
from halogenated p-benzoquinones and aromatic amines, Indian. J.
Chem. Sect. B. Org. Chem. Inc. Med. Chem., 15: 1118-1120.

Petkovich, M., Brand, N., Krust, J. and Chambon, P., 1987, A human
retinoic acid receptor which belongs to the family of nuclear
receptors, Nature, 330: 444-450.

Poellinger, L., Lund, J., Gillner, M., Hansson, L.-A. and Gustafsson,
J.-Å., 1983, Physicochemical characterization of specific and
nonspecific polyaromatic hydrocarbon binders in rat and mouse
liver cytosol, J. Biol. Chem. 2158: 13535-13542.

Poellinger, L., Lund, J., Gillner, M. and Gustafsson, J.-Å., 1985, The
receptor for 2,3,7,8-tetrachlorodibenzo-p-dioxin: similarities and
dissimilarities with the steroid hormone receptors, pp 755-790 in:
"Molecular Mechanism of Steroid Hormone Action", V.K. Moudgil,
ed., Walter de Gruyter, New York.

Poland, A., and Knutson, J.C., 1982, 2,3,7,8-Tetrachlorodibenzo-p-dioxin
and related halogenated aromatic hydrocarbons: examination of the
mechanism of toxicity, Annu. Rev. Pharmacol. Toxicol. 22: 517-
554.

Rannug, A., Rannug, U., Rosenkrantz, H.S., Winqvist, L., Westerholm, R., Agurell, E. and Grafström, A.-K., 1987, Certain photooxidized derivatives of tryptophan bind with very high affinity to the Ah-receptor and are likely to be endogenous signal substances, J. Biol. Chem. , 262: 15442-15427.

Robinson, B., 1963, The Fischer indolisation of cyclohexane-1,4-dione bisphenylhydrazone, J. Chem. Soc., 3097-3099.

Samuels, H.H., Forman, B.M., Horowitz, Z.D. and Zheng-Sheng Ye., 1988, Regulation of gene expression by thyroid hormone, J. Clin. Invest. 81: 957-967.

Sap, J., Munoz, A., Damm, K., Goldberg, Y., Ghysdael, Leutz, A., Beug, H., and Vennström, B., 1986, The c-erb-A protein is a high affinity receptor for thyroid hormone receptor, Nature, 324: 635-640.

Taller, C. and Eichele, G., 1987, Identification and spatial distribution of retinoids in the developing chick limb bud, Nature 327: 625-628.

Thompson, L.L., Weinberger, L., Lebo, R. and Evans, R., 1987, Identification of a novel thyroid hormone receptor expressed in the mammalian central nervous system. Science, 237: 1610-1614.

Tukey, R.H., Hannah, R.R., Negishi, M., Nebert, D.W. and Eisen, H.J., 1982, The Ah locus: correlation of intranuclear appearance of inducer-receptor complex with induction of cytochrome P_1-450 mRNA, Cell, 31: 275-284.

Wattenberg, L.W. 1980. Inhibitors of chemical carcinogens, 1980, pp 517-539 in: "Cancer Achievements, Challenges, and Prospects for the 1980s"., Vol 7, J.H. Burchenal and M.F. Oettegen, eds., Grune and Stratton, New York.

Weinberger, C., Thompson, C.C., Ong, E.S., Lebo, R., Gruol, D.J. and Evans, R., 1986, The c-erb-A gene encodes a thyroid hormone receptor, Nature, 324: 641-646.

Whitlock, J.P., 1986, The regulation of cytochrome P-450 gene expression, Ann. Rev. Pharmacol. Toxicol. 26: 333-369.

Yamamoto, K.R. and Alberts, B.M., 1976, Steroid Receptors: Elements for modulation of eukaryotic transcription, Ann. Rev. Biochem., 45: 722-746.

PROSTANOID METABOLISM AND BIOLOGICALLY ACTIVE PRODUCT FORMATION

David Kupfer

Worcester Foundation for Experimental Biology
222 Maple Avenue
Shrewsbury, MA   01545

INTRODUCTION

This presentation was designed to introduce the reader to the field
of prostanoid metabolism by cytochrome P450 monooxygenases. Therefore,
the description of the biosynthesis of prostanoids carried out by
cyclooxygenase (not a P450 enzyme) was kept brief. To do justice to the
subject of P450 catalyzed metabolism of prostanoids would require an
extensive review of the subject which is not the goal of this
manuscript. Thus, this presentation intends to concentrate on the
metabolism of prostaglandins primarily in the liver. However certain
studies on prostaglandin metabolism by monooxygenases in extrahepatic
tissues which might provide useful comparison with liver findings, as
well as relevance to in vivo situations are also described. Since the
presentation of metabolic studies in extrahepatic tissues was not
intended to be exhaustive, I apologize to those investigators whose
work has been omitted or inadequately covered in this presentation.
Studies on cytochrome P450 catalyzed hydroxylation of arachidonic acid
have been omitted. A comprehensive review of this subject has been
recently published[1]. Additionally the interested reader should
consult original publications on arachidonic acid metabolism by Drs.
Jorge Capdevila (Nashville, TN, USA), Michal Schwartzman (Valhalla, NY,
USA) and Ernst Oliw (Stockholm, Sweden).

A. PROSTANOID BIOSYNTHESIS

The formation of prostanoids is largely controlled by the
availability of the appropriate unsaturated C20 fatty acid (e.g.,
arachidonic acid), which is generated from phospholipids by the action
of phospholipase A2. In turn the released arachidonic acid is oxidized
by the enzyme cyclooxygenase to the endoperoxide hydroperoxide (PGG2)
utilizing molecular oxygen (see Fig. 1). The subsequent steps involve
conversion of PGG2 to PGH2 and its transformation into thromboxane A2
(TXA2), prostacyclin (PGI2) and prostaglandins E2 (PGE2), F2a (PGF2a)
and D2 (PGD2) (Fig 2). The corresponding prostaglandins with a single
double bond PGE1, PGF1a and PGD1 are biosynthesized from eicosatrienoic
acid (a C20 fatty acid with three double bonds).

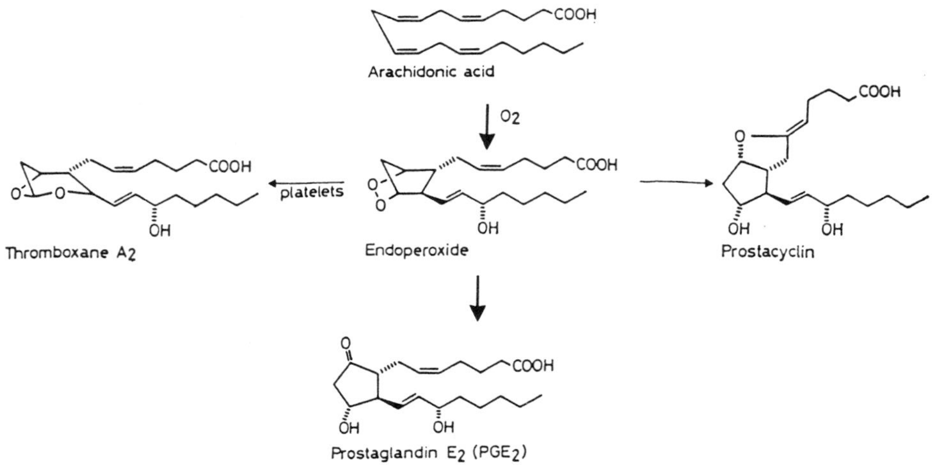

FIGURE 1. PATHWAY OF PROSTANOID BIOSYNTHESIS. The endoperoxide is PGH2. In addition, PGH2 yields PGF2a and PGD2.

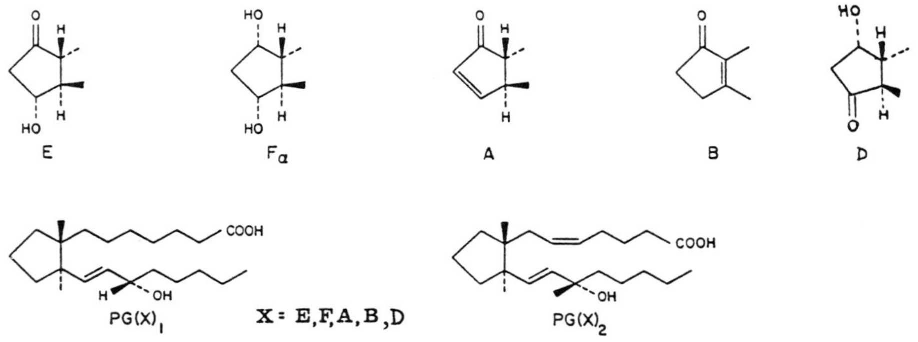

FIGURE 2. STRUCTURES OF PRIMARY PROSTAGLANDINS. PGAs and PGBs are thought not to be endogenous PGs, but are formed during isolation from the corresponding PGEs.

Interestingly, whereas the biosynthesis of prostaglandins does not involve P450s, the conversion of PGH2 to TAX2 and PGI2 require a P-450 mediated catalysis. The latter observations stem from the elegant studies of Ullrich and coworkers[2] who demonstrated that TXA2 and PGI2 are biosynthesized from PGH2 by a novel and unusual class of cytochrome P-450s. These P-450s do not undergo the usual reduction of the heme iron (Fe^{3+} to Fe^{2+}) by NADPH and NADPH-P450 reductase but act in the oxidized Fe^{3+} form as specific isomerases.

B. PROSTANOID METABOLISM

In vivo metabolism

These studies in animals and man involved the metabolism of prostaglandins and the identification of the urinary prostaglandin metabolites[3-6]. The transformation of primary prostanoids into hydroxylated derivatives and into dicarboxylic products in vivo, suggested the likelihood that cytochrome P450 monooxygenases catalyze

294

these reactions. This speculation from in vivo studies provided the impetus for the subsequent extensive experimentations in vitro, which eventually led to the demonstration that microsomal P450 monooxygenases catalyze the hydroxylation of prostanoids.

In vitro metabolism

a) Early studies on hydroxylation of prostaglandins (PGs). The observation in Bengt Samuelsson's laboratory[7,8] that the incubation of prostaglandin A1 (PGA1) with guinea pig liver microsomes yields hydroxylation of PGA1 at the terminal carbon (C_{20}), referred to as w-position and at the penultimate carbon (C_{19}), referred to as w-1 position, suggested that the prostaglandin metabolites observed in vivo are probably derived from an initial monooxygenase catalyzed hydroxylation. By contrast to a facile metabolism of PGA1, these investigators observed that guinea pig liver microsomes did not catalyze the hydroxylation of PGE1. This finding was puzzling, since PGE1 and PGA1 are structurally similar, and hence would be expected to be metabolized by the same enzyme system. Moreover, PGA1 is probably not native to the animal and is merely an artifactually produced, dehydrated product of PGE1, formed during its isolation. These findings raised the important question of which are the enzymes involved in the hydroxylation of the endogenous prostaglandins, i.e., PGEs (PGE1 and PGE2). Additionally, it was also surprising that these investigators observed that rat liver microsomes did not hydroxylate prostaglandins, even though in vivo the rat hydroxylates efficiently PGEs[3,5].

Our interest in identifying endogenous substrates of hepatic and extrahepatic P450s, led us to attempt to resolve the above questions. From the onset, we were struck by the somewhat unorthodox incubation conditions utilized in the above studies. For instance, there was an extremely high level of nicotinamide in those incubations; i.e., nicotinamide was included during the preparation of the microsomes and was also added to the incubation medium. We considered the possibility that the conditions of the above incubations were suboptimal and that nicotinamide has inhibited the hydroxylation of certain prostaglandins much more extensively than that of other prostaglandins. The earlier findings of Schenkman et al.[11] that nicotinamide is highly inhibitory to microsomal monooxygenase catalyzed reactions, supported our suspicions. In fact, studies in our laboratory with guinea pig and rat liver microsomes, in the absence of nicotinamide, demonstrated that both species effectively hydroxylated PGAs (PGA1 and PGA2) and PGEs (PGE1 and PGE2)[9,10]. Additionally, we observed that nicotinamide was indeed a potent inhibitor of hepatic microsomal prostaglandin hydroxylation and hence was deleted from all subsequent incubations. We concluded that because PGA1 is a better substrate than PGE1, PGA1 hydroxylation could be detected even in the presence of nicotinamide. Also, we observed that nicotinamide inhibits much more the w-1 hydroxylation than the w-hydroxylation and therefore in the earlier experiments, using nicotinamide, w-hydroxylation appeared to be more effective than w-1 hydroxylation. However, in the absence of nicotinamide and with saturating concentrations of PGs (0.5-1.0 mM), the w-1 hydroxylation predominated in incubations with liver preparations from guinea pigs, rats and rabbits[9,10,12,13].

Evidence that cytochrome P450s in liver microsomes catalyze the hydroxylation of Prostaglandins

a) Requirement for oxygen, inhibition by carbon monoxide and requirement for NADPH were demonstrated.

b) Inhibition of hydroxylation by classical inhibitors of P450 (SKF525A and metyrapone) and by alternate substrates of P450 was observed.

c) Treatment of animals with inducers of cytochrome P450 affected the in vitro PG-hydroxylation by liver microsomes[10,12,13]. The inducers used were phenobarbital (PB), B-naphthoflavone (BNF) and methylcholanthrene (MC).

Evidence that several P450s in liver microsomes catalyze regioselectively the hydroxylation of prostaglandins

a) Different susceptibility of the two pathways (19- and 20-hydroxylations) to inhibitors of P450 was observed. Invariably, the 19-hydroxylation was found to be much more sensitive to inhibitors than the 20-hydroxylation[12].

b) Inducers of P450 affect the two pathways to a different extent. For instance the treatment of animals with PB altered the ratio of the hepatic microsomal catalyzed hydroxylated products of prostaglandins. There was an increase in the 19-hydroxylation of PGEs and PGAs but only little or no increase in the 20-hydroxylation[12,13]. MC treatment of rats enhanced 19-hydroxylation, but primarily increased the hydroxylation at the w-2 hydroxylation (18-hydroxylation), the latter being a novel site of hydroxylation of prostaglandins[13]. See Figure 3 for sites of hydroxylation. This suggested that a variety of P450s catalyze the 17-, 18-, 19- and 20-hydroxylations.

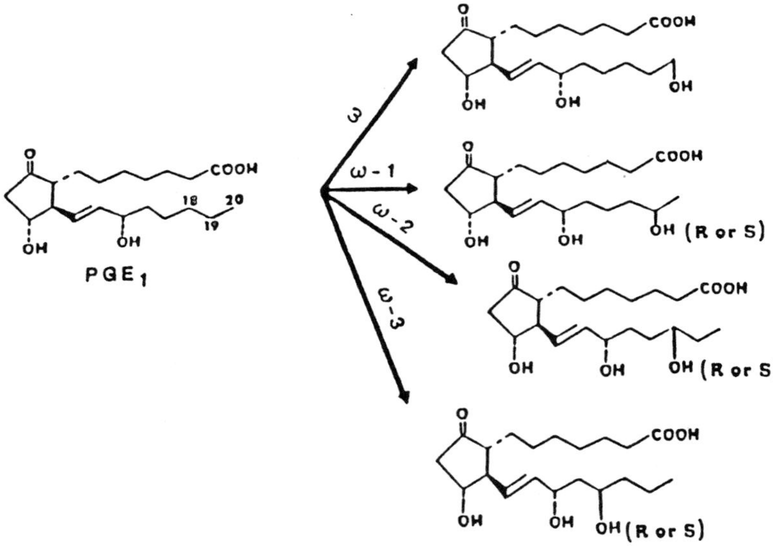

FIGURE 3. PATHWAYS OF HYDROXYLATION OF PROSTAGLANDINS BY CYTOCHROME P450 ISOZYMES

Prostaglandin hydroxylation by microsomes from a variety of tissues and species

Since PGs are ubiquitously found essentially in all tissues examined, it was of interest to determine whether extrahepatic tissues hydroxylate PGs and if so whether indeed cytochrome P450 catalyzes those reactions. The in vitro metabolism of PGAs, PGEs and PGFs by various tissues in a variety of animal species has been studied in several laboratories. TABLE 1 summarizes these findings.

296

TABLE 1. Species and Tissue Specificity in Hydroxylation of Prostaglandins

Species	Tissue	Prostaglandin Hydroxylation[a]		Authors
		19-OH	20-OH	
Guinea Pig	Liver	+	+	Israelsson et al.[7]; Kupfer[9,14,17,18]
	Kidney	+	+	Navarro et al.[15]
	Adrenal	+	+?	Kupfer[16]
Rabbit	Liver	+	+	Kupfer[9,10]; Powell[19,21]
	Kidney (cortex)	+	+	Powell[21]
	Kidney (medulla)	+	-	Powell[21]
	Lung	-	+	Powell[19-21]; Leithauer et al.[30]
	Lung (pregnant)	+	+[b]	Powell[19,20]; Williams et al.[22]; Yamamoto et al.[23]
	Uterus (pregnant)	-	+	Powell[21]; Yamamoto et al.[46]
	Placenta	-	+	Powell[21]; Yamamoto et al.[46]
	Small intestine	+	+	Kusunose[24]
	Colon	trace	+	Kusunose[24,25]
	Brain	trace	trace	Kusunose[24]
Pig (mini)	Kidney	+	+	Okita et al.[27]
Hamster	Liver	+	+	Powell[20]
	Kidney	-	+	Powell[21]
Rat	Liver	+	+	Kupfer & Navarro[14]; Holm et al.[13]; Powell[20]
	Kidney	-	+	Powell[21]
Human	Liver	+	+[c]	Kupfer (unpubl.)
	Placenta	-	+?	Kupfer (unpubl.)
	Seminal vesicles	+	-	Oliw[28,31]
Monkey	Liver	+	+	Oliw[32]
	Seminal vesicles			Oliw et al.[31]
Ram	Seminal vesicles	-	+	Oliw and Hamberg[28]; Oliw et al[29]

[a](+) indicates that a product with identical chromatographic behavior was observed; in many instances the unequivocal identification of the product has not been achieved and there is a possibility that some of the 20-hydroxylated compounds represent a mixture containing 18-hydroxy derivatives. (-) indicates that little or no product was detected. No judgement has been made concerning quantitative values of product.
[b]The level of w-hydroxylation in pregnant rabbit lung microsomes is dramatically higher than in non pregnant rabbit lung.
[c]w-1 hydroxylation is much higher than w-hydroxylation.

Effect of manipulation of liver P-450 monooxygenases on PGs metabolism

a) _Induction_: The first indication that more than one enzyme catalyzes the w and w-1 hydroxylations of prostaglandins by liver microsomes was obtained from our findings that treatments of animals with classical inducers of P-450 monooxygenases, e.g., phenobarbital (PB) and methylcholanthrene (MC) or beta-naphthoflavone (BNF) affect differently the rate of the 20- and 19- hydroxylation. These findings were observed in several species. Additionally, treatment of rats with MC elicited a dramatic induction of a novel route, i.e., that of w-2 hydroxylation.. Whereas the w-1 and w-2 hydroxylations were induced by these compounds, that of the w- was not. Also, in the rabbit ,but not in the rat or human, the affinity for the hepatic w- hydroxylase by PGE1 and PGE2 was much higher (lower Km) than for the w-1 hydroxylase[12] (Kupfer, unpublished).

These findings encouraged further investigations on the nature of the P-450 isozymes catalyzings these hydroxylations. These investigations are depicted below.

Identification of cytochrome P-450 isozymes catalyzing the hydroxylation of prostanoids

Our earlier observation that PB-treatment of rabbits increases the hepatic microsomal w-1 hydroxylation of PGEs and PGAs[12], suggested that P450 LM2, known to be induced by PB in rabbits, is involved in this hydroxylation. However, our initial reconstitution studies with LM2, containing NADPH-P450 reductase and dilauryl phosphatidylcholine and PGE1, did not yield prostaglandin hydroxylation. These findings were frustrating and puzzling to say the least. Subsequently, our studies, in collaboration with Drs Coon and Vatsis, demonstratrated the absolute requirement for cytochrome b_5 for the LM2-catalyzed prostaglandin hydroxylation[33]. These findings prompted a series of investigations with other rabbit liver P450s which are summarized below (see rabbit liver).

Rabbit liver. Cytochrome P450 isozymes isolated from rabbit liver (primarily by Coon and his colleagues (USA)) were named according to their rates of migration on SDS polyacrylamide gel electrophoresis; the P-450 with the fastest migration, lowest relative molecular weight (Mr), was named LM1; those with sequentially increasing Mr were named LM2, 3a, 3b, 3c, 4, 5, 6 and 7 (highest Mr) (TABLE 2). Reconstituted monooxygenase activity composed of the respective P450 isozyme and NADPH-P450 reductase and dilauryl phosphatidylcholine was examined with respect to prostaglandin hydroxylation. It was found that LM2 hydroxylates primarily at w-1 and minimally at the w-; LM3a, 3b and 3c have little or no catalytic activity towards prostaglandin hydroxylation; LM4 hydroxylates solely at the w-1; LM6 hydroxylates at both the w-1 and w-2 and form7 (possibly identical to LM7) hydroxylates solely at the w site (TABLE 2). LM2 and and form 7 _require_ the addition of cytochrome b_5 for significant prostaglandin hydroxylation[33,34,36]. By contrast, LM4 and LM6 do not require b_5.

LM7 ,isolated in Coon's laboratory, appears to be similar to form 7 which we isolated and demonstrated to catalyze the 20-hydroxylation of prostaglandins[34]. Most recently, Kusunose's laboratory isolated a

TABLE 2.   Cytochrome P-450 Isozymes from <u>Rabbit</u> <u>Liver</u>

LM	Mr (K)	Prostaglandin Hydroxylation[a]		
		18-OH	19-OH	20-OH
1[*],[#]	47			
2[*]	48, 51		+++	+
3a[*]	50		+	
3b[*]	52			
3c[*]	53			
4[*]	54-55		+++	
5[**]	57			
6[#]	57	+++	+++	
7[*],[+]	60			+++

Isolated in the laboratories of: [*] Coon; [#] Johnson; [**] Philpot; [+] Kupfer
   There is no report to my knowledge on whether LM1 or 5 catalyze the hydroxylation of prostaglandins. LM 3b and 3c were not active.
   [a]The number of plus (+) signs depicts a rough indication of the relative magnitude of hydroxylating activity; one + representing lowest activity and four +'s the highest activity; For the actual data, the reader should consult the original articles.

P-450 (Mr= 52K) from livers of <u>pregnant</u> rabbits which hydroxylates several prostaglandins at the w and hydroxylates certain fatty acids at the w and w-1 positions[35]. This P450, named $P450_{LPGw}$, is highly similar if not identical to the ω-hydroxylating P450 isolated from lung microsomes of pregnant rabbits[22,23].

   b) <u>Rat liver</u>. The rat liver provided a source of numerous cytochrome P-450 isozymes, which have been widely studied. By contrast to the simplistic nomenclature of P-450s in the rabbit, the naming of isozymes in the rat is confusing, primarily because several groups have generated their own nonenclature (see <u>TABLE 3</u>). The hydroxylation of PGE1 and PGE2 has been examined by our group with MC induced liver microsomes[13]. This study suggested that the induced P450s, P450c (IA1) and/or P450d (IA2) catalyze the increase in the w-2 and w-1 hydroxylation of PGEs observed in MC vs control microsomes. Further, based on our work with antibodies to P-450c and P-450d and on studies with the reconstituted P-450c it was concluded that P-450c hydroxylates PGEs at both the w-1 and the w-2. These findings demonstrated that P450c, known to be structurally related to rabbit LM6, resembles the catalytic activity of reconstituted rabbit LM6 towards prostaglandin hydroxylation[37,38]. However, some differences between P450c and LM6 were noted: whereas, P450c formed a higher ratio of w-2 to w-1 hydroxylated PGs, LM6 was more effective in catalysing the formation of

w-1 than of the w-2 products. The lack of inhibition of the rat
MC-microsome mediated PGE-hydroxylation by antibodies raised against
P450d, indicates that P450d in microsomes <u>does not</u> catalyze
prostaglandin hydroxylation[38]. Nevertheless, since P450d resembles
rabbit LM4 (the latter catalyses w-1 hydroxylation), it would be of
interest to determine whether reconstituted P450d also could
hydroxylate prostaglandins.

Certain constitutive rat liver P-450s were found to catalyze
effectively prostaglandin hydroxylation (collaboration with Dr.
Schenkman's group) (<u>TABLE 3</u>), whereas others were almost inactive[39].
Of particular interest is our finding that female predominant P450
fRLM4 catalyzes the hydroxylation of PGE1 and PGE2 at a **novel** site,
which appears to be w-3. Surprisingly, adult female rat liver
microsomes, known to be rich in fRLM4, did not elicit significant w-3
hydroxylation. Thus, it appears that fRLM4, when present in microsomal
membranes, does not express PG w-3 hydroxylation. Though the mechanism
of this anomaly is currently not understood, it is tempting to
speculate that low affinity of fRLM4 for the limiting levels of
reductase in the microsomes prevents the expression of its catalytic
activity.

c) <u>Rabbit lung</u>. Powell reported that pregnancy or progesterone
treatment of rabbits increases dramatically the w-hydroxylation of PGs
by lung microsomes in the rabbit, but not in the rat or
hamster[19,20]. It is tempting to speculate that this increased lung
activity constitutes a protective mechanism against excessive amounts
of prostaglandins during pregancy. Since lung w-1 hydroxylation was
not affected by this treatment, it appeared that a P-450 with high
regioselectivity for w-hydroxylation was induced by pregnancy.
Additionally, the chronic administration of progesterone to the
nonpregnant rabbit mimicked the effects of pregnancy by stimulating the
lung w-hydroxylase activity. Hence it appeared that the effect by
pregnancy was mediated by progesterone. However, whereas pregnancy
also enhances the w-hydroxylase activity in liver microsomes, treatment
with progesterone had no such effect on the liver indicating that
additional factors, besides progesterone, present in the pregnant
rabbit were required for the induction of the w-hydroxylase. Studies
in the laboratories of Kusunose (Japan) and Masters (USA) demonstrated
that progesterone treatment or pregnancy induced in the lung the
formation of a novel P450, which catalyzes the w-hydroxylation of
PGs[22,23]. Though it appeared that both groups isolated a similar, if
not identical enzyme, some differences with respect to substrate
specificities of the two preparations were noted. In a latter study,
it was observed that pregnancy elevated rabbit lung mRNA which is
translated into protein with the same Mr as the isolated P450 PGs[40].
More recently, Matsubara et al[41] deduced the sequence of this P-450
and demonstrated that it is a unique P-450 with some homology to the
rat liver P-450 induced by clofibrate and which w-hydroxylates lauric
acid.

<u>Monkey and Human studies</u>

In 1966 Hamberg and Samuelsson[42] demonstrated that human seminal plasma
contains large concentration of 19-hydroxy-PGA derivatives. Subsequently it
was shown that PGAs are probably artifacts of isolation of PGEs and it was
concluded that human and monkey semen contain 19-hydroxy-PGE1 and
19-hydroxy-PGE2 and lesser amounts of 19-OH-PGFs[43]. The function of
these 19-OH-PGEs is still not known. However, certain correlation studies
in humans suggest that these compounds might be important to reproductive

TABLE 3. Cytochrome P-450 Isozymes from <u>Rat</u> Liver (Nomenclature and Prostaglandin Hydroxylation).

Nomenclature*					PG Hydroxylase[#]			
					17	18	19	20
Levin	Schenkman	Waxman	Guengerich	Gene Design.				
a	-	3	UT-F	IIA1				
b	PBRLM5	PB-4	PB-B	IIB1				
c	-	BNF-B	BNF-B	IA1		++++	++++	
d	-	ISF-G	ISF-G	IA2				
e	PBRLM6	PB-5	PB-D	IIB2				
f	-	-	-	IIC7				
g	RLM3	-	-	-			+	
h	RLM5	2c	UT-A	IIC11			++++	
i	fRLM4	2d	UT-I	-	++		++	**
j	RLM6	-	-	IIE1				+
k	RLM5a	PB-1	PB-C	IIC6		+	++	+
p	-	PB-2a	PCN-E	IIIA1				
-	-	2a	-	IIIA2				
-	-	6	-	-				
-	RLM2	-	-	-				+

* P-450m and P-450n (49K and 50K) were isolated by Andrew Parkinson's laboratory and are immunochemically related to P-450a[26]. These have not been hitherto characterized with respect to prostaglandin hydroxylation.
# Blank spaces indicate that prostaglandin hydroxylation was not examined.
**Trace amounts.
Results are from Kupfer et al.[39]. The number of pluses, as in Table 2, indicates a rough approximation of the relative rate of the reaction; for actual data the reader should refer to the original article.

success. Though the site of formation of 19-OH-PGEs in vivo has not been established, it was assumed that they are formed from the corresponding PGEs. Our earlier studies suggested that monkey seminal vesicles catalyze the transformation of PGs into chromatographically polar, possibly hydroxylated, metabolites[10]. However, the paucity of tissues at our disposal and the inconsistency of our findings (tissues from some monkeys were not active) prevented the identification of the products. More recently Oliw and coworkers reported that seminal vesicles from monkey and man catalyze the w-1 hydroxylation of PGs, yielding the corresponding 19-OH-PGs[31]. Based on indirect evidence it appears that microsomal P-450 monooxygenase in seminal vesicles catalyzes this reaction. The earlier observations by Kelly and coworkers[44] that species other than primates do not contain 19-OH-PGs, casted doubt on the universality across species, of the involvement of these compounds in reproduction. However, the recent findings by Oliw and Hamberg[28,29] that ram semen contains 20-OH-PGEs and that ram seminal vesicle microsomes catalyze the 20-hydroxylation of PGEs indicate that, in the earlier investigations in the ram, hydroxylated-metabolites other than the 19-OH-PGs were most probably present but merely were not detected. Therefore such studies should be repeated in other species with the intention of exploring for the possible presence in semen of other hydroxylated PGs. Of additional interest is the recent finding by Oliw[45] that human semen contains 18-hydroxy-PGEs. In that regard it is noteworthy that methylcholanthrene treatment of rats dramatically induces the hepatic microsomal 18-hydroxylation of PGEs[13] and that P450c (IA1) is responsible for catalyzing that reaction[38]. It would be of extreme interest to determine whether the 18- and 19 hydroxylated PGEs in human semen are formed by the same P450 in seminal vesicles and whether that P450 resembles liver P450c, structurally and catalytically.

CONCLUSION

The studies on the metabolism of prostaglandins by the various P-450 isozymes indicate that there is considerable regioselectivity of hydroxylation and that the prostaglandins could be metabolized into a variety of hydroxylated prostaglandin derivatives. However, based on in vivo observations, it can be surmized that only certain routes of prostaglandin metabolism, i.e., w- and w-1 hydroxylation are significantly expressed. The reasons for the narrow selection of product formation in vivo is not known. It is possible that the other metabolites are also formed in vivo but have not been hitherto detected or identified. In fact, based on studies by Kelly[44], it has been assumed that only human and monkey seminal fluid contains hydroxylated prostaglandins. However, the recent observation by Oliw et al.[29] that the seminal fluid of the ram contains 20 hydroxy-PGs demonstrated that the above assumption was incorrect and suggests that hydroxylated prostaglandins might be present in seminal fluid of other animals as well. Additionally, it is conceivable that the capability of certain product formation lies in the affinity of a given P-450 isozyme (catalyzing that reaction) for the given prostaglandin and for the reductase and that these factors dictate whether that reaction could be significantly expressed.

The question of whether hydroxylation of prostaglandins represents an activation or an inactivation step, has not been resolved. We observed (collaboration with Dr. John Hildebrandt, Worcester Foundation, Shrewsbury MA, USA) that 19- and 20-OH-PGs exhibit biological activity towards stimulation of human platelet adenylate cyclase, albeit they are less active than the parent prostaglandins[36]. The function of 19-hydroxy-PGs, present in high concentration in human and monkey semen, and of the 20-hydroxy-PGs in ram semen has not been established. Some correlation appears to exist between the levels of hydroxylated PGs and fertility. However, whether high levels of the hydroxylated PGs are required for fertility and whether the presence of extremely low levels of these compounds are the cause of infertility is not known. Further studies are necessary to establish the physiological role of hydroxylated PGs.

ACKNOWLEDGEMENT

The major portion of the work performed in my laboratory presented in this chapter was supported by NIH grants ES00834 from NIEHS and GM 22688 from NIGMS.

REFERENCES

1. F. A. Fitzpatrick, and R. C. Murphy, Pharmacol. Rev. 40:229 (1989).
2. M. Hecker, and V. Ullrich, J. Biol. Chem. 264:141 (1989).
3. K. Svanborg, and M. Bygdeman, Eur. J. Biochem., 28:127 (1972).
4. F. F. Sun, Biochim. Biophys. Acta 348:249 (1974).
5. F. F. Sun, and J. E. Stafford, Biochim. Biophys. Acta 369:95 (1974).
6. C. K. Ellis, M. D. Smigel, J. A. Oates, O. Oelz, and B. J. Sweetman, J. Biol. Chem. 254:4152 (1979).
7. U. Israelsson, M. Hamberg, and B. Samuelsson, Europ. J. Biochem., 11:390 (1969).
8. B. Samuelsson, E. Granstrom, K. Green, and M. Hamberg, Ann. N.Y. Acad. Sci. 180:138 (1971).
9. D. Kupfer, Pharmacol. Ther. 11:469 (1981).
10. D. Kupfer, in: "Hepatic Cytochrome P-450 Monooxygenase System," J. B. Schenkman, and D. Kupfer, eds., pp. 157-187, Pergamon Press, New York (1982).
11. J. B. Schenkman, J. A. Ball, and R. W. Estabrook, Biochem. Pharmacol. 16:1071 (1967).
12. A. D. Theoharides, and D. Kupfer, J. Biol. Chem. 256:2168 (1981).
13. K. A. Holm, R. J. Engell, and D. Kupfer, Arch. Biochem. Biophys. 237:477 (1985).
14. D. Kupfer, and J. Navarro, Life Sci. 18:507 (1976).
15. J. Navarro, D. E. Piccolo, and D. Kupfer, Arch. Biochem. Biophys. 191:125 (1978).
16. D. Kupfer, J. Navarro, G. K. Miranda, and A. D. Theoharides, Arch. Biochem. Biophys. 205:297 (1980).
17. D. Kupfer, J. Navarro, and D. E. Piccolo, J. Biol. Chem. 253:2804 (1978).
18. D. Kupfer, G. K. Miranda, J. Navarro, D. E. Piccolo, and A. D. Theoharides, J. Biol. Chem. 254:10405 (1979).
19. W. S. Powell, and S. Solomon, J. Biol. Chem. 253:4609 (1978).
20. W. S. Powell, J. Biol. Chem. 253:6711 (1978).
21. W. S. Powell, Prostaglandins 19:701 (1980).
22. D. E. Williams, S. E. Hale, R. T. Okita, and B. S. S. Masters, J. Biol. Chem. 259:14600 (1984).
23. S. Yamamoto, E. Kusunose, K. Ogita, M. Kaku, K. Ichihara, and M. Kusunose, J. Biochem. 96:593 (1984).
24. E. Kusonose, M. Kaku, K. Ichihara, S. Yamamoto, I. Yano, and M. Kusunose, J. Biochem. 95:1733 (1984).
25. M. Kaku, E. Kusunose, S. Yamamoto, K. Ichihara, and M. Kusonose, J. Biochem. 97:663 (1985).
26. M. P. Arlotto, D. J. Greenway, and A. Parkinson, Arch. Biochem. Biophys. 270:441 (1989).
27. R. T. Okita, L. K. Parkhill, Y. Yasukochi, B. S. S. Masters, A. D. Theoharides, and D. Kupfer, J. Biol. Chem. 256:5961 (1981).

28. E. H. Oliw, and M. Hamberg, <u>Biochim. Biophys. Acta</u> 879:113 (1986).
29. E. H. Oliw, P. Fahlstadius, and M. Hamberg, <u>J. Biol. Chem.</u> 261:9216 (1986).
30. M. T. Leithauer, D. L. Roerig, S. M. Winquist, A. Gee, R. T. Okita, and B. S. S. Masters, <u>Prostaglandins</u> 36:819 (1988).
31. E. H. Oliw, A. C. Kinn, and U. Kvist, <u>J. Biol. Chem.</u> 263:7222 (1988).
32. E. H. Oliw, <u>J. Biol. Chem.</u> 264:17845 (1989).
33. K. P. Vatsis, A. D. Theoharides, D. Kupfer, and M. J. Coon, <u>J. Biol. Chem.</u> 257:11221 (1982).
34. K. A. Holm, and D. Kupfer, <u>J. Biol. Chem.</u> 260:2027 (1985).
35. Y. Kikuta, E. Kusunose, S. Matsubara, Y. Funae, S. Ymaoka, I. Kubota, and M. Kusunose, <u>J. Biochem.</u> 106:468 (1989).
36. D. Kupfer and K. A. Holm, <u>Drug Metab. Rev.</u> 20(2-4):753 (1989).
37. K. A. Holm, D. R. Koop, M. J. Coon, A. D. Theoharides, and D. Kupfer, <u>Arch. Biochem. Biophys.</u> 243:134 (1985).
38. K. A. Holm, S. S. Park, H. V. Gelboin, and D. Kupfer, <u>Arch. Biochem. Biophys.</u> 269:664 (1989).
39. D. Kupfer, I. Jansson, L. V. Favreau, A. D. Theoharides, and J. B. Schenkman, <u>Arch. Biochem. Biophys.</u> 261:186 (1988).
40. A. S. Muerhoff, D. E. Williams, M. T. Leithauser, V. E. Jackson, M. R. Waterman, and B. S. S. Masters, <u>Proc. Natl. Acad. Sci. USA</u> 84:7911 (1987).
41. S. Matsubara, S. Yamamoto, K. Sogawa, N. Yokotani, Y. Fujii-Kuriyama, M. Haniu, J. E. Shively, O. Gotoh, E. Kusunose, and M. Kusunose, <u>J. Biol. Chem.</u> 262:13366 (1987).
42. M. Hamberg, and B. Samuelson, <u>J. Biol. Chem.</u> 241:257 (1966).
43. P. L. Taylor, and R. W. Kelly, <u>Nature</u> 250:665 (1974).
44. R. W. Kelly, P. L. Taylor, J. P. Hearn, R. V. Short, D. E. Martin, and J. H. Marston, <u>Nature</u> 260:544 (1976).
45. E. H. Oliw, <u>Prostaglandins</u> 35:523 (1988).
46. S. Yamamoto, E. Kusunose, S. Matsubara, K. Ichihara, and M. Kusunose, <u>J. Biochem.</u> 100:175 (1986).

ROLE OF CYTOCHROME P - 450 IN THE ANABOLISM AND CATABOLISM OF ENDOBIOTICS

Hugo Vanden Bossche, Henri Moereels and Paul A.J. Janssen

Janssen Research Foundation
B2340 Beerse, Belgium

INTRODUCTION

Estabrook et al.(1) demonstrated for the first time the participation of a cytochrome P-450 (P450) in the metabolism of endobiotics by proving its role in the steroid C21-hydroxylation reaction, catalyzed by adrenal cortex microsomes. They found that the 21-hydroxylation of 17-hydroxyprogesterone required NADPH and oxygen and was inhibited by carbon monoxide. Furthermore, spectrophotometric measurements showed the presence of an absorption band at 450nm when CO was added to adrenal cortex microsomes reduced with NADPH (1). This study paved the way for innumerable biochemical and endocrinological studies. Now, some 26 years later, a wealth of information is available on the involvement of P450 isozymes in the synthesis of endobiotics in yeasts, fungi, protozoa, plants, invertebrates and vertebrates. Examples of some cellular biosynthetic reactions in which P450 enzymes are essential are listed in Table 1. The metabolism of a wide variety of endogenous compounds is also dependent on P450. Examples are shown in Table 2.
    It is the aim of this paper to review the properties of some of the P450-depentent reactions listed in Tables 1 and 2. The main focus will be on biosynthetic reactions, catalyzed by P450 systems, (Table 1) that are potential targets for a number of compounds of use in medical treatment or plant protection. Since most of the P450 isozymes involved in the catabolism of endobiotics (Table 2) will be discussed in the other chapters of this book, only P450s involved in bile acid synthesis (the most important pathway in the metabolism and excretion of cholesterol), leukotriene and vitamin A metabolism will be described in this chapter.

STEROL SYNTHESIS

    The sterol synthesis pathway can be divided into three segments. The initial step is common to sterol synthesis in fungal, plant and mammalian cells. The first part of the pathway leads from acetyl-CoA, via 3-hydroxy-3-methylglutaryl-CoA (HMG-CoA) and mevalonic acid to squalene (Fig.1). This segment contains the HMG-CoA reductase, the rate-limiting enzyme in the cholesterol biosynthetic pathway. The 2nd part consists of the enzymes involved in the conversion of squalene into lanosterol (fungi and mammalian cells; Fig.1) or cycloartenol (plants). In the 3rd segment of the phytosterol or ergosterol (the main sterol of most fungi) synthesis pathway, cyclo-artenol or lanosterol are first alkylated at C-24 to form 24α-or 24β-methylene or 24α- or 24β-ethylenesterols in plants and 24-methylenedihydrolanosterol in a great number of fungi. These alkylated sterols (plant and fungal microsomes) and lanosterol (*Saccharomyces cerevisiae* and mammalian liver microsomes) form the substrates for the P450-dependent 14α-demethylase ($P450_{14DM}$).

Molecular Aspects of Monooxygenases and Bioactivation of Toxic Compounds
Edited by E. Arınç *et al., Plenum Press, New York,*

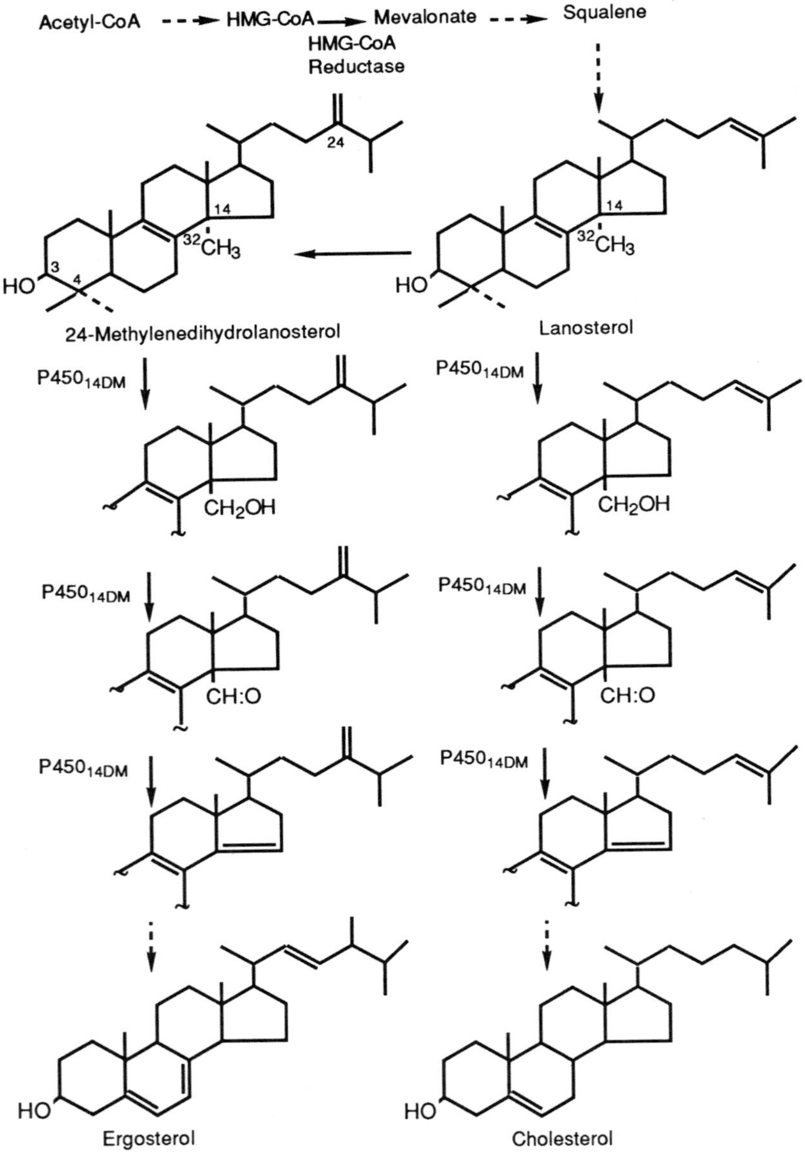

Fig. 1- Synthesis of ergosterol and cholesterol. Dashed arrows= multi-enzyme systems.

Table 1.   Some Cellular Biosynthetic Reactions Catalyzed by P450-dependent Enzymes*

REACTION	P450
STEROL SYNTHESIS	
Lanosterol $\rightarrow$ 4,4-Dimethyl-$\Delta^{8,14,24}$-cholestatrienol	P450$_{14DM}$ [LI]
- - -> Cholesterol	
24-Methylenedihydrolanosterol $\rightarrow$	
4,4-Dimethyl-$\Delta^{8,14,24(28)}$-ergostatrienol	
- - -> Ergosterol	
Obtusifoliol $\rightarrow$ 24-methylenelophenol - - ->	
24-Alkylsterols (phytosterols)	
STEROID HORMONE SYNTHESIS	
Cholesterol $\rightarrow$ Pregnenolone	P450$_{scc}$ [XIA1]
Pregnenolone $\rightarrow$ 17α-Hydroxypregnenolone	P450$_{17\alpha}$ -C17,20
Progesterone $\rightarrow$ 17α-Hydroxyprogesterone	[XVIIA1]
17α-Hydroxy-20-dihydropregnenolone $\rightarrow$ DHEA	
17α-Hydroxy-20-dihydroprogesterone $\rightarrow$ Androstenedione	
Androstenedione $\rightarrow$ Estrone	P450$_{AROM}$
Testosterone $\rightarrow$ Estradiol	[XIXA1]
17α-Hydroxyprogesterone $\rightarrow$ 11-Deoxycortisol	P450$_{C21}$
Progesterone $\rightarrow$ 11-Deoxycorticosterone	[XXIA1]
11-Deoxycortisol $\rightarrow$ Cortisol	P450$_{11\beta}$
11-Deoxycorticosterone $\rightarrow$ Corticosterone	[XIA1]
Corticosterone $\rightarrow$ 18-Hydroxycorticosterone	
18-Hydroxycorticosterone $\rightarrow$ Aldosterone	
11-Deoxycorticosterone $\rightarrow$	
19-Hydroxy,11-deoxycorticosterone	
Androstenedione $\rightarrow$ 19-Hydroxyandrostenedione	
19-Hydroxyandrostenedione $\rightarrow$ 19-Oxoandrostenedione	
19-Oxoandrostenedione $\rightarrow$ 19-Nor-androstenedione	
19-Oxoandrostenedione $\rightarrow$ Estrone	
α-Ecdysone $\rightarrow$ 20-Hydroxyecdisone (β-Ecdysone)	P450$_{C20}$
PROSTAGLANDIN ENDOPEROXIDE REARRANGEMENT	
PGH$_2$ $\rightarrow$ Prostacyclin (PGI$_2$)	P450$_{PGI2}$
PGH$_2$ $\rightarrow$ Thromboxane A$_2$ (TxA$_2$)	P450$_{TxA2}$
ACTIVATION OF VITAMINS	
Vitamin D$_3$ $\rightarrow$ 25-Hydroxyvitamin D$_3$	P450$_{vit.D3C25}$
25-Hydroxyvitamin D$_3$ $\rightarrow$ 1,25-Dihydroxyvitamin D$_3$	P450$_{25OH-}$
	vit.D3C1
GIBBERELLIN SYNTHESIS	
ent-Kaurene $\rightarrow$ ent-Kaurenol $\rightarrow$ ent-Kaurenal $\rightarrow$	P450$_{kaureneC19}$
ent-Kaurenoic acid	

*14DM: 14α-demethylase; scc: side-chain cleavage; 17α: 17α -hydroxylase; C$_{17,20}$: 17,20-lyase; DHEA: dehydroepiandrosterone; AROM: aromatase; C21: 21-hydroxylase; 11β: 11β-hydroxylase; C20: 20-hydroxylase, PGH$_2$: prostaglandin endoperoxide. Figures in square parenthesis are the protein names according to Nebert et al.(3)

Table 2. Some Metabolic Reactions Catalyzed by P450-dependent Enzymes*

REACTION	P450
CHOLESTEROL METABOLISM	
Cholesterol → 7α-Hydroxycholesterol	$P450_{LM}$ 4 II (7, 8)
7α-hydroxycholesterol →	$P450_{LM}$ 4 I (7)
7α,12α-Dihydroxy-4-cholesten-3-one	
26-Hydroxylation	$P450_{C26}$ (9, 10)
PROGESTERONE METABOLISM	
2-Hydroxylation	$P450_{PCN}$ [IIIA](11)
6ß-Hydroxylation	$P450_{3c}$ [IIIA6] (12, 13)
	$P450_{NF}$ (14), $P450_{PCN}$ (11)
15α-Hydroxylation	$P450_{15\alpha}$ (15, 16)
15ß-Hydroxylation	$P450_{PCN}$ (11)
16α-Hydroxylation	$P450_{3c}$ [IIIA6](17)
	$P450_{16\alpha}$ (15)
16ß-Hydroxylation	$P450_{PCN}$ (11)
6ß-HYDROXYPROGESTERONE METABOLISM	
21-Hydroxylation	$P450_{PCN}$ (11)
ANDROSTENEDIONE METABOLISM	
6ß-Hydroxylation	$P450_{NF}$ ($P450_{PB2a}$) (14)
7α-Hydroxylation	$P450_{a}$ ($P450_{PB-3}$)[IIA] (18)
15α-Hydroxylation	$P450_{15\alpha}$ (16)
16ß-Hydroxylation	$P450_{b}$ ($P450_{PB4}$)[IIB1](18)
TESTOSTERONE METABOLISM	
2α-Hydroxylation	$P450_{h}$ ($P450_{PB2c}$)[IIC11](18)
6α-Hydroxylation	$P450_{a}$ ($P450_{PB3}$)[IIA](18)
6ß-Hydroxylation	$P450_{NF}$(14)$P450_{g}$ [IIC] (19)
	$P450_{PCN}$ [IIIA] (12)
7α-Hydroxylation	$P450_{a}$ [IIA](18)
15α-Hydroxylation	$P450_{g}$ (19)
	$P450_{15\alpha}$ (15, 16)
16α-Hydroxylation	$P450_{h}$ (18)
	$P450_{f}$ [IIC7](19)
	$P450_{16\alpha}$ (15)
16ß-Hydroxylation	$P450_{b}$ (18)
17α-Hydroxylation	$P450_{h}$ (18)
5α-ANDROSTANE-3ß,17ß-DIOL METABOLISM	
6α-, 6ß-, 7α-, 7ß-Hydroxylations	P450 in prostate and CNS (20, 21, 22)
ESTRADIOL	
2α-Hydroxylation	$P450_{3c}$ [IIIA6] (6)
	$P450_{h}$ (19)
16α-Hydroxylation	$P450_{16\alpha}$ (15)
LEUKOTRIENE METABOLISM	
LTB_4 → 20-Hydroxy-LTB_4	$P450_{LTB4}$ (23)
VITAMIN A METABOLISM	
Vitamin A- - -> Retinoic acid →	$P450_{4-RA}$ (24)
4-Hydroxyretinoic acid	

*Figures in square parenthesis are the protein names according to Nebert et al.(3). LTB_4: 5(S),12(R)-dihydroxy-6-*cis*-8,10-*trans*-14-*cis*-eicosatetraenoic acid

Cholesterol biosynthesis was apparently carried out even before the divergence of prokaryotes from eukaryotes(2). Thus, the removal of the 14-methyl-group from lanosterol is probably the oldest connection between P450 and cholesterol(2).

P450$_{14DM}$ (belongs to P450 family LI, see ref.3), purified to homogenity from rat liver(4) and *S. cerevisiae*(5) microsomes catalyzes three oxidative steps (Fig.1):

1. the hydroxylation of the C-32-methyl (14α-methyl) group of lanosterol
2. the oxidation of the C-32 alcohol to the C-32 aldehyde
3. the oxidative elimination of the aldehyde as formic acid, resulting in the formation of 4,4-dimethyl-$\Delta^{8,14,24(28)}$-ergostatrienol in fungi, and 4,4-dimethyl-$\Delta^{8,14,24}$-cholestatrienol in *S. cerevisiae* and liver.

The entire dealkylation sequence is catalyzed by a single P450 species. It is of interest to note that 3ß-hydroxylanost-8-en-32-aldehyde, an intermediate in the C-32 demethylation sequence of lanosterol, has been shown to suppress HMG-CoA reductase activity. Studies by Trzaskos et al.(6) suggest that this oxygenated 14α-methyl demethylase intermediate is the endogenously generated modulator of HMG-CoA reductase activity.

The P450$_{14DM}$, purified from rat liver, appeared to be specific in its oxidation toward 14α-methyl sterols with very poor activity toward xenobiotics(6). A modest induction (1.6-fold) of the constitutive P450 isozyme of 14α-methyl sterol oxidase has been found in male Sprague-Dawley rats after 5 days of treatment with 200 mg/kg imidazole(25). Pompon(26), in a study on the functional expression in *S. cerevisiae* of rabbit liver P450$_{LM4}$ and P450$_{LM6}$, used imidazole to stabilize P450$_{LM4}$. Imidazole induced an endogeneous yeast P450 that can be selectively complexed by ketoconazole, a potent inhibitor of the P450$_{14DM}$-dependent ergosterol synthesis in yeast and fungi(5, 27). This investigator did not study the 14α-demethylation catalyzed by this P450. However, P450$_{14DM}$ is the predominant yeast microsomal P450(5), and since the microsomes from the imidazole-treated cells do not show an absorption peak at about 450nm in the presence of 0.5 µM ketoconazole(26), it is tempting to speculate that imidazole also induces or stabilizes P450$_{14DM}$ in *S. cerevisiae*.

The amino acid sequences of the P450$_{14DM}$ from *Candida albicans*(28), *C. tropicalis*(29) and *S. cerevisiae*(30-32) are known. Using a VGAP (alignment with a **v**ariable **gap** penalty) alignment programme (Moereels, De Bie, Tollenaere, J.Comp.-aided Mol.Des., submitted) the *C.albicans* and *C. tropicalis* P450$_{14DM}$ share with the *S. cerevisiae* P450$_{14DM}$ 64.2 and 65.2% identical amino acids(33). P450$_{14DM}$ from *C.albicans* and *C. tropicalis* share 83% identical amino acids(33). The amino acid sequence of the P450$_{14DM}$ from liver is still not published. Since liver and *S. cerevisiae* P450$_{14DM}$ use lanosterol as substrate it would be of interest to see whether their sequences share more identical amino acids than when compared with P450$_{14DM}$ using 24-methylenedihydrolanosterol as substrate.

Yoshida et al.(34) purified a P450 (P450$_{SG1}$) from a nystatin resistant mutant, *S. cerevisiae* SG$_1$ (ATCC 46786, [*mata, his*-1, *erg*11]), which is defective in lanosterol 14α-demethylation. A single nucleotide change resulting in substitution of the glycine-310 residue of P450$_{14DM}$ by an aspartic acid residue was found to have occured in P450$_{SG1}$(32). In this protein the 6th ligand to the heme iron is a histidine residue(32) instead of the hydroxyl group of water or of a serine, tyrosine or a threonine residue, which are the most likely candidates for the 6th ligand in normal P450s.

An analysis of the sterols formed by SG$_1$ grown in PYG-medium, containing 10g polypeptone, 10g yeast extract and 40g glucose per litre, showed the presence of lanosterol (71%), 24-methylene-lanosterol (4%), obtusifoliol (8%), 14-methylfecosterol (2%) and an unidentified 14-methylsterol (12.6%), all 14-methylated sterols. Only a small amount (1.4%) of a 14-desmethylsterol, $\Delta^{7,22}$ergostadienol, was detected (unpublished results).

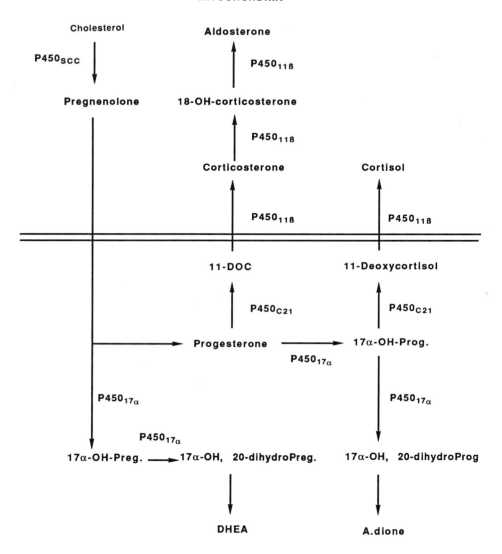

MITOCHONDRIA

Fig. 2- P450-dependent reactions in adrenal mitochondria and microsomes. $P450_{scc}$= side-chain cleavage P450; $P450_{11\beta}$= 11ß-hydroxylase; $P450_{C21}$= 21-hydroxylase; 11-DOC= 11-deoxycorticosterone; Preg.=pregnenolone; Prog.=progesterone; DHEA= dehydroepiandrosterone; A.dione= androstenedione.

These results indicate that P450$_{SG1}$ indeed lost its catalytic activity. The formation of $\Delta^{7,22}$ergostadienol supports the results of Watson et al.(35). These investigators proved that *S. cerevisiae* SG$_1$ is a Δ^{5-6} desaturase mutant. A Chinese hamster ovary cell mutant, AR45, lacking detectable lanosterol 14α-methyl demethylase activity has also been selected(36). This cell line should prove useful in studying regulation of the 14α-demethylase and also the requirements of 14-desmethylsterols for cell proliferation.

As will be discussed in the chapter on " Effects of inhibitors on the P450-dependent metabolism of endogenous compounds in fungi, protozoa, plants and vertabrates" the P450$_{14DM}$ of fungal cells is the target for an important number of antifungal agents.

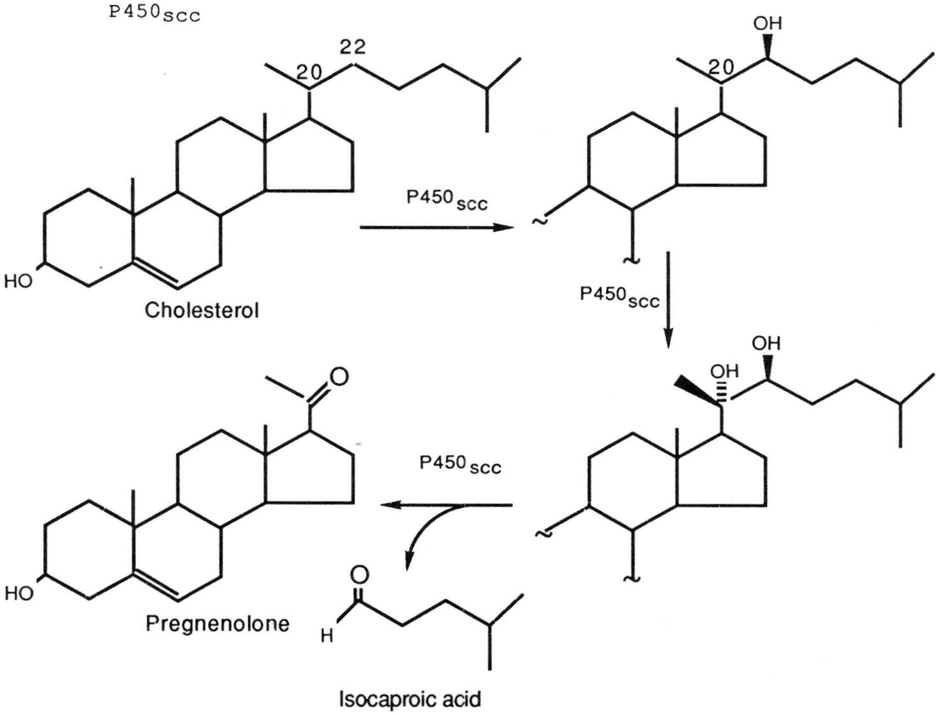

Fig. 3- Cholesterol side-chain cleavage

STEROID HORMONE SYNTHESIS

Mitochondrial P450 Isozymes

Conversion of cholesterol to pregnenolone in mitochondria (Fig.2) is the first rate-limiting step in the synthesis of all steroid hormones. Cholesterol side-chain cleavage (scc) involves three consecutive mono-oxygenation steps and is catalyzed by P450$_{scc}$ (protein name: XIA1)(37-38). The reaction (Fig.3) proceeds via an hydroxylation at the 22-position of cholesterol to yield (22R)-hydroxycholesterol, which is subsequently converted to a dihydroxy intermediate (20R,22R) and finally to pregnenolone and an isocaproic group from the cholesterol side-chain(39).

Protein purification data suggest the existence of a single immunologically identifiable species of $P450_{scc}$(37). Human $P450_{scc}$ is encoded
by a single gene located at chromosome 15(40). Southern blotting of rat
genomic DNA also suggests the presence of a single $P450_{scc}$ gene(41). Rat
cholesterol side-chain cleavage mRNA is found present in ovary, adrenals
and testis, but not in brain, kidney, liver, lung or heart(41). Chung et
al.(40) found that the human $P450_{scc}$ gene is expressed in the placenta in
early and midgestation. The presence of $P450_{scc}$ in the ovaries of rats
during the oestrous cycle, pregnancy, parturition and abortion has been
studied immunohistochemically(42). The luteal cells of cyclic and
pregnant rats were immunostained. $P450_{scc}$ was also expressed in the
interstitial cells of prepubertal and cyclic adult rats, and in the
thecal cells lining the preovulatory follicles.

P450$_{scc}$ is an integral part of the inner mitochondrial membrane.
Reducing equivalents are provided to $P450_{scc}$ via an electron transfer
system localized in the mitochondrial matrix. Electrons from NADPH are
transferred to adrenodoxin reductase which than transfers electrons to
adrenodoxin. This iron-sulphur protein transfers electrons to $P450_{scc}$.
It should be noted that adrenodoxin binding is essential to achieve the
regioselective hydroxylation at the 22R position of the cholesterol side
chain(39). $P450_{scc}$ is a glycoprotein and the sugar moiety may
participate in the binding of adrenodoxin or the subsequent electron
transfer(43). In a recent study Tsubaki et al. (44)showed that either
lysine residue 377 or 381 is essential for adrenodoxin binding. The
region (from residue 369 to residue 381) containing this putative adrenodoxin-binding site is highly homologous to the corresponding region of
bovine mitochondrial cytochrome P-450$_{11ß}$(44). This region overlaps
partly the putative steroid binding place (see under P450$_{11ß}$). cDNA
clones of bovine adrenal cortex $P450_{scc}$ mRNA have been analyzed and the
complete amino acid sequence, including the extrapeptide consisting of 39
amino acids residues at the NH2-terminal has been deduced(45). The
encoded human preP450$_{scc}$ contains 521 amino acids(40). As expected from
the presence of a single $P450_{scc}$ gene in the human genome the testis and
adrenal amino acid sequences are identical, and share 72.7% identical
amino acids with the bovine $P450_{scc}$ sequence(40). Although the
crystallization of bovine $P450_{scc}$ has been reported(46), the three
dimensional structure of this important P450 has still to be solved.

P450$_{11ß}$

Another P450 also known to reside in the inner mitochondrial
membrane is the 11ß-hydroxylase (P450$_{11ß}$). This single enzyme (occurring
only in the adrenals) mediates in the adrenal cortex not only the 11ß-
hydroxylation of 11-deoxycortisol but also the synthesis of aldosterone
(the key mineralocorticoid required for salt retention) from 11-deoxy-
corticosterone (Fig.2)(47). This means that P450$_{11ß}$ has 11ß-hydroxylase,
18-hydroxylase and aldehyde synthase activities. As indicated in Table 1
this P450 also has (in gerbil adrenals) the capacity to carry out the 19-
hydroxylation of 11-deoxycorticosterone (11-DOC) and androstenedione(48).
A P450$_{11ß}$ -dependent conversion of 19-hydroxyandrostenedione to 19-oxo-
androstenedione and further to 19-norandrostenedione (19-oxidase
reaction) has also been described(49). Thus P450$_{11ß}$ seems to possess 10-
demethylase (C10-19-lyase) activity. Suhara and colleagues also
described a P450$_{11ß}$-dependent conversion of androgen to estrogen,
indicating that this multifunctional enzyme catalyzes an aromatase
reaction in the adrenal cortex (49). Aldosterone is produced only in the
zona glomerulosa, while P450$_{11ß}$ exists and produces corticosterone and/or
cortisol in all three zones of the adrenal cortex.

A P450, capable of producing aldosterone from 11-DOC, was purified
from the mitochondria of the zona glomerulosa of adrenal cortex from
sodium-depleted and potassium-repleted rats(50). This P450 was
distinguishable from the P450$_{11ß}$ purified from the zonae fasciculata-

reticularis mitochondria of the same rats. The P450 from the zona
glomerulosa catalyzed the hydroxy-lation of 11-DOC with the formation of
18-hydroxydeoxycorticosterone, 18-hydroxycorticosterone, and a
significant amount of aldosterone. The enzyme from the zonae
fasciculata-reticularis only catalyzed 11ß- and 18-hydroxylation
reactions. Thus at least two forms of $P450_{11ß}$ exist in rat adrenal
cortex mitochondria, but aldosterone synthesis is catalyzed only by the
form present in the zona glomerulosa(50). In bovine adreno-cortical
mitochondria, again two distincts forms of $P450_{11ß}$ have been isolated(51-
52). However, both forms are capable of catalyzing aldosterone
synthesis. As for rat and cattle, two genes, encoding for $P450_{11ß}$, have
also been demonstrated in man(53). Whether both encode for an active
$P450_{11ß}$ or one encodes for a nonfunctional pseudogene is not known. It
is also unknown whether the catalytic activity of the human P450
resembles more that of rat or bovine $P450_{11ß}$.

The entire amino acid sequences of bovine(54) and rat(55) $P450_{11ß}$
have been deduced from the nucleotide sequence. An alignment showed that
these sequences share 59.7% identical amino acids. In rats the $P450_{11ß}$
precursor protein consists of 499 amino acids including an extension
peptide of 24 amino acids at the NH_2-terminus(55). As mentioned above,
the signal peptide of $P450_{scc}$ contains 39 residues. An alignment of both
bovine mitochondrial P450 isozymes shows 38.8% identical amino acids. Of
more interest is that $P450_{scc}$ and $P450_{11ß}$ show two highly conserved
regions. The first region is the heme binding site (the cysteine thought
to serve as the fifth ligand to the heme is in italics):

$P450_{scc}$ ^{454}F G W G V R Q *C* V G R R I A E L E M T L (45)
$P450_{11ß}$ ^{419}F G F G V R Q *C* L G-R R V A E V E M L L (54)

The second region is highly conserved in steroidogenic P450 isozymes
(55). According to Nonaka et al. (55), among the P450s of known
structure, a consensus sequence, L-X-A-X-X-X-E-X-L-R-X-X-P, is found only
in steroidogenic P450s.

SCC- HUMAN	374	V P L L K A S I K E T L R L H P	(40)
SCC- BOVINE	373	V P L L K A S I K E T L R L H P	(45)
SCC- RAT	*	V P L L K A S I K E T L R L H P	(41)
11ß- HUMAN	*	L P L L R A A L K E T L R L Y P	(54)
11ß- BOVINE	362	L P L L R A A L K E T L R L Y P	(54)
11ß- RAT	361	L P L L R A A L K E T L R L Y P	(55)
17α- HUMAN	350	L L L L E A T I R E V L R L R P	(61)
17α- BOVINE	350	L V L L E A T I R E V L R I R P	(56)
17α- PORCINE	*	L V L L E A T I R E V L R F R P	(61)
17α- RAT	349	L L M L E A T I R E V L R I R P	(57)
17α- CHICKEN	353	L P Y L E A T I S E G L R I R P	(71)
C21- HUMAN	342	L P L L N A T I A E V L R L R P	(74)
C21- BOVINE	341	L P L L N A T I A E V L R L R P	(75)
C21- PORCINE	341	L P L L N A T I A E V L R L R P	(73)
C21- MOUSE	334	L P L L M A T I A E V L R L R P	(76)
CONSENSUS		X X X L X A X X X E X L R X X P	(55)

Fig. 4- Alignment of the amino acid sequences of a highly conserved
region in steroidogenic P450s. The consensus sequence (putative steroid
binding site) as proposed by Nonaka et al. (55) is also shown.
*Incomplete sequences. Figures in parenthesis are the references.

Fig. 4 gives the alignment of this conserved region for a number of steroidogenic P450s, mitochondrial as well as microsomal. Nonaka et al. (55), Zuber et al. (56) and Namiki et al. (57) called this region the putative steroid binding site.

Recently, Edwards et al.(58) proposed the tentative α-helices in mammalian cytochromes P450. The location of these α-helices was predicted from their homology with those of cytochrome P450$_{cam}$, [the only P450 of which the three-dimensional structure is known(59,60)] and the likelyhood of the residues within these areas being present within an α-helix. According to the approach used, the putative steroid binding site might be located at helix K.

Microsomal P450 Isozymes

P450$_{17\alpha}$

Cytochrome P-450$_{17\alpha}$ (encoded by gene(s) P450XVII) is found in all steroidogenic tissues, including adrenal cortex, ovarian theca cells and testicular Leydig cells. Chung et al.(61) called the P-450$_{17\alpha}$ enzyme a key branch point in the system of steroid hormone synthesis. Indeed, as 17α-hydroxylase it distinguishes between the synthesis of mineralo-corticoids (e.g. aldosterone) and that of glucocorticoids (cortisol) and as 17,20-lyase it distinguishes between the synthesis of glucocorticoids and C-19 precursors of androgens (Fig. 2).

P450$_{17\alpha}$ has been observed immunohistochemically in pig adrenal cortex, testis and ovary(62). In the adrenal cortex, P-450$_{17\alpha}$ has been found in the zona fasciculata and reticularis but not in the zona glomerulosa(62). In the testis this P450 was present exclusively in Leydig cells and in the ovary, immuno-reactivity was observed exclusively in the theca interna(62). Porcine thecal lyase activity increased as the follicle matured, providing more androgen substrate for the production of estrogen(63).

As with other P450 enzymes from the endoplasmic reticulum, activity is supported by electron transfer from the microsomal flavoprotein, NADPH-P450 reductase. Cytochrome b$_5$ may also be involved in electron transfer. P450$_{17\alpha}$ catalyzes the 17α-hydroxylation of progesterone and pregnenolone. These 17-hydroxylated steroids are converted into 17-hydroxy,20-dihydroprogesterone and 17-hydroxy,20-dihydropregnenolone, respectively, and may undergo scission of the C-17,20 carbon bond (17,20-lyase acivity) to yield androstenedione or dehydroepiandrosterone (DHEA) and formic acid.

Studies of Nakajin and colleagues have shown that in the pig the 17-hydroxylase and 17,20-lyase activities of testicular smooth endoplasmic reticulum result from the action of a single P450(64). The role of a single P450 in catalyzing both reactions has been confirmed for guinea pig(65), porcine(66), bovine(67-68) adrenal, and rat (57,69)and human(70) testis microsomes. Slight differences have been reported in porcine P-450$_{17\alpha}$ isolated from adrenal and testis(64). However, the same P-450$_{17\alpha}$ mRNA is found in human adrenal gland and testis(65). The human(65,73-74) adrenal amino acid sequence is 68.4% identical to the rat testis (57) sequence and 71.1% to the bovine(56,68) adrenal P45017α but only 48.2% to the chicken(71) ovary P450$_{17\alpha}$. From the sequence analysis it could be deduced that there are 75 positions where amino acids are common in bovine and human forms of P450$_{17\alpha}$ but which differ in the rat enzyme(69). This might be at the origin of differences in catalytic properties. Indeed, comparison of the 17α-hydroxylase and the 17,20-lyase activities in COS 1 cells transfected with the bovine, human and rat forms of P450$_{17\alpha}$, shows interesting differences: the rat P450$_{17\alpha}$ is able to convert both 17α-hydroxypregnenolone and 17α-hydroxyprogesterone into their respective androgens, dehydroepiandrosterone (DHEA) and androstenedione, whereas the human and bovine enzymes catalyze only the conversion of 17α-hydroxypregnenolone into DHEA (69). Human $P450_{17\alpha}$ has also been shown to catalyze 16α-hydroxylation of progesterone whereas the rat enzyme cannot(69).

Using the Kyte and Doolittle method, Namiki et al.(57) suggest that the P-$450_{17\alpha}$ of rat testis is anchored to the membrane of the endoplasmic reticulum through two transmembrane regions, one at the NH_2-terminal site (amino acids 2-21) and a second from amino acids 169 up to 186. The two transmembrane regions are also found in the bovine and human P450. In addition four hydrophobic clefts are identified at positions 280-300, 320-360, 400-420 and 465-485 (57). These clefts are also found in bovine and human $P450_{17\alpha}$. The authors (57) also suggest that the second cleft might contain the steroid binding place (an alignment of the sequences of this putative steroid binding domain is shown in Fig. 4).

Deficiency of 17α-hydroxylase is a rare cause of congenital adrenal hyperplasia leading to impaired production of cortisol and/or androgens and estrogens. A detailed investigation of a mutant $P450_{17\alpha}$ protein revealed that the mutation occured to the C-terminal side of the heme binding region(70). The mutant sequence is identical to that of the normal gene except for the four base duplication *CATC* at the position of the codon 479. Consequently the reading frame is altered at this point in the mutant gene leading to a C-terminal sequence which is completely different from that of the normal $P450_{17\alpha}$. Proline 480 is replaced by histidine. In the normal $P450_{17\alpha}$ sequence there are six basic amino acid residues (shown in italics) between amino acids 480 and 508.

Normal: ^{480}P *K V V* F L I D S F *K V K* I *K V R* Q A W *R* E A Q A E G S T

Mutant: ^{480}H P Q G G L S D *R* L F Q S E D Q G A P G L E G S P G

In contrast only one basic amino acid (arginine)is located in this region of the mutant sequence(70). Kagimoto et al.(70) suggest that the reduced positive charge at the C-terminus of the mutant $P450_{17\alpha}$ diminishes its ability to interact with P450 reductase leading to the observed 17α-hydroxylase/17,20-lyase deficiency.

$P450_{C21}$

The microsomal component of the adrenocortical steroidogenic pathway includes next to $P450_{17\alpha}$, $P450_{C21}$. This P450 catalyzes the hydroxylation of progesterone and 17-hydroxyprogesterone at C-21 to yield 11-deoxycorticosterone and 11-deoxycortisol, respectively. As already indicated in the introduction, the 21-hydroxylase was the first to be shown as P450-dependent(1). A deficiency in 21-hydroxylase (about 1 in every 7000 persons) causes congenital adrenal hyperplasia, a disorder of cortisol and aldosterone biosynthesis in females(37). This is one of the most common inborn errors of metabolism.

$P450_{C21}$ is expressed at high levels in adrenal glands, constituting approximately 1% of poly(A^+)RNA(72). The complete sequence of $P450_{C21}$ from pig adrenal microsomes was determined(73). The mature protein consists of 492 amino acid residues. Cysteine$_{427}$ is thought to be the heme binding site. The porcine(69) $P450_{C21}$ amino acid sequence shares with that of the human(74),bovine(75) and mouse(76) P450 79.1%, 87.8%,and 73.3% identical amino acids, respectively.

A single gene product is expressed for steroid 21-hydroxylase in porcine adrenals. Two $P450_{C21}$ genes have been isolated from a human genetic library using a bovine $P450_{C21}$ cDNA(74). Sequence analysis has revealed that these two genes are highly homologous (98%), but that three mutations render one of the genes nonfunctional. According to Higashi et al.(74), tandem arrangement of the pseudo and genuine genes in close proximity could account for the high incidence of $P450_{C21}$ gene deficiency by homologous gene recombination. In a more recent paper, Higashi et al.(77) isolated three steroid 21-hydroxylase B genes from three $P450_{C21}$-deficient patients. Analysis of their structures and functions revealed several base changes in the sequences of the three $P450_{C21}$-B genes as compared to that of the functional B gene. Many of these base changes were identical to those of the inactive $P450_{C21}$-A pseudogene. The three cloned genomic DNAs produced no $P450_{C21}$ activity in COS cells(77). Two of the DNAs were shown to have a point mutation in the second intron, causing aberrant splicing. The other DNA carried three clustered missense mutations in the sixth exon. All these mutations have also been seen in the corresponding site of the $P450_{C21}$-A pseudogene. Based on these data, the investigators suggest the involvement of gene conversion in this genetic disease.

It is of interest that both the 17α-and 21-hydroxylase are using progesterone as substrate. An alignment of the amino acid sequence (Fig. 4) of the human $P450_{17\alpha}$ with that of $P450_{C21}$ showed a well conserved domain (from leucine349 up to proline364 of the $P450_{17\alpha}$ sequence). This domain of the apoprotein of $P450_{17\alpha}$ has been proposed as the putative steroid binding site(Fig. 4).

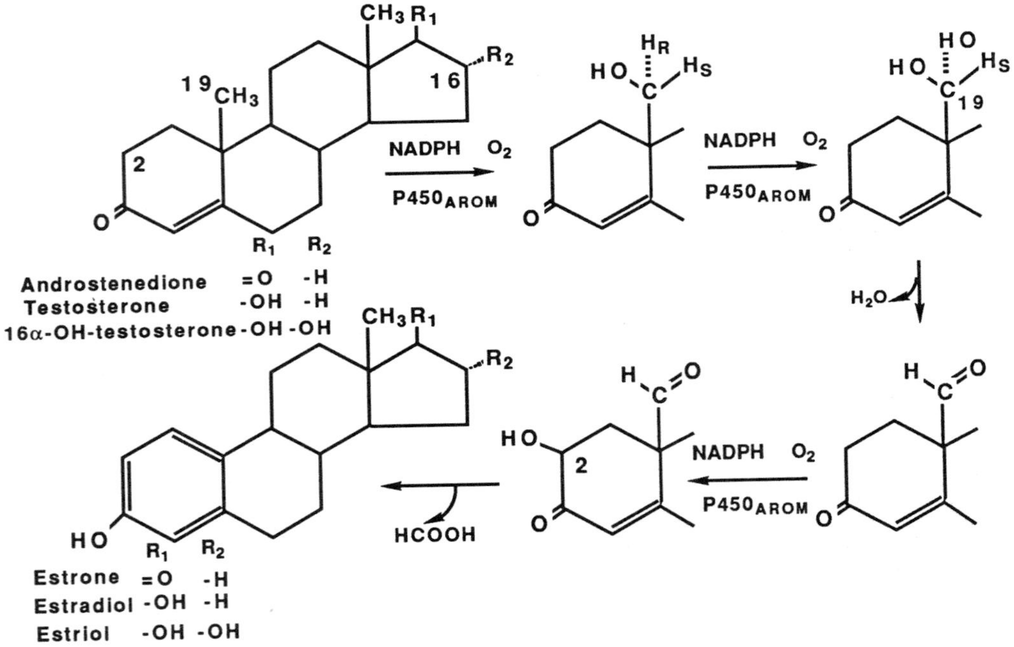

Fig. 5- The reaction sequence catalyzed by $P450_{AROM}$

Aromatisation of androgens into estrogens is one of the most important reactions involved in steroid hormone biosynthesis. The endoplasmic reticulum $P450_{AROM}$ converts androgens (androstenedione, testosterone, 16α-hydroxytestosterone and 19-norandrostenedione) to estrogens by 2 hydroxylations at the C-19 methyl group and a third one at

C-2. This results in the loss of the C-19 methyl group and consequent aromatization of the A ring of the steroid (37,78) (Fig.5). The aromatase enzyme complex is present in the granulosa cells of the ovaries(79), in adipose tissue of both males and females(80-82), in skin fibroblasts (83-84), in the Sertoli cells of the testis(38), in several areas of the brain(85), in placenta(86-88),in human(89) and rat(90) prostate and Dunning R3327H rat prostatic adenocarcinoma(90),and human neoplastic endometrium(91) and breast tumor(82, 92-93). Assays of aromatase in adipose tissue from different quadrants of mastectomy specimens from patients with breast cancer indicate that the activity was always higher in quadrants associated with tumor as compared with non-involved quadrants(94).

Expression of cDNAs encoding respectively the chicken(95) and the human(96) $P450_{AROM}$ in nonsteroidogenic COS-1 monkey kidney cells indicate that the three successive oxidation reactions, necessary to form estrogen from androgen, are catalyzed by a single protein. Recently, Pompon et al.(97) expressed the human placental aromatase in *Saccharomyces cerevisiae*. The results obtained were similar to those reported after expression in COS cells. However the Vmax was about 10 times higher. These investigators proved that yeast P450 reductase is able to transfer electrons from NADPH to the expressed human P450 and that the expressed aromatase activity is negatively regulated by cytochrome b_5.

Using the VGAP alignment programme, the human (96) aromatase sequence shares with the chicken(95) aromatase 71.4% identical amino acids. However the human aromatase shares with bovine adrenal cortex mitochondrial 11ß-hydroxylase ($P450_{11ß}$) and rat testis microsomal 17-hydroxylase ($P450_{17\alpha/17,20-lyase}$) 19.9% and 21.5 identical amino acids only. The alignment of the amino acid sequences of the human $P450_{AROM}$ and human $P450_{17\alpha/17,20-lyase}$ isozymes shows 23.3 % identical amino acids. This little sequence similarity indicates that this enzyme is a unique form of the P450 superfamily.

20-Hydroxylation of Ecdysone ($P450_{ecdysone\ C20}$)

The synthesis of the insect moulting steroid hormone, ecdysone, from cholesterol requires several hydroxylation steps(98). The best studied one is the formation of ß-ecdysone (20-hydroxyecdysone) from α-ecdysone (2ß,3ß,14α,22R, 25-pentahydroxy-5ß-cholest-7-en-6-one). The NADPH-dependent P450 20-monooxygenation of α-ecdysone is catalyzed by mitochondria and microsomes isolated from *Musca domestica* (99). Since only the mitochondrial 20-monooxygenase is regulated by α-ecdysone Srivatsan et al.(99) conclude that only this P450 (s) is involved in the moulting phenomenon.

METABOLISM OF ARACHIDONIC ACID

In the metabolism of arachidonic acid a number of important enzymes are involved, namely cyclooxygenase, the endoperoxide-dependent enzymes, 12-lipoxygenase, 15-lipoxygenase, 5-lipoxygenase, leukotriene synthetic enzymes and cytochrome P450-dependent enzymes. *cis*-Epoxyeicosatrienoic acids (EETs) and certain hydroxyeicosatetraenoic acids (HETEs) originate from a P450 catalytic cycle, the "epoxygenase" pathway (100).
Rat liver microsomal P450 catalyzes the NADPH dependent oxygenation of arachidonic acid to six regioismeric *cis,trans*-dienols, 5-, 8-, 9-, 11-, 12- and 15-HETE(101). While the microsomal P450 catalyzed hydroxylation at carbons 5, 8, 9, 11 and 15 of arachidonic acid proceeds almost without evident enantioselectivity, the formation of 12(R)-hydroxyeicosatetra-enoic acid [12(R)-HETE] is generated with a high degree of enantio-selectivity (81% R vs 19% S)(101). 12(R)-HETE is the predominant stereo-isomer produced by rat liver microsomal P450 (101). 12-HETE is a major proinflammatory metabolite of arachidonic acid in human skin and lesional skin of patients with psoriasis exhibits an increased ability to

synthesize 12-HETE from exogeous arachidonic acid in vitro and contains
high concentrations in vivo (102-103). Studies of Woollard (103)
revealed that 12-HETE, present in the scale of patients with psoriasis,
is not 12(S)-HETE, as formed in platelets, but 12(R)-HETE. Bovine
corneal microsomes incubated with arachidonic acid and a NADPH-generating
system formed two biological active products (104-105). The structures
of these compounds were reported to be 12(R)-hydroxyicosatetraenoic acid
[12(R)-HETE] (104) and 12(R)-hydroxyeicosatrienoic acid. Mechanistic
details on the formation of these monohydroxylated metabolites are still
speculative (100). However, the fact that the formation of these
compounds is dependent on NADPH and inhibited by SKF-525A, antibodies to
human liver arachidonic acid epoxygenase and to NADPH-P450 reductase
(105) proves the involvement of P450.

Other important arachidonate metabolites are thromboxane (TXA_2) and
prostacyclin (PGI_2). The isomerization of prostaglandin-endo-peroxide
(9,11-endo-peroxy, 15-hydroxy arachidonic acid; PGH_2) to prostacyclin and
to thromboxane involves endoperoxide activation(106). PgI_2-synthetase
has been found in e.g. kidney, uterus and lung, but the endothelium of
the aorta is the main source of the enzyme(106). TXA2-synthetase is
present in lung, kidney, umbilical cord and blood platelets(106).
Spectral characteristics of PGI_2-(106-107) and TXA_2 (106-108)- synthetase
indicate that both enzymes are P450 proteins and therefore must contain a
thiolate ligand at the heme iron(107). However, no monooxygenase
activity is associated with their physiological function(107).
Therefore, the name "heme-thiolate enzymes" instead of P450 enzymes has
been proposed(107). From optical and ESR studies on PGI_2- and TXA_2-
synthetase it has been concluded (109) that the ferric enzyme catalyzes
the cleavage of the endoperoxide bond. Studies of Ullrich and colleagues
also prove that PGI_2- synthetase obtains its specificity through an
interaction with the endoperoxide oxygen at C-11, whereas TXA_2-
synthetase interacts with the C-9 endoperoxide oxygen(109).

LEUKOTRIENE METABOLISM

Next to the above discussed P450 isozymes involved in the metabolism
of arachidonic acid, other P450s are needed to metabolize leukotrienes,
again products derived from arachidonic acid. Leukotriene B_4 (LTB_4) is
considered an important mediator of inflammation. It induces chemotaxis
and chemokinesis of polymorphonuclear leukocytes, stimulates the
production of superoxide in neutrophils and eosinophils and can cause
degranulation of polymorphonuclear leukocytes and cellular aggregation.
LTB_4 can be inactivated by hepatic microsomal ω- and ω-1-oxidation to
form 20-hydroxy ω-LTB_4, and 19-hydroxy ω-1-LTB_4. Studies of Romano et
al. (110) suggest that the ω- and ω-1-hydroxylations are mediated by
different isozymes of P450. Hatzelman and Ullrich (111) showed that in
unstimulated human polymorphonuclear leukocytes the LTB_4 ω-monooxygenase
is responsible for the formation of ω-hydroxy-arachidonate. This
indicates that the LTB_4 ω-monooxygenase is not strictly specific for LTB_4
and that some variations in the substrate structure are possible. Recent
studies of Soberman et al. (112) indicate that microsomal $P450_{LTB4}$ from
human polymorphonuclear leukocytes catalyzes three sequential ω-
oxidations leading to the formation of 20-COOH-LTB_4 via 20-hydroxy-LTB_4
and 20-CHO-LTB_4.

VITAMIN D3 ACTIVATION

Vitamin D_3 is hydroxylated in the liver on carbon 25 to produce its
obligatory metabolite 25-OH-vit.D_3(113). In male rat liver this P450-
dependent 25-hydroxylation has been shown to be localized both to the
mitochondrial and the microsomal fraction(114-115). The activity
obtained with the microsomal fraction of male rat liver was less than 15%
of that obtained with the mitochondrial fraction(115). The rate of

conversion was about 3 pmol/mg of mitochondrial protein/ min(115). The microsomal enzyme was purified from male rat liver and characterized as cytochrome P450 ($P450_{vit\ D3-C25}$)(116-117). It is of interest that polyclonal and monoclonal antibodies, prepared against purified male rat liver microsomal $P450_{vit\ D3-C25}$, inhibited both the vitamin D$_3$ 25-hydroxylase and testosterone 16α-hydroxylase activities(118). However, although these two hydroxylase activities are immunochemically indistinguishable, vitamin D$_3$ 25-hydroxylase and testosterone 16α-hydroxylase are different gene products(119).

The microsomal $P450_{vit\ D3-C25}$ was shown to be male specific and hardly detectable in adult female rat liver microsomes(120). This, together with the fact that human liver microsomes are devoid of vitamin D$_3$ 25-hydroxylase activity(121-122), cast doubt about the physiological importance of the liver microsomal enzyme(123). In contrast with the microsomal enzyme, the mitochondrial 25-hydroxylase is 4.6-fold higher in the female rat than in the male rat liver(124). Injection of testosterone into female rats decreased the mitochondrial 25-hydroxylase activity but had no effect on the microsomal enzyme(124). Injection of estradiol valerate to male rats resulted in increased activity of the mitochondrial enzyme while the microsomal activity decreased. These results indicate that sex hormones exert a regulatory control on the mitochondrial vit.D$_3$ 25-hydroxylase(124). Vitamin D$_3$ 25-hydroxylase was purified from female rat liver mitochon-dria(123). This P450 hydroxylates vitamin D$_3$ and 1α-hydroxyvitamin D$_3$ at position 25, but did not show any activity toward xenobiotics. It was slightly inhibited by metyrapone but markedly (81%) in an atmosphere of $CO:O_2:N_2$, 40:20:40(123). Hydroxylation of vitamin D$_3$ is also known to occur in the intestine and kidney(113).

25-Hydroxyvitamin D$_3$, the major circulating form of vitamin D$_3$ must be converted to 1α-25-dihydroxyvitamin D$_3$ (and 24,25-dihydroxyvitamin D$_3$) before full biological activty is achieved(113). This P450-dependent reaction ($P450_{25OHvit.D3C1}$) occurs mainly in kidney mitochondria(113). However, the 1α-hydroxylation step also occurs in human and chicken (embryo) bone cells(125-126), human melanoma cells(127), sarcoid tissue(128), human placenta(129-130) and neonatal human foreskin keratinocytes(122-123). The presence of a 1-hydroxylase in keratinocytes indicates that the skin not only produces vitamin D$_3$ but is also able to produce (abundant quantities) the biologically active form, 1,25-dihydroxyvitamin D$_3$, from 25-hydroxyvitamin D$_3$(131). Exogenous 1,25-dihydroxyvitamin D$_3$, at physiological levels (1 pM to 1 nM), inhibits its production by these cells while stimulating the formation of 24,25-dihydroxy-vitamin D$_3$(132), a metabolite involved in bone ossification(130). Parathyroid hormone(132), cytokines, tumor necrosis factor-α and interferon-γ stimulate the production of 1,25-dihydroxyvitamin D$_3$(133).

Keratinocytes grown in conditions favoring proliferation (0.1 mM calcium) produce more 1,25-dihydroxyvitamin D$_3$ than keratinocytes grown in conditions favoring terminal differentiation (1.2 mM calcium)(133). These studies indicate that the production of 1,25-dihydroxyvitamin D$_3$ varies with the degree of differentiation possibly because of its ability to increase intra-cellular calcium(133). This might play a role in skin diseases, such as psoriasis, that are characterzed by aberrant proliferation and incomplete differentiation.

GIBBERELLIN SYNTHESIS

The gibberellins form a large family of tetracyclic triterpenoid phytohormones. An important part of the gibberellin biosynthetic pathway involves the oxidation of *ent*-kaurene to *ent*-kaurenoic acid. This involves the sequential oxidation of the 19-C methyl group to give *ent*-kaurenol, *ent*-kaurenal and *ent*-kaurenoic acid. All three steps require

NADPH and molecular oxygen and are inhibited by CO, indicating the involvement of a P450(134).

BILE ACID SYNTHESIS

The most important pathway for the metabolism and excretion of cholesterol in mammals is the formation of bile acids(135). The reactions involved in the synthesis of bile acids are very similar to those involved in the hydroxylation, oxidation and conjugation of xenobiotics(135).

The first and major rate-limiting step in the biosynthetic pathway of bile acids in mammalian liver is the 7α-hydroxylation of cholesterol. The microsomal 7α-hydroxylation requires NADPH and oxygen and is inhibited by carbon monoxide and an antibody against NADPH- P450 reductase, indicating that the hydroxylase is a monooxygenase(7-8, 136-137).

Studies of Boström and Wikvall(7) indicate that the liver P450$_{LM4}$ from cholestyramine-treated male rabbits catalyzes not only the 7α-hydroxylation of cholesterol but also the 12α-hydroxylation and 25-hydroxylation of 5ß-cholestane-3α,7α-diol, the 25-hydroxylation of 5ß-cholestane-3α,7α,12α-triol, the 6ß-hydroxylation of testosterone and demethylation of ethylmorphine. Chromatography of electrophoretically homogenous P450$_{LM4}$ on octylamine-Sepharose gave two subfractions, P450$_{LM4}$ I and P450$_{LM4}$ II(7). P450$_{LM4}$ I is inactive in cholesterol 7α-hydroxylation (≤1pmol/nmol P450/min) but catalyzes 12α-hydroxylation and 25-hydroxylation. P450$_{LM4}$ II catalyzes 7α-hydroxylation more effeciently and 12α-hydroxylation and 25-hydroxylation less than the original P450$_{LM4}$ fraction(7). It is of interest that the rate of 12α-hydroxylation was three to four times higher with P450$_{LM4}$ fractions from starved rabbits than with P450$_{LM4}$ fractions from untreated, phenobarbital- or ß-naphthoflavone-treated animals(138). In a recent paper the molecular cloning of cDNA for cholesterol 7α-hydroxylase from rat liver microsomes has been reported (139). The primary structure deduced from the nucleotide sequence of the cDNA indicated that the enzyme constitutes a novel P450 family (139).

Bile acid synthesis by cultured pig hepatocytes increased during the second and third day of culture. This rise was inhibited after addition of conjugated and unconjugated bile acids(140). Kwekkeboom et al. conclude from their studies that feedback regulation of bile acid synthesis is exerted by direct action of bile acids on the hepatocyte(140). This supports earlier findings that bile acid biosynthesis is, at least partly, regulated by feedback inhibition by bile acids returning to the liver via the enterohepatic circulation(135).

Next to the described microsomal P450s, the biosynthesis of bile acids involves in addition a 26-hydroxylation catalyzed by a mito-chondrial fraction of the liver (141). A P450, purified from liver mitochondria of untreated rabbits, catalyzed the 26-hydroxylation of cholesterol, 5-cholestene-3ß,7α-diol, 7α-hydroxy-4-cholesten-3-one, 5ß-cholestane-3α,7α-diol, and 5ß-cholestane-3α,7α,12α-triol (141). This preparation also catalyzed the 25-hydroxylation of vitamin D$_3$ (142). However, further characterization and using a modified purification procedure revealed that different species of liver mitochondrial P450 are involved in the 26-hydroxylation of C$_{27}$-steroids and 25-hydroxylation of vitamin D$_3$ (142). The 26-hydroxylase is also present in pig kidney mitochondria (10). Anderson et al. (9) found that mRNA for the 26-hydroxylating liver mitochondrial P450 is present in duodenum, adrenal, lung and kidney tissues from rabbit and Skrede et al. (143) demonstrated the presence of 26-hydroxylation of C$_{27}$-steroids in human skin fibro-blasts. The structure, as deduced by both DNA sequence analysis of the cDNA and protein sequence analysis, reveals it to be a mitochondrial P450 (9). A signal sequence of 36 residues precedes a coding region of 499 amino acids, predicting a molecular weight of 56,657 for the mature protein (9). This P450 is a member of the P450 gene family XXVI (9). 26-Hydroxycholesterol is a potent suppressor of the HMG-CoA reductase (144). The presence of the 26-hydroxylase in hepatic and non-hepatic tissues suggests that it may be involved in cholesterol homeostasis.

METABOLISM OF RETINOIC ACID

Retinoic acid (RA) the active metabolite of vitamin A (retinol) is rapidly converted to more polar derivatives e.g. 4-hydroxyretinoic acid (145). The involvement of P450 in the 4-hydroxylation by rat liver is indicated by the requirement of oxygen and NADPH, the inhibition by CO and proadifen-HCl (SKF-525A) and by the increase in retinol metabolism after administration of phenobarbital(146-147). The microsomal oxidative metabolism of retinoic acid in hamster liver and intestine also requires oxygen and NADPH and is inhibited by CO (148-151). A P450-dependent retinoic acid metabolism has also been found in rat epidermal microsomes. This metabolism is inhibited by CO, SKF-525A, ketoconazole, but hardly at all by 7,8-benzoflavone (24). Epidermal microsomes from neonatal rats, pretreated topically with retinoic acid or 3-methylcholanthrene shows a 3- to 8-fold increase in retinoic acid metabolism. In contrast, phenobarbital has a much smaller effect (24).

CONCLUSION

A large number of interesting P450-dependent enzymes, which metabolize endogenous substrates, have been shown to exist in a great variety of organs and glands. Of interest is the number of these P450 isozymes present in the skin, indicating that the skin is not only a major site for the metabolism of xenobiotics but that it is also able to synthesize and metabolize key compounds needed for cell proliferation and differentiation.

The great differences observed in the amino acid sequences between P450 isozymes, involved in, for example, steroid biosynthesis, indicate that it should be possible to synthesize selective inhibitors of use in steroid hormone- depending diseases. Of course knowledge of the three-dimensional structure of the eukaryotic isozymes would be more than welcome.

As illustrated, the study of P450s, metabolizing endogenous substrates, was already of help in the description of the molecular basis of some human diseases. This opens new perspectives for early diagnosis and treatment.

REFERENCES

1.   R.W. Estabrook, D.Y. Cooper and O. Rosenthal, The light reversible carbon monoxide inhibition of the steroid C21-hydroxylase system of the adrenal cortex, Biochem. Zeitsch. 338: 741-755 (1963)
2.   D.W. Nebert, D.R. Nelson and R. Feyereisen, Evolution of cytochrome P450 genes, Xenobiotica 19: 1149-1160 (1989)
3.   D.W. Nebert, D.R. Nelson, M. Adesnik, M.J. Coon, R.W. Estabrook,F.J. Gonzalez, F.P. Guengerich, I.C. Gunsalus, E.F. Johnson, B. Kemper, W. Levin, I.R. Phillips, R. Sato and M.R. Waterman, The P450 superfamily: updated listing of all genes and recommended nomenclature for the chromosomal loci, DNA 8: 1-13 (1989)
4.   J. Trzaskos, S. Kawata and J.L.Gaylor, Microsomal enzymes of cholesterol biosynthesis. Purification of lanosterol 14α-methyl demethylase cytochrome P-450 from hepatic microsomes, J. Biol. Chem. 261: 14651-14657 (1986)
5.   Y. Yoshida, Cytochrome P450 of fungi: primary target for azole antifungal agents, in: "Current topics in medical mycology", Vol.2, M.R. McGinnis, ed., Springer Verlag, New York (1988)

6. J.M. Trzaskos, M.F. Favata, R.T. Fischer and S.H. Stam, In situ accumulation of 3S-hydroxylanost-8-en-32-aldehyde in hepatocyte cultures. A putative regulator of 3-hydroxy-3-methyl-glutaryl-coenzyme A reductase activity, J. Biol. Chem. 262: 12261-12268 (1987)

7. H. Boström and K. Wikvall, Hydroxylations in biosynthesis of bile acids. Isolation of subfractions with different substrate specificity from cytochrome P450$_{LM4}$, J. Biol. Chem. 257: 11755-11759 (1982)

8. R. Hansson and K. Wikvall, Purification and properties of chole-sterol 7α-hydroxylase, in: "Cholesterol 7α-hydroxylase (7α-mono-oxygenase)" R. Fears and J.R. Sabine eds., CRC PRESS,INC.,Boca Raton, Florida (1986)

9. S. Anderson, D.L.Davis, H. Dahlbäck, H. Jörnvall and D.W. Russell, Cloning structure and expression of the mitochondrial cytochrome P-450 sterol 26-hydroxylase, a bile acid bio-synthetic enzyme, J. Biol. Chem. 264: 8222-8229 (1989)

10. H. Postlind and K. Wikvall, Evidence for the formation of 26-hydroxycholesterol by cytochrome P-450 in pig kidney mito-chondria, Biochem. Biophys. Res. Commun. 159: 1135-1140 (1989)

11. C. Larroque, R. Lange, P. Maurel, R. Langlois and J.E. van Lier, Rat liver microsomal progesterone metabolism: evidence for differential troleandomycin and pregnenolone 16α-carbonitrile inductive effects in the cytochrome P-450 III family, J. steroid. Biochem. 33: 277-286 (1989)

12. E.F. Johnson, G.E. Schwab and L.E. Vickery, Positive effectors of the binding of an active site-directed amino steroid cytochrome P-4503c, J. Biol. Chem. 263: 17672-17677 (1988)

13. C.L.Potenza, U.R. Pendurthi, D.K. Strom, R.H. Tukey, K.J. Griffin, G.E. Schwab and E.F. Johnson, Regulation of the rabbit cytochrome P-450 3c gene. Age-dependent expression and transcriptional activation by rifampicin, J. Biol. Chem. 264: 16222-16228 (1989)

14. D.J. Waxman, C. Attisano, F.P. Guengerich and D.P. Lapenson, Human liver microsomal steroid metabolism: identification of the major microsomal steroid hormone 6ß-hydroxylase cytochrome P-450 enzyme, Archiv. Biochem. Biophys. 263: 424-436 (1988)

15. N. Harada and M. Negishi, Substrate specificities of cytochrome P-450, C-P-450$_{16\alpha}$ and P-450$_{15\alpha}$ and contribution to steroid hydroxylase activities in mouse liver microsomes, Biochem. Pharmacol. 37: 4778-4780 (1988)

16. R. Lindberg, B. Burkhart, T. Ichikawa and M. Negishi, The structure and characterization of type I P-450$_{15\alpha}$ gene major steroid 15α-hydroxylase and its comparison with Type II P-450$_{15\alpha}$ gene, J. Biol. Chem. 264: 6465-6471 (1989)

17. G.E. Schwab, J.L. Raucy and E.F. Johnson, Modulation of rabbit and human hepatic cytochrome P-450-catalyzed steroid hydroxy-lations by α-naphthoflavone, Mol. Pharmacol. 33: 493-499(1988)

18. D.J. Waxman, A. Ko and C. Walsh, Regioselectivity and stereo-selectivity of androgen hydroxylations catalyzed by cytochrome P-450 isozymes purified from phenobarbital-induced rat liver, J. Biol. Chem. 258: 11937-11947 (1983)

19. D.E. Ryan, S. Iida, A.W. Wood, P.E. Thomas, C.S. Lieber and W. Levin, Characterization of three highly purified cytochromes P-450 from hepatic microsomes of adult male rats, J. Biol. Chem. 259: 1239-1250 (1984)

20. M. Sundin, M. Warner, T. Haaparanta and J.-Å. Gustafsson, Isolation and catalytic activity of cytochrome P-450 from ventral prostate of control rats, J. Biol. Chem. 262: 12293-12297 (1987)

21. M. Wagner, M. Stromstedt, L. Möller and J.-Å. Gustafsson, Distribution and regulation of 5α-androstane-3ß,17ß-diol

hydroxylase in the rat central nervous system, Endocrinology
124:2699-2706 (1989)

22. T. Haaparanta, H. Glaumann and J.-Å. Gustafsson, Characterization
    and endocrine regulation of the cytochrome P-450 dependent
    microsomal hydroxylation of 5α-androstane-3ß,17ßdiol in the rat
    ventral prostate, Endocrinology 114: 2293-2300 (1984)

23. R.J.Soberman, J.P. Sutyak, R.T Okita , D.F. Wendelborn, L.J. Roberts
    and K.F.Austen, (1988) The identification and formation of 20-
    aldehyde leukotriene B4, J. Biol. Chem. 263: 7996-8002 (1988)

24. H. Vanden Bossche, G. Willemsens, P.A.J. Jannsen, (1988) Cytochrome
    P-450-dependent metabolism of retinoic acid in rat skin microso-
    mes: inhibition by ketoconazole, Skin Pharmacol 1: 176-185 (1988)

25. J.M. Trzaskos, W.D. Bowen, A. Shafiee, R.T Fischer and J.L. Gaylor,
    Cytochrome P-450-dependent oxidation of lanosterol in cholesterol
    biosynthesis. Microsomal electron transport and C-32
    demethylation, J. Biol. Chem. 259: 13402-13412 (1984)

26. D. Pompon, cDNA cloning and functional expression in yeast
    Saccharomyces cerevisiae of ß-naphthoflavone-induced rabbit liver
    P-450LM4 and LM6, Eur. J. Biochem. 177: 285- (1988)

27. H. Vanden Bossche, Mode of action of Pyridine, pyrimidine and azole
    antifungals, in: "Sterol biosynthesis inhibitors. Pharmaceutical
    and agrochemical aspects", D. Berg and M. Plempel, eds., Ellis
    Horwood, Chichester, England (1988)

28. M.H. Lai and D.R. Kirsch, Nucleotide sequence oc cytochrome
    P450LIA1 (lanosterol 14α-demethylase) from Candida albicans,
    Nucleic Acids Res., 17: 804 (1989)

29. C. Chen, V.F.Kalb, V.F., T.G.Turi and J.C. Loper, Primary structure
    of the cytochrome P-450 lanosterol 14α-demethylase gene from
    Candida tropicalis, DNA, 7: 617-626 (1988)

30. V.F. Kalb, J.C. Loper, C.R. Dey, C.W. Woods and T.R. Sutter,
    Isolation of a cytochrome P-450 structural gene from Saccharomyces
    cerevisiae, Gene, 45: 237-245 (1986)

31. V,F.Kalb, C.W. Woods, T.G. Turi, C.R. Dey, T.R. Sutter and L.C.
    Loper,Primary structure of the P450 lanosterol demethylase
    gene from Saccharomyces cerevisiae, DNA, 6: 529-537 (1988)

32. N.Y. Ishida, Y. Aoyama, R. Hatanaka, Y. Oyama, S. Imajo, M.
    Ishiguro, T. Oshima, H. Nakazato, T. Noguchi, U.S. Maitra, V.P.
    Mohan, D.B. Sprinson, and Y. Yoshida, A single amino acid
    substitution converts cytochrome P450$_{14}$ DM to an inactive form,
    cytochrome P450$_{SG1}$: complete primary structures deduced from
    cloned DNAs, Biochem. Biophys. Res. Commun.: 155: 317-323 (1988)

33. H. Vanden Bossche, P. Marichal, J. Gorrens, M.-C. Coene, G.
    Willemsens, D. Bellens, I. Roels, H. Moereels and P.A.J.
    Janssen, Biochemical approaches to selective antifungal activity.
    Focus on azole antifungals, Mycoses, in press

34. Y.Yoshida, Y. Aoyama, T. Nishino, H. Katsuki, U.S. Maitra, V.P.
    Mohan and D.B. Sprinson, Spectral properties of a novel
    cytochrome P-450 of a Saccharomyces cerevisiae mutant SG1. A
    cytochrome P-450 species having a nitrogenous ligand trans to
    thiolate. Biochem. Biophys. Res. Commun. 127: 623-628 (1985)

35. P.F. Watson, M.E. Rose and S.L. Kelly, Isolation and analysis of
    ketoconazole resistant mutants of Saccharomyces cerevisiae, J.
    Med. Vet. Mycol. 26: 153-162 (1988)

36. H.W. Chen, D.A. Leonard, R.T. Fischer and J.M. Trzaskos, A
    mammalian mutant cell lacking detectable lanosterol 14α-methyl
    demethylase activity, J. Biol. Chem. 263: 1248-1254 (1988)

37. W.L. Miller, Molecular biology of steroid hormone synthesis,
    Endocrine Rev., 9: 295-318 (1988)

38. M.R. Waterman, M.E. John and E.R. Simpson, Regulation of synthesis
    and activity of cytochrome P-450 enzymes in phydsiological path-
    ways, in: "Cytochrome P-450. Structure, Mechanism, and Function",
    Ortiz de Montellano, Ed., Plenum Press, New York (1986)

39. M. Tsubaki, A. Hiwatashi, Y. Ichikawa, Y. Fujimoto, N. Ikekawa and H. Hori, Electron paramagnetic resonance study of ferrous cytochrome P-450scc-nitric oxide complexes: Effects of 20(R),22(R) dihydrocholesterol and reduced adrenodoxin, Biochemistry 27: 4856-4862 (1988)

40. B.-C. Chung, K.J. Matteson, R. Voutilaine, T.K. Mohandas and W.L. Miller, Human cholesterol side-chain cleavage enzyme, P450scc: cDNA cloning, assignment of the gene to chromosome 15, and expression in the placenta, Proc. Natl. Acad. Sci. USA, 83: 8962-8966 (1986)

41. K.M. McMasters, L.A. Dickson, R.V. Shamy, K. Robischon, G.J. Macdonald and W.R. Moyle, Rat cholesterol side-chain cleavage enzyme (P-450scc) use of a cDNA probe to study the hormonal regulation of P-450scc mRNA levels in ovarian granulosa cells, Gene 57: 1-9 (1987)

42. C. Le Goascogne, N. Sananès, M. Gouézou, E.E. Baulieu and P. Robel, Cell-specific variations and hormonal regulation of immunoreactive cytochrome P-450scc in the rat ovary, J. Reprod. Fert. 85: 61-72 (1989)

43. Y. Ichikawa and A. Hiwatashi, The role of the sugar regions of components of the P-450-linked mixed-function oxidase (monooxygenase) system of bovine adrenocortical mitochondria, Biochim. Biophys. Acta 705: 82- (1982)

44. M. Tsubaki, Y. Iwamoto, A. Hiwatashi and Y. Ichikawa, Inhibition of electron transfer from adrenodoxin to cytochrome P-450scc by chemical modification with pyridoxal 5'-phosphate: identification of adrenodoxin-binding site of cytochrome P-450scc, Biochemistry 28: 6899-6907 (1989)

45. K. Morohashi, Y. Fugii-Kuriyama, Y. Okada, K. Sogawa, T. Hirose, S. Inayama and T. Omura, Molecular cloning and nucleotide sequence of cDNA for mRNA of mitochondrial cytochrome P-450 (SCC) of bovine adrenal cortex, Proc. Natl. Acad. Sci. USA 81: 4647-4651 (1984)

46. Y. Iwamoto, M. Tsubaki, A. Hiwatashi and Y. Ichikawa, Crystallization of cytochrome P-450scc from bovine adrenocortical mitochondria, FEBS Lett. 233: 31-36 (1988)

47. K. Yanagibashi, M. Haniu, J.E. Shively, W.H. Shen and P. Hall, The synthesis of aldosterone by the adrenal cortex. Two zones (fasciculata and glomerulosa) possess one enzyme for 11ß-, 18-hydroxylation and aldehyde synthesis, J. Biol. Chem, 261, 3556-3562 (1986)

48. T.D. Drummond, J.I. Mason, J.L. McCarthy (1988) Gerbil adrenal 11ß- and 19-hydroxylating activities respond similarly to inhibitory or stimulatory agents: two activities of a single enzyme. J. Steroid Biochem. 29: 641-648 (1988)

49. K. Suhara, K. Ohashi, K. Takahashi and M. Katagiri, Aromatase and nonaromatizing 10-demethylase activity of adrenal cortex mito-chondrial P-450₁₁ß., Archiv. Biochem. Biophys. 267: 31-37 (1988)

50. T. Ogishima, F. Mitani and Y. Ishimura, Isolation of aldosterone synthase cytochrome P-450 from zona glomerulosa mitochondria of rat adrenal cortex, J. Biol. Chem. 264: 10935-10938 (1989)

51. S. Kirita, K.-I. Morohashi, T. Hashimoto, H. Yoshioka, Y. Fujii-Kuriyama and T. Omura, Expression of two kinds of cytochrome P-450(11ß) mRNA in bovine adrenal cortex, J. Biochem 104: 683-686 (1988)

52. T. Ogishima, F. Mitani, Y. Ishimura, Isolation of two distinct cyto-chromes P-450₁₁ß with aldosterone synthase activity from bovine adrenocortical mitochondria, J. Biochem., 105: 497-499 (1989)

53. H. Globerman, A. Rösler, R. Theodor, M.I. New and P.C.White, An inherited defect in aldosterone biosynthesis caused by a mutation in or near the gene for steroid 11-hydroxylase, N. Eng. J. Med., 319: 1193-1197 (1988)

54. S.C. Chua, P. Szabo, A. Vitek, K.-H. Grzeschik, M. John and P.C. White, Cloning of cDNA ancoding steroid 11ß-hydroxylase (P450c11). Proc. Natl. Acad. USA 84: 7193-7197 (1987)

55. Y. Nonaka, N. Matsukawa, K.-I. Morohashi, T. Omura, T. Ogihara, H. Teraoka and M. Okamoto, Molecular cloning and sequence analysis of cDNA encoding rat adrenal cytochrome P-450$_{11ß}$, FEBS Lett. 255: 21-26 (1989)

56. M.X. Zuber, M.E. John, T. Okamura, E.R. Simpson, and M.R. Waterman, Bovine adrenocortical cytochrome P-450$_{17\alpha}$, regulation of gene expression by ACTH and elucidation of primary sequence, J. Biol. Chem. 261: 2475-2482 (1986)

57. M. Namiki, M. Kitamura, E. Buczko and M.L. Dufau, Rat testis P-450$_{17\alpha}$ cDNA: the deduced amino acid sequence, expression and secondary structural configuration, Biochem. Biophys. Res. Commun. 157: 705-712 (1988)

58. R.J. Edwards, B.P. Murray, A.R. Boobis and D.S. Davies, Identification and location of α-helices in mammalian cytochromes P-450, Biochemistry 28: 3762-3770 (1989)

59. T.L. Poulos, B.C. Finzel, I.C. Gunsalus, G.G. Wagner and J. Kraut, The 2.6-Å crystal structure of Pseudomonas putida cytochrome P-450, J. Biol. Chem. 260: 16122-16130 (1985)

60. T.L. Poulos, B.C. Finzel and A.J. Howard, High-resolution crystal structure of cytochrome P450$_{CAM}$, J. Mol. Biol. 195: 687-700 (1987)

61. B.-C. Chung, J. Picado-Leonard, M. Haniu, M. Bienkowski, P.F. Hall, J. E. Shively and W. L. Miller, Cytochrome P450c17 (steroid 17α-hydroxylase/17,20 lyase): cloning of human adrenal and testis cDNAs indicates the same gene is expressed in both tissues, Proc. Natl. Acad. Sci. USA 84: 407-411 (1987)

62. H. Sasano, J.I. Mason and N. Sasano, Immunohistochemical analysis of cytochrome P-450 17α-hydroxylase in pig adrenal cortex, testis and ovary, Mol. Cell. Endocrinol. 62: 197-202 (1989)

63. S.A. Tonetta and M. Hernandez, Modulation of 17α-hydroxylase/C17,20-lyase activity in porcine theca cells, J. steroid. Biochem. 33: 263-270 (1989)

64. S. Nakajin, Shively, P.-M. Yuan and P.F. Hall, Microsomal cytochrome P-450 from neonatal pig testis: two enzymatic activities (17α-hydroxylase/C17,20-lyase) associated with one protein, Biochemistry 20: 4037-4042 (1981)

65. S. Kominami, K. Shinzawa and S. Takemori, Purification and some properties of cytochrome P-450 specific for steroid 17α-hydroxylation and C_{17}-C_{20} bond cleavage from guinea pig adrenal microsomes, Biochem. Biophys. Res. Commun. 109: 916-921 (1982)

66. S. Nakajin, M. Shinoda, M. Haniu, J.E. Shively and P.F. Hall, C_{21} steroid side chain cleavage enzyme from porcine adrenal microsomes. Purification and characterization of the 17α-hydroxylase/C17,20-lyase cytochrome P-450, J. Biol. Chem. 259: 3971-3976 (1984)

67. M.X. Zuber, E.R. Simoson and M.R. Waterman, Expression of bovine 17α-hydroxylase cytochrome P-450 cDNA in nonsteroidogenic (COS 1) cells, Science 234: 1258-1261 (1986)

68. C.R. Bhasker, B.S. Adler, A.Dee, M.E. John, M.X. Zuber, R. Ahlgren, X. Wang, E.R. Simpson and M.R. Waterman, Structural characterization of the bovine CYP17 (17α-hydroxylase) gene, Archiv. Biochem. Biophys. 271: 479-487 (1989)

69. H.R. Fevold, M.C. Lorence, J.L. McCarthy, J.M. Trant, M. Kagimoto, M.R. Waterman and J.I. Mason, Rat P450$_{17\alpha}$ from testis: characterization of a full-lenght cDNA encoding a unique steroid

hydroxylase capable of catalyzing both Δ^4- and Δ^5-steroid-17,20-lyase reactions, Mol. Endocrinol. 3: 968-975 (1989)

70. M. Kagimoto, J.S.D. Winter, K. Kagimoto, E.R. Simpson and M.R. Waterman, Structural characterization of normal and mutant human steroid 17α-hydroxylase genes: molecular basis of one example of combined 17α-hydroxylase/17,20 lyase deficiency, Mol. Endocrinol. 2: 564-570 (1988)

71. H. Ono, M. Iwasaki, N. Sakamoto and S. Mizuno, cDNA cloning and sequence analysis of a chicken gene expressed during the gonadal development and homologous to mammalian cytochrome P-450c17, Gene 66: 77-85 (1988)

72. K. L. Parker, B.P. Schimmer, D.D. Chaplin and J.G. Seidman, Characterization of a regulatory region of the steroid 21-hydroxylase gene, J. Biol. Chem. 261: 15353-15355 (1986)

73. M. Haniu, K. Yanagibashi, P.F. Hall and J.E. Shively, Complete amino acid sequence of 21-hydroxylase cytochrome P-450 from porcine adrenal microsomes, Arch. Biochem. Biophys. 254: 380-384 (1987)

74. Y. Higashi, H. Yoshioka, M. Yamane, O. Gotoh and Y. Fujii-Kuriyama, Complete nucleotide sequence of two steroid 21-hydroxylase genes tandemly arranged in human chromosome: a pseudogene and a genuine gene, Proc. Natl. Acad. Sci. USA 83: 2841-2845 (1986)

75. H. Yoshioka, K.-I. Morohashi, K. Sogawa, M. Yamane, S. Kominami, S. Takemori, Y. Okada, T. Omura and Y. Fujii-Kuriyama, Structural analysis of cloned cDNA for mRNA of microsomal cytochrome P-450(C21) which catalyzes steroid 21-hydroxylation in bovine adrenal cortex, J. Biol. Chem. 261: 4106-4109 (1986)

76. D.D. Chaplin, L.J. Galbraith, J.G. Seidman, P.C. White and K.L. Parker, Nucleotide sequence analysis of murine 21-hydroxylase genes: mutations affecting gene expression, Proc. Natl. Acad. Sci. USA 83: 9601-9605 (1986)

77. Y. Higashi, A. Tanae, H. Inoue, T. Hiromasa and Y. Fujii-Kuriyama, Aberrant splicing and missense mutations cause steroid 21-hydroxylase[P-450(C21)] deficiency in humans: possible gene conversion products, Proc. Natl. Acad. Sci. USA 85: 7486-7490 (1988)

78. J.T. Kellis and L.E. Vickery, The active site of aromatase cytochrome P-450. Differential effects of cyanide provide evidence for proximity of heme-iron and carbon-19 in the enzyme-substrate complex, J. Biol. Chem. 262: 8840-8844 (1987)

79. G.J. Hickey, S. Chen, M.J. Besman, J.E. Shively, P.F. Hall, D. Gaddy-Kurten and J.S. Richards, Hormonal regulation, tissue distribution, and content of aromatase cytochrome P450 messenger ribonucleic acid and enzyme in rat ovarian follicles and corpora lutea: relationship to estradiol biosynthesis, Endocrinology 122: 1426-1436 (1988)

80. C.J. Newton, D.L. Samuel and V.H.T. James, Aromatase activity and concentrations of cortisol, progesterone and testosterone in breast and abdominal adipose tissue, J. Steroid. Biochem. 24: 1033-1039 (1986)

81. C.R. Mendelson, J.C. Merrill, M.P. Steinkampf and E.R. Simpson, Regulation of the synthesis of aromatase cytochrome P-450 in human adipose stromal and ovarian granulosa cells, Steroids 50: 51-59 (1987)

82. D.W. Killinger, E. Pewrel, D. Daniilescu, L. Kharlip and M.E. Blackstein, Aromatase activity in the breast and other peripheral tissues and its therapeutic regulation, Steroids 50, :523-536 (1987)

83. M. Fujimoto, G,D,Berkovitz, T.R. Brown and C.J. Migeon, Time-dependent biphasic response of aromatase to dexamethasone in cultured human skin fibroblasts, J. Clin. Endocrinol. Met. 63: 468-474 (1986)

84. A. Chabab, J.-C. Nicolas and C. Sultan, Aromatase activity in human skin fibroblasts: characterization by an enzymatic method, J. steroid. Biochem. 25: 157-163 (1986)

85. N.J. MacLusky, A.S. Clark, F. Naftolin and P.S. Goldman-Rakic, Estrogen formation in the mammalian brain: possible role of aromatase in sexual differentation of the hippocampus and neocortex, Steroids 50:459-474 (1987)

86. J.T. Kellis Jr. and L.E. Vickery, Purification and characterization of human placental aromatase cytochrome P-450, J. Biol. Chem. 262: 4413-4420 (1987)

87. P.F. Hall, S. Chen, S. Nakajin, M. Shinoda and J.E. Shively, Purification and characterization of aromatase from human placenta, Steroids 50: 37-49 (1987)

88. D.D. Hagerman, Human placenta estrogen synthetase (aromatase) purified by affinity chromatography, J. Biol. Chem. 262: 2398-2400 (1987)

89. Y. Kaburagi, M.B. Marino, R.Y. Kirdani, J.P. Greco, J.P. Karr and A.A. Sandberg, The possibility of aromatization of androgen in human prostate, J. steroid Biochem. 26: 739-742 (1987)

90. S.A. Marts, G.M. Padilla and V. Petrow, Aromatase activity in microsomes from rat ventral prostate and Dunning R3327H rat prostatic adenocarcinoma, J. steroid Biochem. 26: 25-29 (1987)

91. L. Tseng, J. Mazella, M.IFunt, W.J. Mann and M.L. Stone, Preliminary studies of aromatase in human neoplastic endometrium, Obstet. Gynecol.63: 150-154, 1984.

92. V.H.T. James, J.M. McNeill, L.C. Lai, C.J. Newton, M.W. Ghilchik, and M.J. Reed, Aromatase activity in normal breast tumor tissues: in vivo and in vitro studies, Steroids 50: 270-279 (1987)

93. R.J. Santen, Potential clinical role of new aromatase inhibitors, Steroids 50, 575-593 (1987)

94. W.R. Miller and J. O'Neill, The importance of local synthesis of estrogen within the breast, Steroids 50: 537-548 (1987)

95. M.J. McPhaul, J.F. Noble, E.R. simpson, C.R. Mendelson and J.D. Wilson, The expression of a functional cDNA encoding the chicken cytochrome P-450arom (aromatase) that catalyzes the formation of estrogen from androgen, J. Biol. Chem. 263: 16358-16363 (1988)

96. C.J. Corbin, S. Graham-Lorence, M.McPhaul, J.I. Mason, C.R. Mendelson and E.R. Simpson, Isolation of a full-lenght cDNA insert encoding human aromatse system cytochrome P-450 and its expression in nonsteroidogenic cells, Proc. Natl. Acad. Sci. USA 85: 8948-8952 (1988)

97. D. Pompon, R.Y.-K. Liu, M.J. Besman, P.-L. Wang, J.E. Shively and S. Chen, Expression of human placental aromatase in Saccharomyces cerevisiae, Mol. Endo. 3:1477-1487 (1989)

98. C.F. Wilkinson, Role of mixed-function oxidase in insect growth and development, in: "Bioregulators for pest control", P.A. Hedin, ed., American Chemical Society, Washington, D.C.(1985)

99. J. Srivatsan, T. Kuwahara and M. Agosin, The effect of alpha-ecdysone and phenobarbital on the alpha-ecdysone 20-monoooxyenase of house fly larva. Biochem. Biophys. Res. Commun. 148: 1075-1080 (1987)

100. F.A. Fitzpatrick and R.C. Murphy, Cytochrome P-450 metabolism of arachidonic acids: formation and biological actions of "epoxy-genase"-derived eicosanoids, Pharmacol. Rev. 40: 229-241 (1989)

101. J. Capdevilla, P. Yadagiri, S. Manna and J.R. Falck, Absolute confi-guration of the hydroxyeicosatetraenoic acids (HETEs) formed during catalytic oxygenation of arachidonic acid by microsomal cytochrome P-450, Biochem. Biophys. Res. Commun. 141: 1007-1011 (1986)

102. S. Hammarström, M. Hamberg, B. Samuelson, E.A. Duell, M. Stawiski and J.J. Voorhees, Increased concentrations of nonesterified arachidonic acid, 12 L-hydroxy-5,8,10,14-eicosatetraenoic acid,

prostaglandin E_2, and prostaglandin $F_{2\alpha}$ in epidermis of psoriasis, Proc. Natl. Acad. Sci. 72: 5130-5134 (1975)

103. P.M. Woollard, Stereochemical difference between 12-hydroxy-5,8,10,14-eicosatetraenoic acid in platelets and psoriatic lesions, Biochem. Biophys. Res. Commun. 136: 169-175 (1986)

104. M.L. Schwartzman, M. Balzy, J. Masferrer, N.G. Abraham, J.C. McGiff and R.C. Murphy, 12(R)-hydroxyicosatetraenoic acid: a cytochrome P-450-dependent arachidonate metabolite that inhibits Na^+, K^+-ATPase in the cornea, Proc. Natl. Acad. Sci. 84: 8125-8129(1987)

105. R,C. Murphy, J.R. Falck, S. Lumin, P. Yadagiri, J.A. Zirrolli, M. Balazy, J.L. Masferrer, N.G. Abraham and M.L. Schwartzman, 12(R)-hydroxyeicosatrienoic acid: a vasodilator cytochrome P-450-dependent arachidonate metabolite from the bovine corneal epithelium, J. Biol. Chem. 263: 17197-17202(1988)

106. V. Ullrich and H. Graf, Prostacyclin and thromboxane synthase as P-450 enzymes, TIPS 5:352-355 (1984)

107. M. Hecker, W.J. Baader, P. Weber and V. Ullrich, Thromboxane synthase catalyses hydroxylations of prostaglandin H2 analogs in the presence of iodosylbenzene, Eur. J. Biochem. 169: 563-569 (1987)

108. M. Haurand and V. Ullrich, Isolation and characterization of thromboxane synthase from human platelets as a cytochrome P-450 enzyme, J. Biol. Chem. 260: 15059-15067 (1985)

109. M. Hecker and V. Ullrich, On the mechanism of prostacyclin and thromboxane A2 biosynthesis, J. Biol. Chem. 264: 441-150 (1989)

110. M.C. Romano, R.D. Eckardt, P.E. Bender, T.B. Leonard, K.M. Straub and J.F. Newton, Biochemical characterization of hepatic microsomal leukotriene B4 hydroxylases, J. Biol. Chem. 262: 1590-1595(1987)

111. A. Hatzelmann and V. Ullrich, The ω-hydroxylation of arachidonic acid by human polymorphonuclear leukocytes, Eur. J. Biochem. 173: 445-452(1988)

112. R.J. Soberman, J.P. Sutyak, R.T. Okita, D.F. Wendelborn, L.J. Roberts and Austen, The identification and formation of 20-aldehyde leukotriene B4, J. Biol. Chem. 263: 7996-8002(1988)

113. H.F. DeLuca and H. Schnoes, Vitamin D: recent advances, Ann. Rev. Biochem. 52: 411-439 (1983)

114. I. Björkhem and I. Holmberg, Assay and properties of a mitochondrial 25-hydroxylase active on vitamin D3, J. Biol. Chem. 253: 842-849 (1978)

115. I. Björkhem and I. Holmberg,On the 25-hydroxylation of vitamin D3 in vitro studied with a mass fragmentographic technique, J. Biol. Chem. 254: 9518-9524 (1979)

116. S. Andersson, I. Holmberg and K. Wikvall, 25-Hydroxylation of C_{27}-steroids and vitamin D3 by a constitutive cytochrome P-450 from rat liver microsomes, J. Biol. Chem. 258: 6777-6781 (1983)

117. S.-I. Hayashi, M. Noshiro and K. Okuda, Purification of cytochrome P-450 catalyzing 25-hydroxylation of vitamin D3 from rat liver microsomes, Biochem. Biophys. Res. Commun. 121: 994-1000 (1984)

118. S.-I. Hayashi, T. Omura, T. Watanabe and K. Okuda, Immunochemical evidence for the catalysis of vitamin D3 25-hydroxylation and testosterone 16α-hydroxylation by homologous forms of cytochrome P-450 in rat liver microsomes, J. Biochem. 103: 853-857 (1988)

119. S.-I. Hayashi, K.-I. Morohashi, H. Yoshioka, K. Okuda and T. Omura, Expression of a rat liver microsomal cytochrome P-450 catalyzing testosterone 16α-hydroxylation in Saccharomyces cerevisiae: vitamin D3 25-hydroxylase and testosterone 16α-hydroxylase are distinct forms of cytochrome P-450, J. Biochem. 103: 858-862 (1988)

120. S. Andersson and H. Jörnvall, Sex differences in cytochrome P-450-dependent 25-hydroxylation of C27-steroids in rat liver microsomes, J. Biol. Chem. 261: 16932-16936 (1986)

328

121. K. Saarem and J.I. Pedersen, 25-Hydroxylation of 1α-hydroxyvitamin
     D-3 in rat and human liver, Biochim. Biophys. Acta 840: 117-126
     (1985)
122. K. Saarem, S. Bergseth, H. Oftebro and J. I. Pedersen, Subcellular
     localization of vitamin D3 25-hydroxylase in human liver, J. Biol.
     Chem. 259: 10936-10940 (1984)
123. O. Masumoto, Y. Ohyama and K. Okuda, Purification and character-
     ization of vitamin D 25-hydroxylase from rat liver mitochondria,
     J. Biol. Chem. 263: 14256-14260 (1988)
124. K. Saarem and J.I. Pedersen, Sex differences in the hydroxylation of
     cholecalciferol and of 5ß-cholestane-3α,7α,12α-triol in rat liver,
     Biochem. J. 247: 73-78 (1987)
125. R.T. Turner, J.E. Puzas, M.D. Forte, G.E.Lester, T.K. Gray, G.A.
     Howard and D.J. Baylink, In vitro synthesis of 1,25-
     dihydroxychole-calciferol and 24,25-dihydroxycholecalciferol by
     isolated carvarial cells, Proc. Natl. Acad. Sci. USA, 77: 5720-
     5724 (1980)
126. G.A. Howard, T. Russell, R.T. Turner, D.J. Sherrard and D.J.
     Baylink, Human bone cells in culture metabolize 25-hydroxyvitamin
     D3 to 1,25-dihydroxyvitamin D3 and 24,25-dihydroxyvitamin D3, J.
     Biol. Chem. 256: 7738-7740 (1981)
127. T.L. Frankel, R.S. Mason, P. Hersey, E. Murray and S. Posen, The
     synthesis of vitamin D metabolites by human melanoma cells, J.
     Clin. Endocrinol. Metab. 57: 627-631 (1983)
128. G.L. Barbour, J.W. Coburn, E. Slatopolsky, A. W. Norman and R.L.
     Horst, Hypercalcemia in an anephric patient with sarcoidosis:
     evidence for extrarenal generation of 1,25-dihydroxyvitamin D, N.
     Engl. J. Med. 305: 440-443 (1981)
129. J.A. Whitsett, M. Ho, R.C. Tsang, E.J. Norman and K.G. Adams,
     Synthesis of 1,25-dihydroxyvitamin D3 by human placenta in vitro,
     J. Clin. Endicrinol. Metab. 53: 484-488 (1979)
130. Y. Weisman, A. Harell, S. Edelstein, M. David, Z. Spirer and A.
     Golander, 1α,25 Dihydroxyvitamin D3 and 24,25 dihydroxyvitamin D3
     in vitro synthesis by human decidia and placenta, Nature 281: 317-
     319 (1979)
131. D.D. Bikle, M.K. Nemanic, J.O. Whitney and P.W. Elias, Neonatal
     human foreskin keratinocytes produce 1,25-dihydroxyvitamin D3,
     Biochemistry 25: 1545-1548 (1986)
132. D.D. Bikle, M.K. Nemanic, E. Gee and P. Elias, 1,25-Dihydroxyvitamin
     D3 production by human keratinocytes: kinetics and regulation, J.
     Clin. Invest. 78: 557-566 (1986)
133. S. Pillai, D.D. Bikle and P.M. Elias, Vitamin D and epidermal
     differentiation: evidence for a role of endogenously produced
     vitamin D metabolites in keratinocyte differtiation, Skin
     Pharmacol. 1: 149-160 (1988)
134. J.E. Graebe, Gibberellin biosynthesis and control, Ann Rev Physiol
     38: 419-465 (1987)
135. I. Björkhem, Mechanism of bile acid biosynthesis in mammalian liver,
     in: "Sterols and bile acids" H. Danielson and J. Sjövall, eds.,
     Elsevier, Amsterdam, (1985)
136. H. Danielsson and K. Wikvall, Evidence for a specific cytochrome P-
     450 with short half-life catalyzing 7α-hydroxylation of
     cholesterol, Biochem. Biophys. Res Commun. 103: 46-51 (1981)
137. W.R. Rush and R. Fears, Requirements for cytochrome P-450: evidence
     from metabolic studies and the response to xenobiotics, in:
     "Cholesterol 7α-hydroxylase (7α-mono-oxygenase)" R. Fears and J.R.
     Sabine eds., CRC PRESS,INC.,Boca Raton, Florida (1986)
138. R. Hansson and K. Wikvall, Hydroxylations in biosynthesis of bile
     acids. Cytochrome P-450 LM4 and 12α-hydroxylation of 5ß-
     cholestane-3α,7α-diol, Eur. J. Chem. 125: 423-429 (1982)

139. M. Noshiro, M. Nishimoto, K.-I. Morohashi and K. Okuda, Molecular cloning of cDNA for cholesterol 7α-hydroxylase from rat liver microsomes. Nucleotide sequence and expression, FEBS Lett. 257: 97-100(1989)

140. J. Kwekkeboom, E.M. van Voorthuizen, H.M.G. Princen and H.J. Kempen, Feedback inhibition of bile acid synthesis in cultured pig hepatocytes, Biochem. Biophys. Res. Commun., 155: 850-856 (1988)

141. K. Wikvall, Hydroxylations in biosynthesis of bile acids. Isolation of a cytochrome P-450 from rabbit liver mitochondria catalyzing 26-hydroxylation of C27-steroids, J. Biol. Chem. 259: 3800-3804(1984)

142. H. Dahlbäck, Characterization of the liver mitochondrial cytochrome P-450 catalyzing the 26-hydroxylation of 5ß-cholestane-3α,7α,12α-triol, Biochem. Biophys. Res. Commun. 157: 30-36(1988)

143. S. Skrede, I. Björkhem, A.A. Kvittingen and M.S. Buchmann, Demonstration of 26-hydroxylation of C27-steroids in human skin fibroblasts, and a deficiency of this activity in cerebrotendinous xanthomatosis, J. Clin. Invest. 78: 729-735 (1986)

144. F.R. Taylor, S.E. Saucier, E.P. Shown, E.J. Parish and A.A. Kandutsch, Correlation between oxysterol binding to a cytosolic binding protein in the repression of hydroxymethylglutaryl coenzyme A reductase, J. Biol. Chem. 259: 12382-12387(1984)

145. A.B. Roberts and H.F. DeLuca, Pathways of retinoic acid metabolism in the rat, Biochem. J. 102: 600-605(1967)

146. M.A. Leo, S. Iida and C.S. Lieber, Retinoic acid by a system reconstituted with cytochrome P-450, Arch. Biochem. Biophys. 234: 305-311(1984)

147. M.A. Leo and C.S. Lieber, New pathway for retinol metabolism in liver micrisomes, J.Biol. Chem. 260: 5228-5231(1985)

148. A.B. Roberts, Microsomal oxidation of retinoic acid in hamster liver, intestine, and testis, Ann. N.Y. Acad. Sci. 359: 45-53(1981)

149. A.B. Roberts, M.D. Nichols, D,L. Newton and M.B. Sporn, In vitro metabolism of retinoic acid in hamster intestine and liver, J. Biol. Chem. 254: 6296-6302(1979)

150. A.B. Roberts, C.A. Frolik, M.D. Nichols and M.B. Sporn, Retinoid-dependent induction of the in vivo and in vitro metabolism of retinoic acid in tissues of the vitamin A-deficient hamster, J. Biol. Chem. 254: 6303-6309(1979)

151. C.A. Frolik, In vitro and in vivo metabolism of all-trans and 13-cis-retinoic acid in the hamster, Ann. N.Y. Acad. Sc. 359: 37-44(1981)

INTERACTIONS OF ESTROGENIC PESTICIDES WITH CYTOCHROME P450

David Kupfer

Worcester Foundation for Experimental Biology
222 Maple Avenue
Shrewsbury, MA 01545

This presentation examines a variety of questions concerning the interactions of chlorinated hydrocarbon pesticides with cytochrome P450 monooxygenases. In particular, it intends to discuss those pesticides which exhibit estrogenic activity in mammals. An important question concerns whether the estrogenic activity of these pesticides resides in the parent compound or whether estrogenicity of these compounds is primarily due to their respective metabolites. In the latter case, it is anticipated that the alteration in the activity of the enzymes catalyzing the formation of these metabolites would have a pronounced effect on the estrogenic activity of the parent compound. Hence it would be important to identify the enzymes (P450s) catalyzing their formation. These aspects will be examined in some detail.

The estrogenic activity of a given compound is usually detected *in vivo* by a variety of 'estrogenic endpoints', the most common of which is an increase in uterine weight. In attempting to determine whether a compound is an estrogen or a proestrogen, we realized that it would be extremely difficult to obtain unequivocal resolution of that question from *in vivo* studies. The reason is because it is impossible to completely inhibit the *in vivo* metabolism of a given compound by administering a specific inhibitor of the enzyme(s) catalyzing that metabolism, without introducing some degree of toxicity; also, in such experiments we found that reproducibility of results was not consistent. Therefore, we developed an *in vitro* method to determine whether a compound is a proestrogen or an estrogen. This method is described below.

Additionally, we observed that certain chlorinated hydrocarbon pesticides can undergo metabolic activation *in vitro*, yielding reactive intermediates which bind irreversibly, apparently covalently, to microsomal proteins. This suggested that the metabolism of these compounds might yield toxic manifestation and should be a cause of concern. Also, we obtained evidence that the reactive intermediates may bind to certain cytochrome P450 isozymes resulting in inactivation of their catalytic activities. Lastly, the effects of metabolism on the interactions of these pesticides with the estrogen receptor and the possible biological consequences of such interactions, will be discussed.

Molecular Aspects of Monooxygenases and Bioactivation of Toxic Compounds
Edited by E. Arınç *et al., Plenum Press, New York,*

Discovery of the estrogenic activity of chlorinated hydrocarbon pesticides.

It has been known for about 40 years that DDT exhibits estrogenic activity in animals. In 1950, Burlington and Lindeman demonstrated estrogenic activity of DDT in roosters[1]. These observations were not always confirmed in subsequent studies by other investigators[2]. Presumably because of those reasons, these findings lay dormant for a number of years until in 1968-1971 Welch, Levin and Conney (USA) and Bitman and Cecil (USA) independently demonstrated that o,p'DDT (usually present as approximately 15% contaminant of DDT preparations) is the major estrogenic component in commercial DDT preparation[3-5] (FIGURE 1 structures). Based on these findings, it appears conceivable that the lack of observable estrogenic activity of DDT, in some of the earlier studies, may have been due to the use of purer DDT preparations containing much lower levels of o,p'DDT.

In subsequent studies, Conney's group proposed that o,p'DDT is not an estrogen per se, but merely is a **proestrogen** and that its metabolite(s) are estrogenic. This conclusion was based on their observation that the administration to female rats of carbon

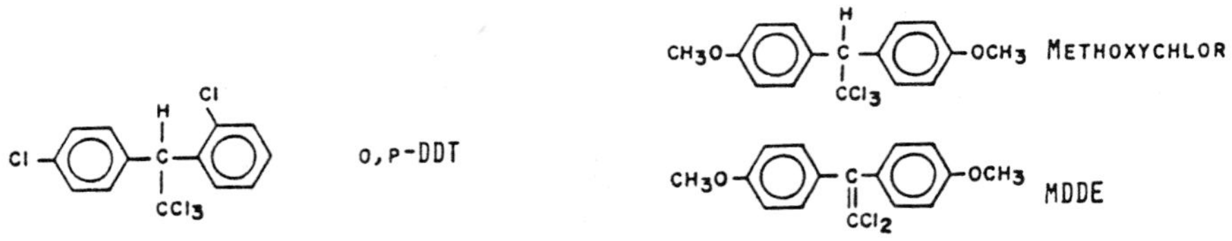

FIGURE 1. Structures of DDT derivatives and homologs

tetrachloride, a known inactivator of hepatic P450 monooxygenases, inhibited the estrogenic activity of o,p'DDT but enhanced that activity of administered estradiol[5]. By contrast, studies by other investigators indicate that o,p'DDT is active per se. For instance, several groups demonstrated that o,p'DDT binds to the estrogen receptor in rat uterine cytosol and human breast and uterine tumors[6-9] and that o,p'DDT exhibits estrogenic activity *in vitro* by causing the elevation of the so called *induced protein*[10]. Also, it has been observed that the phenolic metabolites of o,p'DDT are equipotent or only slightly more potent estrogens than o,p'DDT[40]. Lastly, other investigators were unable to demonstrate that carbon tetrachloride inhibits the estrogenic activity of o,p'DDT and potentiates the estrogenic activity of estradiol[11,12]. Thus the weight of the evidence indicates that o,p'DDT is an estrogen per se. Nevertheless, it is conceivable that o,p'DDT is also a 'proestrogen' and that certain of its hitherto unidentified metabolites are more potent estrogens than o,p'DDT. If the latter proves affirmative, it might explain the reason for the contrasting conclusions reached in the different laboratories concerning whether o,p'DDT is an estrogen or a proestrogen.

The above mentioned uncertainty with respect to whether o,p'DDT is an estrogen per se, prompted us to design experiments to try to resolve that question. Also, we attempted to determine whether estrogenic "biodegradable" pesticides, currently employed to substitute for DDT which has been banned in the industrially developed countries, are estrogens or proestrogens. Methoxychlor is an example of such a biodegradable pesticide which demonstrates estrogenic activity *in vivo* and the question we asked was whether methoxychlor is an estrogen or a proestrogen.

Determination of whether chlorinated hydrocarbons and related compounds are estrogens or proestrogens

To examine that question, several approaches were used:

    a)   Examination of the effects of metabolism by hepatic monooxygenase on estrogenic activity of these and related compounds.

    b)   Determination of whether contaminants are estrogenic or proestrogenic?

To develop the above procedures, we primarily utilized methoxychlor and its estrogenic contaminants [MDDE and chlorotrianisene (TACE)] as model compounds:

Metabolism of methoxychlor by hepatic microsomal monooxygenase

Earlier observations demonstrated that methoxychlor is metabolized by sequential demethylation, yielding mono-demethylated (mono-OH-M) and bis-demethylated (bis-OH-M)[13-15]. Most recently, we demonstrated that bis-OH-M is further hydroxylated to a trihydroxy metabolite (tris-OH-M), which is a catechol[16] (**FIGURE 2**). The identification of the metabolites of methoxychlor was achieved by gas chromatography/mass spectrometry and by HPLC/mass spectrometry. Further, the identification of the tris-OH-M as a catechol was achieved by a relatively novel technique referred to as hydrodynamic voltammetry[17].

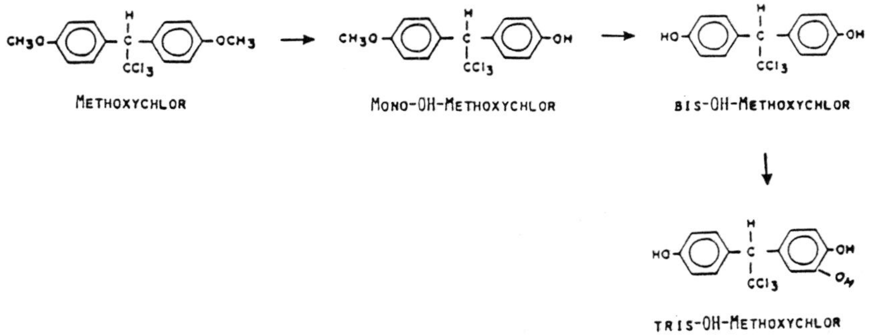

FIGURE 2. Metabolism of methoxychlor by rat liver microsomes

Additionally, we observed that methoxychlor undergoes metabolic activation by the liver microsomal monooxygenase yielding a reactive intermediate which binds covalently to microsomal proteins. These studies will be further discussed under the section called "Covalent Binding"

Metabolism of Chlorotrianisene (TACE) by the hepatic microsomal monooxygenase

Our interest in the metabolism of TACE was initiated when we realized that TACE (**FIGURE 3**) is an estrogenic contaminant of methoxychlor[18]. Interestingly, TACE has been used as a long acting therapeutic estrogen for the treatment of postpartum breast engorgement and for prostatic cancer.

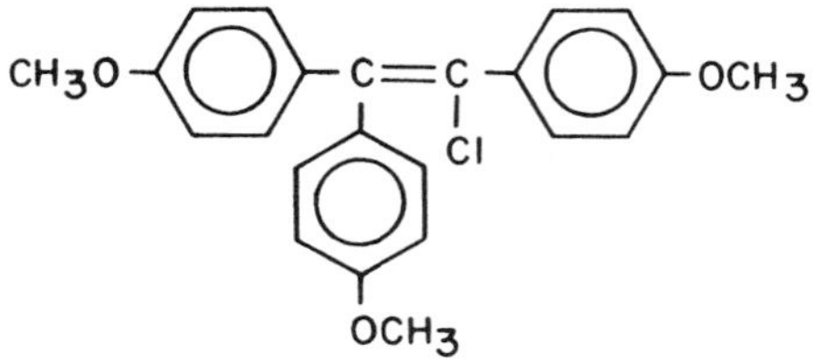

FIGURE 3.   Chlorotrianisene (TACE)

It has been proposed that TACE is a proestrogen. However, most recent studies indicate that TACE is also an **antiestrogen**[19]. In that regard it resembles other triphenylethylene antiestrogens, like tamoxifen, in exhibiting both partial agonistic and antagonistic estrogenic effects. This raised the question whether the antiestrogenic activity of TACE is due to its metabolites; if so, TACE would be classified as a **proantiestrogen**.

The metabolism of TACE *in vitro* was studied by Reunitz and his colleagues[20,32]. Using rabbit and rat liver microsomes, these investigators demonstrated that TACE, like methoxychlor, undergoes demethylation to the phenolic monodemethylated (mono-OH-T) and bis-demethylated TACE (bis-OH-M). It is not known, however, whether a tridemethylated TACE (tris-OH-T) is also formed.

Additionally, we found that TACE undergoes metabolic activation and the reactive intermediate binds covalently to microsomal proteins[21]. A more extensive discussion of this phenomenon is presented below under Covalent Binding.

Metabolism of MDDE, a methoxychlor contaminant, by the hepatic microsomal monooxygenase

MDDE (FIGURE 1) is both a metabolite and a contaminant of methoxychlor[18]. We observed that MDDE is metabolized by the hepatic monooxygenase into mono- and bis-demethylated products[22,23]. Also, MDDE undergoes metabolic activation yielding covalent binding to microsomal proteins (see below under Covalent Binding).

METHOD TO DETERMINE WHETHER A COMPOUND IS A PROESTROGEN

It is generally accepted that for a given compound to elicit estrogenic activity, the compound has to bind to the estrogen receptor (ER) in the target tissues. In turn, this binding exhibits a redistribution of the ER between the nuclear and cytosolic compartments, yielding an increase in the nuclear estrogen receptor (ERn) and a decrease in the cytosolic receptor (ERc). This phenomenon occurs both *in vivo* within a brief period (1-3 hrs) after the administration of the estrogen and *in vitro* following the incubation of the compound with isolated uteri. Recent evidence, however, indicates that the presence of the estrogen receptor in the cytosol and the estrogen mediated "translocation" of the receptor from the cytosol to the nucleus are an artifact of homogenization of the tissues and represents a misinterpretation of the actual mechanism involved[24,25]. Namely, the preponderance of the evidence indicates that the receptor normally resides in the cell nucleus bound to chromatin DNA. Also, it appears that in its unliganded form, in the absence of estrogen, the

334

receptor exhibits low affinity for the nuclear chromatin. Thus, during homogenization, the unliganded receptor is released from the nuclei into the cytosol. By contrast, in the presence of the ligand, the nuclear receptor acquires high affinity for the DNA and hence homogenization does not cause its dissociation from the chromatin. The high affinity of the receptor for the chromatin is thought to be generated by the estrogen, causing the release of a 90 KDa heat-shock protein (hsp90)[26,27]. The loss of hsp90 apparently reveals a high affinity binding site for DNA.

Nevertheless, independent of the actual mechanism of the estrogen-mediated receptor dynamics, we utilized this characteristic of the estrogen receptor in order to develop an *in vitro* procedure for examining whether a compound is a proestrogen[22,28]. This procedure monitors both the metabolism of the compound and its subsequent "estrogenic activity"; i.e., effects on the subcellular localization of the receptor. Briefly, this assay combines in the same vessel a metabolizing system (liver microsomal monooxygenase) and an estrogen detecting system (uterine ER). A compound is considered to be a proestrogen if during incubation with uteri it requires an <u>active</u> microsomal monooxygenase (i.e., presence of NADPH) for eliciting an increase in uterine ERn and a decrease in ERc. By contrast, a compound is an estrogen per se, if an increase in ERn occurs in the absence of an active monooxygenase ,i.e., presence of liver microsomes <u>without</u> NADPH or when <u>heat-inactivated</u> microsomes are employed[22,28].

Are o,p'DDT, methoxychlor or MDDE proestrogens?

The classical assay using binding to the uterine cytosolic ERc (competition with [^{3}H]-estradiol), demonstrated that whereas o,p'DDT exhibits significant binding, methoxychlor does not. In fact, incubation of o,p'DDT with rat uteri, in the absence of liver microsomes, elevated ERn and diminished ERc. However, methoxychlor did not exhibit that activity (**TABLE 1**). This suggested that o,p'DDT has intrinsic estrogenic activity and that methoxychlor has no such estrogenic activity[12]. By contrast, using the above *in vitro*

TABLE 1   The effect of incubation of o,p'DDT and methoxychlor with rat uteri (liver microsomes absent) on the distribution of the uterine estrogen receptor in cytosol (ERc) and nuclei (ERn).

Compound[a]	fmol of /uterus[b]		
	ERc	ERn	ERc + ERn
o,p'DDT	244[*]	303[**]	547
--	377	197	574
Methoxychlor	271	177	448
--	241	169	410

a   10 μM of o,p'DDT or methoxychlor; -- (control) indicates compounds were left out.
b ^{3}H-E$_2$ bound in an exchange assay.
* $p \leq 0.05$;   ** $p \leq 0.001$

method, we established that methoxychlor is a proestrogen; namely, there was a requirement for active liver monooxygenase for methoxychlor to cause a decrease in ERc and an increase in ERn (TABLE 2). However, o,p'DDT behaved like an estrogen per se; i.e., o,p'DDT produced an

TABLE 2   EFfect of incubation of methoxychlor or MDDE with rat uteri and rat hepatic microsomes supplemented with NADPH on the distribution of uterine ER.

Compound, 2 μM	Microsomes[a]	fmol/uterus[b]		
		ERc	ERn	ERc + ERn
Methoxychlor	Inactive	769	83	852
Methoxychlor	Active	559*	309*	868
MDDE	Inactive	960	111	1071
MDDE	Active	300*	492*	792

a   Active = microsomes plus NADPH; inactive = microsomes boiled for 10 min plus NADPH; similar results were obtained with active microsomes lacking NADPH.
b ^{3}H-E$_2$ bound in an exchange assay.
* $p \leq 0.05$

increase in ERn in the absence of active microsomal monooxygenase and the addition of active monooxygenase did not enhance significantly this effect (not shown). As expected from the above findings, mono- and bis-hydroxy metabolites of methoxychlor were found to exhibit estrogenic activity per se (TABLE 3). MDDE was found to be a proestrogen and

TABLE 3   Effect of incubation of demethylated metabolites of methoxychlor and MDDE with rat uteri in the <u>absence</u> of liver microsomes on ERc and ERn.

Compound	fmol/uterus		
	ERc	ERn	ERc + ERn
Mono-OH-M (2.0 μM)	233*	424*	657
-	702	68	770
Bis-OH-M (0.2 μM)	386*	236*	622
-	585	67	655
Mono-OH-MDDE (2.0 μM)	477*	460*	937
-	809	78	887
Bis-OH-MDDE (0.2 μM)	393*	413*	806
-	839	90	929

* $p \leq 0.05$

From Bulger and Kupfer[22]

its demethylated metabolites were shown to be estrogenic. Further, we demonstrated that whereas MDDE is not estrogenic its mono- and bis-demethylated phenolic products (mono-OH-MDDE and bis-OH-MDDE, respectively) are strongly estrogenic (**TABLE 3**)[22,23]. This indicates that MDDE is a proestrogen.

COVALENT BINDING OF METHOXYCHLOR AND OF ITS CONTAMINANTS AND METABOLITES

a) <u>Methoxychlor</u>. In addition to demethylation and hydroxylation, methoxychlor undergoes metabolic activation by the hepatic P450 monooxygenase, yielding covalent binding to microsomal proteins[29,30]. Treatment of rats with phenobarbital increases the covalent binding of methoxychlor by about 10 fold (**TABLE 4**). Antibodies raised against

TABLE 4   Effect of inducers of P450 on the covalent binding of methoxychlor equivalents to rat liver microsomal proteins.

	pmoles bound/min/mg protein	
Treatment		Treated/Control
Phenobarbital	271 ± 21^a	10
Control (H_2O)	27 ± 5	
Methylcholanthrene	26 ± 4	1
Control (corn oil)	31 ± 2	

Values are mean $\pm$ S.E.

^{a} $p \leq 0.001$

the major P450 induced by phenobarbital (i.e., P450b) strongly inhibited covalent binding, indicating that P450b (P450IIB1) and/or P450e (450IIB2) are involved in catalyzing the formation of the reactive intermediate (**TABLE 5**). The question was raised whether the formation of the reactive intermediate necessitates demethylation or whether the pathway of covalent binding involves a different pathway. This question is discussed below under "<u>Structure of the Reactive Intermediates of Methoxychlor</u>".

b) <u>MDDE</u>. We observed that MDDE, like methoxychlor, undergoes a monooxygenase-mediated activation, yielding covalent binding[23,35].

c) <u>TACE (chlorotrianisene)</u>. TACE is metabolized by demethylation to the mono- and bis-demethylated derivatives[20,32]. We demonstrated that TACE undergoes metabolic activation, however by contrast to methoxychlor, phenobarbital does not stimulate its covalent binding. Surprisingly, methylcholanthrene (MC) stimulates about 60 fold the covalent binding of TACE (**TABLE 6**). MC does not appear to enhance demethylation[21], suggesting that demethylation is not essential for covalent binding. It appears that the major portion of the covalent binding of TACE in MC microsomes is catalyzed by P450c (IA1), but not by P450d (IA2). The evidence is based on inhibition of covalent binding by antibodies to P450c/d, but not by antibodies to P450d, and that reconstituted P450c catalyzes effectively covalent binding of TACE (**TABLE 7,8**)[31].

TABLE 5   Effect of monoclonal antibodies (MAb) against P450b on covalent binding of methoxychlor equivalents to liver microsomal proteins.

Treatment of Rats[c]	MAb[a]	Covalent Binding (% of control = no MAb)
Phenobarbital	-	100
Phenobarbital	+[b]	150
Phenobarbital	+	19
--	-	100
--	+[b]	68
--	+	90

[a]   Anti-P450b at a ratio of MAb/microsomes of approximately 2.5:1. MAb were kindly provided by Drs. Gelboin and Park (NIH, Bethesda, MD). - indicates no MAb.

[b]   Equivalent amount of nonspecific MAb instead of anti-P450b.

[c]   Phenobarbital (37.5 mg/kg) was given i.p. twice daily for 4 days;-- indicates rats received a similar regimen, albeit only water.

TABLE 6   Effect of inducers of P450s on the covalent binding of chlorotrianisene (TACE) equivalents to rat liver microsomal proteins.

Treatment	pmoles bound/min/mg protein	Treated/Control
Phenobarbital	17 ± 2	1
Control (H_2O)	15 ± 2	
Methylcholanthrene	$877 \pm 34^{*}$	60
Control (corn oil)	14 ± 2	

Values are mean $\pm$ S.E.

* $p \leq 0.001$

Data abstracted from Juedes et al.[21].

STRUCTURE OF REACTIVE INTERMEDIATES OF METHOXYCHLOR AND TACE AND OF THEIR PROTEIN ADDUCTS

a) Reactive Intermediate of Methoxychlor (M*).   Indirect evidence suggests that M* is a free radical. The evidence is based on the observations that antioxidants/free radical scavengers and spin trappers, at concentrations which do not inhibit the overall metabolism of methoxychlor to phenolic derivatives, strongly inhibit methoxychlor covalent binding. However, an ESR study did not provide support for M* being a free radical; there was no indication of spin adduct formation between M* and a spin trapper (PBN or TMPO). The lack of spin adduct,

TABLE 7   Effects of antibodies (Ab) to P450c/d and to P450d on the
covalent binding of TACE to liver microsomal proteins from
methylcholanthrene-treated rats.

Ratio (Ab/microsomal proteins)	Ab[a]	Covalent Binding (% of Control)
-	-	100
0.25	Anti-c/d	97
0.50	"	69
1	"	45
5	"	34
0.25	Anti-d	126
0.50	"	126
1	"	118
5	"	90

[a] Anti-P450-c/d (MAb 1-7-1) was provided by Drs. Gelboin and Park
(NIH, Bethesda, MD); anti-P450d was provided by Dr. Paul Thomas
(Rutgers University, NJ).

Data from Juedes and Kupfer[31].

TABLE 8   Covalent binding of TACE in reconstituted cytochrome P450c
system.

	nmol bound/30 min.
Complete system	20.5
Complete minus NADPH	0.4
Complete minus P450c	0 4
Complete minus reductase	0.4

Complete system: 0.1 μM P450c, 0.1 μM NADPH-P450 reductase,
1.0 mg BSA and 2 mM NADPH; ^{3}H-TACE (100 μM).

Data from Juedes and Kupfer[31].

however, could be due to the brief half-life of M* and/or due to
competing rapid reactions between M* and proteins in microsomes.   The
nature of the "active oxygen" participating in activation of
methoxychlor was examined.   Indirect evidence indicates that superoxide
anion is involved: a) The copper complex of diisopropyl salicylate
(DIPS), a superoxide dismutase mimetic agent, strongly inhibited
covalent binding of methoxychlor[33]; however, the zinc DIPS complex,
which has no superoxide dismutase activity, did not inhibit covalent
binding and b) methoxychlor is metabolized in aprotic medium by
superoxide, losing a chlorine[34], indicating that methoxychlor could be
metabolized by superoxide and suggesting the formation of a free radical
intermediate in that reaction.

The nature of adduct formation between M* and proteins was investigated. Our studies suggest that the adduct is formed by the reaction of M* with the the cysteine sulfhydryl. This is based on the observation of a strong inhibition of covalent binding of methoxychlor to proteins by compounds containing free sulfhydryl groups, but not by those containing blocked sulfhydryls[29]. Additionally, we demonstrated that the treatment of the M*-protein adduct with Raney-nickel (known to cleave carbon-sulfur bonds), released a significant portion of the radioactivity derived from M*[30]. Our observation that the incubation of radiolabeled methoxychlor with rat liver microsomes and N-acetylcysteine or cysteine increases several fold the formation of a radiolabeled water soluble product, indicates that M* could react readily with the sulfhydryl containing compounds. The question of whether demethylation of methoxychlor is essential for covalent binding was addressed, with dual-labeled methoxychlor (^3H in the methoxyl and ^{14}C in the ring). Evidence indicates that demethylation is not essential for covalent binding of methoxychlor to microsomal proteins[35]. These findings demonstrate that the primary reactive intermediate is not a catechol, generated by demethylation and hydroxylation. This observation together with the above findings on the involvement of superoxide in generation of M* invites a speculation that M* is a free radical, formed by modification of the side chain by the homolytic cleavage of the C-Cl bond or of the C-H bond. If indeed the latter is the case, the biphenyl methine free radical could be stabilized by resonance involving the para methoxyls. Studies are in progress to identify the adduct formed between M* and N-acetylcysteine or cysteine. These studies should provide information concerning the structure of M*.

b) Reactive intermediate of TACE (T*). Attempts to identify the T* involved a similar approach to that used with methoxychlor. Like M*, T* appears to be a free radical. Also, it seems that T* is not a demethylated product, since treatment of rats with methylcholanthrene (MC), which dramatically increases the covalent binding of TACE, does not elevate significantly the demethylation of TACE. However, because of unavailability of a dual-labeled TACE, the more direct experiments with a dual labeling approach have not been performed. T* appears to react with sulfhydryl groups. Again, this supposition is based on indirect evidence. Among the 21 natural amino acids tested, only cysteine inhibited covalent binding of TACE to proteins (Juedes and Kupfer, unpublished). Similarly N-acetylcysteine and glutathione strongly inhibit covalent binding of TACE. Interestingly, T* appears to be relatively long lived, since it reacts avidly with added albumin[21,31]. This indicates that the reactive intermediate of TACE might be able to survive in transport and react with proteins distant from its site of formation.

DO METHOXYCHLOR METABOLITES REACT WITH AND INACTIVATE P450s?

Indirect evidence suggests that the reactive intermediate of methoxychlor binds covalently to microsomal P450s[30]. The evidence is based on the observation that the incubation of ^{14}C-methoxychlor with liver microsomes and NADPH yields radiolabeled protein bands on SDS-polyacrylamide gel electrophoresis which exhibit Mr values between 45K and 60K. However, there was no decrease in the level of spectrally measured microsomal P450s (CO complex of dithionite reduced P450), indicating that the heme environment crucial for CO complex formation was not significantly altered. More recently, we observed that the 2-hydroxylation of estradiol is inhibited by micromolar concentrations of methoxychlor and that this inhibition was increased by the prior incubation with methoxychlor in a time dependent manner[36]. These findings suggest that methoxychlor is a suicide inactivator, possibly a mechanism-based inhibitor, of certain P450s. However, it is not known yet whether the inhibition of 2-hydroxylation of estradiol by methoxychlor is not due to strong affinity and hence apparently irreversible, noncovalent, binding of methoxychlor metabolites.

DOES TACE METABOLISM AFFECT THE UTERINE ESTROGEN RECEPTOR?

If indeed as previously suggested, TACE is a proestrogen then its incubation with uteri and liver microsomes supplemented by NADPH, would be expected to result in diminished ERc and elevation of ERn. However, surprisingly, the incubation of TACE with rat uteri and NADPH did not elevate ERn, but instead the sum total of the ERc plus ERn was markedly diminished (**TABLE 9**)[37]. These results suggest that the diminished level of receptor (assayed by binding of ^{3}H-estradiol) was due to masking of the estrogen binding site on the receptor, by a TACE metabolite. It is tempting to speculate that the masking of the estrogen binding site was due to covalent binding of T* to the estrogen receptor. Support for such a mechanism could be derived from our observation that T* can diffuse from the microsomes into the media and react avidly with albumin[31], suggesting that T* is relatively stable and could diffuse into the uteri and bind to the ER. However, other interpretations of these results are also possible. Nevertheless, independent of the actual mechanism for this diminished receptor level, it would be of extreme interest to determine whether the receptor is actually inactivated by TACE metabolism and in turn whether this could explain the mechanism of the antiestrogenic action of TACE.

TABLE 9. Effect of incubation of TACE with liver microsomes (with or without NADPH) in the presence of rat uteri on the distribution of estrogen receptor in uterine cytosol (ERc) and nuclei (ERn).

NADPH	fmol/uterus[a]		
	ERc	ERn	Σ ERc + ERn
-	610 ± 43	96 ± 11	707
+	$204 \pm 45^*$	138 ± 7	342^*

Each incubation contained TACE (0.2 μM) and rat liver microsomes (1 mg protein) and uterine strips from 6 rats.
[a] ^{3}H-E_2 bound in an exchange assay.
* $p \leq 0.05$.

REMAINING QUESTIONS AND CONCLUSIONS

It would be important to establish whether the formation of protein adducts with the estrogenic pesticides causes cellular or organ toxicity. The structures of the reactive intermediates of methoxychlor and TACE require identification. Further the adducts which appear to be formed with sulfhydryl containing compounds (glutathione and cysteine) and with proteins should be characterized and their potential cellular toxicity examined.

It appears that with each of the compounds which we examined there are two main pathways of metabolism: one leading to demethylated estrogenic/antiestrogenic derivatives and the other leading to reactive intermediates and covalent binding (**FIGURE 4**). It would be important to establish where the two pathways diverge, which P450s catalyze the two pathways and how each pathway is regulated. Such knowledge will permit the determination of how the 'decision making' process permits the selection of one versus the other pathway.

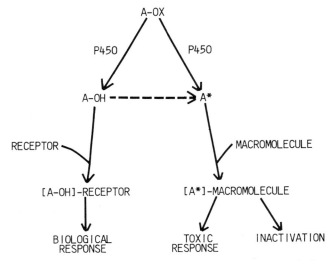

"DECISION MAKING" IN METABOLIC ROUTING

FIGURE 4 A-OX represents a substrate with a methoxyl (e.g., methoxychlor or TACE). A* = Reactive intermediate. Macromolecule = protein or DNA. Receptor = Estrogen Receptor, in the case of methoxychlor and TACE.

An important aspect of our findings is the observation that TACE metabolism causes a decrease in assayable levels of the estrogen receptor. Currently, it is not known whether this is due to the masking of the estrogen binding site on the receptor or due to diminished level of intact receptor. The question needing resolution is the determination of the nature of the change in the estrogen receptor and whether this change represents activation or deactivation of the receptor. If indeed this change represents inactivation of the receptor by covalent binding of the reactive intermediate of TACE, then this finding would constitute a breakthrough in our understanding of the mechanism of antiestrogen action of TACE and possibly that of other triphenylethylene antiestrogens.

The finding that a variety of environmental compounds interact with the estrogen receptor and that metabolism controls that interaction, is important. Animals and probably humans could be susceptible to detrimental effects of exposure to estrogens, particularly during the early periods of development. The promiscuity of the estrogen receptor towards unusual structures ,such as chlordecone (kepone)[38,39], is puzzling. One possible explanation is that the receptor can undergo an induced fit and hence accommodate highly diverse structures. Currently it is not possible to predict which compounds will be estrogenic. Similarly, there is no available analysis, based on strictly structural considerations, which could indicate with high degree of certainty whether a given pollutant will be an estrogen or a proestrogen. Therefore, to my knowledge, our assay is the only test currently available for resolution of such questions. However, refinement of that assay is needed.

It has been established that environmental factors produce changes in the composition of hepatic and extra-hepatic isozymes of cytochrome P450 in the human population and in the animal kingdom . The induction and repression of certain isozymes of P450 are known to be subject to environmental influences. It is likely that certain environmental influences could elicit the formation of novel P450s which have not been hitherto uncovered and which could produce qualitative and quantitative changes in metabolic routing of the exogenously absorbed environmental compounds and of endogenous substances. Such changes could dictate the metabolic routing and consequently the biological consequences of potentially harmful environmental pollutants. These considerations indicate that further studies should be encouraged to shed light on the general subject of interrelationships of environmental exposure with metabolic influences and how these factors interact.

REFERENCES

1. H. Burlington, and V. F. Lindeman, Proc. Soc. Exp. Biol. Med. 74:48 (1950).
2. A. L. Fisher, H. H. Keasling, and F. W. Schueler, Proc. Soc. Exp. Biol. Med. 81:439 (1952).
3. W. Levin, R. M. Welch, and A. H. Conney, Fed. Proc., 27:649 (1968).
4. J. Bitman, H. C. Cecil, S. J. Harris, and G. F. Fries, Science 162:371 (1968).
5. R. M. Welch, W. Levin, and A. H. Conney, Toxicol. Appl. Pharmacol. 4:358 (1969).
6. J. A. Nelson, Biochem. Pharmacol. 23:447 (1974).
7. M. S. Forster, E. L. Wilder, and W. L. Heinrichs, Biochem. Pharmacol. 24:1777 (1975).
8. D. Kupfer, and W. H. Bulger, Pest. Biochem. Physiol. 6:561 (1976).
9. D. Kupfer, and W. H. Bulger, Res. Commun. Chem. Pathol. Pharmacol. 16:451 (1977).
10. G. M. Stancel, J. S. Ireland, V. R. Mukku, and A. K. Robison, Life Sci. 27:1111 (1980).
11. J. A. Nelson, R. F. Struck, and R. James, J. Toxic. Environ. Hlth. 4:325 (1978).
12. D. Kupfer, and W. H. Bulger in: Estrogens in the Environment (J. A. McLachlan, ed), Elsevier/North Holland, New York, pp. 239-263 (1980).
13. I. P. Kapoor, R. L. Metcalf, R. F. Nystrom, and G. K. Sangha, J. Agric. Food Chem. 18:1145 (1970).
14. J. A. Nelson, R. F. Struck, and R. James, Pharmacologist 18:247 (1976).

15. W. H. Bulger, R. M. Muccitelli, and D. Kupfer, Biochem. Pharmacol. 27:2417 (1978).
16. D. Kupfer, W. H. Bulger, and A. D. Theoharides, Res. Chem. Toxicol. (in press).
17. M. Hamilton, and P. T. Kissinger, Drug Metab. Dispos. 14:5 (1986).
18. C-J. H. Lamoureux, and V. J. Feil, J. Assoc. Off. Anal. Chem. 63:1007 (1980).
19. C. A. Powers, M. A. Hatala, and P. J. Pagano, Mol. Cell. Endocrinol. 66:93 (1989).
20. P.C. Reunitz, and M. M. Toledo, Biochem. Pharmacol. 30:2203 (1981).
21. M. J. Juedes, W. H. Bulger, and D. Kupfer, Drug. Metab. Dispos. 15:786 (1984).
22. W. H. Bulger, V. J. Feil, and D. Kupfer, Mol. Pharmacol. 27:115 (1985).
23. D. Kupfer, and W. H. Bulger, Federation Proc. 46:1864 (1987).
24. W. J. King, and G. L. Greene, Nature 307:745 (1984).
25. W. V. Welshons, M. E. Lieberman, and J. Gorski, Nature 307:747 (1984).
26. G. Redeuilh, B. Moncharmont, C. Secco, and E-E. Baulieu, J. Biol. Chem. 262:6969 (1987).
27. J. J. Dougherty, R. K. Puri, and D. O. Toft, J. Biol. Chem. 259:8004 (1984).
28. D. Kupfer, and W. H. Bulger, Life Sci. 25:975 (1979).
29. W. H. Bulger, J.E. Temple, and D. Kupfer, Toxicol. Appl. Pharmacol. 68:367 (1983).
30. W. H. Bulger, and D. Kupfer, Drug Metab. Dispos. 17:487 (1989).
31. M. J. Juedes, and D. Kupfer, Drug Metab. Dispos. (in press).
32. P.C. Reunitz, Drug Metab. Dispos. 6:631 (1978).
33. D. Kupfer, W. H. Bulger, and F. J. Nanni, Biochem. Pharmacol. 35:2775 (1986).
34. J. L. Roberts, T. S. Calderwood, and D. T. Sawyer, J. Am. Chem. Soc. 105:7691 (1983).
35. W. H. Bulger, and D. Kupfer, Biochem. Pharmacol. (in press).
36. C. Mani, and D. Kupfer, Fed. Proc., abstract (in press).
37. D. Kupfer, and W. H. Bulger, FEBS Letters (in press).
38. W. H. Bulger, R. M. Muccitelli, and D. Kupfer, Mol. Pharmacol. 15:515 (1979).
39. B. Hammond, B. S. Katzenellenbogen, N. Krauthammer, and J. McConnell, Proc. Natl. Acad. Sci. USA 76:6641 (1979).
40. J. Bitman, H. C. Cecil, S. J. Harris, and V. J. Feil, J. Agric. Food Chem. 26:149 (1978).

EFFECTS OF INHIBITORS ON THE P450-DEPENDENT METABOLISM OF ENDOGENOUS

COMPOUNDS IN FUNGI, PROTOZOA, PLANTS AND VERTEBRATES

Hugo Vanden Bossche, Patrick Marichal, Gustaaf Willensens
and Paul A.J. Janssen

Janssen Research Foundation
B2340 Beerse, Belgium

INTRODUCTION

A great number of the present antifungal agents belong to the class
of nitrogen heterocycle derivatives. Examples are the pyrimidine
antifungals, triarimol , fenarimol and nuarimol, the pyridine derivative,
buthiobate , the imidazoles, miconazole ,clotrimazole , econazole,
imazalil , tioconazole, bifonazole, sulconazole and ketoconazole and
finally the triazole antifungals, azaconazole, propiconazole,
terconazole, fluconazole, itraconazole and saperconazole. All these
antifungal agents have been shown to inhibit ergosterol synthesis (the
main sterol in most yeasts and fungi) by interacting with the cytochrome
P450-dependent 14α-demethylation ($P450_{14DM}$) of lanosterol in e.g.
Saccharomyces cerevisiae or of 24-methylenedihydrolanosterol in most
fungal cells (for reviews see refs. 1-4).
It is not surprising that nitrogen heterocycles affect cytochromes
P450 (P450). Indeed in 1972, Wilkinson et al.(5) described a long list
of imidazole derivatives as potent inhibitors of P450-dependent reactions
in liver microsomes. Each of the imidazoles investigated exhibited a
type II difference spectrum with a peak at 430-431 nm and a trough at
390-393 nm. This suggests that the unhindered nitrogen (N_3 in the
imidazole ring) binds to the catalytic heme iron atom at the site
occupied by the exchangeable sixth ligand (6).
It is of great interest that a number of the 14α-demethylase inhibitors
(DMIs) listed above show much higher affinity for the fungal $P450_{14DM}$ than
for P450 isozymes of the hosts (1).
It is the aim of this paper to review the interaction, of some of
the above listed antifungals, with ergosterol synthesis in fungal cells
and P450-dependent metabolic systems in the host. Compounds that
interfere with P450 isozymes in steroid hormone biosynthesis, and plant
growth regulators are also included.

14α-DEMETHYLASE INHIBITORS

Ragsdale and Sisler (7-9) were the first to prove that the
pyrimidine derivative, triarimol (Fig.1), inhibited ergosterol synthesis
in *Ustilago maydis*, a fungal pathogen of maize (blister smut). Triarimol
is effective in controlling powdery mildews. In triarimol-treated
sporidia of *U. maydis* reduced levels of ergosterol coincided with the
accumulation of 24-methylenedihydrolanosterol, obtusifoliol and 14α-
methyl-8,24(28) ergosterol, indicating an inhibition of the 14α-
demethylase (10). Studies of Kato et al.(reviewed in ref. 11) showed
that the pyridyl derivative, buthiobate (Fig. 1), has a similar

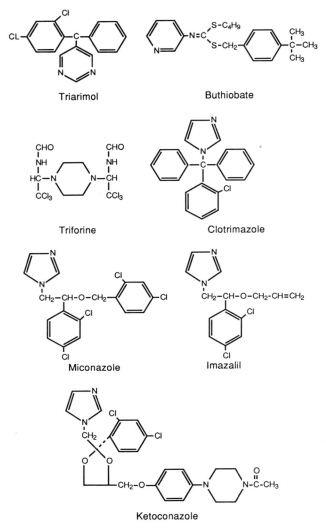

Fig. 1- Chemical structures of some ergosterol biosynthesis inhibitors

mode of action as triarimol. Aoyama and Yoshida(12) proved that buthiobate specifically bound to $P450_{14DM}$ from *Saccharomyces cerevisiae* and inhibited lanosterol 14α-demethylation in a reconstituted system consisting of $P450_{14DM}$ and NADPH-P450 reductase, both purified from *S. cerevisiae* microsomes. The addition of buthiobate to $P450_{14DM}$ in the oxidized form caused a type II spectral change (12). Compounds, causing this kind of spectral changes, bind to the 6th coordination position of the heme iron. The spectral change was saturated when one mole of buthiobate was bound to one mole of P450, indicating that buthiobate formed ono-to-one complexes with oxidized $P450_{14DM}$ (12). Addition of $Na_2S_2O_4$ to the buthiobate-$P450_{14DM}$ complex changed the absorption spectrum to that of a buthiobate-ferrous $P450_{14DM}$ complex; brief bubbling of carbon monoxide converted the buthiobate complex to the reduced -CO compound, showing maximum absorption at 447 nm (12). This suggests that the pyridine fungicide has lower affinity for the heme iron complex than CO. As shown by Wilkinson et al.(5, 13) the imidazole-liver P450 complex was also easily replaced by CO. However, the latter investigators used relatively small 1-arylimidazoles. Therefore it is possible that these compounds bind only slightly to the apoprotein moiety of the cytochrome.

Buthiobate is not inhibiting the Δ^{22}-desaturation of ergosta-5,7-dien-3ß-ol, the last step in ergosterol synthesis (14). This hydrocarbon dehydrogenase requires NADPH and molecular oxygen, and is inhibited by CO and metyrapone. This indirect evidence suggests that the reaction is catalyzed by a P450 ($P450_{22-DS}$) (15). Further studies showed that the Δ^{22}-desaturation is not blocked by rabbit antibodies against $P450_{14DM}$, indicating that $P450_{22-DS}$ is different from $P450_{14DM}$ (15).

Buthiobate only partially inhibited the O-deethylation of 7-ethoxycoumarin by rat liver microsomes and neither the benzphetamine N-demethylation nor the p-nitrosoanisole O-demethylation were inhibited (16). These studies certainly prove that buthiobate is a selective inhibitor of the 14α-demethylase. However, buthiobate is an almost equipotent inhibitor of the 14α-demethylases in both *S. cerevisiae* and rat liver microsomes (16).

An interesting ergosterol biosynthesis inhibitor is the piperazine derivative, triforine (Fig.1). Triforine differs from triarimol and buthiobate in having its heterocyclic nitrogen atoms substituted. Triforine inhibits ergosterol synthesis with concomitant accumulation of 14-methylated sterols, similar to those found after treatment of fungi with triarimol (17). Triarimol-resistant mutants of *Cladosporium cucumerinum* showed cross resistance to triforine.(11). However, triforine did not block the synthesis of ergosterol from $[^{14}C]$-mevalonate in a cell-free system of *S. cerevisiae* (18). This suggests that in intact cells triforine is metabolized to an active form. An other possibility is that triforine is inactive against the *S. cerevisiae* $P450_{14DM}$, which uses lanosterol instead of 24-methylenedihydrolanosterol as substrate. As mentioned before, 24-methylenedihydrolanosterol is the substrate for $P450_{14DM}$ of most fungal cells.

The introduction of imidazole and triazole antifungals has improved the treatment of plant pathogens and the imidazoles, miconazole (Fig.1) and clotrimazole (Fig. 1) ameliorated topical antifungal treatment. However topical treatment is often inconvenient and several forms of dermatomycosis and systemic mycoses are not controlled by conventional drug therapy. Therefore the introduction of the orally active imidazole derivative, ketoconazole (Fig. 1), represented a major important advance in antifungal chemotherapy. As will be discussed further on, ketoconazole was also of great help in the study of cytochrome P-450 isozymes.

The inhibition of ergosterol synthesis in *U. avenae* by the triazole fungicide, triadimefon (19) together with the effect of triarimol (20) and N-dodecylimidazole (21) on cholesterol biosynthesis in rat liver led

Matolscy et al. (22) to propose sterol biosynthesis as a potential target for antifungal action of the triazole antifungals.

Ergosterol synthesis is not only inhibited by the pyrimidine, pyridine and triazole antifungals of use in plant protection, also imidazole derivatives, active against a long list of yeasts and fungi, pathogenic to plants, animals or man, have been found to interfere with ergosterol synthesis and specially with $P450_{14DM}$. Examples are clotrimazole (23), N-dodecyl-imidazole (24), econazole (23), imazalil (Fig. 1) (25), ketoconazole (26). miconazole (27), parconazole (28), tioconazole (23), all imidazole derivatives. Examples of triazole derivatives are etaconazole (29), propiconazole (30), terconazole (31) and fluconazole (32).

As expected from the studies of Wilkinson et al. (5), all these azoles have been found to yield type II spectra and/or to compete with carbon monoxide for binding to the sixth coordination position of the reduced heme iron in P450(s) present in yeast or fungal microsomes. Examples are the azole agricultural fungicides, azaconazole, penconazole, propiconazole and imazalil (4) and the medicinal azole antifungals, miconazole, clotrimazole, bifonazole, ketoconazole (4), parconazole and terconazole (25).

Using a purified preparation of *S. cerevisiae* $P450_{14DM}$, Yoshida and Aoyama (33) proved that ketoconazole, as 1-methylimidazole, interacted with the heme iron of $P450_{14DM}$ at the sixth coordination position. Titration of $P450_{14DM}$ with ketoconazole indicated the stoichiometric binding of the fungicide to this P450 with very high affinity. Upon addition of hydrosulfite, the ketoconazole-$P450_{14DM}$ complex was reduced to the corresponding ferrous complex. Addition of CO to the reduced complex showed a slow replacement of ketoconazole by CO (32). Similar results were obtained by using microsomal preparations of *Candida albicans* (34). Furthermore, this reactivity of the reduced azole-$P450_{14DM}$ complex was not affected by replacing the imidazole moiety in ketoconazole by a triazole ring (34).

The low reactivity of the ketoconazole-$P450_{14DM}$ complex with CO and the fact that the imidazole can be replaced by a triazole moiety, indicate that ketoconazole not only binds to the heme iron, but, that the large N-1 substituent of ketoconazole also binds to the apoprotein moeity and that the composition of this non-ligand part determines the interaction with the apoprotein. Indeed, this interaction was much less with the much smaller buthiobate molecule and the 1-methylimidazole-$P450_{14DM}$ complex was promptly converted to the reduced CO-complex (33). It is obvious that compounds that bind simultaneously to hydrophobic domains of the protein and to the heme iron are more active than those that bind to the iron atom only. This is illustrated by the fact that imidazole (binding to the 6th coordination position of the heme iron only) is a weak inhibitor of the P450-dependent aldrin epoxidation in rat liver microsomes (IC_{50}-value=3.6mM), whereas phenylimidazole is a much more potent inhibitor (IC_{50}-value= 1.5 µM)(35).

Yoshida and Aoyama also proved that ketoconazole inhibited lanosterol 14α-demethylase activity of a reconstituted system consisting of $P450_{14DM}$ and NADPH-P450 reductase, purified from *S. cerevisiae* microsomes (2, 33). The inhibition was linearly dependent on the amount of ketoconazole and reached 100% when an equal amount of ketoconazole to $P450_{14DM}$ was added. This further proves that ketoconazole inactivates the 14α-demethylase system by forming a stoichiometric complex with $P450_{14DM}$. That azole antifungals are extremely potent inhibitors of the 14α-demethylase system in yeasts and fungi can also be deduced from its effects on ergosterol synthesis by intact cells. For example, when ketoconazole was added to exponentially growing *Candida albicans* or *Aspergillus fumigatus*, 50% inhibition of ergosterol synthesis was reached after one hour of contact with 4 nM and 58 nM, respectively (36). With the topically active imidazole derivative, miconazole, 50% inhibition of ergosterol synthesis, by exponentially growing *C. albicans*, was already reached at 0.7 nM (27).

Selectivity

Topically applied antifungals could interfere with P450 isozymes present in the skin. For example, clotrimazole has been shown to inhibit the P450-dependent aryl hydrocarbon hydroxylase (AHH) activity in human hair follicles, a readily available source of human epithelial tissue (37). Fifty per cent inhibition (IC_{50}-value) was achieved at about 1 μM. Topically applied clotrimazole also inhibited AHH activity in epidermal microsomes prepared from neonatal rats pretreated with solvent (acetone), coal tar, aroclor or 3-methylcholanthrene (38). The IC_{50}-values were 0.12, 0.16, 0.26 and 0.24 μM, respectively (38). Clotrimazole was also found to be in vitro an inhibitor of epoxide hydrolase activity in rat epidermal microsomes, with a 50% inhibition at 0.1 mM (38). Furthermore, topical application of clotrimazole to the skin of Balb/c mice substantially increased the latent period for the develpoment of skin tumors by 3-methylcholanthrene (38). Miconazole and ketoconazole inhibit (IC_{50}-values are 10 μM and 0.65 μM, respectively) the P450-dependent metabolism of retinoic acid in epidermal microsomes from neonatal rats (39).

Topical application of ketoconazole, at doses of 1, 5 and 10 mg/kg, on neonatal Wistar rats, 1h before application of retinoic acid (1 mg/rat) results in a dose-dependent inhibition of retinoic acid metabolism by epidermal microsomes isolated 24h later (39). Topical administration of ketoconazole to the scalp of human volunteers decreased in some of them the P450-dependent 7-ethoxyresorufin-O-deethylase activity in human hair roots (40). However, following daily oral administration of 200 mg ketoconazole (5 days), a substantial decrease (from 2.73±0.71 to 0.71±0.26 pmol resorufin/30min/20 human hair roots) occurred (40). These results indicate that some azole antifungals might be useful in inhibiting P450-dependent reactions in the skin that are able to transform harmless compounds into carcinogens. These results also focus attention on the fact that orally active antifungal agents whose activity is baised on interaction with the P450-dependent 14α-demethylation, should be tested for their effects on mammalian and plant P450 isozymes.

It is obvious that the first candidate to be tested is the $P450_{14DM}$ involved in cholesterol synthesis. To obtain 50% inhibition of the 14α-demethylation of lanosterol in phytohemagglutinin (PHA) stimulated human peripheral lymphocytes, 10^{-7} M of ketoconazole is needed (3). When macrophages (J774G8) were incubated for 48h in the presence of $[^{14}C]$-acetate and ketoconazole, 50% inhibition of the incorporation of ^{14}C into cholesterol was obtained at 1.5 μM (41). Fifty % inhibition of cholesterol synthesis in subcellular fractions of male rat liver is reached at 2 μM ketoconazole and 6 μM miconazole (42). It should be noted that 50 % inhibition of ergosterol synthesis by a subcellular fraction of C. albicans was observed at 0.07 μM and 0.09 μM (42) of ketoconazole and miconazole, respectively. Ergosterol synthesis inhibition in macro-phages infected with *Leishmania mexicana mexicana* amastigotes [*Leishmania* is known to synthesize ergosterol (43)] was achieved at 0.2 μM (41). As shown above, ketoconazole inhibits ergosterol synthesis in intact *C.albicans* cells for 50% at 4nM, and complete depletion of ergosterol is achieved at 7.5 nM (44). Yoshida and colleagues,using the already mentioned reconstituted system found that ketoconazole inhibited the 14α-demethylation of 24,25-dihydrolanosterol and of 32-hydroxy-24,25-dihydrolanosterol [i.e. the product of the hydroxylation of the C-32 methyl (14α-methyl group) of lanosterol] but the inhibitory effect on the removal of the 32-hydroxy-methyl group was weaker(45). This suggests that in the yeast reconsti-tuted system, ketoconazole preferentially inhibits the hydroxylation step and not so much the oxidation of the C-32 alcohol to a C-32 aldehyde or the oxidative elimination of the aldehyde as formic acid. Trzaskos et al. (46) using hepatic microsomes, found that both miconazole and keto-conazole promote oxysterol accumulation (the major oxysterol accumulating

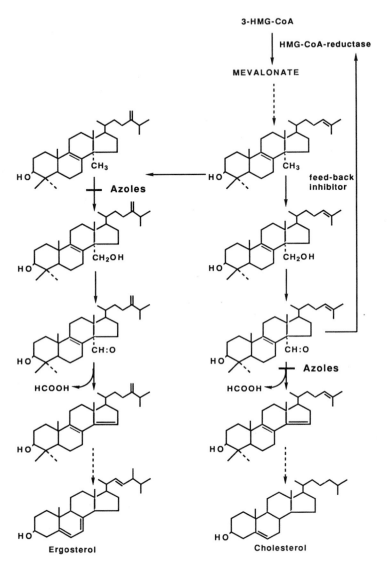

Fig. 2- Inhibition by azole antifungals (Azoles) of ergosterol (yeast and fungi) and cholesterol (mammalian cells) synthesis. 3-HMG-CoA= 3-hydroxy-3-methylglutaryl-coenzyme A. Dashed arrows=multi-enzyme systems.

was the C-32-aldehyde) prior to complete inhibition of lanosterol de-methylation. Fifty % inhibition of lanosterol 14α-demethylation was achieved at 0.9 and 0.4μM miconazole and ketoconazole, respectively.

These studies (46) indicate that, miconazole and ketoconazole at concentrations higher than those needed to inhibit lanosterol 14α-demethylase in yeast, inhibit, the lyase activity (i.e. the last step in the 14-demethylation system) instead of the hydroxylation step in yeast (Fig.2). This suggests that P450s doing the same job in different species are not necessarily identical. However, Trzaskos and Henry(47), studying the effects of the triazole derivative, flusilazole, found that the affinity of this triazole for a partly purified P450 from *S. cerevisiae* microsomes is similar to that for a partly purified P450 from rat liver microsomes. However, when they compared the binding of flusilazole to P450(s) in the microsomes, the affinity for the yeast P450(s) was almost 24-times higher. Studying the effect on the lanosterol 14α-demethylation it was found that the hepatic microsomal preparation was almost 100 times less sensitive than the yeast preparation (47). From these studies, Trzaskos and Henry conclude that the apparent increase in susceptibility of fungal preparations observed upon exposure to azole antifungals is due in part to a protective effect afforded by other susceptible P450 isozymes present in mammalian preparations. However, Trzaskos and Henry (47) used *S. cerevisiae* only. Although the P450$_{14DM}$ of both rat liver and *S. cerevisiae* might be quite similar it can not be excluded that P450$_{14DM}$ of other yeasts and fungi differ from the hepatic or *S. cerevisiae* counterpart. As shown in the chapter on "Role of cytochrome P-450 in the anabolism and catabolism of endobiotics" (this book) the P450$_{14DM}$ of *S. cerevisiae* shares with those of *C. albicans* and *C. tropicalis* 64.2 and 65.2 identical aminoacids, whereas the P450$_{14DM}$ of both candidas share 83% identical amino acids. Furthermore, ketoconazole is a much more potent inhibitor of ergosterol synthesis by *C. albicans* than by *S. cerevisiae*, both grown for 24h in a casein hydrolysate-yeast extract-glucose medium (27). Fifty per cent inhibition was achieved at 2.1 nM and 330 nM, respectively (unpublished results).

STEROID SYNTHESIS

Ketoconazole is an oral antifungal imidazole with a broad spectrum of antifungal activity. The effective daily single dose of ketoconazole for most fungal infections is 200 mg. The incidence of side effects is low. However, in 1981, gynecomastia was reported in two patients during treatment with 200mg ketoconazole daily (48) and in a few patients receiving high multiple doses (600-1200 mg/day) of the antifungal (49). The observation of De Felice et al.(48) triggered a multidisciplinary study of the endocrinological applications of ketoconazole and was at the onset of in-depth studies on mammalian P450 isozymes. Indeed, this rare side effect suggested that in such patients ketoconazole affected the estrogen: androgen ratio (49). No clear reduction in the plasma estradiol levels at 600 mg ketoconazole per day was seen (49). Measuring the testosterone serum levels, a dose related reversible decrease was observed (49-56).
The finding of a marked but transient decrease in plasma testosterone levels was the start for a series of studies to determine the site of action of ketoconazole in the androgen biosynthetic pathway. Testosterone synthesis was studied in a rat testicular subcellular fraction (S10-fraction) containing the cytosol and microsomes (57-59).

With pregnenolone as substrate, a 50% inhibition of androgen synthesis was achieved after a 3 h incubation period in the presence of 5 μM ketoconazole (58). This inhibition coincided with an accumulation of 17α-hydroxy,20-dihydropregnenolone (Fig.3). Maximum accumulation was

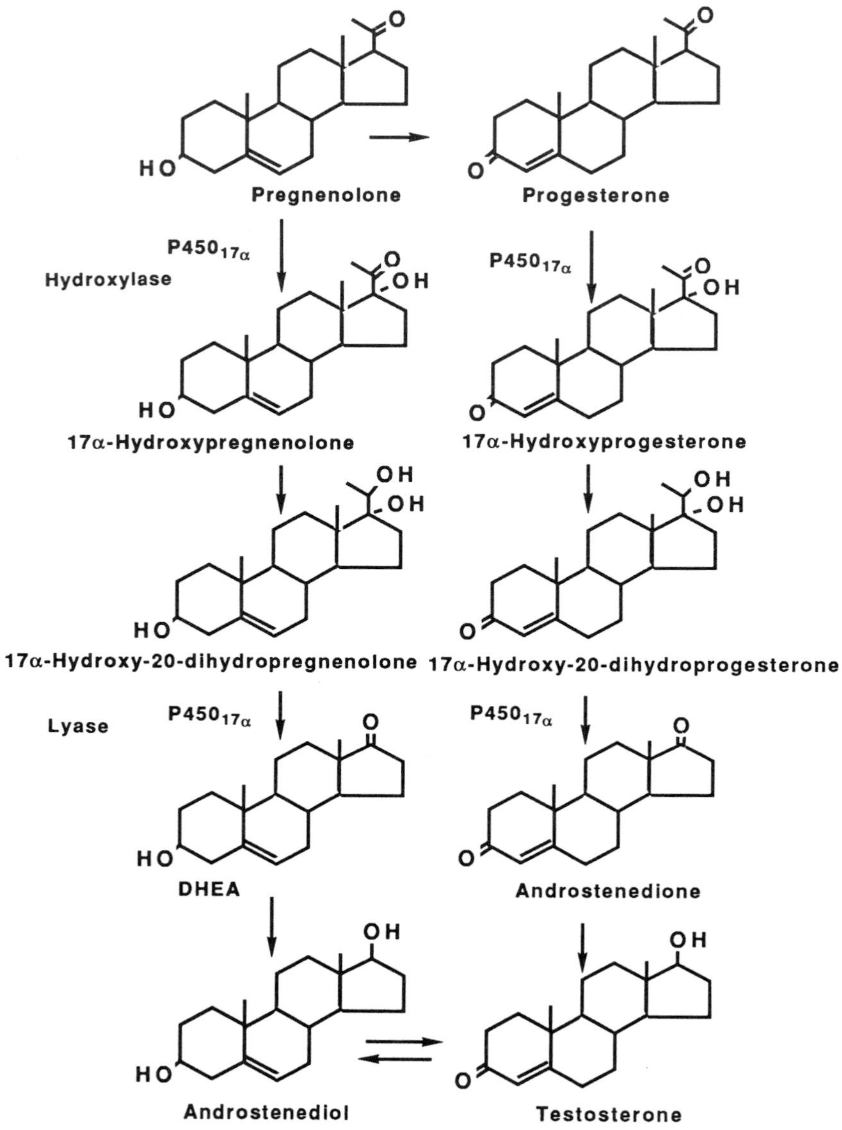

Fig. 3– The synthesis of androgens from pregnenolone and progesterone. $P450_{17\alpha}$= 17α-hydroxylase/17,20-lyase. DHEA= dehydroepiandrosterone.

reached at 0.5μM. At concentrations >1μM an accumulation of pregnenolone was observed with a concomitant decrease in 17α-hydroxy,20-dihydropregnenolone (58). Further studies revealed that ketoconazole inhibited the conversion of 17α-hydroxy,20-dihydroprogesterone into androstenedione by rat testicular microsomes, 50% inhibition was achieved at 0.2 μM (57, 59). These studies prove that ketoconazole is an inhibitor of the 17,20-lyase.

Studies by Rajfer et al.(60), using minced decapsulated testes from Sprague-Dawley rats, indicated that the ketoconazole induced inhibition of androgen synthesis originated from an inhibitory effect on the 17,20-lyase only (60). Similar results were reported by Kan et al.(61). Higashi et al.(62) showed that ketoconazole inhibited the activities of steroid 17α-hydroxylase and 17,20-lyase in rat and human testis, in human testis the investigators also found inhibition of the 16α-hydroxylation of progesterone. In rat testis the 17,20-lyase was almost 4-times more sensitive to ketoconazole than the 17α-hydroxylase (62).

This difference in sensitivity is of interest since it has been shown that the microsomal 17-hydroxylase and 17,20-lyase activities result from the action of a single P450 ($P450_{17\alpha}$) (63-64, see also "Role of cytochrome P450 in the anabolism and catabolism of endobiotics", this book).

The 17,20-lyase reaction that leads to the cleavage of a carbon-carbon bond is chemically more complex than the 17α-hydroxylation reaction (65). Studies of Yanase et al. (65) showed that deletion of a phenylalanine in the N-terminal region of human cytochrome $P450_{17\alpha}$ results in the partial combined 17α-hydroxylase/17,20-lyase deficiency. However, 17α-hydroxylase activity of this mutant protein measured in intact COS 1 cells was less than 37% of that observed upon expression of the wild-type enzyme, whereas 17,20-lyase activity of the mutant was less than 8% of that observed with the normal enzyme (65). This might indicate that the 17,20-lyase may require a more rigid active side and is thus more altered by the missing phenylalanine residue. These results also suggests that conformational changes in $P450_{17\alpha}$ are needed to hydroxylate pregnenolone or progesterone and to bind 17α-hydroxy-20-dihydropregnenolone or 17α-hydroxy-20-dihydroprogesterone. The P450-17α-hydroxy-20-dihydropregnenolone or P450-17α-hydroxy-20-dihydroprogesterone complex seems to be more apt to bind ketoconazole.

Ketoconazole also inhibits androgen synthesis when added to bovine adrenal microsomes, 50% inhibition was achieved at 0.4 µM. This inhibition coincided with an accumulation of pregnenolone and 11-deoxycorticosterone (54). At higher concentrations, ketoconazole also inhibited the synthesis of 11-deoxycortisol coinciding with a further increase in 11-deoxycorticosterone. The accumulation of the latter mineralocorticoid indicates that ketoconazole interferes much less with the 21-hydroxylase. In rat adrenal microsomes a K_i value of 5 µM was found (62). This is a 12.5 times higher value as found for the rat testicular 17,20-lyase (62). Ketoconazole also had little or no effect on 21-hydroxylase activity in human adrenal microsomes (66). This is surprising since $P450_{17\alpha}$ and $P450_{C21}$ show a well conserved domain (from aspartic acid$_{345}$ up to proline$_{367}$ of the $P450_{17\alpha}$ sequence). This domain of the apoprotein of the rat $P450_{17\alpha}$ has been proposed as the putative steroid binding site (63) and is present in the human and bovine $P450_{17\alpha}$ (see "Role of cytochrome P450 in the anabolism and catabolism of endobiotics", this book).

Ketoconazole inhibits the 17α-hydroxylase and 17,20-lyase in human adrenal microsomes (66). Ayub and Levell found that the human adrenal 17,20-lyase was about twice more sensitive to ketoconazole than the 17α-hydroxylase (67). These investigators also compared the activity of a series of imidazole antifungals on steroidogenic enzymes of human adrenal microsomes. The order of decreasing inhibitory potency, as determined from values for both 17α-hydroxylase and 17,20-lyase activities, was: bifonazole> clotrimazole> ketoconazole> miconazole> econazole> isoconazole> tioconazole (67). The high activity of bifonazole corresponds with its high affinity for the microsomal P450 of piglet testis. Indeed, 50% inhibition of CO binding was already achieved at 0.07 µM, whereas 1.2 µM was needed with *C. albicans* microsomes (4). An effect of ketoconazole (400 mg every 8 hours) on the synthesis of adrenal steroids was also proven in fourteen, previously untreated patients with stage-D metastatic prostate cancer (52). Next to a striking decrease in testosterone plasma levels a decline in androstenedione and dehydroepiandrosterone,which are mainly of adrenal origin, was noted (52). As expected from the in vitro results, the decrease in androgen levels coincided with a rise in 17α-hydroxy-progesterone and progesterone plasma levels.

The fact that ketoconazole interferes with testicular as well as adrenal androgen synthesis has made it a good candidate for the treatment of androgen dependent prostate carcinoma (54-56,68-72). At the high dose (400 mg every 8 hours) used the principal side effect is gastric discomfort with nausea (55, 72) which resolves with administration of antinauseants or upon discontinuation of the drug (73).

From all these studies it can be deduced that ketoconazole, at high doses, is effective both clinically and endocrinologically in the treatment of metastatic prostate carcinoma, but its use is limited by gastric discomfort. However, these investigations provided us with a battery of P450 systems to evaluate activity and predict possible toxicity and have triggered a multidisciplinary study to open new possibilities in medical treatment.

A product of these studies is R 75251. This imidazole derivative (Fig. 4) is devoid of antifungal activity but shares with ketoconazole the effects on P450 isozymes involved in testicular androgen synthesis (74). R 75251 also inhibits retinoic acid metabolism in epidermal microsomes of neonatal rats, 50% inhibition of the 4-hydroxylation of retinoic acid is reached at 2 μM (unpublished results). Up to 10μM, R75251 has no effect on $P450_{14DM}$, $P450_{scc}$ or $P450_{C-21}$ (21-hydroxylase) catalyzed reactions (74). R75251 is, compared to ketoconazole, a much more potent inhibitor of the placental aromatase; 50% inhibition was achieved at 2.9 nM (74).

Oral administration of this imidazole derivative to male dogs demonstrates that it lowers testosterone and androstenedione plasma levels to castrate levels (R. De Coster, personal information). Values close to castrate levels are found in male volunteers treated with a single dose of 300 mg (J. Bruynseels, personal communication). This non-steroid, at the moment under study in patients with progressive prostate cancer, has in contrast with ketoconazole, perfect gastric tolerance (L. Dennis, personal communication).

It has already been mentioned that ketoconazole is a poor inhibitor of the aromatase, this in contrast to the topically used imidazole antifungal, econazole. Econazole (Fig. 4) is a potent inhibitor of the aromatase in human placental microsomes (75) and of the human placental $P450_{AROM}$ expressed in transfected Cos-1 cells (76). In the latter cells an IC_{50} value of 0.04 μM was found (76). Using human placenta microsomes and $[4-^{14}C]$-androst-4-ene-3,17 dione (40 nM) as substrate (25 μg of microsomal protein; 6 min of incubation at 37°C) we reached 50% inhibition with 1.8 nM econazole and 5.9 nM miconazole, whereas 3 μM of ketoconazole was needed (unpublished results). Mason et al.(75) compared the inhibitory effects of econazole with those of imazalil, miconazole, prochloraz, clotrimazole, ketoconazole, and aminoglutethimide on the human placenta microsomal aromatase. Fifty per cent inhibition was achieved at 0.03, 0.15, 0.6, 0.7, 1.8, 60 and 45 μM, respectively.

The knowledge gathered with azole antifungals and the study of the different P450 systems paved the way to the development of a new non-steroidal aromatase inhibitor, R 76173. This triazole derivative (Fig. 4) is in contrast with the above mentioned azoles, devoid of any effect on the $P450_{14DM}$-dependent ergosterol or cholesterol synthesis. R76713 is a potent inhibitor of the aromatase in human placental microsomes. Fifty per cent inhibition of the conversion of $[4-^{14}C]$ androst-4-ene-3,17-dione into estrogens by human placental microsomes is reached at 2.7 nM i.e. at concentrations 520 and 1549 times lower than those needed for aminoglutethimide and ketoconazole, respectively (74).

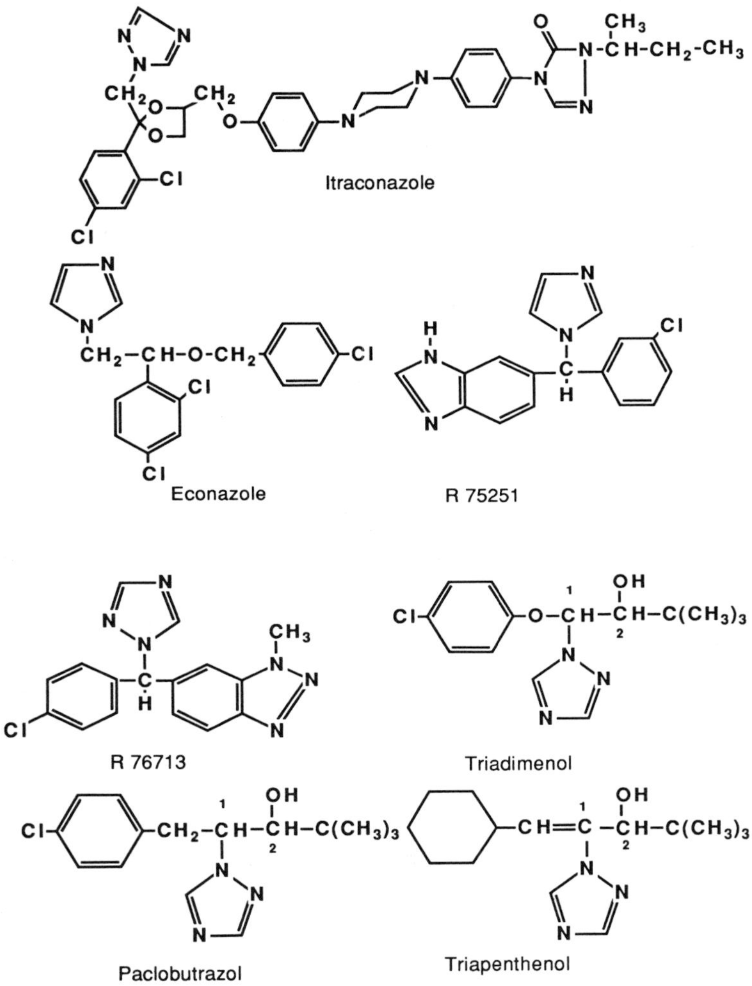

Fig. 4— Chemical structures of itraconazole, econazole, R 75251, R 76713 and of sterol synthesis inhibitors with fungicidal and plant growth regulator activity

The IC50-value corresponds very well with the 3 nM found for rat granulosa cells(77). In vivo, a single oral dose of 0.05 mg/kg lowered plasma estradiol levels of rats primed with pregnant mare's serum gonadotropin by more than 90% (77).

In contrast with ketoconazole, R 76713 is devoid of effects on the cholesterol side chain cleavage (P450$_{scc}$). At 10µM it has a slight effect on adrenal androgen and 11-deoxycortisol synthesis, suggesting that R 76713 is a poor inhibitor of both the 17-hydroxylase and the 17,20-lyase. A 30 % decrease in cortisol synthesis is observed at 10µM only [with ketoconazole 50% inhibition is reached at 0.4 µM (74)]. The 17-hydroxylase-17,20-lyase system in rat testis microsomes is more than 300 times less sensitive than the aromatase in the microsomal fraction of human placenta. Furthermore, R 76713 is a poor inhibitor of the regio- and stereoselective P450-dependent hydroxylations of steroids by rat liver microsomes into estrogens by human placental microsomes is reached at 2.7 nM i.e. at concentrations 520 and 1549 times lower than those needed for aminoglutethimide and ketoconazole, respectively (74). The IC50-value corresponds very well with the 3 nM found for rat granulosa cells(77). In vivo, a single oral dose of 0.05 mg/kg lowered plasma estradiol levels of rats primed with pregnant mare's serum gonadotropin by more than 90% (77).

Both, R 75251 and R 76713, are good examples to prove that P450 systems can be exploited in the construction of compounds that may become important tools in medical treatment.

ITRACONAZOLE

Although ketoconazole is a much more potent inhibitor of the P450-dependent ergosterol synthesis than of for example P450$_{17\alpha}$, an antifungal free of effects on mammalian P450s was welcome. The P450s studied were of help in the selection of itraconazole (Fig.4). This orally active triazole antifungal shares with ketoconazole its broad spectrum of activity. It is, as compared with ketoconazole, much more active against *Aspergillus* spp., *Microspurum canis* and *Sporothrix schenckii*, and is highly active in meningeal cryptococcosis(78-79). Itraconazole is, as compared with ketoconazole, a much more potent inhibitor of ergosterol synthesis in *Aspergillus fumigatus*. For example, in *A. fumigatus* grown for 24h in Sabouraud broth, more than 125 times more ketoconazole than itraconazole is needed to reach 50% inhibition of ergosterol synthesis (36).

After 24h of growth of *A. fumigatus* in Sabouraud broth the intracellular pH was about 5.4, at this pH ketoconazole's imidazole nitrogen is 91% protonated whereas the triazole nitrogen of itraconazole is for 2% protonated only (36). Therefore, the lower activity of ketoconazole against *A. fumigatus* might originate partly from protonation of its imidazole nitrogen at the intracelular pH measured especially at the hyphal apex (34, 36). Indeed this nitrogen should be unprotonated to bind to the heme iron in P450$_{14DM}$. Obviuosly, potency is also affected by the uptake and transport. When cells were collected from cultures of *A. fumigatus* containing [^{3}H]itraconazole or [^{3}H]ketoconazole, that had been added at a concentration of 5 nM or 0.1 µM (the cultures having been incubated for 24h before and 1h after the addition of the azoles), 3 to 8.6 times more itraconazole than ketoconazole was found in the cells (34).

Itraconazole also differs from ketoconazole by the fact that it has almost no effect on P450-dependent reactions in testis microsomes, adrenal mitochondria and microsomes (4, 57, 80-81). Up to 10 µM, itraconazole is devoid of inhibitory effects on the human placental aromatase, the 1-hydroxylation of 25-hydroxyvitamin D3 and it does not effect the regio- and stereospecific hydroxylations of testosterone (34). Itraconazole has no effect on the metabolism of retinoic acid by epidermal microsomes from neonatal rats (39). At concentrations as high as ≥10µM, 50% inhibition was observed of the *p*-nitroanisole *-O-*

demethylase, N,N-dimethylaniline-N-demethylase or aniline hydroxylase in untreated-, 3-methylcholanthrene- or phenobarbital-induced liver microsomes from male Wistar rats (82).

Itraconazole, inhibits (IC_{50}-value = 0.95 μM) cholesterol synthesis in mouse macrophages (J 774 G8)(41). Fifty % inhibition of cholesterol synthesis in subcellular fractions of male rat liver is reached at 0.7 μM itraconazole (83). It should be noted that 50 % inhibition of ergosterol synthesis by a subcellular fraction of C. albicans was reached at 0.08 μM of itraconazole (83) and that at 0.005 μM it abolished ergosterol synthesis in intact C. albicans (80).

High concentrations of itraconazole have an effect on testosterone synthesis by dispersed rat testicular cells. Without stimulation, 50% inhibition was found at 4.5 μM itraconazole. In the presence of human chorionic gonadotropin (hCG), 50% inhibition was achieved at 1.6 μM (80). However, the production of the precursors 17α-hydroxyprogesterone and progesterone was affected in the same way as that of testosterone. This indicates that the effect of high concentrations of itraconazole might originate from an effect on cholesterol synthesis, and is not related to interference with the 17α-hydroxylase and/or 17,20-lyase. Indeed 50% inhibition of cholesterol synthesis in Leydig cells was observed after 4 h of incubation in the presence of 1 μM itraconazole (unpublished results).

Itraconazole also inhibits ergosterol synthesis in *Leishmania mexicana mexicana* (41). Fifty % inhibition is achieved at 0.15 μM (41).

GIBBERELLIN SYNTHESIS

During the development of fungicides it was noted that some of the 14-demethylase inhibitors (DMI fungicides) caused stunting plants (84). This observation was the start of research programmes with the objective of developing plant growth regulators. Examples are paclobutrazol (Fig. 4), uniconazol and triapenthenol (Fig. 4) (84). The side effect observed is due, at least partly, to inhibition of gibberellin biosynthesis. Gibberellins are plant hormones involved in elongation growth of plants. Thus, inhibition of this hormone synthesis should result in growth retardation.

The target of these plant growth regulators is the P450-dependent oxidation of *ent*-kaurene to *ent*-kaurenoic acid (Fig. 5). It is of interest that the 1R,2R enantiomer of paclobutrazol is a more potent inhibitor of ergosterol synthesis than the 1S,2S enantiomer, whereas the latter is a much better inhibitor of gibberellin synthesis (85).

A comparison of triadimenol (Fig. 4) and paclobutrazol demonstrates that minor structural changes can result in major differences in biological activities, target sites and stereochemical discriminations (85). Instead of the 1R,2R enantiomer of paclobutrazol, the 1S,2R-enantiomer of triadimenol shows the highest fungicidal activity (85). Triadimenol seems to be inactive in gibberellin synthesis, however, the 1S,2R-enantiomer of triadimenol inhibits the synthesis of phytosterols. The 1R,2S- enantiomer of triadimenol also is an inhibitor of plant sterol synthesis but lacks fungicidal activity (85). Similar studies done with triapenthenol showed that the S -enantiomer causes growth inhibition by inhibiting gibberellin synthesis while the R -enantiomer inhibits growth at high concentrations only.

CONCLUSION

The examples, discussed in this paper, demonstrate that P450 systems can be exploited in the construction not only of important antifungals that are devoid of endocrinological effects, but also of compounds that may become important tools in chemotherapeutic and agrochemical applications.

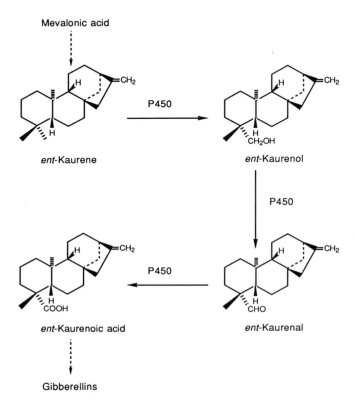

Fig. 5- The P450-dependent synthesis of gibberellins from mevalonate. Dashed arrows represent multi-enzyme steps.

REFERENCES

1. H. Vanden Bossche, Mode of action of pyridine, pyrimidine and azole antifungals, in: "Sterol biosynthesis inhibitors. Pharmaceutical and agrochemical aspects", D. Berg and M. Plempel, eds., Ellis Horwood, Chichester, England (1988)
2. Y. Yoshida, Cytochrome P450 of fungi: primary target for azole antifungal agents, in: "Current topics in medical mycology", Vol.2, M.R. McGinnis, ed., Springer Verlag, New York (1988)
3. P.A.J. Janssen and H. Vanden Bossche, Mode of action of cytochrome P-450 monooxygenase inhibitors. Focus on azole derivatives, Arch. Pharm. Chem. 15: 23-40(1987)
4. H. Vanden Bossche, P. Marichal, J. Gorrens, D. Bellens, H. Verhoeven, M.-C. Coene, W. Lauwers and P.A.J. Janssen, Interaction of azole derivatives with cytochrome P-450 isozymes in yeast, fungi, plants and mammalian cells, Pestic. Sci. 21: 289-306(1987)
5. C.F. Wilkinson, K. Hetnarski and T.O. Yellin, Imidazole derivatives- A new class of microsomal enzyme inhibitors, Biochem. Pharmacol. 21: 3187-3192(1972)
6. P.R. Ortiz de Montellano and N.O. Reich, Inhibition of cytochrome P-450 enzymes, in: "Cytochrome P-450 structure, mechanism and biochemistry", P.O.Ortiz de Montellano, ed., Plenum Press, New York (1986)
7. N.N. Ragsdale and H.D. Sisler, Inhibition of ergosterol synthesis in Ustilago maydis by the fungicide triarimol, Biochem. Biophys. Res. Commun. 46: 2048-2053(1972)
8. N.N. Ragsdale and H.D. Sisler, Mode of action of triarimol in Ustilago maydis, Pesticide Biochem. 3: 20-29(1973)
9. N.N. Ragsdale, Specific effects of triarimol on sterol synthesis in Ustilago maydis, Biochim. Biophys. Acta 380: 81-96(1975)
10. H.D. Sisler and N.N. Ragsdale, Fungotoxicity and growth regulation involving aspects of lipid biosynthesis, Neth. J. Plant Path. 83 (suppl.1): 81-91(1977)
11. T. Kato. Sterol biosynthesis in fungi, a target for broad spectrum fungicides, in:"Chemistry of plant protection. 1. Sterol biosynthesis, inhibitors and anti-feeding compounds", G. Haug and H. Hoffmann, eds., Springer Verlag, Berlin (1986)
12. Y. Aoyama, Y. Yoshida, S. Hata, T. Nishino and H. Katsuki, Buthiobate: a potent inhibitor for yeast cytochrome P-450 catalyzing 14α-demethylation of lanosterol, Biochem. Biophys. Res. Commun. 115: 642-647(1983)
13. C.F. Wilkinson, K. Hetnarski and L.J. Hicks, Substituted imidazoles as inhibitors of microsomal oxidation and insecticide synergists, Pest. Biochem. Physiol. 4: 299-312(1974)
14. S. Hata, T. Nishino, H. Katsuki, Y. Aoyama and Yoshida, Y., Two species of cytochrome P-450 in ergosterol biosynthesis of yeast, Biochem. Biophys. Res. Commun., 116: 162-166 (1983)
15. S. Hata, T. Nishino, M. Komori and H. Katsuki, Involvement of cytochrome P-450 in Δ^{22}-desaturation in ergosterol biosynthesis in yeast, Biochem. Biophys. Res. Commun., 103: 272-277(1981)
16. Y. Yoshida, Y. and Y Aoyama, Y., Effects of buthiobate, a fungicide, on cytochrome P-450 of rat liver microsomes, J. Pharmacobiol. Dyn. 8: 432-439(1985)
17. J.L. Sherald, N.N. Ragsdale and H.D. Sisler, Similarities between the systemic fungicides triforine and triarimol, Pestic. Sci. 4: 719-727(1973)
18. T. Kato, Biosynthetic processes of ergosterol as the target of fungicides, in: "Pesticide chemistry, human welfare and the environment: mode of action and toxicology", S. Matsunaka, D.H. Hutson and S.D. Murphy, eds., Pergamon Press, Oxford (1982)

19. H. Buchenauer, Hemmung der Ergosterinbiosynthese in _Ustilago avenae_ durch Triadimefon und Fluorotrimazol, _Z. Pflanzenkr. Pflanzenschutz_, 83: 363-367(1976)

20. K.A. Mitropoulos, G.F. Gibbons, C.M. Connell and R.A. Woods, Effect of triarimol on cholesterol biosynthesis in rat liver subcellular fractions, _Biochem. Biophys. Res. Commun_. 71: 892-900(1976)

21. S.D. Atkins, B. Morgan, K.H. Baggaley and J. Green, Isolation of 2,3-oxidosqualene from the liver of rats treated with 1-dodecylimidazole, a novel hypocholesteremic agent, _Biochem. J_. 130: 153-157(1972)

22. G. Matolsky, M. Kovács, M. Tüske and B. Tóth, Studies on the antifungal action of potential steroid inhibitors, _Neth. J. Plant Pathol_. 83 (Suppl. 1): 39-47(1977)

23. M.S. Marriott, Inhibition of sterol biosynthesis in _Candida albicans_ by imidazole-containing antifungals, _J. Gen. Microbiol_. 117: 253-255(1980)

24. M.J. Henry and H.D. Sisler, Effects of miconazole and dodecylimidazole on sterol biosynthesis in _Ustilago maydis_, _Antimicrob. Ag. Chemother_. 15: 603-607(1979)

25. H. Vanden Bossche, W. Lauwers, G. Willemsens, P. Marichal, F. Cornelissen and W. Cools, Molecular basis for the antimycotic and antibacterial activity of N-substituted imidazoles and triazoles. Inhibition of isoprenoid biosynthesis, _Pestic. Sci_. 15: 188-198(1984)

26. H. Vanden Bossche, G. Willemsens, W. Cools, F. Cornelissen, W.F. Lauwers and J.M. Van Cutsem, In vitro and in vivo effects of the antimycotic drug ketoconazole on sterol synthesis, _Antimicrob. Ag. Chemother_. 17: 922-928(1980)

27. H. Vanden Bossche, G. Willemsens, W. Cools, W.F. Lauwers and L. Le Jeune, Biochemical effects of miconazole on fungi. II. Inhibition of ergosterol biosynthesis in _Candida albicans_, _Chem. Biol. Interact_. 21: 59-78(1978)

28. G.W. Pye and M.S. Marriott, Inhibition of sterol C14-demethylation by imidazole containing antifungals, _Sabouraudia_ 20: 325-329(1982)

29. J.D. Weete, M.S. Sancholle and C. Montant, Effects of triazoles on fungi:II.lipid composition of _Taphrina deformans_, _Biochim. Biophys. Acta_ 753: 19-29(1983)

30. E. Ebert, J. Gaudin, W. Muecke, K. Ramsteiner, C. Vogel and H. Fuhrer, Inhibition of ergosterol biosynthesis by etaconazole in _Ustilago maydis_, _Z. Naturforsch_. 38c: 28-34(1983)

31. G. Cauwenbergh and H. Vanden Bossche, Terconazole pharmacology of a new antimycotic agent, _J.Reprod. Med_., 34: 588-592(1989)

32. M.S. Marriott and K. Richardson, The discovery and mode of action of fluconazole, _in_: "Recent trends in the discovery and development and evaluation of antifungal agents", R.A. Fromtling, ed., J.R. Prous Science Publishers, Barcelona(1987)

33. Y. Yoshida and Y. Aoyama, Interaction of azole fungicides with yeast cytochrome P-450 which catalyzes lanosterol 14α-demethylation, _in_: "In vitro and in vivo evaluation of antifungal agents", K. Iwata and H. Vanden Bossche, eds., Elsevier Science Publishers, Amsterdam (1986)

34. H. Vanden Bossche, P. Marichal, J. Gorrens, H. Geerts and P.A.J. Janssen, Mode of action studies. Basis for the search of new antifungal agents, _Ann. N.Y. Acad. Sci_. 544: 191-207(1988)

35. T.D. Rogerson, C.F. Wilkinson and K. Hetarski, Steric factors in the inhibitory interaction of imidazoles with microsomal enzymes, _Biochem. Pharmacol_. 26: 1039-1042(1977)

36. H. Vanden Bossche, P. Marichal, H. Geerts and P.A.J. Janssen, The molecular basis for itraconazole's activity against _Aspegillus fumigatus_, _in_: "_Aspergillus_ and aspergillosis", H. Vanden Bossche, D.W.R. Mackenzie and G. Cauwenbergh, eds., Plenum Press, New York (1988)

37. H. F. Merk, H. Mukhtar, I. Kaufmann, M. Das and D.R. Bickers, Human hair follicle benzo[a]pyrene and benzo[a]pyrene-7,8-diol metabolism: effect of exposure to a coal tar-containing shampoo, J. Invest. Dermatol., 88: 71-76(1987)

38. H. Mukhtar, B.J. Del Tito, M. Das, E.P. Cherniack, A.D. Cherniack and D.R. Bickers, Clotrimazole, an inhibitor of epidermal benzo(a)pyrene metabolism and DNA binding and carcinogenicity of the hydocarbon, Canc. Res. 44: 4233-4240(1984)

39. H. Vanden Bossche, G. Willemsens, P.A.J. Janssen, (1988) Cytochrome P-450-dependent metabolism of retinoic acid in rat skin microsomes: inhibition by ketoconazole, Skin Pharmacol 1: 176-185 (1988)

40. H.F. Merk, H. Mukhtar, B. Schutte, I. Kaufmann, M. Das and D.R.Bickers, 7-Ethoxyresorufin-O-deethylase activity in human hair roots: a potential marker for toxifying species of cytochrome P-450 isozymes, Biochem. Biophys. Res. Commun., 148: 755-761(1987)

41. D.T. Hart, W.J. Lauwers, G. Willemsens, H. Vanden Bossche and F.R. Opperdoes, Perturbation of sterol biosynthesis by itraconazole and ketoconazole in *Leishmania mexicana mexicana* infected macrophages, Mol. Biochem. Parasitol. 33: 123-134(1989)

42. G. Willemsens, W. Cools and H. Vanden Bossche, Effects of miconazole and ketoconazole on sterol synthesis in a subcellular fraction of yeast and mammalian cells,. in "The Host Invader Interplay" , H. Vanden Bossche, ed., Elsevier/ North Holland Biomedical Press, Amsterdam (1980)

43. J.D. Berman, L.J. Goad, D.H. Black and G.G. Holz Jr., Effects of ketoconazole on sterol synthesis by *Leishmania mexicana mexicana* amastigotes in murine macrophage tumor cells, Mol. Biochem. Parasitol. 20: 85-92(1986)

44. H. Vanden Bossche, G. Willemsens, P. Marichal, W. Cools and W. Lauwers, The molecular basis for the antifungal activity of N-substituted azole derivatives. Focus on R 51211, in: "Mode of action of antifungal agents", Cambridge University Press, Cambridge (1984)

45. Y. Aoyama, Y. Yoshida, Y. Sonoda and Y. Sato, Metabolism of 32-hydroxy-24,25-dihydrolanosterol by purified cytochrome P-450$_{14DM}$ from yeast. Evidence for contribution of the cytochrome to whole process of lanosterol 14α-demethylation, J. Biol. Chem. 262: 1239-1243(1987)

46. J.M. Trzaskos, R.T. Fisher and M.F. Favata, Mechanistic studies of lanosterol C-32 demethylation. Conditions which promote oxysterol intermediate accumulation during the demethylation process, J. Biol. Chem. 261: 16937-16942(1986)

47. J.M. Trzaskos and M.J. Henry, Comparative effects of the azole-based fungicide flusilazole on yeast and mammalian lanosterol 14α-methyl demethylase, Antimicrob. Ag. Chem. : 33, 1228-1231(1989).

48. R. De Felice, D.G. Johnson and J.N. Galgiani, Gynecomastia with ketoconazole, Antimicrob. Ag. Chemother. 19: 1073-1074(1981)

49. A. Pont, P.L. Williams, S. Azhar, R.E. Reirz, Ketoconazole blocks testosterone synthesis, Arch. Intern. Med. 142: 2137-2140(1982)

50. Th. Schürmeyer and E. Nieschlag, Ketoconazole-induced drop in serum and saliva testosterone, Lancet, Nov. 13, 1098-1099, 1982.

51. R.J. Santen, H. Vanden Bossche, J. Symoens, J. Brugmans and R. De Coster, Site of action of low dose ketoconazole on androgen biosynthesis in men, J. Clin. Endocrinol. Metab. 57: 732-736(1983)

52. R. De Coster, I. Caers, M.-C. Coene, W. Amery, D. Beerens and C. Haelterman, Effects of high dose ketoconazole therapy on the main plasma testicular and adrenal steroids in previously untreated prostatic cancer patients, Clin. Endocrinol. 24: 657-664(1986)

53. D. Feldman, Ketoconazole and other imidazole derivatives as inhibitors of steroidogenesis, Endocrine Rev. 7: 409-420(1986)

54. H. Vanden Bossche, R. De Coster and W. Amery, Pharmacological and clinical uses of ketoconazole, in: "Pharmacology and clinical uses of inhibitors of hormone secretion and action," B.J.A. Furr and A.E. Wakeling, eds., Baillière Tindall, London (1987)

55. W.K. Amery, R. De Coster and I. Caers, Ketoconazole: from an antimycotic to a drug for prostate cancer, Drug Develop. Res. 8: 299-307(1986)

56. J. Trachtenberg and J. Zadra, Steroid synthesis inhibition by ketoconazole: sites of action, Clin. Invest. Med. 11: 1-5(1988)

57. H. Vanden Bossche, W. Lauwers, G. Willemsens and W. Cools, The cytochrome P-450 dependent C17,20-lyase in subcellular fractions of the rat testis: differences in sensitivity to ketoconazole and itraconazole, in : "Microsomes and Drug Oxidations", A.R. Boobis, J. Caldwell, F. de Matteis and C.R. Elcombe, eds., Taylor and Francis, London (1985)

58. H. Vanden Bossche, W. Lauwers, G. Willemsens and W. Cools, Ketoconazole an inhibitor of the cytochrome P-450 dependent testosterone biosynthesis, in "Therapeutic Principles in Metastatic Prostatic Cancer. Progress in Clinical and Biological Research", Vol. 185A, F.H. Schroeder and B. Richards, eds., Alan R. Liss, New York (1985)

59. W. Lauwers, L. Le Jeune, H. Vanden Bossche and G. Willemsens, Identification of $17\alpha,20\alpha$-dihydroxyprogesterone in testicular extracts after incubation with ketoconazole, Biomed. Mass. Spectro. 12: 296-301(1985)

60. J. Rajfer, S. Sikka, H.W. Xie and R.S. Swerdloff, Effect of ketoconazole on steroid production in rat testis, Steroids 46: 867-881(1985)

61. P. Kan, M.A. Hirst and D. Feldman, Inhibition of steroidogenic cytochrome P-450 enzymes in rat testis by ketoconazole and related imidazole anti-fungal drugs, J. Steroid. Biochem. 23: 1023-1029(1985)

62. Y. Higashi, M. Omura, K. Suzuki, H. Inano and H. Oshima, Ketoconazole as a possible universal inhibitor of cytochrome P-450 dependent enzymes: its mode of inhibition, Endocrinol. Japan 34: 105-115(1987)

63. M. Namiki, M. Kitamura, E. Buczko and M.L. Dufau, Rat testis P-$450_{17\alpha}$ cDNA: the deduced amino acid sequence, expression and secondary structural configuration, Biochem. Biophys. Res. Commun. 157: 705-712 (1988)

64. H.R. Fevold, M.C. Lorence, J.L.McCarthy, J.M. Trant, M. Kagimoto, M.R. Waterman and J.I. Mason, Rat P$450_{17\alpha}$ from testis: characterization of a full-lenght cDNA encoding a unique steroid hydroxylase capable of catalyzing both Δ^4- and Δ^5-steroid-17,20-lyase reactions, Mol. Endocrinol. 3: 968-975 (1989)

65. T. Yanase, M. Kagimoto, S. Suzuki, K. Hashiba, E.R. Simpson and M.R. Waterman, Deletion of a phenylalanine in the N-terminal region of human cytochrome P-$450_{17\alpha}$ results in partial combined 17α-hydroxylase/17,20-lyase deficiency, J. Biol. Chem. 264: 18076-18082 (1989)

66. R.M. Couch, J. Muller, Y.S. Perry and J.S.D. Winter, Kinetic analysis of inhibition of human adrenal steroidogenesis by ketoconazole, J. Clin. Endocrinol. Met. 65: 551-554 (1987)

67. M. Ayub and M.J. Levell, Inhibition of human adrenal steroidogenic enzymes in vitro by imidazole drugs including ketoconazole, J. steroid. Biochem. 32: 515-524(1989)

68. J. Trachtenberg, N. Halpern and A. Pont, Ketoconazole a novel and rapid treatment for advanced prostatic cancer, J. Urol. 130: 152-153(1983)

69. J. Trachtenberg, Ketoconazole therapy in advanced prostatic cancer, J. Urol. 132: 61-63(1984)

70. J. Trachtenberg, Ketoconazole therapy in advanced prostatic cancer, J. Urol. 137: 959(1987)

71. W. Heyns, A. Drochmans, E. vander Schueren and F. Verhoeven, Endocrine effects of high dose ketoconazole therapy in advanced prostatic cancer, Acta Endocrinol. 110: 276-283(1985)

72. J.A. Witjes, P.F. del Moral, A.D.H. Geboers, L. Baert, J. Casselman, M. Stragier and F.M.J. Debruyne, Ketoconazole high dose in the treatment of metastatic prostate carcinoma, Akt. Urol. 20: 29-32(1989)

73. T. Eichenberger, J. Trachtenberg, P. Toor and A. Keating, Ketoconazole: a possible direct cytotoxic effect on prostate carcinoma cells, J. Urol. 141: 190-191(1989)

74. H. Vanden Bossche, G. Willemsens, D. Bellens, I. Roels and P.A.J. Janssen. From 14α-demethylase inhibitors in fungal cells to androgen and oestrogen biosynthesis inhibitors in mammalian cells, Biochem. Soc. Trans.18: 10-13(1990)

75. J.I. Mason, B.R. Carr and B.A. Murry, Imidazole antimycotics: selective inhibitors of steroid aromatization and progesterone hydroxylation, Steroids 50: 179-189(1987)

76. C.J. Corbin, S. Graham-Lorence, M. McPhaul, J.I. Mason, C.R. Mendelson and E.R. Simpson, Isolation of a full-lenght cDNA insert encoding human aromatase system cytochrome P-450 and its expression in nonsteroidogenic cells, Proc. Natl. Acad. Sci. 85: 8948-8952(1988)

77. W. Wouters, R. De Coster, M. Krekels, J. van Dun, D. Beerens, C. Haelterman, A. Raeymaekers, E. Freyne, J. Van Gelder, M. Venet and P.A.J. Janssen, R 76713, a new specific non-steroidal aromatase inhibitor, J. steroid Biochem. 32: 781-788(1989)

78. G. Cauwenbergh and J. Van Cutsem, Role of animal and human pharmacology in antifungal drug design, Ann. N.Y. Acad. Sci. 544: 264-269(1988)

79. J. Van Cutsem, F. Van Gerven and P.A.J. Janssen, The in vitro and in vivo antifungal activity of itraconazole, in: "Recent trends in the discovery,development and evaluation of antifungal agents", R.A. Fromtling, R.A., ed., J.R. Prous publishers, S.A., Barcelona (1987)

80. H. Vanden Bossche, D. Bellens, W. Cools, J. Gorrens, P. Marichal, H. Verhoeven, G. Willemsens, R De Coster, D. Beerens, C. Haelterman, M.-C. Coene, W. Lauwers and L. Le Jeune, Cytochrome P-450: target for itraconazole, Drug Develop. Res. 8: 287-298(1986)

81. H. Vanden Bossche, Itraconazole: a selective inhibitor of the cytochrome P-450 dependent ergosterol biosynthesis, in: "Recent trends in the discovery and development and evaluation of antifungal agents", R.A. Fromtling, ed., J.R. Prous Science Publishers, Barcelona(1987)

82. K. Lavrijsen, J. Van Houdt and D. Thijs, Interaction of miconazole, ketoconazole and itraconazole with rat liver microsomes, Xenobiotica 17: 45-47(1987)

83. H. Vanden Bossche, G. Willemsens, P. Marichal, W. Cools and W. Lauwers, The molecular basis for the antifungal activities of N-substituted azole derivatives. Focus on R 51 211, in: "Mode of Action of Antifungal Agents",A.P.J. Trinci and J.F. Ryley, eds., Cambridge University Press, Cambridge (1984)

84. K. Lürssen, (1988) Triazole plant growth regulators: effects and mode of action, in: "Sterol biosynthesis inhibitors. Pharmaceutical and agrochemical aspects", D. Berg and M. Plempel, eds., Ellis Horwood Ltd, Chichester, England (1988)

85. W. Köller, Isomers of sterol synthesis inhibitors: fungicidal effects and plant growth regulator activities, Pestic. Sci. 18: 129-147(1987)

MECHANISM OF METABOLIC ACTIVATION OF NITROIMIDAZOLES

Anthony Y. H. Lu and Peter G. Wislocki

Merck Sharp & Dohme Research Laboratories
Rahway, NJ 07065 USA

INTRODUCTION

5-Nitroimidazoles are highly effective therapeutic agents (1) against a variety of anaerobic bacteria and protozoa (Table 1). For this reason, the medicinal use of these agents has been explored in humans and in food-producing animals. For example, metronidazole (1-hydroxyethyl-2-methyl-5-nitroimidazole) is used in human therapy because of its proven activity against trichomoniasis as well as its prophylactic and therapeutic efficacy against a wide range of gram positive anaerobic bacteria (2, 3). Ronidazole [1-methyl-5-nitroimidazole-2-yl)-methylcarbamate] is a highly effective drug for the treatment of turkey blackhead and swine dysentery. However, as a class, 5-nitroimidazoles are genotoxic. Various studies have demonstrated the critical importance of the reduction of the nitro group in their cytotoxicity, mutagenicity, carcinogenicity and interaction with DNA (2-5). To investigate the mechanism of metabolic activation of 5-nitroimidazoles, the covalent binding of ronidazole to proteins has been extensively studied in rats.

In Vivo Metabolism and Disposition of Ronidazole

When ronidazole labeled with ^{14}C at various positions is administered to rats, radioactive residues are widely distributed in all tissues, with the highest levels found in liver and muscle (6). The feces generally contain 40-50% of the dose and urine contains 30-40% of the labeled drug. Little or no ronidazole can be detected after several days in all tissues, indicating that the drug is extensively metabolized. Analysis of the liver and muscle samples after two-day treatment shows that a great portion of the total radioactivity in these tissues is protein-bound.

In Vitro Metabolic Activation and Protein Binding of Ronidazole

Since protein-bound residues are abundant in tissues of ronidazole-treated rats, *in vitro* studies are carried out to investigate the mechanism of metabolic activation and protein

Molecular Aspects of Monooxygenases and Bioactivation of Toxic Compounds
Edited by E. Arınç *et al., Plenum Press, New York,*

Table 1. Structure and Therapeutic Use of Some 5-Nitroimidazoles

Compound	Structure	Therapeutic Use
Metronidazole (Flagyl)		Anti-protozoal (Trichomoniasis, giardiasis, amebiasis) Anti-bacterial (Vincent's Disease)
Ornidazole		Trichomoniasis, Amebiasis, Meningitis
Dimetridazole		Turkey histomoniasis Swine dysentery
Ronidazole		Turkey histomoniasis, Swine dysentery

binding of ronidazole. These studies are aimed to provide the following information.

A. The enzyme activating system - The generation of reactive intermediate from ronidazole and the formation of ronidazole-protein adducts require NADPH and anaerobic conditions, suggesting that nitro reduction is essential for metabolic activation (7). Covalent-binding is totally inhibited at high oxygen concentration. Inhibition studies suggest that the cytochrome P450-containing electron transport chain is involved in the metabolic activation of ronidazole, but purified NADPH-cytochrome P450 reductase by itself can carry out this reaction (8). Although xanthine oxidase can also activate ronidazole to produce protein-bound metabolite, its physiological role may be limited since allopurinol, an inhibitor of xanthine oxidase, does not inhibit covalent binding catalyzed by liver homogenates. Other inhibitor studies indicate that neither cytosolic DT-diaphorase, alcohol dehydrogenase, nor aldehyde oxidase are involved in the activation of ronidazole to protein-bound metabolite(s).

B. Protein target - The reactive metabolite generated from nitro reduction of ronidazole binds nonspecifically to a large number of proteins in the incubation system (8). Covalent binding does not appear to be related to protein function. Thus, bovine serum albumin is readily alkylated, and the radioautography of protein bands obtained from microsomal incubation shows that many proteins are alkylated, including cytochrome P450 and NADPH-cytochrome P450 reductase.

C. The nucleophile target on proteins - The primary target of protein alkylation is the cysteine thiol group (9). This is supported by the finding that covalent binding is inhibited by cysteine, but not by other amino acids. Furthermore, covalent binding of ronidazole to bovine serum albumin is strongly inhibited when the thiol groups of the albumin are specifically blocked with methyl methanethiosulfonate but restored when the blocking group is reductively removed.

D. Chemical model - The chemical reduction of the nitro group of ronidazole in aqueous solution is known to result in extensive fragmentation of the imidazole ring, producing simple molecules such as ammonia, carbon dioxide, cyanide, methylamine, and acetamide. However, when ronidazole is reduced anaerobically by dithionite in the presence of excess amounts of cysteine, the isolated cysteine adducts retain the intact imidazole nucleus (Fig. 1). This study also indicates that cysteine can be attached to the reduced ronidazole at either the 2-methylene or the C-4 position (10).

E. Structure of bound metabolite - Using ronidazole radiolabeled at different positions for covalent binding studies (9, 11), it is established that the major ronidazole-protein adduct retains the imidazole nucleus but loses the carbamoyl group and the proton at the C-4 position (Table 2.) Structure-activity studies suggest that cysteine is added primarily at the 2-methylene position. No evidence is obtained for the dicysteine protein adduct formation.

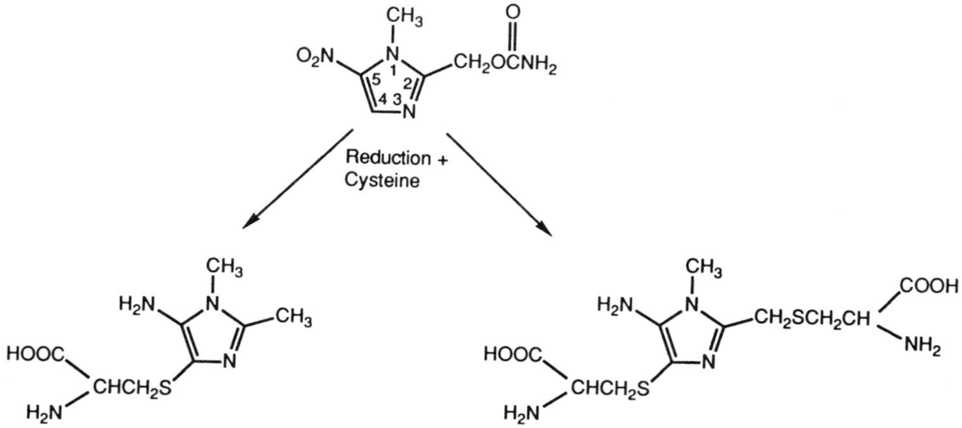

Figure 1. Mono- and di-cysteine ronidazole adduct
formation following anaerobic reduction
of ronidazole by dithionite in the
presence of excess amounts of cysteine.

Table 2. In Vitro Covalent Binding of Ronidazole and
Derivatives to Rat Liver Microsomal Proteins

Compound	Label Position	% of Maximum Protein Binding
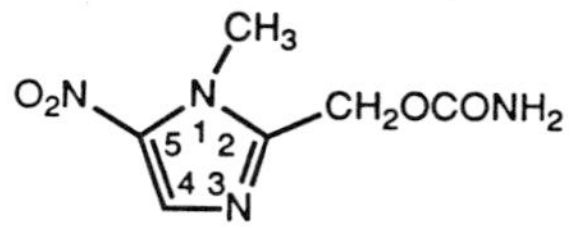	$4,5\text{-}^{14}C$ $2\text{-}^{14}CH_2$ $1\text{-}N\text{-}^{14}CH_3$ $O^{14}CONH_2$	100 91 92 15
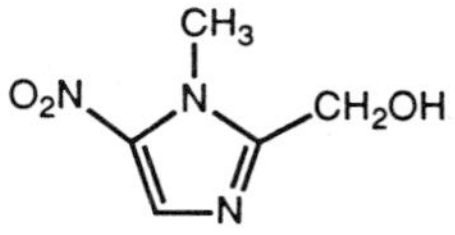	$2\text{-}^{14}CH_2OH$	15
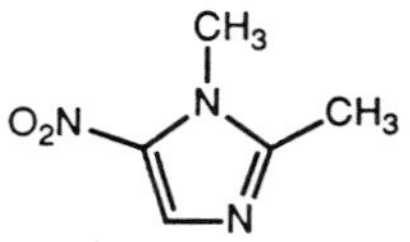	$1\text{-}N\text{-}^{14}CH_3$	15
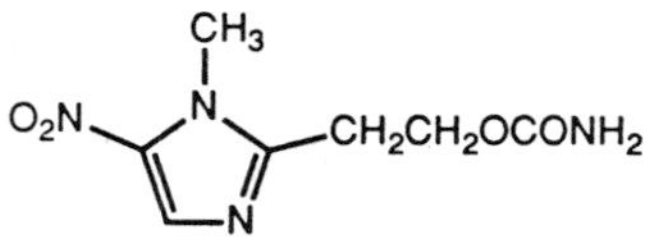	$1\text{-}N\text{-}^{14}CH_3$	19
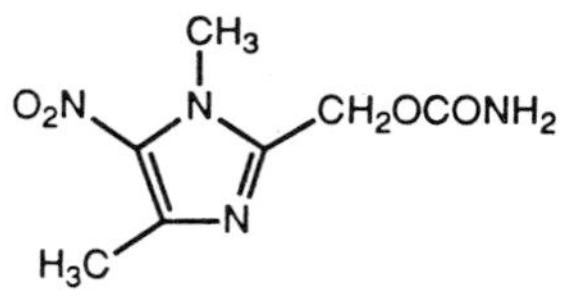	$1\text{-}N\text{-}^{14}CH_3$	15

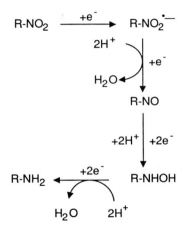

Figure 2. The stepwise reduction of a
nitro group.

Since nitro reduction involves a 6-electron stepwise process (Fig. 2), the nitrogen oxidation state involved in covalent binding is investigated. In this experiment, ronidazole is reduced by stoichiometric amounts of dithionite under anaerobic conditions (12). Two moles of dithionite are required for the complete reduction of one mole of ronidazole, the maximum release of the 4-proton and the maximum covalent binding to proteins, suggesting that the 4-electron product, hydroxylamine, is the reactive species.

F. Kinetics - In a series of experiments, it is established that virtually all of the ronidazole metabolized results in the formation of the reactive metabolite via a mechanism involving the release of the 4-proton (Table 3). For every 20 molecules of reactive metabolite formed, only one molecule alkylates proteins whereas the other 19 molecules are presumably degraded in water to generate ring fragment products.

G. Mechanism of activation - Based on these data, a mechanism for ronidazole activation is proposed (Fig. 3). In this scheme, ronidazole is shown to undergo 4-electron enzymatic reduction to the hydroxylamine (II) which then activates the C-4 position to nucleophilic attack by either protein thiols or water to produce IIIa and IIIb. Subsequent loss of the 4-proton and carbamic acid yields a Michael-like acceptor capable of reacting with thiol nucleophiles at the 2-methylene positions (Vb) or with water producing Va. The formation of Vb is preferred since the expected product, carboxymethylcysteine, has been identified from acid hydrolysates of the protein adduct. Even though the addition of a protein cysteine residue to the 2-methylene carbon may represent the major pathway adduct formation, the extent of this pathway in protein alkylation has not been established.

In Vivo Protein Binding of Ronidazole in Rats

A. Multiple label study - Ronidazole labeled at different positions is administered in rats. Six hours after dosing, liver and muscle tissues are collected for protein-bound residue analysis. This study establishes that *in vivo*, protein bound residues also retain the intact imidazole nucleus but have lost the tritium at the C-4 position (13).

B. Acid hydrolysis - Despite extensive efforts, attempts to isolate the cysteine adduct from proteins have been unsuccessful. Alternative approaches are used to characterize the acid-hydrolyzed fragments. HPLC analysis of the *in vitro* (microsomal incubation) and *in vivo* (six-hour rat liver and muscle) protein-bound samples obtained from 2-methylene ^{14}C-labeled ronidazole gives identical radiochromatographic profiles (13). In addition, acid hydrolysis of the *in vitro* and *in vivo* samples generates similar amounts of methylamine from N-methyl-labeled ronidazole, carboxymethyl cysteine from 2-methylene-labeled ronidazole, and oxalic acid from 4,5-ring-labeled ronidazole (13). These results demonstrate that the mechanism of metabolic activation and covalent binding of ronidazole is identical for the *in vitro* microsomal system and the intact animal.

Mechanism of Mutagenicity of Ronidazole

Using the Ames' test, the mutagenicity of ronidazole and its structural analogs is examined both in the presence and in the absence of the S-9. Table 4 shows that the presence of a nitro group (most important), a carbamoyl group, and an unsubstituted C-4 position are essential for the full mutagenic activity of ronidazole (14). The amine derivatives are totally inactive. This structural requirement for mutagenicity shares the same structural features required for covalent binding, suggesting that the same reactive intermediate may be responsible for the mutagenicity and protein binding of ronidazole.

Mechanism Based Structure Modifications

Based on the knowledge of the mechanism of ronidazole metabolic activation, attempts are made to design non-mutagenic 5-nitroimidazoles. Since the C-4 position is critical for the generation of the reactive intermediate and the mechanism of activation, substituent at the C-4 position is incorporated in the molecule. Interestingly, some 5-nitroimidazoles containing a substituent at the C-4 position have greatly decreased mutagenic activity but still retain good antiprotozoal activity (15). For example, 1,2-dimethyl-4-(2-hydroxyethyl)-5-nitroimidazole shows good antitrichomonal activity *in vitro* but possessing only 0.28% of the mutagenicity of ronidazole. Thus, it appears that the reactive species of the 5-nitroimidazoles responsible for the antiprotozoal activity is different from those responsible for the mutagenicity and covalent binding of this class of compound.

Table 3. Stoichiometry in NADPH-Dependent, Metabolism-Mediated Covalent Binding of Ronidazole to Microsomal Proteins

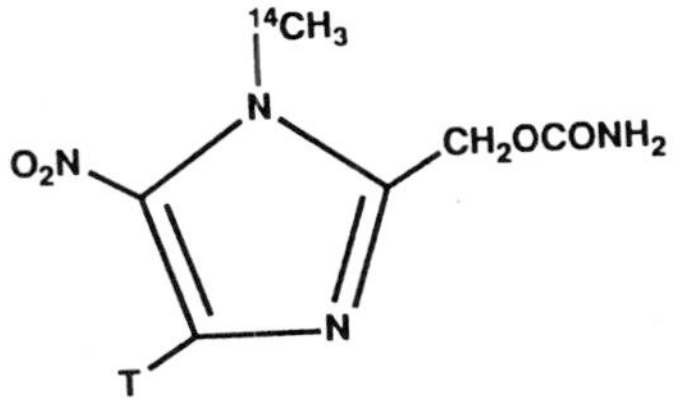

$$\frac{\text{RONIDAZOLE METABOLIZED}}{^{14}\text{C-COVALENT BINDING}} = 20$$

$$\frac{\text{TOTAL REACTIVE INTERMEDIATE FORMED}}{^{14}\text{C-COVALENT BINDING}} = 18$$

$$\frac{\text{TRITIUM RELEASED}}{^{14}\text{C-COVALENT BINDING}} = 17$$

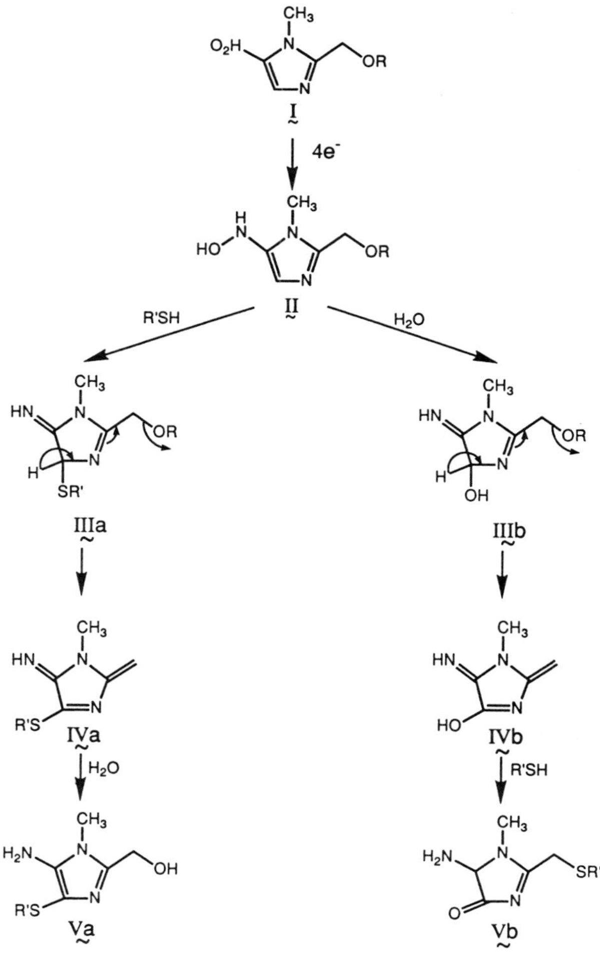

Figure 3. Proposed mechanism for ronidazole activation.

Table 4. Relative Mutagenic Activities
of Ronidazole and Derivatives

Compound	% of Activity

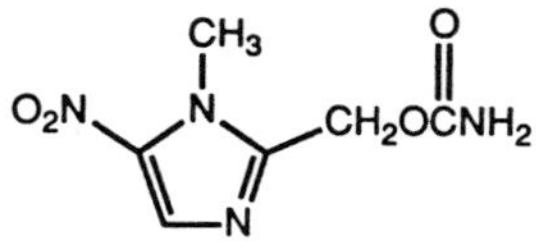

100

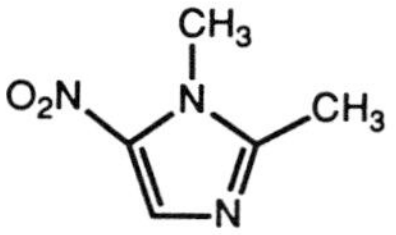

10

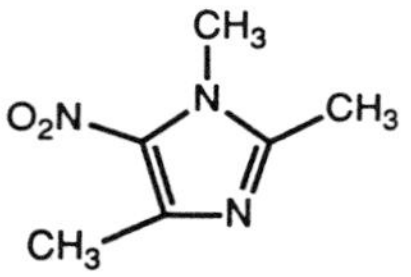

1

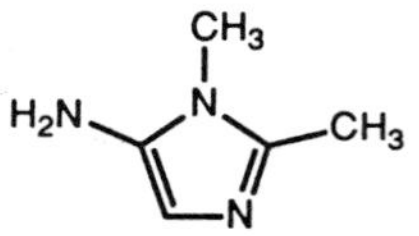

0

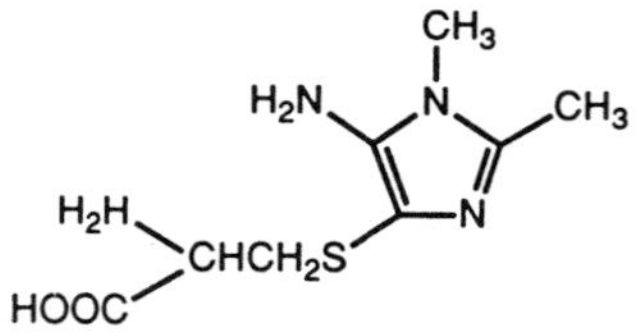

0

Acknowledgment

We wish to thank Terry Rafferty for her help in preparing this manuscript.

References

1. J. F. Rossignol, H. Maisonneuve and Y. W. Cho, Nitroimidazoles in the treatment of trichomoniasis, giardiasis and amebiasis, Int. J. Clin. Pharmacol. Ther. Toxicol. 22:63 (1984).
2. D. I. Edwards, Mechanisms of selective toxicity of metronidazole and other nitroimidazole drugs, Br. J. Vener. Dis. 56:285 (1980).
3. M. Muller, Mode of action of metronidazole on anaerobic bacteria and protozoa, Surgery 93:165 (1983).
4. M. Muller, Reductive activation of nitroimidazoles in anaerobic microorganisms, Biochem. Pharmacol. 35:37 (1986).
5. P. Goldman, R. L. Koch, T. C. Yeung, E. J. T. Chrystal, B. B. Beaulieu, M. A. McLafferty and G. Sudlow, Comparing the reduction of nitroimidazoles in bacteria and mammalian tissues and relating it to biological activity, Biochem. Pharmacol. 35:43 (1986).
6. F. J. Wolf, R. Alvaro, L. R. Chapin, M. L. Green, D. E. Wolf, F. R. Koniuszy, J. J. Steffens and T. A. Jacob, Tissue residue studies with ronidazole: effect of label site on total radioactivity content of rat tissues, J. Agric. Food Chem. 32:706 (1984).
7. S. B. West, P. G. Wislocki, K. M. Fiorentini, R. Alvaro, F. J. Wolf and A. Y. H. Lu, Drug residue formation from ronidazole, a 5-nitroimidazole. I. Characterization of *in vitro* protein alkylation, Chem.-Biol. Interactions 41:265 (1982).
8. S. B. West, P. G. Wislocki, F. J. Wolf and A. Y. H. Lu, Drug residue formation from ronidazole, a 5-nitroimidazole. II. Involvement of microsomal NADPH-cytochrome P450 reductase in protein alkylation *in vitro*, Chem.-Biol. Interactions 41:281 (1982).
9. G. T. Miwa, S. B. West, J. S. Walsh, F. J. Wolf and A. Y. H. Lu, Drug residue formation from ronidazole, a 5-nitroimidazole. III. Studies on the mechanism of protein alkylation *in vitro*, Chem.-Biol. Interactions, 41:297 (1982).
10. P. G. Wislocki, E. S. Bagan, W. J. A. VandenHeuvel, R. W. Walker, R. F. Alvaro, B. H. Arison, A. Y. H. Lu and F. J. Wolf, Drug residue formation from ronidazole, a 5-nitroimidazole. V. Cysteine adducts formed up reduction of ronidazole by dithionite or rat liver enzymes in the presence of cysteine, Chem.-Biol. Interactions, 49:13 (1984).
11. G. T. Miwa, R. Wang, R. Alvaro, J. S. Walsh and A. Y. H. Lu, The metabolic activation of ronidazole to reactive metabolites by mammalian, cecal bacterial and *T. foetus* enzymes, Biochem. Pharmacol. 35:33 (1986).
12. G. L. Kedderis, L. S. Argenbright and G. T. Miwa, Mechanism of reductive activation of a 5-nitroimidazole by flavoproteins: model studies with dithionite, Arch. Biochm. Biophys. 262:40 (1988).

13. G. T. Miwa, R. F. Alvaro, J. S. Walsh, R. Wang and A. Y. H. Lu, Drug residue formation from ronidazole, a 5-nitroimidazole. VII. Comparison of protein-bound products formed *in vitro* and *in vivo*, Chem.-Biol. Interactions, 50:189 (1984).
14. P. G. Wislocki, E. S. Bagan, M. M. Cook, M. O. Bradley, F. J. Wolf and A. Y. H. Lu, Drug residue formation from ronidazole, a 5-nitroimidazole. VI. Lack of mutagenic activity of reduced metabolites and derivatives of ronidazole. Chem.-Biol. Interactions 49:27 (1984).
15 J. S. Walsh, R. Wang, E. Bagan, C. C. Wang, P. Wislocki and G. T. Miwa, Structural alterations that differentially affect the mutagenic and antitrichomonal activities of 5-nitroimidazoles, J. Med. Chem. 30:150 (1987).

BENZENE METABOLISM

Robert Snyder and Suzanne Pirozzi Chatterjee

Department of Toxicology and Pharmacology
Rutgers The State University of New Jersey
Piscataway, New Jersey

Exposure to benzene in industry has been a problem for over a century and today it has become a problem for the entire population with the prevalence of benzene in gasoline. The first reports of the metabolites of benzene in urine included the finding of phenol (1); hydroquinone and catechol (2); trans, trans-muconic acid (3) and phenylmercapturic acid (4). The availability of [^{14}C] after WWII led R.T. Williams to suggest to D.V. Parke that he should synthesize radiolabeled benzene and study its

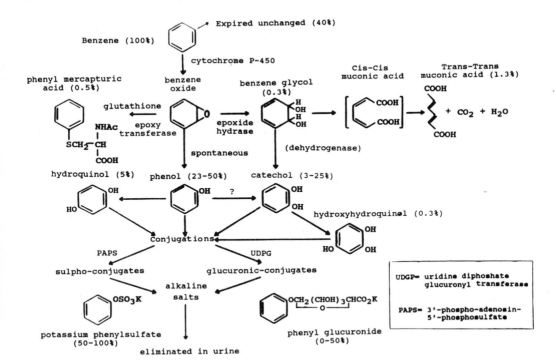

Figure 1. Initial In Vivo metabolic pathways for proposed for benzene based on the early studies of benzene metabolism. (Taken from Ref. 7 and reproduced with permission.)

Molecular Aspects of Monooxygenases and Bioactivation of Toxic Compounds
Edited by E. Arınç *et al., Plenum Press, New York,*

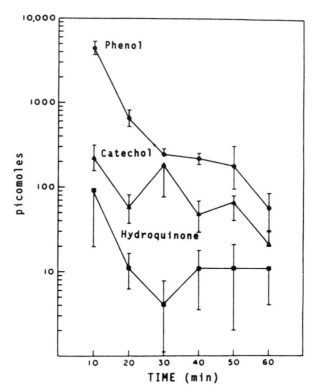

Figure 2. Radioactive metabolites recovered from blood following the introduction of [^{14}C]benzene into bone marrow. (Taken from Ref. 19 and reproduced with permission.)

metabolism in rabbits. The result was the first quantitative evaluation of the metabolites of benzene (5,6). About 20% of the radioactivity could be recovered as phenol, 5% as hydroquinone, approximately 2% as trans, trans-muconic acid, 1% as catechol, and 0.5% as phenylmercapturic acid. Based on these studies the metabolic pathway seen in Fig. 1 was proposed and with slight modification was the accepted pathway for many years. Thus, benzene is hydroxylated via cytochrome P-450 (8) to yield benzene oxide (9,10). Benzene oxide can spontaneously rearrange to form phenol which can then be further hydroxylated via cytochrome P-450 (11,12) to yield hydroquinone, catechol and trihydroxybenzene. Although suggested to be a metabolite by Parke and Williams (6) the formation of 1,2,4-trihydroxybenzene has not been well studied. These phenolic metabolites appear in the urine principally as conjugates of glucuronic acid or ethereal sulfate.

It is generally accepted that benzene must undergo metabolic activation to produce toxicity (13-15). The data suggest that the first and principle site of benzene metabolism is the liver; liver microsomes were reported to metabolize benzene to phenol (8,16,17). Furthermore, partial hepatectomy, coinciding with marked reduction in benzene metabolism, correlated with protection against benzene toxicity (15). Sawahata and Neal (11) demonstrated that liver microsomes also metabolize phenol to hydroquinone and catechol. The metabolites of benzene then enter the blood stream and subsequently accumulate in the bone marrow (18), the major site of benzene toxicity. Irons et al. (19) reported that the infusion of benzene into the bone marrow results in benzene metabolism (Fig.2); however, the relative capacity of the bone marrow to metabolize benzene appears to be much less than the liver.

376

Table 1. Inhibition of [³H]Benzene Metabolism in Mouse Liver
Microsomes   CO[a]

Gas phase		CO:O₂ ratio	% Control activity
%O₂	%CO		
4.0	0	0	100
4.0	2.0	0.5	65
4.0	4.0	1.0	34
4.4	9.5	2.16	25
3.6	31.0	8.6	11
4.1	49.2	12.0	8

[a]The data represents mean values from 3 experiments. (Taken
from Ref. 8 and reproduced with permission.)

Table 2. ³H-Benzene Metabolism in Rabbit Bone Marrow Microsomes

Sample	pmoles/mg protein/45 min			
	(H)	(C)	(P)	(X)
Zero Time	14.7	1.0	18.9	10.9
Air - TPNH	11.2	4.3	18.2	9.3
Air + TPNH	13.5	6.0	89.0	24.8
CO - TPNH	14.7	4.4	19.0	9.4
CO + TPNH	12.5	5.7	34.0	13.5

[³H]Benzene (3.17 C_i/mmole - 145 μM) was incubated with rabbit bone
marrow microsomes (10 mg/incubation - 0.025 nmole P-450/mg) in the
presence or absence of a TPNH generating system (3.8 mM TPN, 5.9 mM
G6P and 10 units G-6PD) in a total volume of 2.0 ml in septum sealed
vials for 45 min at 37°C. The CO/O_2 ratio in gassing experiments was
9:1. $HClO_4$ was added to terminate the reaction and the supernatant
fraction was removed. After extraction of metabolites into ether the
concentration of metabolites were determined by   HPLC.   H =
hydroquinone;   C = catechol; P = phenol; X = unidentified metabolite.
(Taken from Ref. 20 and reproduced with permission.)

The evidence indicates a role for cytochrome P-450 in the metabolic
activation of benzene in microsomes of mouse liver (Table 1) (8) and rat
bone marrow (Table 2) (20). Benzene hydroxylation has been postulated to
occur via either of two mechanisms.   Jerina et al. (9) suggested that
hydroxylation of benzene occurred via the intermediate formation of an
epoxide (benzene oxide) which could rearrange to form phenol, conjugate
with GSH to form a premercapturic acid or become hydrated via epoxide
hydrolase to yield benzene dihydrodiol. Alternatively, it has been
suggested that radical mechanisms may play a role in benzene metabolism
(11). Ingelman-Sundberg et al. (21, 22) have proposed that cytochrome P-
450, functioning in an uncoupled manner, could produce hydroxyl radicals
which could directly form phenol (Fig. 3). Gorson and Coon (23) have
suggested that benzene hydroxylation might occur via a free radical
mediated mechanism at low concentrations of benzene, whereas the epoxide
pathway predominates at higher concentrations of benzene. These studies

were performed in vitro using reconstituted mixed function oxidase systems. If free radical hydroxylation of benzene is a significant metabolic pathway in vivo biphenyl would be an expected metabolite. As yet, neither biphenyl, nor any of its known metabolites, have been identified in the urine of humans or animals exposed to benzene.

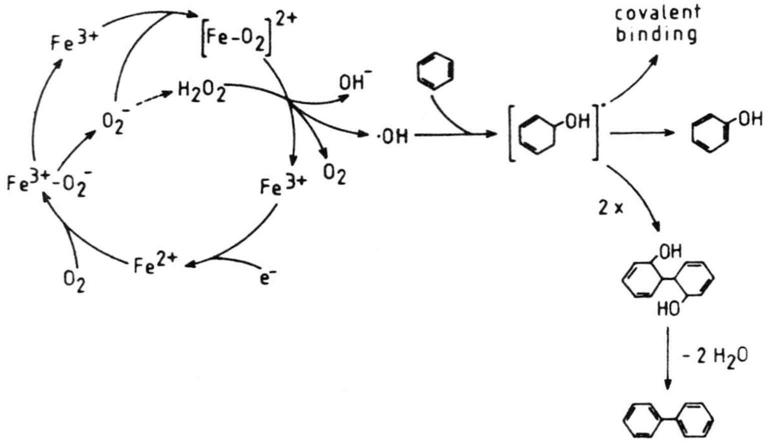

Figure 3. Proposed scheme for the metabolic activation of benzene and related compounds. (Reproduced from Ref. 21 with permission.)

Table 3. Benzene Hydroxylation by Purified Rabbit P450 Isozymes

Isozyme	Product formed (nmole/min/nmole P450)		
	Phenol	Hydroquinone	Total
2	3.66	0.17	3.83
2 (0.25 mM)	0.64	0.03	0.67
3a	4.60	0.38	4.98
3a (0.3 mM)	2.48	0.60	3.08
3b	0.45	0.17	0.62
3c	0.19	0.17	0.36
4	0.40	0.12	0.52
6	0.60	0.12	0.72

Reaction mixtures usually contained 2.0 mM benzene (except for isozymes 2 and 3a which utilized 0.25 mM and 0.3 mM, respectively, as noted in the Table), 0.1 μM P450, 0.3 μM P450 reductase, 30 μg/ml lipid, 0.4 μM b_5, 0.1 M potassium phosphate buffer (pH 6.8 for isozyme 3a and pH 7.4 for the other isozymes), 1.0 mM ascorbate, and 1.0 mM NADPH. The reactions were run for 10 min at 37°C and quenched by the addition of ice-cold ether. The products were determined by HPLC as described under Methods. The values represented the means of two to three separate experiments each run in duplicate. Reproduce from Ref. 29 with permission.

The microsomal metabolism of benzene can be stimulated by some enzyme inducers. Benzene metabolism can be induced by phenobarbital and benzene in rabbit and rat liver (16), by DMSO in rat liver (17) and by benzene in mouse liver (8). Saito et al. (24) demonstrated that benzene, but not any of its hydroxylated metabolites, induced benzene metabolism in rat liver. Post and Snyder (25,26) showed that the apparent Km for benzene metabolism induced by benzene in rat liver was lower by an order of magnitude than the Km for phenobarbital-induced benzene metabolism and that the pH optima of the resulting enzymes in microsomes were different, suggesting that at least two forms of cytochrome P-450 could hydroxylate benzene.

Table 4. Phenol Hydroxylase Activity of Purified Rabbit P450 Isozymes

	Hydroquinone formed (nmol/min/nmol P450)	
Isozyme	$-b_5$	$+b_5$
2	0.06	0.56
3a	9.00	15.4
3b	0.29	0.63
3c	0.04	0.39
4	0.16	0.78
6	0.31	0.81

The reaction mixtures contained 0.1 μM P450, 0.3 μM P450 reductase, 0.4 μM b_5 when indicated, 30 μg/ml lipid, 0.1 M potassium phosphate buffer (pH 6.8 for 3a and pH 7.4 for the others), 1 mM ascorbate, 0.1 mM phenol, and 1.0 mM NADPH to initiate the reaction. After a 10-min incubation at 37°C the reactions were quenched with ether and hydroquinone was determined by HPLC as described under Methods. The only product observed was hydroquinone. The rates represent the means of two to four experiments each run in duplicate. (Reproduce from Ref. 29 with permission.)

Cytochrome P-450IIE1 appears to be the most effective enzyme in the metabolism of benzene (27-30). Koop (29) demonstrated the effectiveness of rabbit liver cytochrome 3a (equivalent to IIE1) in metabolizing both benzene (Table 3) and phenol (Table 4). Both the phenobarbital-induced cytochrome P-450 IIB1 and the isoniazide-induced cytochrome P-450 IIE1 metabolize benzene; the metabolism is enhanced by the addition of cytochrome b_5 (Table 5). In addition to stimulating overall metabolism, cytochrome b_5 increases the relative proportion of hydroquinone formed from benzene by cytochrome P-450 IIE1 (Table 5). Investigations of the metabolism of benzene have also focused on the fate of the major metabolite, phenol. The metabolism of phenol by rat liver microsomes occurs according to first order kinetics at various concentrations of phenol (Fig. 4) (12). Both benzene and phenobarbital induced phenol metabolism although benzene was more effective and resulted in the production of a greater amount of hydroquinone.

Table 5. The Effect of Cytochrome B_5 on Benzene Metabolism by
Cytochromes P450IIB1 and P450IIE1

Cytochromes	Benzene	- cytochrome B_5			+ cytochrome B_5		
		%P	%HQ	total nmoles formed	%P	%HQ	total nmoles formed
P450IIB1	0.32 mM	100	0	0.22	100	0	0.50
	1.0 mM	90	10	0.78	94	6	1.60
	4.0 mM	95	5	5.88	95	5	10.60
	12.0 mM	95	5	13.60	96	4	21.16
P450IIE1	20 μM	84	16	0.19	60	40	1.16
	80 μM	92	8	0.50	72	28	2.85
	320 μM	94	6	1.26	86	14	5.44
	800 μM	94	6	2.55	91	9	8.21

Reaction mixtures contained, 0.1 nmol cytochrome P-450, 1200 units NADPH cytochrome P-450 reductase, 15 ug dilauroylphosphatidylcholine and 1.0 mM NADPH in a total volume of 0.25 ml. [^{14}C]benzene was used in all experiments and cytochrome b_5, when present, was added in equimolar concentration to the cytochrome P-450. Subsequent to stopping the reaction (10 min for P450IIE1; 20 min for P450IIB1), an aliquot of the reaction mixture was filtered and spiked with unlabelled standards. Metabolites were separated by HPLC and quantitated by liquid scintillation counting.

The finding of muconic acid in urine following benzene treatment led Parke and Williams (5,6) to suggest that the benzene ring might be opened. The possibility that the benzene ring might be opened to yield toxic metabolites such as muconaldehyde was proposed by Goldstein et al. (31) and demonstrated in vitro by Latriano et al. (32). The mechanism for the formation of muconic acid from muconaldehyde has been suggested to occur via a two-step oxidation by aldehyde dehydrogenases (33). Muconaldehyde has also been demonstrated to be reduced to a dialcoholic product (muconal) by alcohol dehydrogenase (34). It is clear that further studies are needed to elucidate the routes of muconaldehyde metabolism.

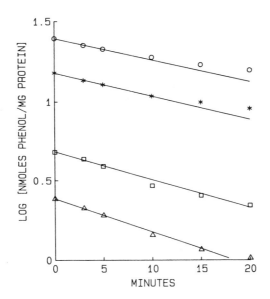

Figure 4.   Disappearance of [^{14}C]phenol in rat liver microsomal incubations. Varying concentrations of [^{14}C]phenol (0.005 mM, 0.01 mM, 0.03 mM, 0.05 mM) were incubated with microsomes in the presence of a NADPH-generating system for the indicated time periods. The residual [^{14}C]phenol was determined after separation of hydroquinone and catechol metabolites. Each data point represents the mean of two experiments each consisting of duplicate samples. The lines were drawn on the basis of linear regression analysis. (Reproduce from Ref. 12 with permission.)

Currently, the potential pathways of benzene metabolism suggest a central role of benzene oxide (Fig. 5). The fate of benzene oxide would appear to determine the degree to which toxic metabolites are produced. Whereas low substrate concentration of benzene might lead directly to the formation of phenol, higher concentrations, which we know are more toxic, drive the reaction toward the formation of the epoxide. Since phenol is the major metabolite of benzene it is likely that the non-enzymatic rearrangement of the oxide is the most rapid reaction which benzene oxide

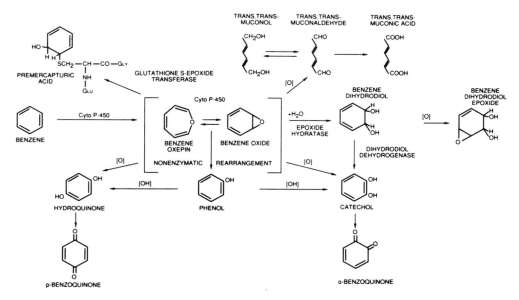

Figure 5.   The potential pathways of benzene metabolism.

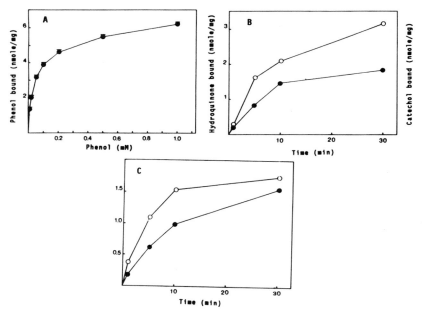

Figure 6.   The effect of phenol concentrations (A) on the covalent binding
of metabolically activated phenol to microsomal proteins and
the time courses of the covalent binding of hydroquinone (B)
and catechol (C) to microsomal proteins. Closed circles
represent incubations with and open circles without NADPH.
Incubations were described in Materials and Methods except that
phenol was replaced with 6 μM hydroquinone or 6.5 μM catechol
as needed. (Reproduce from Ref. 37 with permission.)

Table 6. Effect of Myeloperoxidase on Covalent Binding of [^{14}C]Phenol and
Metabolites to Microsomal Protein In Vitro

Condition	Covalent binding[a] (pmole phenol equivalents bound/min/ 0.7 mg microsomal protein)
Myeloperoxidase (0.17 unit), H$_2$O$_2$ (2.2 mM)	1534.1 ± 128.1
Myeloperoxidase (0.17 unit), H$_2$O$_2$ (4.4 mM)	1474.4 ± 129.9
Myeloperoxidase[b](0.17 unit), H$_2$O$_2$ (8.8 mM)	1458.6 ± 57.2
Myeloperoxidase (0.17 unit), H$_2$O$_2$ (8.8 mM), ascorbate (1 mM)	70.6 ± 20.1[c]
H$_2$O$_2$ (4.4 mM)[d]	114.1 ± 32.1[c]

[^{14}C]Phenol (1mM) was incubated with 0.7 mg of washed hepatic microsomal protein isolated from PB-pretreated guinea pigs and sodium phosphate buffer (100 mM, pH 7.4). The incubations was carried out for 5 min at 37°C. The total volume was 1 ml.
[a] Values are means ± standard deviation, N=4; 1 unit=1 umole of guaiacol oxidized/min/mg; guinea pig myeloperoxidase=1.3 units/mg.
[b] Control.
[c] Significantly different from the control (p < 0.01).
[d] Data from Table 3.
(Reproduced from Ref. 38 with permission.)

undergoes. The dihydrodiol, muconic acid, and the premercapturic acid are formed in lesser quantities probably because they depend upon enzymatic reactions whose rates are determined by substrate concentration and the affinity of the enzyme for the substrate. Since benzene oxide rearranges rapidly to form phenol, very little benzene oxide is available as substrate for either epoxide hydrolase or glutathione transferase which might explain why those reactions do not yield larger quantities of metabolites. Subsequent hydroxylation of phenol to yield hydroquinone, catechol, or trihydroxybenzene may occur in the liver or in the bone marrow.

The production of biological reactive intermediates from benzene leads to covalent binding of these metabolites to cellular macromolecules such as proteins and nucleic acids. The binding of benzene metabolites to DNA is discussed in the next chapter. The metabolic activation of benzene leads to metabolites which covalently bind to proteins in liver, brain, kidney, spleen, fat (35), muscle, and bone marrow (36). Incubations of benzene metabolites, phenol, hydroquinone, and catechol with liver microsomes (Fig. 6) also lead to covalent binding to protein in NADH-independent reactions (37). The data suggest that peroxidases play a significant role in the metabolism of benzene to its phenolic metabolites in the bone marrow. Smart and Zannoni (38) demonstrated the role of myeloperoxidase in metabolic activation of phenol (Table 6) and Schlosser et al. (39) showed that prostaglandin H synthetase can also play a role in the metabolic activation of phenol and hydroquinone. In contrast, Smart and Zannoni have also reported that DT-diaphorase (38) and ascorbic acid (38, 40) reduced covalent binding of phenol by liver microsomes. It appeared that the production of reactive metabolites of these phenolic compounds (e.g., semiquinones and quinones) depends upon the oxidative state of the organ. Smart and Zannoni (38) have postulated that benzene toxicity may be a result of covalent binding of the oxidized forms of phenol, catechol, or hydroquinone within the bone marrow and the limited ability of the bone marrow to keep these metabolites in their reduced state could explain the selective toxicity of benzene. They have suggested that benzene is not a liver toxin because of the high level of reductive enzymes (DT-diaphorase, quinone reductases, carbonyl reductase, etc...). Wermeth et al. (41) have provided additional evidence to support this concept by demonstrating that carbonyl reductase in man may act to reduce quinones to hydroquinones and thereby exert provide a mechanism which can protect against benzene toxicity..

If benzene metabolites can be retained in their reduced form in the liver, i.e., as phenolic compounds, they can be conjugated with ethereal sulfate or glucuronic acid (5) to compounds which are thought to be detoxication products. It is tempting to speculate that some part of the toxicity of benzene may be mediated by deconjugation products in peripheral organs but such pathways have yet to be established for benzene metabolites.

ACKNOWLEDGMENT

The results from the author's laboratory have been supported by NIH grant ES-02931.

References

1. O. Schultzen and B. Nauynyn, Uber das verhalten des kohlenwasserstoffes, Arch. Anat. Physio. 1:349 (1867).
2. M. Nencki and P. Giacosa, Uber die oxydation der aromatischen kohlenwasserstoffe in turkoyser, Z. Physiol. Chem. 4:325 (1880).
3. M. Jaffe, Cleavage of the benzene ring in the organism. I. The excretion of muconic acid in the urine after ingestion of benzene. Z. Physiol.Chem. 62:58 (1909).
4. S. Zbarsky and L. Young, The conversion of benzene to phenyl-mercapturic acid in the rat, J. Biol. Chem. 151:487 (1943).
5. D. Parke and R.T. Williams, Studies on detoxification. The metabolism of benzene: (a) the formation of phenylglucuronide and phenylsulphuric acid from ^{14}C benzene; (b) the metabolism of ^{14}C phenol, Biochem J. 55:337 (1953).
6. D. Parke and R.T. Williams, Detoxication XLIV. Metabolism of benzene containing ^{14}C benzene, Biochem. J. 54:231 (1954).
7. G.M. Rusch, B.K. Leong, and S. Laskin, Benzene Metabolism, J. Toxicol. Environ. Health Suppl 2:23 (1977).
8. L. Gonasun, C.M. Witmer, J. Kocsis, and R. Snyder, Benzene metabolism in mouse liver metabolism, Toxicol. Appl. Pharmacol. 26:398 (1973).
9. D. Jerina, J. Daley, B. Witkop, P. Zaltzman-Nirenberg and S. Udenfriend, Role of arene oxide-oxepin system in the metabolism of aromatic substrates. I. In vitro conversion of benzene oxide to premercapturic acid and dihydrodiol, Arch. Biochem. Biophys. 128:176 (1968).
10. A. Tunek, K. Platt, P Bentley, and F. Oesch, Microsomal metabolism of benzene to species irreversibly binding to microsomal protein and the effects of modification of this metabolism, Molec. Pharmacol. 14:920 (1978).
11. T. Sawahata and R. Neal, Biotransformation of phenol to hydroquinone and catechol by rat liver microsomes, Mol. Pharmacol. 23:453 (1983).
12. S. Gilmour, G. Kalf, and R. Snyder, Comparison of the metabolism of benzene and its metabolite phenol in rat liver microsomes in "Biological Reactive Intermediates III: Mechanisms of Action in Animal Models and Human Diseases", J.J. Kocsis, D.J. Jollow, C.M. Witmer, J.O. Nelson, and R. Snyder eds., Plenum Press, New York (1986), p.223.

13. K. Nomiyama, Studies on the poisoning by benzene and its homologues. Oxidation rate of benzene and benzene poisoning, Med. J. Shinshu. Univ. 7:41 (1962).
14. R. Snyder, L.S. Andrews, E.W. Lee, C.M. Witmer, M. Reilly, and J.J. Kocsis, Benzene metabolism and toxicity in "Biological Reactive Intermediates", D.J. Jollow, J.J. Kocsis, R. Snyder, H. Vainio, Plenum Press, New York (1977), p.286.
15. D. Sammet, E. Lee, J.J. Kocsis, and R. Snyder, Partial heptatectomy reduces both metabolism and toxicity of benzene, J. Toxicol.. Environ. Health. 5:785 (1979).
16. R. Snyder, F. Uzuki, L. Gonasun, E. Bromfeld , and A. Wells, The metabolism of benzene in vitro, Toxicol. Appl. Pharmacol. 11:346 (1967).
17. J.J. Kocsis, S. Harkaway, M.C. Santoyo, and R. Snyder, Dimethyl sulfoxide: Interactions with aromatic hydrocarbons, Science 160:427 (1968).
18. W.F. Greenlee, E.A. Gross, and R.D. Irons, Relationship between benzene toxicity and disposition of ^{14}C-labeled benzene metabolites in the rat, Chem. Biol. Interact. 33:285 (1981).
19. R. Irons, J. Dent, T. Baker, and D. Rickert, Benzene is metabolized and covalently bound in bone marrow in situ, Chem. Biol. Interact. 30:241 (1980).
20. L. Andrews, H. Sasame, and J.R. Gillette, ^{3}H-benzene metabolism in rabbit bone marrow, Life Sci. 25(7):567 (1979).
21. M. Ingelman-Sundberg and A.L. Hagbjork, On the significance of the cytochrome P-450-dependent hydroxyl radical-mediated oxygenation mechanism.
22. I. Johansson and M. Ingelman-Sundberg, Hydroxyl radical-mediated, cytochrome P-450-dependent metabolic activation of benzene in microsomes and reconstituted enzyme systems from rabbit liver. J. Biol. Chem. 258:7311 (1983).
23. L.D. Gorsky and M.J. Coon, Evaluation of the role of free hydroxyl radicals in the cytochrome P-450-catalyzed oxidation of benzene and cyclohexanol, Drug Metab. Disp. 13:169 (1985).
24. F.U. Saito, J.J. Kocsis, and R. Snyder, Effect of benzene on hepatic drug metabolism and ultra structure. Toxicol. Appl. Pharmacol. 26:209 (1973).
25. G.B. Post and R. Snyder, Effect of enzyme induction on microsomal benzene metabolism, J. Toxicol. Environ. Health. 11:811 (1983).
26. G.B. Post and R. Snyder, Fluoride stimulation of microsomal benzene metabolism, J. Toxicol. Environ. Health. 11:799 (1983).
27. P. Baune, J. Flinois, E. LePrevost, and J. Leroux, Influence of ethanol and benzene on cytochrome P-450 fractions in rat liver microsomes, Drug Metab. Dispos. 11:499 (1983).
28. I. Johansson and M. Ingelman-Sundberg, Benzene metabolism by ethanol-, acetone-, and benzene-inducible cytochrome P-450 (IIE1) in rat and rabbit liver microsomes, Cancer Res. 48:5387 (1988).
29. D.R. Koop, C.L. Laethem, and G.G. Schnier, Identification of ethanol-inducible P-450 isozyme 3a (P-450IIE1) as a benzene and phenol hydroxylase, Toxicol. Appl. Pharmacol. 98:278 (1989).
30. T.A. Chepiga, C.S. Yang, and R. Snyder, Benzene metabolism by two purified, reconstituted rat hepatic mixed function oxidase systems, Toxicologist 10(1):128 (1990).
31. B.D. Goldstein, G. Witz, J. Javid, M. Amoruso, T. Rossman, and B. Wolder, Muconaldehyde, a potential toxic intermediate of benzene metabolism in "Biological Reactive Intermediates II. Part A", R. Snyder, D.V. Parke, J.J. Kocsis, D. Jollow, G.G. Gibson, and C.M. Witmer, Eds., Plenum Press, New York, 1982, p. 331.
32. L. Latriano, B.D. Goldstein, and G. Witz, Formation of muconaldehyde, an open-ring metabolite of benzene, in mouse liver microsomes: an additional pathway for toxic metabolites, Proc. Natl. Acad. Sci. 83:8356 (1986).

33. T.A. Kirley, B.D. Goldstein, W.M. Maniara, and G. Witz, Metabolism of trans,trans-muconaldehyde, a microsomal hematotoxic metabolite of benzene, by purified yeast aldehyde dehydrogenase and a mouse liver soluble fraction, Toxicol. Appl. Pharmacol. 100:360 (1989).

34. D. Ross, P. Holzner, and D.R. Petersen, Hepatic metabolism and toxicity of trans-trans muconaldehyde, Toxicologist 10(1):185 (1990).

35. R. Snyder, E.W. Lee, and J.J. Kocsis, Binding of labeled benzene metabolites to mouse liver and bone marrow, Res. Comm. Chem. Pathol. Pharmacol. 20(1):191 (1978).

36. S. Longacre, J. Kocsis, and R. Snyder, Influence of strain differences in mice on the metabolism and toxicity of benzene, Toxicol. Appl. Pharmacol. 60:398 (1981).

37. H. Wallin, P. Melin, C. Schelin, and B. Jergil, Evidence that covalent binding of metabolically activated phenol to microsomal proteins is caused by oxidized products of hydroquinone and catechol, Chem. Biol. Interact. 55:335 (1985).

38. R.C. Smart and V.G. Zannoni, DT-diaphorase and peroxidase influence the covalent binding of the metabolites of phenol, the major metabolite of benzene, Mol. Pharmacol. 26:105 (1984).

39. M.J. Schlosser, R.D. Shurina, and G.F. Kalf, Metabolism of phenol and hydroquinone to reactive products by macrophage peroxidase or purified prostaglandin H synthetase, Environ. Health Perspect. 82:229 (1989).

40. R.C. Smart and V.G. Zannoni, Effect of ascorbate on covalent binding of benzene and phenol metabolites to isolated tissue preparations, Toxicol. Appl. Pharmacol. 77:334 (1985)

41. B. Wermuth, K. Platts, A. Seidel, and F. Oesch, Carbonyl reductase provides the enzymatic basis of quinone reduction in man, Biochem. Pharmacol. 35:1277 (1986).

MECHANISMS OF BENZENE TOXICITY

Suzanne Pirozzi Chatterjee and Robert Snyder

Department of Pharmacology and Toxicology
Rutgers The State University of New Jersey
Piscataway, New Jersey

INTRODUCTION

Benzene is a ubiquitous pollutant in the environment and in many workplaces. It is known to produce aplastic anemia, leukemia, chromosomal damage, and immunotoxicity. This discussion will consider mechanisms by which benzene produces decrements in bone marrow function, alters the production of cytokines, and impairs the hematopoietic system. Methods of evaluating bone marrow function such as cell counting techniques, colony forming units assays, cytokine assays, and the $[^{59}Fe]$ uptake technique will be summarized. Mechanisms involving covalent binding of reactive metabolites to protein and DNA which can explain the occurrence of aplastic anemia will be discussed. Implications for benzene induced leukemia and the relationship to chromosome damage will be examined. Current interest in interactions among benzene metabolites leading to toxic effects observed with benzene will be analyzed. Finally, areas where new research is needed will be proposed.

HEMATOTOXICITY (APLASTIC ANEMIA)

For many years benzene has been known to perturb hemopoiesis. Selling (1) and Weiskotten (2) in 1916 demonstrated that injection of rabbits with benzene produced cytopenia and in 1920 Weiskotten et al. (3) reported cytopenia in rabbits exposed to benzene by inhalation. Since then cytopenia has been produced in a variety of animals using various routes of administration (4-7). The severe depression in the levels of circulating blood cells in benzene-treated animals which result in pancytopenia and aplastic anemia ultimately reflects the damage to the bone marrow, a major site of hemopoiesis (8) and the target organ for benzene toxicity (9-11).

To understand the toxic effects of benzene on bone marrow it is first necessary to briefly review our current understanding of hemopoiesis, regulation of bone marrow, and the techniques used to study the system. Currently the favored model for hemopoiesis which fits most of the experimental evidence proposes that the hemopoietic stem cell performs a self-renewing function by giving rise to new stem cells, thereby preventing the stem cell population from dwindling, while also generating progenitor cells that are committed to differentiate via defined lineages that result in the production of mature blood cells (Fig.1)(12). The decision of the

Molecular Aspects of Monooxygenases and Bioactivation of Toxic Compounds
Edited by E. Arınç *et al., Plenum Press, New York,*

387

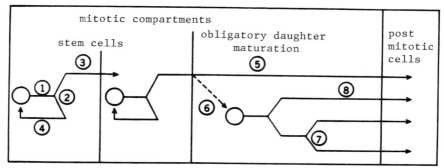

Fig. 1.   Patterns of mitosis.   In stem cell mitotic
compartments, following DNA synthesis (1) and mitosis
(2), daughter cells either may mature and enter the
next compartment (3) or may retain the characteristics
of the mother cell (4). Upon leaving the stem cell
compartment or the last of a concatenated group of
compartments, the daughter may mature without further
mitosis (5) or may undergo further mitosis (6) of a
type in which both daughters must mature with (7) or
without (8) repetitive subsequent mitoses. Taken from
Ref. 12 and reproduced with permission.

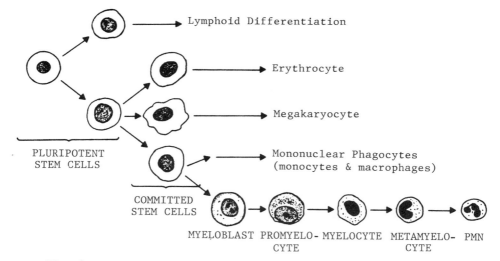

Fig. 2.   Scheme of stem cell differentiation and granulopoietic
maturation. Taken from Ref. 16 and reproduced with
permission.

cycling stem cell to self-renew or produce committed pluripotent progeny is a random or stochastic event. Once the pluripotent cells commit to differentiation there is a progressive restriction in the genetic expression during proliferation such that the progeny are progressively restricted in both the differentiation pathway, i.e., in going from pluripotent to oligopotent and eventually to monopotent cells, and in their proliferative potential (13-16). (See Fig. 2.)

Although the stem cell can differentiate along any cell lineage, it is the stromal cells found closely associated with stem cells (17) that modulate hemopoiesis by providing a favorable environment that sustains and regulates proliferation and differentiation (18-22). Stromal cells consist of a number of cell types with various functions (Table 1). In addition to supplying a matrix where stem cells may lodge and proliferate, stromal cells synthesize and present hemopoietic growth factors or cytokines to stem cells to induce their proliferation and differentiation (23). (See references 24-26 for a review of cytokines.) The question of which of the cells in the stroma play a key role in the regulation of hemopoiesis has not been demonstrated unequivocally; however some investigators suggest that this cell may be the macrophage (22, 27-29). Humoral regulators also modulate hemopoiesis but this aspect of regulation is beyond the scope of this discussion.

Table 1. Bone Marrow Stromal Cells and their Function in the Maintenance of the Microenvironment

Bone Marrow Stromal cells	Function
Endosteal cells	Stem cell for the renewal of marrow stroma Synthesis of cytokines
Fibroblasts	Production of collagen, reticulum, and mucopolysaccharides Synthesis of cytokines Possible induction of differentiation and commitment
Reticuloendothelial cells and monocytes/ macrophages	Synthesis of cytokines Phagocytosis Possible induction of differentiation and commitment
Fat cells	Replacement for bone marrow stromal cells which were destroyed
Sinusoids	Trapping of hemopoietic cells Regulating maturation of blood cells Exchange of metabolites Means of allowing matured blood cells to enter the circulation Maintenance of a microenvironment capable of supporting cell growth
Bone	Possible production of cytokines Support

Table 2. In Vivo Effects of Recombinant Growth Factors

Factor	Species	Status	Effects measured
Murine IL-3	Mouse	Normal	↑ Cycling of stem cells ↑ Number of progenitor cells ↑↑ Cellularity in spleen ↑ (moderate) Circulating granulocytes
Human GM-CSF	Monkey	Normal or pancytopenic	↑ Numbers of circulating granulocytes ↑ Reticulocytes
Human G-CSF	Monkey	Normal or myelosuppressed	↑ Numbers of circulating neutrophils Shortened recovery time after cytotoxic treatment
Human erythropoietin	Human	Anemia of renal disease	↑ Hemoglobin ↑ Sense of well being

Where measured, the effects were found to be dose-dependent. There was no reported toxicity (except for some hypertension after erythropoietin). Phase I clinical trials with CSFs are presently taking place in several centers. (Taken from Ref. 37 and reproduced with permission).

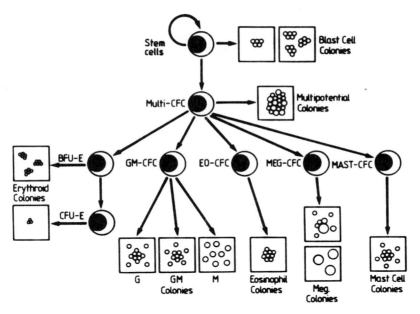

Fig. 3.   Schematic diagram of the stem and progenitor cells
generating the mature blood cells, with their acronyms
and a representation of the general features of the
colonies formed in vitro by these cells. Taken from
Ref. 36 and reproduced with permission.

The model of hemopoiesis and its regulation was developed as a result
of using a variety of techniques such as colony forming unit (CFU) assays,
long term bone marrow culture (LTBMC), and [^{59}Fe] incorporation.
Demonstration of stem cells was first shown in the spleen colony forming
assay in which lethally irradiated recipient mice were reconstituted with
bone marrow cells. Individual stem cells first seed in the spleen and then
begin to proliferate to form nodules which contained cells of all
hemopoietic lineages, i.e., colony forming unit-spleen (CFU-S) (30). The
development of a clonal in vitro culture system using semisolid medium by
Pluznik and Sach (31) and Bradley and Metcalf (32) demonstrated the
proliferation and differentiation of progenitor cells called colony forming
cells (CFC) into mature colonies of all lineages known as colony forming
units in culture (CFU-C) [Fig. 3.] (33-36). The addition of exogenous
growth factors or cytokines in culture induces hemopoietic cells to
proliferate and differentiate along the various blood cell lineages (Fig.
4). Cytokines have been proposed to be genuine regulators of the various
cell populations in vivo based on indirect evidence or by analogy. Recently
recombinant technology has made cytokines available for in vivo studies and
many of the in vitro effects of cytokines have also been observed in vivo
[Table 2] (37).

The LTBMC developed by Dexter and associates (38-40) provided an in
vitro model of hemopoiesis. A marrow derived adherent cell layer is
established in liquid culture and three weeks later an inoculum of fresh
bone marrow cells is added. During propagation of the culture a non-
adherent population appears which is subsequently assessed for stem and
progenitor cells during the culture period using colony forming unit
assays.

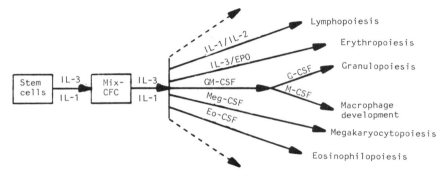

Fig. 4. Growth factors involved in the development of the various blood cell
lineages. IL = interleukin; EPO = erythropoietin; G = granulocyte;
M = macrophage; Meg = megakaryocyte; Eo = eosinophil; CSF = colony-
stimulating factor. Taken from Ref. 37 and reproduced with
permission.

The uptake of [^{59}Fe] into circulating erythrocytes in vivo has been used
as a measure of erythropoiesis (41-44). Figure 5 displays the scheme of
maturation and amplification of the pluripotent stem cell to the circulating
erythrocyte (16). Pronormoblasts, normoblasts, and reticulocytes, which are
morphologically distinct cells, appear successively at 24, 48, 72 hrs after
re-initiation of red cell development by erythropoietin (45). Hemoglobin is
synthesized only in cells between the pronormoblast and reticulocyte stage
of differentiation (46) and only these cells will incorporate iron into
hemoglobin. Therefore, the erythroid cell series can be divided into three
classes: cells that can utilize iron for hemoglobin synthesis but do not
proliferate (reticulocytes), cells that utilize iron for hemoglobin
synthesis and proliferate (normoblasts and pronormoblasts), and cells that
do not utilize iron for hemoglobin synthesis but can proliferate and commit
to differentiation along the erythroid pathway (pluripotent stem cell and
erythropoietin responsive stem cell).

The effect of an agent on each compartment of erythropoiesis can be
measured by allowing each of the erythrocyte precursors to mature to the
reticulocyte stage after treatment. The size of the reticulocyte pool is
then measured by administering [^{59}Fe] which rapidly disappears from the
circulation and is delivered to the marrow (47) where reticulocytes
incorporate iron for hemoglobin synthesis. The peripheral blood is then
sampled for [^{59}Fe] uptake 24 hr later at which time the label in the blood
represents the incorporation of iron in the hemoglobin during the period
when matured reticulocytes are released into the blood. The size of the
reticulocyte pool, from the maturation of each cellular compartment, would
then indicate if any damage occurred to any of the earlier precursor cell
types as a result of exposure to an agent (Fig. 6). Evaluation of various
cytotoxic agents whose mechanisms of action is known (48, 49) has
demonstrated the usefulness of using ferrokinetics in evaluating chemically
induced erythropoietic toxicity (Fig 7). Vinblastine which arrests mitosis
at metaphase and produces damage to the hemopoietic stem cell severely
reduced the [^{59}Fe] uptake at both 24 and 48 hr, whereas at 72 hr the
inhibition was less severe. Methotrexate and cytarabine, which are specific
S-phase cycle inhibitors, reduced iron uptake at 24 and 48 hr, whereas, at
72 and 96 hr [^{59}Fe] uptake returned to the control value. Cyclophosphamide
and busulfan which are toxic to both proliferating and nonproliferating
cells reduced [^{59}Fe] uptake at all times.

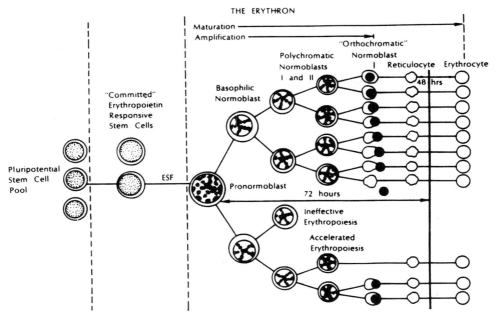

THE ERYTHRON

Fig. 5.   The erythron. The erythropoietic organ and circulating
red cell mass constitute the erythron. Depicted here
are the composition of the erythron and the maturation
and amplification of the red cell series.
Pluripotential stem cells are recruited into the pool
of committed erythropoietin-responsive stem cells.
Under the stimulus of erythropoietin (ESF), these
committed cells differentiate into pronormoblasts.
Normal maturation occurs by four mitotic divisions
during which the pronormoblast matures to the
reticulocyte by a continual differentiation process,
passing through three arbitrarily defined stages: the
basophilic normoblast, polychromatophilic normoblasts
1 and 2, and the "orthochromatic" normoblast, which
by ejection of its nucleus emerges as a reticulocyte.
Amplification results from maturational divisions
such that each pronormoblast results in a potential
16 reticulocytes. The maturation process requires a
period of approximately 72 h. In the ensuing 48 h,
the circulating reticulocyte matures into an
erythrocyte. Ineffective erythropoiesis, with failure
of erythroblast maturation and intramedullary cell
destruction, and accelerated erythropoiesis, with
skipped maturational divisions, are also illustrated.
Taken from Ref. 16 and reproduced with permission.

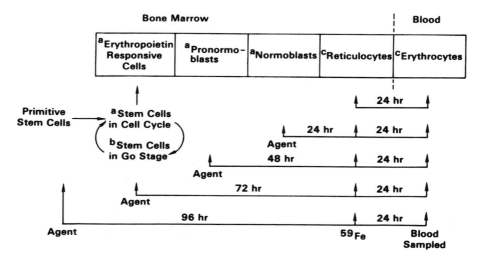

Fig. 6.   Developmental sequences of cells involved in erythropoiesis experimental scheme for determining the effect of agent on each cellular compartment: (a) proliferating cells; (b) nonproliferating cells but can be activated to proliferating cells; (c)nonproliferating mature cells. Taken from Ref. 49 and reproduced with the permission.

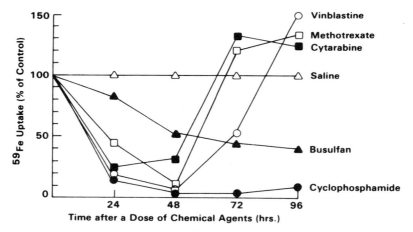

Fig. 7.   Twenty-four hour [^{59}Fe] uptake in mice measured at different time intervals following a single IP dose of cytarabine (150 mg/kg), methotrexate (20 mg/kg), vinblastine (2 mg/kg), cyclophosphamide (200 mg/kg), and a single sc dose of busulfan (320 mg/kg). Values represent the mean of 10 mice per group and were significantly different in all cases except those at 24 hr after busulfan, 72 hr after methotrexate, and 96 hr after cytarabine; 72 hr after cytarabine and also at 96 hr after methotrexate and vinblastine the significance of the differences was $p < 0.05$ whereas in all other cases the differences were $p < 0.001$. Taken from Ref. 49 and reproduced with permission.

Evaluating benzene's effect on blood cell production using the previously described techniques has shown that benzene disrupts hemopoiesis at several stages. Exposure to benzene at levels which induce anemia result in elevated levels of splenic hemosiderin which suggests that circulating erythrocytes may be directly hemolyzed by benzene or that benzene may affect erythropoietic production by yielding short-lived red blood cells (7,50). We have demonstrated that benzene reduced the incorporation of [^{59}Fe] into circulating red blood cells of mice (48,49,51). Figure 8 shows that iron uptake was not affected after 1, 12, or 72 hr whereas iron uptake was reduced after 24 and 48 hr at 440 and 2200 mg/kg doses. These results provide evidence that after a short term treatment benzene does not interfere with iron incorporation or maturation of reticulocytes nor does it affect nonproliferating cells (stem cells at G_o state). Benzene does, however, interfere with proliferation of pronormoblasts and normoblasts.

The metabolites, phenol, catechol, and hydroquinone have also been shown, although to a lesser degree than benzene, to reduce [^{59}Fe] uptake during erythrocyte production (52). Further work performed in our laboratory has demonstrated that administration of either p-benzoquinone or muconaldehyde inhibited erythropoiesis (53). Irons et al. (54) has demonstrated that after repeated exposure, peripheral lymphocytes and differentiating bone marrow cell precursors were the most sensitive cell populations. Benzene treatment induced marrow cells to enter the cell cycle; however maturation was arrested prior to the mitotic stage preventing the cells from completing the cell cycle.

Benzene affects other hemopoietic cell types besides morphologically recognizable cells in the bone marrow. Uyeki et al. (55) and Gill et al. (56) reported that multiple exposures to benzene (4000-5000 ppm) significantly reduced the number of CFU-S. Lower levels of benzene (100-400 ppm, 6 hr/day, 5 day/ week for 5 to 9 days) also reduced the number of stem cells (57-59). Progeny of stem cells, i.e., GM-CFU-C, have also been affected by exposure to benzene (55,57,59). Individual metabolites of benzene, phenol, hydroquinone, and benzene dihydrodiol have only a slight affect on bone marrow cellularity and the ability of colony forming cells (CFC) to proliferate (60). Recently benzene concentrations at or in the range of occupational exposure levels were shown to reduce the ability of CFU-GM (61) and CFU-E (62) to proliferate in vitro.

Bone marrow stromal cells have also been affected by benzene and its metabolites. Lethally irradiated, benzene-exposed mice could not resume normal hemopoiesis after injection with untreated normal bone marrow cells whereas incubation of normal cells with benzene prior to injection in lethally irradiated animals resulted in normal hemopoiesis (63). Stromal cell layers derived from benzene-treated mice (58,64) and phenol-treated mice (64) have reduced abilities to support granulocyte/macrophage colony forming cells. In vitro the metabolites of benzene have been shown to inhibit the growth of the stromal cell layer. The relative potency of the benzene metabolites is: hydroquinone > benzoquinone> benzenetriol > catechol > phenol. The ability of this layer to support the growth of G/M-CFC was inhibited by benzoquinone and hydroquinone at low doses and catechol and benzenetriol at high doses (65). Recent studies have demonstrated that the toxic effects of benzene on the bone marrow can be prevented by coadministration of indomethacin (66,67) suggesting that prostaglandin H synthetase may play a role in benzene-induced toxicity. However, the coadministration of indomethacin with phenol and hydroquinone, did not prevent the metabolite-induce toxicity (67).

The development of established animal models which mimic the effects in humans has provided the means to study the mechanisms involved in benzene induced aplastic anemia. Molecular mechanisms which could explain benzene toxicity include covalent binding of benzene metabolites to cellular macromolecules such as DNA or RNA or by direct alkylation of protein sulfhydryl (SH) groups. Irons et al. (68) demonstrated that hydroquinone, but not phenol or catechol, inhibited polymerization of tubulin to a greater extent than colchicine (Fig. 9). Tubulin possesses nucleophilic SH groups that bind guanosine triphosphate (GTP) which in turn stabilizes tubulin for further polymerization to microtubules necessary for spindle

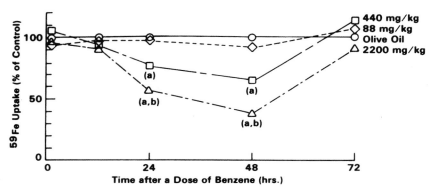

Fig. 8.   Effect of benzene on 24-hr erythrocyte utilization in mice measured at different time intervals following single sc dose of benzene as a 50% (v/v) benzene-olive oil solution: 88, 400, and 2200 mg/kg. Control group received olive oil only. Values represent the mean of more than 11 mice per group (11 to 26); (a) significantly different from control (p < 0.001); (b) significantly different from 440 mg/kg dose (p < 0.01). Taken from Ref. 49 and reproduced with permission.

formation during cell division (69). Hydroquinone and p-benzoquinone interfere with GTP-tubulin binding by alkylating SH groups (68,70). Another protein which is inhibited by benzene metabolites, hydroquinone and p-benzoquinone but not phenol or catechol, is mitochondrial DNA polymerase gamma (Fig. 10) (71). The binding of [14C]hydroquinone to the polymerase was prevented by N-ethylmaleimide, and unlabeled hydroquinone or p-benzoquinone suggesting that these metabolites bind to the same sulfhydryl group thereby inactivating DNA polymerase gamma and inhibiting DNA replication. Thus at least two types of molecules essential for cell division are inhibited by benzene metabolites via a reaction at the protein level.

Benzene can be converted to reactive species which covalently bind to DNA. Lutz and Schlatter (72) administrated [^{14}C] or [^{3}H] labeled benzene to rats and demonstrated covalent binding of benzene metabolites to liver DNA (Table 3). Gill et al. (73) later showed that radiolabeled metabolites of benzene covalently bind to hemopoietic cell nuclear DNA. In vitro RNA synthesis was inhibited by benzene in mitochondria from liver or bone marrow indicating the capacity of the mitochondria to bioactivate benzene to metabolites (74) [Fig. 11]. Rushmore et al. (75) identified DNA adducts using rabbit bone marrow or rat liver mitoplasts in which the DNA was labeled with [^{3}H]dGTP and then exposed to [^{14}C]benzene in vitro. Enzymatic hydrolysis of the DNA to deoxynucleosides and subsequent chromatography on Sephadex LH-20 columns yielded seven guanine nucleoside adducts from the

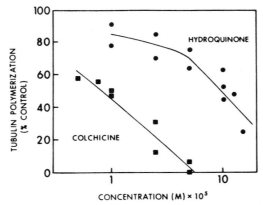

Fig. 9.   Concentration-response curves for colchicine and hydroquinone inhibition of microtubule polymerization. Inhibition of polymerization was calculated as % of control (ordinate) after 10 min incubation at 37°C for different concentrations (abscissa) of colchicine (■) or hydroquinone (●). Taken from Ref. 68 with permission.

hydrolysate of rabbit bone marrow cell mitochondrial DNA whereas six guanine nucleoside adducts were identified in the hydrolysate of DNA from rat liver mitoplast DNA (Fig. 12). When [^{3}H]ATP was used instead of labeled GTP two deoxyadenosine adducts were formed. The addition of the metabolites, p-benzoquinone or 1,2,4-benzene-triol to the incubation isotopically diluted the amount of labeled benzene metabolites which covalently bound to mtDNA (Fig 13). p-Benzoquinone and 1,2,4-benzenetriol competed with benzene labeled metabolites from the mitochondria suggesting it is possible that the mitochondrial metabolites could be p-benzoquinone or 1,2,4 benzenetriol. Recently we have reported the formation of adducts of the bone marrow from benzene-treated rats using [^{32}P]postlabeling technique (53). Figure 14 shows some of the possible structures of the DNA adducts which form in vivo when exposed to benzene (76).

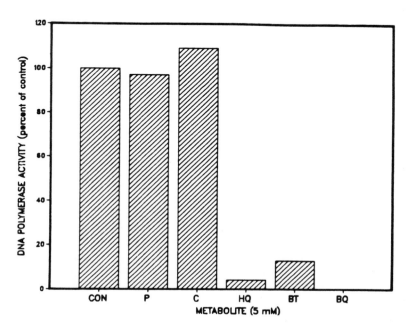

Fig. 10.   The effect of metabolites on the activity of DNA
polymerase gamma. Enzyme activity was assayed in the
presence or absence of 5 mM each of phenol (P),
catechol (C), hydroquinone (HQ), 1,2,4-benzenetriol
(BT) and p-benzoquinone (BQ) for 1 h at 37°C, under
conditions described in Methods. Control activity was
3.4 pmol/ug enzyme. Each value represents the means
of triplicate determinations which agreed within 5%.
Taken from Ref. 71 and reproduced with permission.

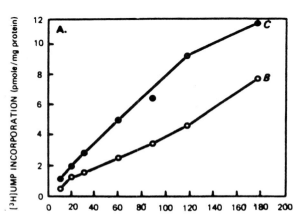

Fig. 11. RNA synthesis in vitro in liver mitochondria from partially hepatectomized-benzene treated rats. Mitochondria were incubated with [^{3}H]UTP under optimal conditions for RNA synthesis as described in Methods. Each point represents the mean of duplicate experiments which agreed within 5%. (•) C, control, RNA synthesis in mitochondria from 24 hr regenerating rat liver; (O) B, RNA synthesis in mitochondria from animals treated in vivo 10 hr post-hepatectomy with 2200 mg benzene/kg body weight ip. Taken from Ref. 74 and reproduced with permission.

Table 3. Binding of Radioactive Benzene and Tritiated Water to Rat Liver DNA

Line No.	Description	Units	Tritiated benzene		Carbon-14 benzene		Control non-covalent binding	Control tritiated water	
1.	Experiment	No	1	2	3	4	5	6	7
2.	Total benzene administered	mg	19.7	19.7	21.2	21.2	44		
3.	Total radioactivity	mCi	12.2	12.2	1.0	1.0	0.53	3.4	3.7
4.	Period of exposure	h	19	10.5	9.25	11.5	2	26	23
5.	Radioactivity in whole liver	dpm/mg	8200	11900	1400	820	2.4×10^5	3100	3200
6.	Radioactivity of isolated DNA	cpm	545	496	112	98	23.3	96	126
7.	Background of equal volume buffer	cpm	29.4	22.5	29.0	29.0	21.6	26.6	29.4
8.	Specific activity of DNA	dpm/mg	834	821	41.6	37.0	3.4	122	162
9.	Radioactivity bound to DNA calculated as umole benzene/mole DNA phosphate		2.40	2.36	1.57	1.40			

Taken from Ref. 72 and reproduced with permission.

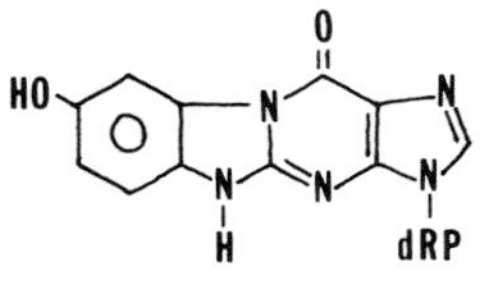

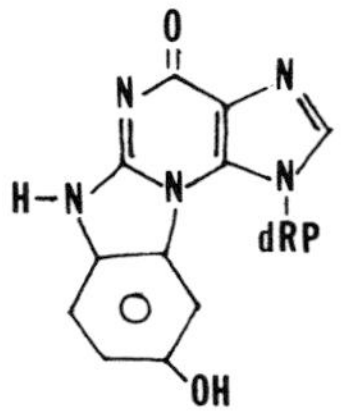

dGMP-5'
dGMP-3'
dGDP-3',5'

Deoxyguanosine nucleotide adducts

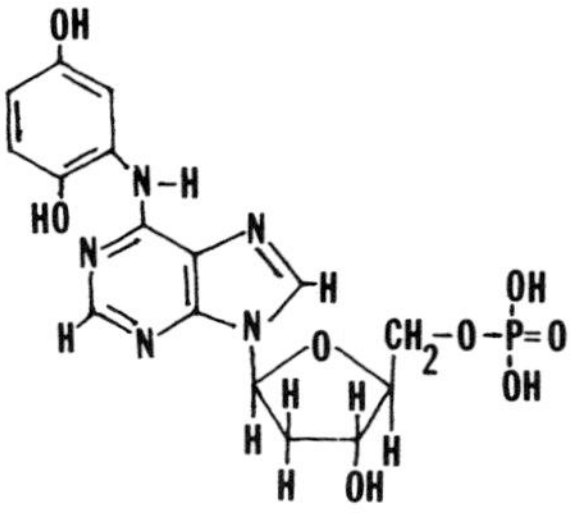

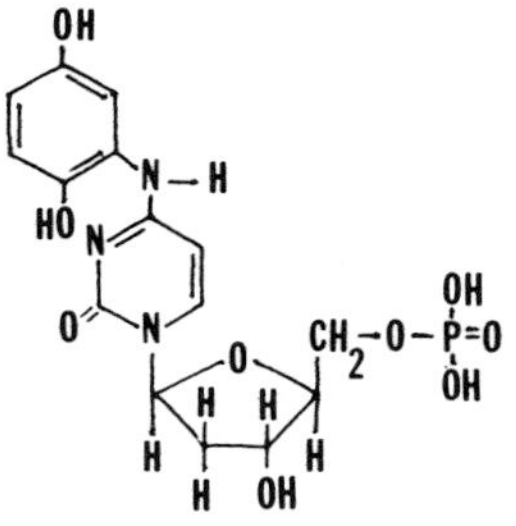

Deoxyadenosine 5' monophosphate adduct   Deoxycytidine 5'monophosphate adduct

Fig. 12.   Sephadex LH-20 column chromatography of an enzymatic
            hydrolysate of [^{3}H]dGMP-labeled [^{14}C]benzene-modified
            mtDNA. The DNA from 50 mg of rabbit bone marrow cell
            mitoplasts   or   100   mg   of   rat   liver   mitoplasts
            prelabeled in vitro with [^{3}H]-dGTP and exposed to
            [^{14}C]benzene   under   optimal   conditions   for   mtRNA
            synthesis in vitro, was isolated by isopycnic
            centrifugation in a CsCl gradient as described under
            Methods. The labeled mtDNA was heat denatured and
            completely hydrolyzed to deoxyribonucleosides by
            treatment with a series of nucleases, as described
            under Methods, and the hydrolysate was then applied
            to a Sephadex LH-20 column. The column (0.8 x20 cm)
            was treated first with 30 ml of 10 mM (NH$_4$)$_2$CO$_3$ (pH
            8.9) to elute the unmodified nucleosides and then
            with 35% ethanol containing 10 mM (NH$_4$)$_2$CO$_3$ to elute
            the nucleoside-adducts. Fractions (1.0 ml) were
            collected and the radioactivity measured in a liquid
            scintillation spectrometer. Panel A: column profile
            showing the nucleosides present in a hydrolysate of
            [^{3}H]dGMP-labeled [^{14}C]benzene-modified rabbit bone
            marrow cell mtDNA; (•), [^{3}H], (O), [^{14}C].   Panel B:
            column profile showing the nucleosides present in a
            hydrolysate of [^{3}H]dGMP-labeled [^{14}C]benzene-modified
            rat liver mtDNA; (•), [^{3}H]; (O), [^{14}C]. Taken from
            Ref. 75 and reproduced with permission.

401

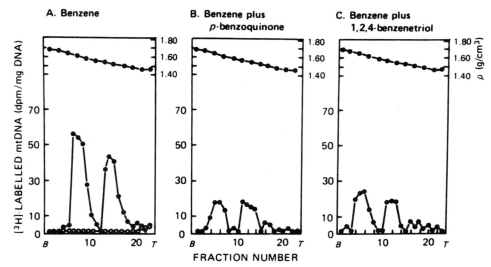

Fig. 13.   The effect of the putative benzene metabolites on the
formation of [³H]benzene-derived adducts in mtDNA in
vitro. Mitoplasts from rabbit bone marrow cells were
incubated in the presence of equal concentrations of
[³H]benzene and p-benzoquinone or [³H]benzene and
1,2,4-benzenetriol for 4 h under optimal conditions
for mtRNA synthesis in vitro as described under
Methods. At the end of the incubation, the mitoplasts
were recovered by centrifugation, lysed in a buffer
containing 2% SDS, and digested with pronase and RNase
and the [³H]-labeled mtDNA was then subjected to
isopycnic centrifugation in a CsCl-ethidium bromide
gradient for 72 h at 34,000 rev./min at 20°C in a
Spinco SW 50.1 rotor. Fractions (200 ul) were
collected and the radioactivity determined as
described under Methods. Panel A: rabbit bone marrow
cell mitoplasts incubated in the presence of
[³H]benzene (5 mM, 3000 dpm/nmol) (●) [³H]-labeled
mtDNA fractions 4-8, closed circular DNA; fraction 9-
12, linear DNA. (O), effect of DNase treatment. Panel
B: rabbit bone marrow cell mitoplasts incubated
simultaneously with [³H]benzene (5 mM, 3000 dpm/nmol)
and 5 mM p-benzoquinone. Panel C: rabbit bone marrow
cell mitoplasts incubated simultaneously with
[³H]benzene 5 mM, 3000 dpm/nmol) and 5 mM 1,2,4-
benzenetriol. B, bottom of gradient; T, top of
gradient. Taken from Ref. 75 and reproduced with
permission.

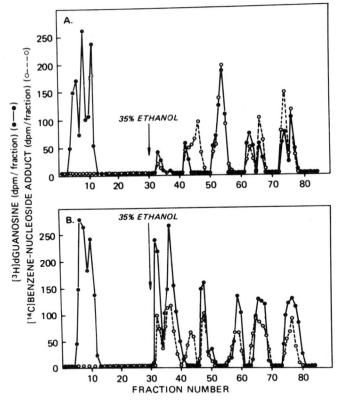

Fig. 14.   Possible structures of adducts.

LEUKEMOGENESIS

Although benzene exposure results in pancytopenia or one of its
cytopenic variants a small percentage of humans with cytopenia eventually
exhibit aplastic anemia or leukemia. Acute myeloblastic leukemia or one of
its variants is the form of leukemia most often observed (77-81). For many
years attempts to demonstrate leukemia in animals exposed to benzene have
given negative results (7,82,83). Snyder C.A. et al.(84) exposed C57Bl/6
mice to 300 ppm benzene for 6 hr/day x 5 days/week for life and observed
hemopoietic neoplasms such as lymphoma and myeloid leukemia. However, the
incidence of leukemia was too low to conclusively demonstrate that benzene
induces myeloid leukemia.   Benzene has been reported to produce solid
tumors in rat zymbal gland, an organ not present in humans (85,86). The
difficulty of demonstrating leukemia in mice during chronic benzene
treatment may be caused by the fact that benzene inhibits proliferation of
leukemic cells.

Cronkite and associates (87-90) have developed a model for benzene-
induced leukemia which circumvented this problem. C57Bl/6 or CBA/Ca male
mice were exposed to benzene 6 hr/day x 5 day/week for 16 weeks after which
exposure ceased. The mice were then held for lifetime study. These animals
displayed a significant increase in leukemia/lymphoma compared to sham-
exposed mice.

Previously the absence of a model for benzene induced leukemia has been a major obstacle in elucidating the mechanism of benzene leukemogenesis. The model developed by Cronkite and associates may help to demonstrate key events, i.e., chemical modification of DNA and or gene expression, responsible for benzene leukemogenesis and possibly establish a dose-relationship between genomic alterations resulting from exposure to benzene and the occurrence of leukemia.

CLASTOGENESIS

The difficulties inherent in establishing a reliable and reproducible model for benzene-induced leukemia have not hampered the investigation of benzene as a chromosome-damaging agent. A common feature of chemical carcinogens is their interaction with DNA. Many laboratories employed the Ames Salmonella test for mutagens using benzene and have reported that benzene is non-mutagenic (91,92). Benzene has been known to induce genetic toxicity in humans. Forni et al. (93) performed chromosome studies on humans who had been exposed to benzene during their lifetime. A significant increase in stable and unstable chromosome aberrations were observed. Stable chromosome aberrations were still present many years after cessation of exposure to benzene; in some cases abnormal clone formation in peripheral blood lymphocytes was reported (92,94).

In vivo animal studies have reported benzene-induced genetic effects such as chromosome and chromatid aberrations, sister chromatid exchange (SCE) and micronuclei. Tice et al. (95) observed significant increases in SCE in bone marrow cells taken from mice exposed to 28 ppm benzene for four hours. This is of particular interest since the recommended workplace exposure standard is 1 ppm over an 8 hr day as a time weighted average. Benzene had also been reported to induce micronuclei in bone marrow polychromatic erythrocytes (96,97). Luke et al. (98) reported the induction of micronuclei in polychromatic erythrocytes in the peripheral blood of mice exposed to benzene. Coadministration of indomethacin with benzene reduced the formation of micronuclei in peripheral polychromatic erythrocytes (99). The induction of micronuclei in bone marrow polychromatic erythrocytes by benzene has been correlated with urinary metabolites, muconic acid, hydroquinone, and phenyl glucuronide. Administration of hydroquinone induced micronuclei whereas phenol and catechol did not. In contrast to benzene none of these metabolites yielded the benzene urinary metabolite muconic acid (100). These results suggest that the benzene metabolite trans, trans-muconaldehyde, which is thought to be the metabolic precursor to urinary muconic acid (101), might play a role in micronuclei formation. A comparison of the occurrence of SCE in peripheral blood lymphocytes and micronuclei in bone marrow polychromatic erythrocytes in mice and rats exposed to benzene at 0-1000 ppm for 6 hr demonstrated that SCE and micronuclei can be induced by benzene in both species (Table IV) (102). These cumulative results suggest that it might be possible to use urinary metabolites of benzene or SCE in peripheral blood lymphocytes as biomarkers of exposure.

IMMUNOTOXICITY

Benzene was one of the first chemicals demonstrated to yield an immunotoxic response. Early in this century benzene-exposed rabbits were shown to have an increased susceptibility to pneumonia (103,104) and tuberculosis (105). The impact of benzene on the inflammatory reaction and on antibody production was manifested in studies which showed that benzene treatment caused decreases in red cell lysins and agglutinins for killed typhoid bacilli and inhibition of the production of antibacterial antibodies (106,107). Smolik and associates (108-110) demonstrated alterations in serum immunoglobulins and complement levels in benzene-

404

exposed Polish workers, indicating that benzene could induce auto-immunity and allergy. Rozen et al. (111) reported that short term inhalation exposure in mice at or near occupational exposure levels significantly depressed the proliferation of bone marrow-derived B cells, responsible for antibody production, and thymic T cells, responsible for cell-mediated immunity. Benzene was shown to alter other parameters of cell-mediated immunity such as lowering host resistance to infectious agents (112) and virus-induced tumors (113).

Recently bone marrow phagocytes have been demonstrated to be activated following benzene treatment of mice (114). These phagocytes produced more hydrogen peroxide in response to the phorbol ester tumor promotor, TPA, than phagocytes from control animals. Although activated phagocytes are known to destroy foreign antigens via production of highly reactive and toxic oxygen intermediates, such as superoxide anion, hydrogen peroxide, and hydroxyl radicals (115), it is possible that these same reactive intermediates may contribute to bone marrow injury (116,117). It is, therefore, important to understand the role played by the immunological processes in preventing carcinogenesis since the leukemic activity of benzene may be attributed in part to benzene's alteration of the immune response.

INTERACTIONS OF BENZENE METABOLITES

Recent studies of benzene metabolism and toxicity have focussed on the interactions of the metabolites of benzene. The addition of phenol to incubations containing horseradish peroxidase, hydroquinone, and hydrogen peroxide stimulated the metabolism of hydroquinone (Fig. 15) (118) and the formation of p-benzoquinone. Covalent binding of hydroquinone increased in the presence of phenol when horseradish peroxidase was replaced with bone marrow cell protein (Table V) (119). In vivo coadministration of phenol and hydroquinone could reproduce benzene-induced myelotoxicity (Fig. 16) suggesting that phenol may stimulate hydroquinone metabolism to p-benzoquinone in bone marrow. We have demonstrated that coadministration of phenol with hydroquinone or catechol potentiates the inhibition of erythropoiesis (53). Witz et al. (120) has shown that the benzene metabolite, muconaldehyde, reduced red blood cell count and bone marrow cellularity in CD-1 mice. We have reported that coadministration of very low doses of benzoquinone and muconaldehyde was the most effective treatment in inhibiting erythrocyte production (53). These results strongly suggest that benzene toxicity may be the result of interactions of benzene metabolites in bone marrow cells.

CONCLUSION

It is clear from this discussion that many questions remain unanswered: What events determine whether the outcome of benzene exposure will be aplastic anemia or leukemia? Does it depend on dose, genetics, interactions with other chemicals, or some other parameter yet to be determined? What are the best biomarkers of benzene exposure? Can accurate biomarkers for exposure help to predict toxic effects? What levels of exposures would provide an environment with minimum risk of benzene toxicity or leukemogenesis? Continued mechanistic studies may provide answers to these questions.

ACKNOWLEDGEMENT

The results from the author's laboratory have been supported by NIH grant ES-02931.

Table 4. The Induction of Cytogenetic Damage in Male DBA/2 Mice and Sprague-Dawley Rats

Benzene (ppm/hr)	Number of animals	SCEs/metaphase[a]	Number of micronuclei/1,000 PCEs[a]
DBA/2 mice			
0	10 (SCE), 11 (micronuclei)	5.9 ± 0.2 (0.14)[b]	2.1 ± 0.3
10.3 ± 0.9	5	7.6 ± 0.2 (0.17)*,**	9.0 ± 0.6*
100.1 ± 1.6	5	9.5 ± 0.2 (0.23)*,**	20.3 ± 0.7*
991.2 ± 5.9	5	13.8 ± 0.9 (0.37)*,**	28.1 ± 0.8*
Sprague-Dawley rats			
Experiment 1			
0	10	8.6 ± 0.1 (0.17)	2.0 ± 0.1
10.0 ± 0.2	5	10.4 ± 0.2 (0.27)*,**	6.2 ± 0.4*
30.0 ± 0.3	5	11.1 ± 0.1 (0.23)*,**	7.6 ± 0.4*
Experiment 2			
0	20	8.2 ± 0.1 (0.13)	1.7 ± 0.2
0.1 ± 0.01	5	8.2 ± 0.1 (0.24)	1.6 ± 0.3
0.3 ± 0.02	5	8.2 ± 0.1 (0.19)	2.0 ± 0.3
1.0 ± 0.02	5	9.1 ± 0.2 (0.28)*	4.4 ± 0.3*
3.0 ± 0.05	5	10.5 ± 0.2 (0.25)*,**	4.8 ± 0.4*

[a]Mean ± standard error among animals within a group.

[b]The numbers in parentheses denote the standard error among cells within a group.

*Significantly different from the concurrent controls (P < 0.05) using Student's t test (one-sided).

**Significantly different from the concurrent controls (P < 0.05) using Mann-Whitney U test (one-sided).

(Taken from Ref. 102 and reproduced with permission.)

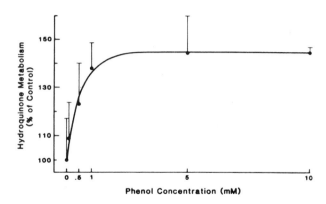

Fig. 15. Effects of various phenol concentrations on hydroquinone metabolism. Incubation conditions are as described under Materials and Methods. The data represent the means and standard deviations of three experiments. The 0 and 0.1 phenol treatment groups differ significantly ($p < 0.05$; paired t test). Taken from Ref. 118 and reproduced with the permission.

Table 5. Effect of Phenol on the Covalent Binding of ^{14}C-Hydroquinone to Mass Bone Marrow Cell Protein in the Presence of Hydrogen Peroxide

Conditions[a]	Binding
	nmol/mg protein
[14]C-Hydroquinone (0.01), cells	0.7 ± 0.2[b]
[14]C-Hydroquinone (0.01 mM), cells, H_2O_2 (0.1 mM)[c,d]	9.6 ± 0.9
Plus phenol (0.01 mM)	10.6 ± 0.4
Plus phenol (0.1 mM)	12.1 ± 1.5[b]
Plus phenol (1 mM)	14.8 ± 0.6[b]

Values represent means $\pm$ SD of four separate experiments using bone marrow cells from different mice. Bone marrow cells were obtained as described in the text.
[a]Cell density was 1.5×10^6/ml. Incubations were performed for 5 minutes at 37 C.
[b]Significantly different to value obtained using cells, hydroquinone and peroxide, $p < 0.01$. Statistical analyses were performed by one way analysis of variance followed by a Dunnetts t test for multiple comparisons to a single control.
[c]Autoclaved cell controls resulted in binding of 0.38 nmol of hydroquinone equivalents/mg of protein under these conditions.
[d]Binding of ^{14}C-phenol under these conditions was 16% of the value obtained using ^{14}C-hydroquinone.
(Taken from ref. 119 and reproduced with permission.)

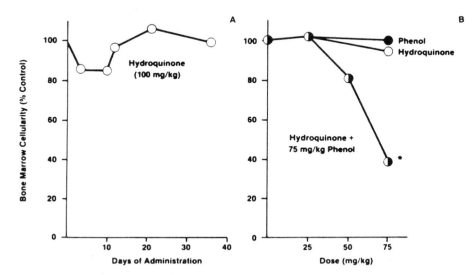

Fig. 16.   Influence of phenol on suppression of bone marrow
cellularity by hydroquinone. B6C3F1 male mice, 8-10
weeks of age, were administered phenol, hydroquinone,
or both intraperitoneally at the doses indicated,
twice daily, 6 hr apart. (A) Hydroquinone alone (O)
over 36 days of administration. (B) Hydroquinone (25-
75 mg/kg) in the presence of phenol (75 mg/kg)(◑);
phenol alone (•) after 12 days of administration.
Each point represents the mean femoral cellularity of
five or six animals expressed as a percentage of
control (untreated) mice which averaged 1.63 x 10^7,
1.02 x 10^6 SE) nucleated cells per femur. Asterisk
indicates significant difference from controls (p <
0.001). Taken from Ref. 118 and reproduced with
permission.

REFERENCES

1. L. Selling, Benzol as a leukotoxin. Studies on degeneration regeneration of blood and hematopoietic organs, <u>Johns Hopkins Hospital Rep</u>. 17:83 (1916).

2. H.G. Weiskotten, S.C. Shwartz, H.S. Steinsland, The action of benzol on blood and blood forming tissues, <u>J. Med. Res.</u> 35:63 (1916).

3. H.G. Weiskotten, G.B.F. Gibbs, E.O. Boggs, and E.R. Templeton, The action of benzol. VI. Benzol vapor leucopenia (rabbit), <u>J. Med. Res.</u> 41:425 (1920).

4. V. Hough and S. Freeman, Relative toxicity of commercial benzene, toluene, and xylene, <u>Fed. Proc.</u> 3:20 (1944).

5. M. Wolf, V. Rowe, D. McCollister, R. Hollingsworth, F. Oyen Toxicological studies of certain alkylated benzenes and benzene, <u>A. M. A. Arch. Ind. Health.</u> 14:387 (1956).

6. W. Diechmann, W. MacDonald, E. Bernal, The hemopoietic tissue toxicity of benzene vapors, <u>Toxicol. Appl. Pharmacol.</u> 5:201 (1978).

7. C.A. Snyder, B. Goldstein, A. Sellakumar, S. Nohnan, I. Bromberg, M. Erlichman, and S. Laskin, Hematotoxicity of inhaled benzene to Sprague-Dawley rats and AKR mice at 300 ppm, <u>J. Toxicol. Environ. Health.</u> 4:605 (1978).

8. M. Tavassoli and J.M. Yoffey, "Bone Marrow: Structure and Function", Alan R. Liss Inc., New York (1983).

9. L.S. Andrews, E.W. Lee, C.M. Witmer, J.J. Kocsis, and R. Snyder, Effects of toluene on metabolism, disposition, and hematopoietic toxicity of [^{14}C]benzene, <u>Biochem. Pharmacol.</u> 26:293 (1977).

10. L. Andrews, H.A. Sasame, J.R. Gillette, [^{3}H]Benzene metabolism in rabbit bone marrow, <u>Life Sciences</u> 25:567 (1979).

11. D.E. Rickert, T.S. Baker, J.S. Bus, C.S. Barrow, and R.D. Irons, Benzene disposition in the rat after exposure by inhalation, <u>Toxicol. Appl. Pharmacol.</u> 49:417 (1979).

12. D.R. Boggs, Hemostatic regulatory mechanisms of hematopoiesis, <u>Ann. Rev. Physiol.</u> 28:39 (1966).

13. A.P. Korn, R.M. Hinckelman, F.P. Ottensmeyer, and J.E. Till, Investigations of stochastic model of hemopoiesis, <u>Exp. Hematol.</u> 1:362 (1973).

14. M. Ogawa, P.N. Porter, T. Nakahata, Renewal and commitment to differentiation of hematopoietic stem cells (An Interpretive Review), <u>Blood</u> 61:823 (1983).

15. T. Nakahata, A.J. Gross, M. Ogawa, A stochastic model of self-renewal and commitment to differentiation of the primitive hemopoietic stem cells in culture, <u>J. Cell. Physiol.</u> 113:455 (1982).

16. W.J. Williams, E. Beutler, A.J. Erslev, and R.W. Rundles, "Hematology", 2nd ed., McGraw-Hill Book Company, New York 1977.

17. M.A. Lichtman, The ultrastructure of the hematopoietic environment of the marrow: a review, <u>Exp. Hematol.</u> 9:391 (1981).

18. J.J. Trentin, Determination of bone marrow stem cell differentiation by stromal hemopoietic inductive microenvironment (HIM), <u>Am. J. Pathol.</u> 65:621 (1971).

19. M. Tavassoli, Studies on hemopoietic microenvironment, <u>Exp. Hematol.</u> 3:213 (1975).

20. N.S. Wolf, The hemopoietic microenvironment, <u>Clin. Haematol.</u> 8:469 (1979).

21. M. Tavassoli and A. Friedenstein, Hemopoietic stromal micro-environment, <u>Am. J. Hematol.</u> 15:195 (1983).

22. T.D. Allen and T.M. Dexter, The essential cells of the hemopoietic microenvironment, <u>Exp. Hematol.</u> 12:517 (1984).

23. I.N. Rich, A role for the macrophage in normal hemopoiesis. I. Functional capacity of bone-marrow-derived macrophages to release hemopoietic growth factors, Exp. Hematol. 14:738 (1986).

24. S.C. Clark and R. Kamen, The human hematopoietic colony-stimulating factor, Science 236:1229 (1987).

25. D. Metcalf, Haemopoietic growth factors. Med. J. Austr. 148:516 (1988).

26. A. Billiau, J. Van Damme, G. Opdenakkar, W.E. Fibbe, J.H.F. Falkenburg, and J. Content, Interleukin 1 as cytokine inducer, Immunobiol. 172:323 (1986).

27. I.N. Rich, A role for the macrophage in normal hemopoiesis. II. Effect of varying physiological oxygen tensions on the release of hemopoietic growth factors from bone-marrow-derived macrophages in vitro, Exp. Hematol. 14:746 (1986).

28. G.C. Bagby, Production of multilineage growth factors by hematopoietic stromal cells: Intercellular regulatory network involving mononuclear phagocytes and interleukin-1, Blood Cells 13:147 (1987).

29. E.P. Cronkite, Analytical review of structure and regulation of hemopoiesis, Blood Cells 14:313 (1988).

30. J.E. Till and E.A. McCulloch, A direct measure of the radiation sensitivity of normal mouse bone marrow cells, Radiation Res. 14:213 (1961).

31. D.H. Pluznik and L. Sach, The cloning of normal mast cells in tissue culture, J. Cell. Comp. Physiol. 66:319 (1965).

32. T.R. Bradley and D. Metcalf, The growth of mouse bone marrow cells in vitro, Aust. J. Biol. Med. Sci. 44:287 (1966).

33. J.R. Stephenson, A.A Axelrad, D.L. McLeod, and M.M. Shreeve, Induction of colonies of hemoglobin synthesizing cells by erythropoietin in vitro, Proc. Natl. Acad. Sci. USA 68:1542 (1971).

34. D. Metcalf, G.T.V. Nossal, N.L. Warner, J.F. Miller, T.E. Mandel, J.E. Layton, G.A. Gutman, Growth of B-lymphocyte colonies in vitro, J. Exp. Med. 142:1534 (1975).

35. E. Gerassi and L. Sach, Regulation of the induction of colonies in vitro by normal human lymphocytes, Proc. Natl. Acad. Sci. USA 73:4546 (1976).

36. D. Metcalf, "The Molecular Control of Blood Cells", Harvard University Press, Cambridge (1988).

37. B.I. Lord and N.G. Testa, The hemopoietic system: structure and regulation in "Hematopoiesis: Long-Term Effects of Chemotherapy and Radiation", Nydia G. Testa and Robert Peter Gale eds., Marcel Dekker, New York (1988).

38. T.M. Dexter, T.D. Allen, L.G, Lajtha, R. Schofield, and B.I. Lord, Stimulation of differentiation of hemopoietic cells in vitro, J. Cell Physiol. 82:461 (1973).

39. T.M. Dexter, T.D. Allen, and L.G. Lajtha, Conditions controlling the proliferation of haemopoietic stem cells in vitro, J. Cell. Physiol. 91:335 (1977).

40. T.M. Dexter, N.G. Testa, T.D. Allen, T. Rutherford, and E. Scholnick The regulation of haemopoesis in long term bone marrow cultures, IV. Molecular and Cell Biological Aspects of erythropoiesis, Blood 58:699 (1981).

41. T.J. Haley "Nuclear Hematology", Academic Press, New York (1965).

42. C.F. Baxter, E.H. Belchar, E.B. Harris, and L.F. Lamerton, Anaemia and erythropoiesis in the irradiated rat: an experimental study with particular reference to techniques involving radioactive iron, Br. J. Haematol. 1:86 (1955).

43. R.J. Cole and J. Paul, The effects of erythropoietin on haem synthesis in mouse yolk sac and cultured foetal liver cells, J. Embryol. Exp. Morph. 15:245 (1966).

44. W.R. Bruce and E.A. McCulloch E.A., The effects of erythropoietic stimulation on the hemopoietic colony-forming cells of mice, _Blood_ 23:216 (1964).

45. E. Filmanowicz and C.W. Gurney, Studies on erythropoiesis. XVI. Response to a single dose of erythropoietin in polycythemic mice. _J. Lab. Clin. Med._ 57:65 (1961).

46. B.J. Payne, H.B. Lewis, T.E. Murchinson, and E.A. Hart, Hematology of laboratory animals, in: "Handbook of Laboratory Science, VIII," E.C. Melby, Jr.,and N.H. Altman, Eds. CRC Press, Cleveland (1976).

47. T. Bothwell, R. Charlton, J. Cook, and C. Finch "Iron metabolism in man", Blackwell Scientific Publication, Oxford (1979).

48. E.W. Lee, J.J. Kocsis, and R. Snyder, Acute effect of benzene on [^{59}Fe] incorporation into circulating erythrocytes, _Toxicol. Appl. Pharmacol._ 27:431 (1974).

49. E.W. Lee, J.J. Kocsis, and R. Snyder, The use of ferrokinetics in the study of experimental anemia, _Env. Health Perspect._ 39:29 (1981).

50. C.A. Snyder, B. Goldstein, A. Sellakumar, and R.E. Albert, Evidence for hematotoxicity and tumorigenesis in rats exposed to 100 ppm benzene, _Am. J. Ind. Med._ 5:429 (1984).

51. E.W. Lee, J.J. Kocsis, and R. Snyder, Dose-dependent inhibition of [^{59}Fe] incorporation into erythrocytes after a single dose of benzene, _Res. Commun. Chem. Pathol. Pharmacol._ 5:547 (1973).

52. L.E. Bolcsak and D.E. Nerland, Inhibition of erythropoiesis by benzene and benzene metabolites, _Toxicol. Appl. Pharmacol._ 69:363 (1983).

53. R. Snyder, E. Dimitriadis, R. Guy, P. Hu, K. Cooper, H. Bauer, G. Witz, and B.D. Goldstein, Studies on the mechanism of benzene toxicity, _Environ. Health Perspect._ 82:31 (1989).

54. R.D. Irons, H.D. Heck, B.J. Moore, and K.A. Muirhead, Effects of short-term benzene administration on bone marrow cell cycle kinetics in rats, _Toxicol. Appl. Pharmacol._ 51:399 (1979).

55. E.M. Uyeki, A.E. Askar, D.W. Shoeman, and T.U. Bisel, Acute toxicity of benzene inhalation to hemopoietic precursor cells, _Toxicol. Appl. Pharmacol._ 40:49 (1977).

56. D. Gill, V. Jenkins, R. Kempen, and S. Ellis, The importance of pluripotent stem cells in benzene toxicity, _Toxicology_ 16:163 (1980).

57. J.D. Green, C.A. Snyder, J. Lobue, B.D. Goldstein, and R.E. Albert, Acute and chronic dose/response effects of inhaled benzene on multipotential hemopoietic stem (CFU-S) and granulocyte/macrophage progenitor (GM CFU-C) cells in CD-1 mice, _Toxicol. Appl. Pharmacol._ 58:492 (1981).

58. K. Harigaya, M.E. Miller, E.P. Cronkite, and R.T. Drew, The detection of _in vivo_ liquid bone marrow cultures, _Toxicol. Appl. Pharmacol._ 60:346 (1981).

59. E.P. Cronkite, T. Inoue, A.L. Carsten, M.E. Miller, J.E. Bullis, and R.T. Drew, Effects of benzene inhalation on murine pluripotent stem cells, _J. Toxicol. Environ. Health_ 9:441 (1982).

60. A. Tunek, T. Olofsson,and M. Berlin, Toxic effects of benzene and benzene metabolites on granulopoietic stem cells and bone marrow cellularity in mice, _Toxicol. Appl. Pharmacol._ 59:149 (1981).

61. K. Toft, T. Olofsson, A. Tunek, and M. Berlin, Toxic effects on mouse bone marrow caused by inhalation of benzene, _Arch. Toxicol._ 51:295 (1982).

62. K. Baarson, C.A. Snyder, and R.E. Albert, Repeated exposures of C57B16 mice to 10 ppm inhaled benzene markedly depressed erythropoietic colony formation, _Toxicol. Lett._ 20:337 (1984).

63. V.N. Frash, B.G. Yushov, A.V. Karaulov, and V.L. Skuratov, Mechanism of action of benzene on hematopoiesis. Investigation of hematopoietic stem cells, _Bull. Exp. Biol. Med._ 82:985 (1976).

64. K.W. Gaido and D. Weirda, Modulation of stromal cell function in DBA/2J and B6C3F1 mice exposed to benzene or phenol, Toxicol. Appl. Pharmacol. 81:469 (1985).

65. K.W. Gaido and D. Weirda, In vitro effects of benzene metabolites on mouse bone marrow stromal cells, Toxicol. Appl. Pharmacol. 76:45 (1984). 66. K.W. Gaido and D. Wierda, Suppression of bone marrow stromal cell function by benzene and hydroquinone is ameliorated by indomethacin, Toxicol. Appl. Pharmacol. 89:378 (1987).

67. S.J. Pirozzi, M. Schlosser, and G.F. Kalf, Prevention of benzene induced myelotoxicity and prostaglandin synthesis in bone marrow of mice by inhibitors of prostaglandin H synthetase, Immunopharmacol. 18:39 (1989).

68. R.D. Irons and D.A. Neptun, Effects of principle hydroxy-metabolites of benzene on microtubule polymerization, Arch. Toxicol. 45:297 (1980).

69. K. Mann, M. Giesel, H. Fasold, W. Haase, Isolation of native microtubules from porcine brain and characterization at the GTP binding sites, FEBS Lett. 92:45 (1974).

70. R.W. Pfiefer and R.D. Irons, Alteration of lymphocyte function by quinones through sulfhydryl-dependent disruption of microtubule assembly, Int. J. Immunopharmacol. 5:463 (1983).

71. C. Schwartz, R. Snyder, and G. Kalf, The inhibition of mitochondrial DNA replication in vitro by metabolites of benzene, hydroquinone and p-benzoquinone, Chem-Biol. Interact. 53:327 (1986).

72. W.K. Lutz and C.H. Schlatter, Mechanism of the carcinogenic action of benzene:irreversible binding to rat liver DNA, Chem-Biol. Interact. 18:241 (1977).

73. D.P. Gill and A. Ahmed, Covalent binding of [^{14}C]benzene to cellular organelles and bone marrow nucleic acids, Pharmacol. 30:1127 (1981).

74. G.F. Kalf, T.R. Rushmore, and R. Snyder, Benzene inhibits RNA synthesis in mitochondria from liver and bone marrow, Chem-Biol. Interact. 42:353 (1982).

75. T. Rushmore, R. Snyder, and G. Kalf, Covalent binding of benzene and its metabolites to DNA in rabbit bone marrow mitochondria in vitro, Chem.-Biol. Interact. 49:133 (1984).

76. L. Jowa, G. Witz, R. Snyder, S. Winkle and G. Kalf, Synthesis and characterization of deoxyguanine-benzoquinone adducts, J. Appl. Toxicol. 10:47 (1990).

77. E.C. Vigliani and G. Saita, Benzene and leukemia, N. Engl. J. Med. 271:872 (1964).

78. M. Aksoy, H. Dincal, T. Akgun. S. Erdem, and G. Dincol, Hematological effects of chronic benzene poisoning in 217 workers, Br. J. Ind. Med. 28:296 (1971).

79. M. Aksoy, S. Erdem, and G. Dinol, Leukemia in shoe workers exposed chronically to benzene, Blood 44:837 (1974).

80. E. Vigliani and A. Forni, Benzene and leukemia, Environ. Res. 11:122 (1976).

81. B.D. Goldstein, Hematotoxicity in humans in benzene toxicity, a critical evaluation, J. Toxicol. Environ. Health, Suppl. 2:69 (1977).

82. J.-L. Amiel, Essai negtif d'induction de leuce mies chez les souris par le benzene, Rev. Fr. Etud. Clin. Biol. 5:198 (1960).

83. J. Ward, J. Weisberger, R. Yamamoto, T. Benjanub, C. Brown, and E. Weisberger, Long term effects of benzene in C57Bl/6N mice, Arch. Environ. Health 30:22 (1975).

84. C.A. Snyder, B.D. Goldstein, A.R. Sellakumar, I. Bromberg, S. Laskin, and R.E. Albert, The inhalation toxicology of benzene:incidence of hematopoietic neoplasms and hematotoxicity in AKR/J and C57B1/6J mice, Toxicol. Appl. Pharmacol. 54:323 (1980).

85. C. Maltoni and C. Scarnato, First experimental demonstration of the carcinogenic effects of benzene, Med. Lav. 5:352 (1981).

86. C. Maltoni, G. Cotti, L. Valgimigli, and A. Mandrioli, Zymbal gland carcinomas in rats following exposure to benzene by inhalation, Am. J. Ind. Med. 3:11 (1982).

87. E.P. Cronkite, J.E. Bullis, T. Inoue, and R.T. Drew, Benzene inhalation produces leukemia in mice, Toxicol. Appl. Pharmacol. 75:358 (1984).

88. E.P. Cronkite, R.T. Drew, T. Inoue, and J.E. Bullis, Benzene hematotoxicity and leukemogenesis, Am. J. Ind. Med. 7:447 (1985).

89. E.P. Cronkite, Benzene hematotoxicity and leukemogenesis, Blood Cells 12:129 (1986).

90. E.P. Cronkite, Chemical leukemogenesis: benzene as a model, Hematol. 24:2 (1987).

91. B.J. Dean, Recent findings on the genetic toxicology of benzene, toluene, xylenes, and phenol, Mut. Res. 154:153 (1985).

92. B.J. Dean, Genetic toxicology of benzene, toluene, xylene and phenols, Mut. Res. 47:75 (1978).

93. A.M. Forni, A. Capellini, E. Pacifico, and E.C. Vigliani, Chromosome changes and their evolution in subjects with past exposure to benzene, Arch. Environ. Health 23:385 (1971).

94. H.G.S. Van Raalte and P. Grasso, Hematological myelotoxic, clastogenic, carcinogenic, and leukemogenic effects of benzene, Regulat. Toxicol. Pharmacol. 2:153 (1982).

95. R.R. Tice, T.F. Vogt, and D.L. Costa, Cytogenetic effects of inhaled benzene in murine bone marrow in Genotoxic Effect of Airborne Agents, Environ. Sci. Res. 25:257 (1982).

96. M. Hite, M. Pecharo, I. Smith, and S. Thornton, The effect of benzene in micronucleus test, Mut. Res. 77:149 (1980).

97. M.A. Diaz, J. Reiser, and J. Diez, Studies on benzene mutagenesis, I. The micronucleus test, Experientia 36:297 (1980).

98. C.A. Luke, R.R. Tice, and R.T. Drew, Duration and regimen induced micronuclei in the peripheral blood of mice exposed chronically to benzene, Environ. Mutagen.7 Suppl.(3):29, (1985).

99. S.J. Pirozzi, J.F. Renz, and G.F. Kalf, The prevention of benzene-induced genotoxicity in mice by indomethacin, Mut. Res. 222:291 (1989).

100. M.M. Gad-El Karim, V.M.S. Ramanujam, and M.S. Legator, Correlation between the induction of micronuclei in bone marrow by benzene exposure and the excretion of metabolites in urine of CD-1 Mice, Toxicol. Appl. Pharmacol. 85:464 (1986).

101. T.A. Kirley, B.D. Goldstein, W.M. Maniara and G. Witz, Metabolism of trans,trans-muconaldehyde, a microsomal hematotoxic metabolite of benzene, bu purified yeast aldehyde dehydrogenase and a mouse liver soluble fraction, Toxicol. Appl. Pharmacol. 100:360 (1989).

102. G.L. Erexson, J.L. Wilmer, W.H. Steinhagen, and A.D. Kligerman, Induction of cytogenic damage in rodents after short-term inhalation of benzene, Environ. Mut. 8:29 (1986).

103. M.N. Winternitz and A.D. Hirschfelder, Studies upon experimental pneumonia in rabbits: Part I-III, J. Exp. Med. 17:657-664, 1913.

104. A.D. Hirschfelder and M.C. Winternitz, Studies upon experimental pneumonia in rabbits. Part IV. Is there a parallelism in trypanocidal and pneumoccoccidal action of drugs?, J. Exp. Med.17:666 (1913).

105. W.C. White and A.M. Gammon, The influence of benzol inhalation of experimental pulmonary tuberculosis in rabbits, _Trans. Assoc. Amer. Physicians_ 29:332 (1914).

106. W.A. Camp and E.A. Baumgartner, Inflammatory reaction in rabbits with a severe leukopenia, _J. Exp. Med._ 22:174 (1915).

107. L. Hektoen, The effects of benzene on the production of antibodies, _J. Infect. Dis._ 19:69 (1916).

108. R. Smolik, K. Grzybek-Hryncewicz, A. Lange, and W. Zatonski, Serum complement levels in workers exposed to benzene, toluene, and xylene, _Int. Arch. Arbeitsmed._ 31:243 (1973).

109. A. Lange, R. Smolik, W. Zatonski, and J. Syzmanska, Serum immunoglobulin levels in workers exposed to benzene, toluene, and xylene, _Int. Arch. Arbeitsmed._ 32:37 (1973).

110. A. Lange, R. Smolik, W. Zatonski, and H. Glazman, Leukocyte agglutinins in workers exposed to benzene, toluene, and xylene, _Int. Arch. Arbeitsmed._ 31:45 (1973).

111. M.G. Rozen, C.A. Snyder, and R.E. Albert, Depression in B- and T-lymphocytes mitogen-induced blastogenesis in mice exposed to low concentrations of benzene, _Toxicol. Lett._ 20:343 (1984).

112. G.J. Rosenthal and C.A. Snyder, Modulation of the immune response to Listeria monocytogenes by benzene inhalation, _Toxicol. Appl. Pharmacol._ 80:502 (1985).

113. G.J. Rosenthal and C.A. Snyder, Inhaled benzene reduces aspects of cell-mediated tumor surveillance in mice, _Toxicol. Appl. Pharmacol._ 88:35 (1987).

114. D.L. Laskin, L. MacEachern, and R. Snyder, Activation of bone marrow phagocytes following benzene treatment of mice, _Environ. Health Perspect._ 82:75 (1989).

115. D.O. Adams and T.A. Hamilton, The cell biology of macrophage activation, _Ann. Rev. Immunol._ 2:283 (1984).

116. B.M. Babior, Oxidants from phagocytes. Agents of defense and destruction, _Blood_ 64:959 (1984).

117. P.A. Ceruitti, Prooxidant states and tumor promotion, _Science_ 227:375 (1985).

118. D.A. Eastmond, M.T. Smith, and R.D. Irons, An interaction of benzene metabolites reproduces the myelotoxicity observed with benzene exposure, _Toxicol. Appl. Pharmacol._ 91:85 (1987).

119. V. Subrahmanyam, A. Sadler, E. Suba, and D. Ross, Stimulation of _in vitro_ bioactivation of hydroquinone by phenol in bone marrow cells, _Drug Metab. Disp._ 17:348 (1989).

120. G. Witz, S.R. Gondi, and B.D. Goldstein, Short-term toxicity of trans, trans muconaldehyde, _Toxicol. Appl. Pharmacol._ 80:511 (1985).

HEMATOTOXICITY, LEUKEMOGENECITY AND CARCINOGENECITY OF

CHRONIC EXPOSURE TO BENZENE

Muzaffer Aksoy

Department of Biology
Research Institute for Basic Sciences
TÜBİTAK, Gebze 74, Kocaeli, Turkey

INTRODUCTION

As it is known benzene (C_6H_6) is the so called parent hydrocarbon of the aromatic group, the resonant cyclic compound consisting only of carbon and hydrogen. Benzene stems not only from industrial sources but it is present in natural products etc. On the other hand, incomplete combustion of all natural organic substances causes the formation of aromatic hydrocarbons. In this respect, tobacco smoke contains benzene in small amounts (1). Benzene is an excellent solvent for numerous solids. In addition, benzene is used in industry as a starting material in the production of numerous organic chemicals. Also, public exposure to benzene present in automobile gasoline ranging between 0.8 and 5 per cent has raised increased concern about this chemical agent as a significant environmental pollutant. Benzene is absorbed via the lungs and approximately 40 or 50 per cent of it is retained. Due to the high lipid solubility, benzene accumulates in fat and nervous system, and about 40-50 per cent of benzene with some carbon dioxide is excreted unchanged from the lungs. Furthermore, as Sherwood and Carter (2) showed that even after 24 hours a small percentage of benzene is exhaled unchanged. As it is known the major organ system implicated for benzene metabolism are the liver and the bone marrow. After inhalation, apart from insignificant traces of benzene in urine, most of the absorbed benzene remaining is ultimately metabolized into derivates which are more water soluble and thus more removable by the kidneys. The fact that the benzene metabolizing system is located in the cell microsomal fraction of the liver is demonstrated by differential centrifugation of the liver homogenate. The mixed function oxidases (mono-oxidases) in the endoplasmotic reticulum of the parenchym cells perform epoxidation of the nucleus to form arene oxides which decompose spontaneously to form phenol. Benzene oxide or epoxide, an unstable chemical which can attach to DNA or RNA giving cause to mutation. All these problems were discussed by Snyder (3) in the foregoing paper. Furthermore, recently Uyeki et al. (4) and later other investigators (5,6) showed experimentally in mice that high doses of benzene such as 4680 or 4000 ppm cause rapid decrease of colony forming cells (CFC). These experiments (4,5,6) and that of Post et al. (7) disclosed that benzene inhibits the self-renewal of pluropotent stem cells and that of the cells of microenvironment and by this phenomenon causes aplastic anemia. Furthermore, benzene is a suspected leukemogenic agent since the first description of a case of leukemia due to this chemical by Le Noire and Claude (8) in 1897. Recently Maltoni and Scarnato (9) showed in rats and mice that intake of high doses of benzene causes malignant tumours in several tissues. If an agent is toxic to the pluropotent stem cell or to microenvironment, it can cause an alteration in two important capabilities of the colony forming cells. These are: 1) Self renewal, and 2) Differentiation to produce a variety of lineage progenitor cells (10). In other words, the longterm maintenance of self-renewal and differentiation. Any block or disturbance in these capabilities will cause either aplastic anemia (a block in self-renewal) or leukemia (a block in differentiation).

MATERIAL AND METHODS

All individuals with blood dsycrasies due to chronic benzene exposure and the laboratory methods used are described in the following papers (11,12,13).

Hematological Disorders Caused by Chronic Exposure to Benzene

Chronic benzene toxicity predominately associated with hematotoxicity is known since description of 9 cases of aplastic anemia due to this chemical agent by Santesson in 1897 (14). The findings of chronic benzene poisoning are mainly leucopenia, thrombocytopenia and pancytopenia. As an example of chronic benzene toxicity, it will be interesting to summarize briefly the results of an investigation among 217 workers with this chemical toxicity in Istanbul during the period of 1969 and 1970 (11). In 51 of these workers (23.5%) hematological abnormalities due to benzene toxicity were detected. Leucopenia was present in 9.7%, thrombocytopenia in 1.84%, leucopenia associated with thrombocytopenia in 4.6%, pancytopenia in 2.76%, acquired pseudo-Pelger Huet anomaly in 0.46%, lymphocytosis in 0.46%, giant platelets in 0.46%, eosinophilia in 2.3% and basophilia in 0.46% (11). The maximum levels for eosinophil and basophil in this study were 8% and 2% respectively. The concentration of benzene in the workplaces ranged between 30 and 210 ppm. The period of exposure to benzene ranged between 3 months and 17 years. Curiously in a worker with normal hematological data of this study, an acute erythroleukemia developed 4 years later (11,15). Except leucopenia, thrombocytopenia and pancytopenia there are numerous blood and laboratory changes due to chronic benzene exposure. These are briefly: decrease in the leucocyte osmotic resistance, decreased phagocytic function of neutrophiles, reduced glucogen content and inhibited activity of preoxidase of neutrophiles etc. (16). According to Mosczynski and Lisewitz (17), in individuals with chronic benzene exposure there is a decrease in T lymphocyte count without showing a change in their function. As to the presence of macrocystosis in red blood cells in chronic benzene toxicity, according to the present author it does not have to accompany chronic benzene toxicity (16). On the other hand, there are very few reports on the state of bone marrow in chronic benzene toxicity before the development of aplastic anemia. The state of the bone marrow in chronic benzene toxicity was studied in 11 patients by the present author and his associates (11). Four with leucopenia associated with thrombocytopenia, one with lymphocytosis, one with pseudo-Pelger Huet anomaly and five with pancytopenia which were diagnosed during a survey among healthy appearing workers. No remarkable changes were observed in two of them. One bone marrow was hypercellular with maturation arrest in the erythroid and myeloid series, whereas the second bone marrow was slightly hypercellular with maturation arrest in granulocytic series. In 2 workers, the bone marrow was normocellular, and in 2 slightly hypocellular with maturation arrest either in the erythroid or myeloid series. In another 3, the bone marrow showed marked vacuolization as observed in some cases of chloramphenicol-induced pancytopenia. On the other hand, animal experiments with benzene performed by Biancacchio and Farmerielle (18) showed that there is considerable increase in urinary coproporphyrine level. Furthermore, Kahn and Muzyka (19) observed in 74 workers with chronic exposure to benzene that there were complaints such as frequent headaches, easy fatiguebility and disturbances in sleep and memory. Also 60 per cent of the workers had increased delta-aminolevulinic acid in erthrocytes and 30% of them also had increased coproporphyrine in urine. On the other hand, the same investigators showed experimentally that chronic benzene exposure causes changes in the levels of delta-aminolevulinic acid, porphobilinogen and coproporphyrine in the central nervous system. According to Kahn and Muzyka (19) the disorders of porphyrine metabolism due to chronic exposure to benzene toxicity take place before appearance of hematological abnormalities. On the other hand, Lange et al. (20) who studied immunoglobulin level in workers chronically exposed to benzene, toluene and xylene and they found that serum IgG and IgA levels were decreased. Despite this, IgM levels were either normal or slightly elevated. Similarly Smolik et al. (21) determined serum complement levels in the workers with benzene, toluene and xylene exposure. Most of the complement levels were lower than the mean value of the control group. According to the investigators (21), these findings showing the changes in immunoglobulin levels suggest the involvement of immunologic factors in the pathologic mechanism of chronic exposure to benzene and its homologs. On the other hand, Sobcszyk et al. (22) performed EEG recordings in female workers with chronic benzene exposure following the occurrence of hematological changes and increased phenol excretion. In 40 per cent of these workers there were abnormal findings such as changes in temporal regions, paroxysmal burst and generalized changes. According to the investigators these changes in bioelectric activity of the brain may be considered an index of chronic benzene toxicity.

Skin and Chronic Benzene Toxicity

The recent studies performed Suston et al. (23) showed that a worker exposed 150 times to a rubber solvent containing 0.5% of benzene will absorb approximately 6 mg of benzene daily. This represents a significant addition to 14 mg of benzene estimated to be retained in the body following inhalation of benzene at a concentration of 1 ppm. If one considers that the skin of tire workers is frequently cracked and fissured almost assures that the penetration of benzene contacting the skin will be easier and greater than would be expected for the intact skin.

According to Suston et al. (23) these results should be evaluated carefully from the standpoint of health.

APLASTIC ANEMIA DUE TO CHRONIC BENZENE EXPOSURE

The total bone marrow depresion that is mostly seen in advanced cases of chronic benzene toxicity will exhibit itself as pancytopenia in the peripheral blood reflecting mostly the picture of classical aplastic anemia. Despite this, there are some laboratory findings which are rarely seen in idiopathic type or in this hematologic disorder caused by other agents. We shall discuss these findings later.

Frequency

Aplastic anemia is a rare disorder in Western countries. It is between 1.4 and 3.1 x 10^6 per year (24). According to Böttiger (25) the incidence of this hematologic disorder is at least 4 or 5 times more common in the East than in the West. Furthermore the same investigator has considered that the differences in the prevalence of aplastic anemia between West and East has multifactorial explanation. In this respect except for the possible role of genetic difference in susceptibility he accuses the other exogenous factors such as occupational hazards including benzene, high incidence of hepatitis and the use of chloramphenicol. On the other hand, using an indirect and possibly an inaccurate method, the present author calculated the incidence of aplastic anemia in Turkey as 9.3 per million. From the standpoint of the incidence of aplastic anemia, Turkey is similar to the Eastern countries. Actually, following the control of the use of benzene in the workplaces in Turkey, the annual number of aplastic anemia due to this chemical decreased rapidly. Despite this, in Turkey benzene like chloramphenicol is still one of the main factors playing role in the development of aplastic anemia. As an example, the present author and his associates (26) performed a study on 108 patients with this hematological disorder during the period of 1973 and 1982 in Istanbul. In this study the most important possible causal agent for the development of aplastic anemia was benzene which accounted for 25 patients (23.1%) of the series. Furthermore, in 1988 we have studied a fatal case of aplastic anemia due to benzene exposure in Istanbul. The patient was a shoeworker and the solvent used in the workplace contained 8.3% benzene determined by gas-chromatography. Furthermore, in a study on the benzene contents of the solvents, thinners etc. used in the workplaces in Istanbul and vicinities during the period between 1983 and 1985, the present author and his associates (12) established that the 76.4% of the solvents and 19.1% of the thinners contained benzene more than 1% ranging between 1.1 and 7.6% (mean:2.8%).

Clinical and Hematological Findings of Aplastic Anemia due to Chronic Benzene Exposure

Characterized by pancytopenia, the clinical and hematological picture of aplastic anemia due to benzene in a whole is not different from that of classical aplastic anemia, idiopathic or due to different chemicals or other agents. Despite this, there are some findings which are absent in this hematologic disorder caused by other agents. These are briefly:

418

mild bilirubinemia, changes in osmotic fragility, shortened erythrocyte survival time, increased fecal urobilinogen and mild reticulocytosis (16). These findings are not found in all cases of aplastic anemia due to chronic benzene exposure. A mild bilirubinemia with a range of 1.3 mg to 1.8 mg/100 ml was present in 8 out of 28 patients with aplastic anemia due to chronic benzene exposure. Furthermore, mild or moderate increase in either before or after incubation as performed by the quantitative osmotic fragility test was present in 13 out of 20 pancytopenic patients with aplastic anemia due to chronic benzene exposure (16,27,28). As it is known lactate dehydrogenase activity in aplastic anemia is in normal range. Contrary to this, in some cases of pancytopenia due to chronic benzene exposure, this enzyme can increase (16,28). Relative lymphocytosis is usually a characteristic finding of aplastic anemia. Contrarily, in benzene mediated aplastic anemia absolute lymphopenia is not rare. In our series, it was observed in 24 out of 32 patients with benzene exposure (16,28). On the other hand, there are no symptoms and findings relating to nervous system in aplastic anemia, either idiopathic or due to different agents. Despite this, in some cases of aplastic anemia due to to chronic benzene exposure, there are some findings relating to neurological system. Although as early as in 1967 Truhaut (29) indicated in his report relating benzene, the possibility of long-term effects of this chemical agent on the nervous system such as polyneuritis of the lower extremities, there are very few reports on this subject. In 1982 Baslo and Aksoy (30) performed a neurological, electromyographical and motor conduction velocity examinations in 6 patients with aplastic anemia due to this chemical. In addition, sensory conduction velocities were measured in 3 patients. Neurological abnormalities such as global atrophy and decreased sensory vibration of lower extremities (in one case), distal latency lengthening of median nerve (in one case), decrease in the sensory conduction velocities of the lower extremities (in one case) were found. There was a certain relationship between the presence of neurological abnormalities and the period of exposure.

The findings of the bone marrow examination obtained in chronic benzene toxicity are extremly variable ranging from complete aplasia to a highly hyperplastic bone marrow (16,31). In terminal stage, bone marrow can be found fully acellular. According Mallory et al. (31) the effects of benzene on the bone marrow will vary from individual to individual and are mostly indepedent of the length of exposure. Mostly, more than one type of faci can be encountered in the bone marrow. Furthermore, Moeschlin and Speck (32) studied the bone marrow of animals poisonened by benzene by autoradiography with ^{3}H-thymidine and found that the results varied from animal to animal. From 19 bone marrows studied, 4 were very hypoplastic, 6 hypoplastic, 4 were hyperplastic and 5 normocellular. No correlation was found between cellularity of the bone marrow and the duration of exposure to benzene. The present author and Erdem (33) performed a follow-up study in 44 pancytopenic patients with chronic exposure to benzene. Only 21 patients had hypocellular bone marrow, it was normocellular in 13 patients and hypercellular in 8. On the other hand, there was a clear relationship between the types of cellularity of the bone marrow and the outcome including the development of leukemia.

Out of 21 patients with hypocellular bone marrow, 11 died
(52.4%) and in 5 patients (23.8%) leukemia developed later.
Contrary to this, only 1 out of 13 (7.7%) pancytopenic
patients with normocellular bone marrow died. On the other
hand 4 out of 8 patients with hypercellular bone marrow
recovered completely (50%). Two died from complications of
aplastic anemia such as bleeding and infection. In one of
the remaining two patients, leukemia developed and in the
second 8 years after complete recovery the occurrence of
myeloid metaplasia was the cause of death. Erbengi and the
present author (34) performed ultrastructual studies in 4
patients with chronic benzene toxicity. Ultrastructural
findings were in accordance with the changes in the bone
marrow observed by light microscopy. Additioanlly, there
was an increase in plasma cells with maturation arrest in
the later phases of erythroid series accompanied with
phagocytic activity of the reticulum cells. This study showed
hyperactivation in the marrow of the one case and depression
in the erythroid elements of the second patient. These
different findings may give the impression that the effect of
benzene on the bone marrow may change from patient to patient.

Benzene-Hepatitis-Aplastic Anemia Syndrome

As it is well-known, viral hepatitis is one of the agents
which may cause aplastic anemia. Recently we have observed a
severe case of aplastic anemia ten months after the appearance
of hepatitis B (16). The propositus was a 32 year old technician
working for 6 years in a petroleum plant in Eastern Thrace. A
sample of the product disclosed 2.2 per cent of benzene. Ten
months ago he had hepatitis B for a duration of 1 month. At
that time he was not anemic and there were no signs of
hemorrhagic diathesis. He never used chloramphenicol.
Approximately 10 months later, the signs of aplastic anemia such
as epistaxis, lassitude, anemia etc appeared. He was
hospitalized and there was a severe pancytopenia associated with
very hypoplastic bone marrow. Proteinogram disclosed
hypogammaglobulinemia. The liver tests were in normal range.
HBs Ag and HBe Ag were negative and anti-HBs, anti-HBe,
anti-HBc were positive. The clinical and hematologic picture of
the propositus was similar to those of chloramphenicol-
hepatitis-aplastic anemia syndrome described by Hodykinson (35).
Following the failure of oxymethalone therapy a trial with
anti-lymphocytic serum, the patients clinical and hematological
picture improved considerably. He is in a complete remission at
the present. We called this syndrome "Benzene-hepatitis-
aplastic anemia syndrome". We are strongly in favor of proposing
that viral hepatitis acted as contributary factor in the
development of severe aplastic anemia associated with
chronic benzene exposure. Therefore, we propose that the
individuals who might have been exposed to benzene, should
be protected from viral hepatitis by measures such as
vaccination etc.

On the other hand, even rarely benzene may cause myeloid
metaplasia and also paroxysmal nocturnal hemoglobinuria. In
our series of hematological disorders due to chronic benzene
toxicity we have observed 3 patients with myeloid metaplasia
and 2 cases of paroxysmal hemoglobinuria (16). In one of these
cases myeloid metaplasia developed 9 years following aplastic
anemia due to chronic benzene exposure (16,33).

Cytogenetic Abnormalities due to Chronic Benzene Exposure

Since 1964 both human and other experimental studies
performed by numerous investigators particularly by Pollini and
Colombi (36) Vigliani and Forni (37), Though and Court-Brown
(38) and Erdoğan and Aksoy (39) showed that benzene can induce a
significant rate of numerical and structural abnormalities in
somatic cells, both peripheral lymphocytes or bone marrow cells.
Chromosomal aberations may be stable (Cs) or unstable (Cu). On
the other hand, the cytogenetic study by Picciani (40) on
workers exposed to benzene less than 10 ppm, has revealed an
increase in chromosomal aberration rates compared to that of
controls. Furthermore, experimental studies performed by some
investigators such as Kissling and Speck (41), Morimoto (42),
Philip and Jensen (43), Styles and Richardson (44) and Erexson
et al. (45) demonstrated increased chromosomal aberrations by
benzene.

LEUKEMOGENECITY AND CARCINOGENECITY OF CHRONIC BENZENE TOXICITY

Since the first description of a case of acute leukemia due
to chronic benzene exposure in 1897 by Le Noire and Claude (8)
this chemical agent has been strongly suspected as a
leukemogenic and possibly carcinogenic agent. The second and
more carefully studied case report of this type of hematologic
malignancy namely that of acute lymphoblastic leukemia due to
this chemical was reported by Delore and Borgomono in 1928 (46).
Following these two patients, numerous case reports and small
and large groups of the patients associated with chronic benzene
exposure have been published. These are: Vigliani and Forni (47)
in Italy, Goguel et al. (48) in France, Tareef et al. (49) in
USSR, Aksoy et al. (13,15,16,50) in Turkey and Yin et al. (51)
in China and the groups surveyed by Browning (52). On the other
hand, until 1974 the benzene experiments on animals have not
yielded satisfactory results that would lead to a ready
acceptance of leukemogenic effect of benzene. Considering the
failure of animal experiments in producing leukemia by benzene,
the present author (15) decided to perform an epidemiologic
study among shoeworkers exposed chronically to benzene in the
period of 1967 and 1974 in Istanbul.

The Incidence of Leukemia among Shoe, Slipper and Handbag Workers Exposed Chronically to Benzene in Istanbul

Prior to 1955 and 1960 benzene was used infrequently in
Turkey including Istanbul (15). An overwhelming majority of
shoes, slippers and handbags were manufactured in small and
mostly unhygienic workshops where usually less than 10 persons
were employed. At that time shoeworkers were preparing the glue-
adhesives by processing rubber in gasoline. As they experienced
that the benzene containing glue-adhesives were extremely
practical and cheaper, they replaced their adhesives with the
new products sold in the market. Approximately since 1961,
numerous cases of chronic benzene toxicity such as leucopenia,
thrombocytopenia and pancytopenia or namely aplastic anemia
among these workers have been noted (11,16,50). In addition,
again from 1960 to 1972 the present author and his associates
investigated 46 cases of benzene mediated aplastic
anemia (16,28,33). Among them, 34 workers were from small shoe-

manufacturing workplaces. During these years, in 35 working environments, great majority in Istanbul and primarily shoe-workplaces, the concentration of benzene measured by a gas detector (Drager) ranged between 150 and 210 ppm and on rare occasion reached to 650 ppm (11,16,28,33,50).

In addition, the benzene content of 98 adhesives and thinners available between 1970 and 1972 in Istanbul determined by gas chromatography was 9% and 88% (33). Towards 1967, we started to investigate leukemic shoeworkers associated with benzene exposure and performed an epidemiological study among 28 500 shoe, slipper and handbag workers in Istanbul during the period from 1967 to 1974 (15,16,50). During this period 31 leukemic individuals were detected among the group described above. The crude incidence of leukemia in these shoeworkers was 13.59 per 100 000 which is significantly higher than the incidence of 6 per 100 000 in general population (15,16,50). The peak incidence of leukemia among shoeworkers living in Istanbul was recorded in 1973 and it was 24.5/100 000 (Table 1) (13,16,50). Our calculation of the incidence of leukemia was probably underestimated, since we were not able to include the leukemic shoeworkers who may have been admitted to other hospitals in Istanbul other than the one we studied.

Decline of Leukemia after Phase-out of Benzene in Istanbul

The number of leukemic shoeworkers in Istanbul during the period of 1974 and 1975 decreased to the level of 1969 and 1970 and none were recorded in the following 3 years (13,16). (Table 1) Between 1979 and 1988 we have observed 22 new cases of leukemia with chronic benzene exposure, only 3 of them were shoeworkers living in Istanbul (13,16,55). The decline in the annual occurence of leukemia in Istanbul may be due to the prohibition and gradual discontinuation of benzene use in this city starting with 1969 (13,16).

On the other hand, reappearance of leukemia after 1979 in Istanbul and the other Turkish cities may be attributed either to the variation of the interval between the occurrence of leukemia and exposure or to the continuing use of solvents and thinners containing benzene (13,16).

Occurrence of Two Acute Leukemia Cases in a Tire Cord Fabric

In a six year period we have observed 2 cases of acute leukemia, one lymphoblastic and the second myeloblastic in a modern tire cord fabric where 550 workers were employed (13,16). The working conditions were usually good and workplaces were properly ventilated. The concentration of benzene determined by gas chromatography in one place of this plant was 110 ppm and of the solvents used had benzene content of 5%. The incidence of leukemia in this fabric was 60.6 per 100 000 which is significantly higher than the incidence 6 per 100 000 in the general population (13).

Development of Leukemia in Pancytopenic Patients

It is often noted that leukemia with chronic benzene exposure often develops following a period of bone marrow depression. This fact is emphasized by several authors including

422

Table 1. Annual Number of Leukemic Shoe-Workers
in Istanbul Between 1967 and 1978

Years	No.of Leukemic Shoe-Workers
1967	1
1968	1
1969	3
1970	4
1971	6
1972	5
1973	7
1974	4
1975	3
1976	0
1977	0
1978	0

Vigliani and Saita (53), Forni and Vigliani (54) and Browning
(52). In our series comprising 59 leukemic patients with chronic
benzene exposure, a preceding pancytopenic period was present
in 14 patients of which 5 with acute myeloblastic leukemia, 5
with preleukemia, 2 with erythroleukemia, 1 with acute
myelomonocytic leukemia and 1 acute with undifferentiated
leukemia (16,55). Interestingly this period was present in 50
per cent of the patients with preleukemia. The interval between
the onset of the preceding period and that of the manifestation
of luekemia varied from 6 months to 6 years (13,16,55). Quite
often the clinical and hematological findings of the
pancytopenia improved considerably or the condition even
disappeared completely, but despite this leukemia developed
later.

Genetic Factors

As it is known, there are numerous data suggesting that
genetic factors or host factors may play a role in the
development of leukemia (56). Several families in which two or
more cases of leukemia of acute or chronic type, mostly lymphoid
occurred, showed this in the same or successive generations
(57). In addition to this, there is a frequent occurrence of
leukemia in identical twins (58). As can be seen from Table 2,
in our series, in six leukemic patients associated with chronic
benzene exposure, a familial connection was established
(13,16,50). Two were uncle and nephew and two were cousins. The
father of the fifth leukemic patient, a 65 year old shoemaker
with a long history of benzene exposure, died in a hospital in
Istanbul with a diagnosis of myelosclerosis, but reevaluation of
the case report by the present author showed that this patient
with a longterm exposure to benzene had acute leukemia of an
unidentified type (16,50). The father of the 7th patient with
chronic lymphoid leukemia associated with chronic benzene
exposure had never been exposed to this chemical. But he used
for a long time saccharine and colchicine. The role of these two
drugs in the etiology of chronic leukemia is unknown. These 6
luekemic patients among first and second degree relatives
constituted 10.2 per cent of the leukemic patients

Table 2. The Genetic Relationship Between 7 Patients with
Chronic Benzene Toxicity

Case	Age	Dur.of Exp. (Year)	Occupation	Types of Leukemia	Genetic Relationship
1	43	6	Shoeworker	AML	The Paternal Uncle of Case 2
2	24	4	Shoeworker	ALL	The Nephew of Case 1
3	36	7	Shoeworker	Preleu.	Maternal Cousin of Case 4
4	48	3	Painter	AML	Maternal Cousin of Case 3
5	36	15	Shoeworker	A.Ery.L.	Son of Case 6
6	65	35	Shoeworker	A.Un.L.[b]	Father of Case 5
7	43	2	Owner of a wall-paper printing shop	CLL	Son of Case 8
8[a]	77		Owner of a stationary shop	CLL	Father of Case 7

[a]He used 10 or 20 tablettes of saccharin 1-2 tablettes colchicine daily for a long time.

[b]A.Un.L.:Acute Undifferentiated Leukemia

with benzene exposure (16,50,55). In these 6 patients, the development of leukemia was possibly due to the presence of genetic or intrinsic factors, in addition to benzene as an environmental determinant. This suggestion is in accordance with the view that leukemia may be due to combination of various intrinsic and extrinsic factors (56).

Individual Susceptibility

Individual or family susceptibility first suggested by Reifschneider (59) has been considered as an important factor in the incidence of chronic benzene toxicity. Two of our patients associated with aplastic anemia due to chronic benzene toxicity, one fatal, were brothers (60). Also two pancytopenic patients with chronic exposure to benzene, one fatal were cousins (61). Similarly the son of one our patients with aplastic anemia due to chronic benzene toxicity showed leucopenia following a short time exposure to this chemical (16,61). As a very illustrative example of individual susceptibility, it is convenient to present the following family with chronic benzene toxicity. A 38 year old man and his 32 year old wife were engaged in manufacturing whistles (13,16). For this purpose they dipped the plastic material into an open liquid containing 88.42% of benzene and 9.25 of toluene. Following exposure to benzene for a period of 6 months, a severe aplastic anemia developed in the wife. Contrarily, in her husband, following an exposure of 14 years all hematological data were within normal limits. The cause of the variation concerning individual susceptibility is obscure. According to Browning (52) the most probable explanation lies in the innate difference in the potency of different individuals to carry out the metabolic detoxification. Some individuals may be congenitally deficient in the capacity to detoxify benzene or particularly its metabolites. Individual susceptibility is not accepted by all investigators dealing with benzene problem. Despite this, we are strongly inclined to believe in the important role of individual susceptibility in the development of chronic benzene toxicity.

Other Epidemiological Studies in the Workers with Chronic Benzene Exposure

As early as in 1938 Vigliani and his associates including Penati, Saita and Forni (37,47,53,63,64) published several case studies on leukemia associated with chronic benzene exposure in Milan and Pavia. Occupational benzene exposure were identified in retrogravure plants and cottage shoe-factories. Duration of exposure was between 3 and 24 years. Benzene concentration ranged between 25 and 600 ppm. In 1976 Vigliani calculated the risk of leukemia for workers in Milan and Pavia to be at least 20 times higher than that of the general population (64). In 1977 Infante et al. (65) reported the results of an OSHA proposed epidemiological study on cohorts of the rubber workers exposed chronically to benzene in the course of manufacturing pliofilm at 2 Ohio rubber plants in the period between 1940 and 1949 and followed the group for vital health status up to 1975. In comparison with 2 control groups, a significant excess of leukemia was found in the benzene exposed workers (65). On the other hand, there was a five-fold excess of death from myeloid and monocytic leukemia combined. According to Rinsky et al. (66) when the data obtained in the above mentioned study were analyzed by the length of employement, a significant excess in leukemia was observed among workers engaged for five or more years, but not among workers less than five years. Recently, Yin et al. (51) in China performed a retrospective study on 28 460 workers in the period from 1972 to 1981. Thirty leukemic workers with chronic benzene exposure and four cases in the control group were found. The leukemia mortality rate was 14/100 000 persons in benzene cohort and 2/100 000 in the control group. On the other hand, the data obtained by Yin et al. (51) and those of the present author and his associates (13,15) are similar. Furthermore, in Yin et al. (51) study, the average period of exposure to benzene was 11.4 years. This period was 10.2 years in the studies of the present author and his associates (14,15,16,50).

Distribution of the Types of Leukemia in Chronic Benzene Toxicity

It is matter of fact that there is a significant difference relating to the distribution of the types of leukemia in several groups of this hematologic malignancy due to chronic exposure to benzene. In one group acute types of leukemia predominate whereas in the other series chronic types of leukemia takes the

Table 3. The Distribution of the Types of Leukemia in 59
Patients with Chronic Benzene Toxicity

Types of Leukemia	Exposed		Non-Exposed	
	No	%	No	%
AML	23	39	8	16
ALL	5	8.5	13	26
Preleukemia	8	13.5	1	2
A.Erythroleukemia	10	17	2	4
A.Myelomonocytic L.	5	8.5	3	6
A.Promyelocytic L.	1	1.7		
A.Undifferentiated L.	1	1.7		
Hairy Cell Leukemia	1	1.7		
CML	3	5	10	20
CLL	2	3.4	13	26

most important place. As can be seen in Table 3, in our series,
53 leukemic individuals, namely 89.8 per cent, had various forms
of acute leukemia. Contrary to this, in the unexposed group the
percentage of leukemia was 46%. On the other hand, in a study
among 3715 hematologic patients admitted to the hematology
section of Istanbul Medical School, 695 patients had leukemia
(26). Among them, 396 (57%) patients had acute types of luekemia
and 299 (43%) showed the findings of chronic leukemia.
Therefore, the low incidence of chronic types of leukemia in
our series of leukemic patients with benzene exposure is
interesting. This result is in accordance with those of Vigliani
and Forni (47), Infante et al. (65) and Yin et al. (51). Acute
types of leukemia consisted of 82.7% of Vigliani and Forni (47)
and 85.5% of the series described by Infante et al. (65).
Recently Yin et al. (51) in China obtained similar results,
namely 76.6% of their leukemic workers with chronic benzene
exposure showed symptoms of acute types of leukemia. Contrarily,
there are several reports describing the frequent occurrence of
chronic types of leukemia such as chronic lymphoid and myeloid
leukemia in chronic benzene toxicity. These types of leukemia
were present in 47.8% in Goguel et al. series (48). Similarly in
the group surveyed by Browning (52), chronic types of leukemia
were recorded as 39.3%. Recently the present author (16,67)
tried to explain differences concerning epidemiological data in
the patients with acute and chronic types of leukemia in chronic
benzene toxicity. These are: 1. Differences in the content of
benzene used. In cases of acute leukemia, the percentage of
benzene was high, ranging between 9 and 88%. Contrarily, the
benzene content of the solution used by a case with chronic

lymphoid leukemia was present as low as 2.8% (67). **2.** The
adhesives which were used by nearly all the workers with acute
leukemia in our series contained only benzene not the other
homologs of this chemical as toluene and xylene. Contarily, the
solvent used in the workplaces of one patient with chronic
lymphoid leukemia contained 95% toluene and 2.8% benzene (67).
Another patient with chronic lymphoid leukemia was later exposed
to different chemicals (67). Similarly the patient with hairy
cell leukemia (67), later worked in the place where polystyrene
and polyethylene were used. **3.** The great majority of the
individuals with acute leukemia in our series (13,16,67) was
exposed to high concentration of benzene ranging 150 and 210 ppm
during all working hours. Contrarily, five out of six patients
with chronic leukemia were exposed to benzene intermittently and
a short time during their work (16,67). Considering these data
we suggest the striking difference in the types of leukemia due
to chronic benzene toxicity in our series and the series of some
other investigators (48,49,52) may be partially explained by the
exposure levels and the presence of the other homologs of
benzene such as toluene and xylene and other chemicals (16,67).

Malignant Lymphoma and Benzene

Although there are few reports showing the possible role
of benzene in the development of different types of malignant
lymphoma associated with chronic benzene exposure, there is no
doubt about the causal relationship between this chemical and
malignant lymphoma. In 1947 Bousser et al. (68) reported a case
of lymphosarcoma occurred 3.5 years after the cesation of
benzene exposure. Since then a few cases of malignant lymphoma
with chronic benzene exposure appeared in the medical
literature (16). Furthermore Wirtschafter and Bichell (69)
showed in rats that tissue responses should be observed in lymph
nodes, spleen, thymus and bone marrrow of rats after a single
injection of benzene. Also there was an increase of macrophages
in the lymph nodes. In addition, the spleen disclosed marked
proliferation of the reticuloendothelial cells of the pulp with
morphological alterations. Recently, Irons et al. (70) showed
that the effects of benzene administration in mice on lymphocyte
function are observed at doses of the compound that produce
little or no measurable differences in the number of circulating
cells. In 1974 the present author and his associates (71)
published six cases of Hodgkin's disease with chronic benzene
exposure for a period ranging between 1 and 28 years and a mean
of 11 years. Despite the lack of statistical data, we suggested
that chronic benzene exposure might play a causal role in the
development of Hodgkin's disease. In the following years we have
studied seven more cases of malignant lymphoma (13,16). All were
chronically exposed to benzene for 1 to 28 years. In 1979 Vianna
and Polan (72) performed a comparative study on the mortality
rates of different types of malignant lymphoma among male
workers in different occupations with exposure to benzene and or
other coal-tar fractions and a non-exposed group. According to
the investigators, the results were consistent with the
possibility that chronic benzene exposure might be important in
the etiology of different types of malignant lymphoma. On the
other hand, Norseth et al. (73) performed a study of cancer
incidence among rubber workers in Norway. They found an
increased risk of bladder cancer, lymphoma and leukemia in the
footwear department. According to the investigators, a benzene
based glue was used in the footwear department until 1940 and

possibly in small amounts for a few years thereafter. Furthermore there are several studies showing the high mortality rate of malignant lymphoma among pathologists, chemists and persons handling chemicals containing benzene (74).

Multiple Myeloma and Chronic Benzene Exposure

There are few reports showing the possible role of chronic benzene exposure in the development of multiple myeloma. Torres et al. (75) cited 2 patients with IgG myeloma associated with chronic benzene exposure. Among our series of malignancies associated with chronic exposure to benzene, we studied 6 cases of multiple myeloma (13,16,76). Only in one of these patients there was a short period of pancytopenia with hypoplastic bone marrow. Severe myeloma nephropathy and Bence Jones proteinuria developed rapidly. The remaining 5 patients had different types of multiple myeloma with chronic benzene exposure. De Coufle et al. (77) performed a historical cohort mortality study comprising 259 employees of a chemical plant where benzene has been used in large quantities. According to De Coufle et al. (76), the results of this study are consistent with the leukemongenic effect of occupational exposure to benzene and they also raise the possibility that multiple myeloma could also be linked to benzene. Rinsky et al. (78) performed a study of epidemiological risk assessement among a total of 1165 men with at least 1 ppm/day of cumulative exposure to benzene between the period of 1940 and 1965. There was a statistically significant increase in death from leukemia (9 observed vs. 2.7) expected and from multiple myeloma (4 observed vs. 1 expected). According to this study, three of the four deaths from multiple myeloma that were observed in this cohort occurred in the group with the lowest cumulative exposure to benzene, and 4 persons with multiple myeloma had an exceptionally long latency period for hematologic cancers (20 years).

Lung Cancer and Chronic Benzene Toxicity

In 1976 the present author (79) considering the 5 individuals with lung cancer exposed chronically to benzene, suggested a causal relationship between this chemical and lung cancer. In our series of malignancies associated with chronic benzene exposure we have studied 6 cases of lung cancer (13,16). The ages of patients ranged between 31 and 57 with a mean 45 and duration of exposure 8 and 35 (mean : 15). Only 3 showed mild hematologic abnormalities such as leucopenia or lymphopenia. Because benzene available in Turkey did not contain benzo-pyrene as determined by ultraviolet spectrophotometry (33), lung cancer in these six cases cannot be attributed to this carcinogenic agent. Benzene is mainly absorbed via the lungs and about 40% of the absorbed portion is exhaled unchanged and after 24 hours a small percentage of it can be detected in the expired air (2). Therefore, a carcinogenic effect of benzene in the lung is possible. Because our patients were also smokers, a possible role of smoking in the development of lung cancer should also be considered.

References

1. J. R. Newsome, V. Norman and C. H. Keith, Vapor phase analysis to tobacco smoke, Tobacco Smoke Sci. 9:102 (1965).

2. R. J. Sherwood and F. G. Carter, The measurement of occupational exposure to benzene vapour, Ann. Occup. Hyg. 13:125 (1970).

3. R. Snyder, Benzene metabolism, in: "Molecular Aspects of Monooxygenases and Bioactivation of Toxic Compounds", E. Arınç ed., Plenum, London (1990).

4. E. Uyeki, A. Ashkar, A. Shoeman and T.Bisel, Acute toxicity of benzene inhalation to hemopoietic precursor cells,Toxicol. Apll. Pharmacol. 40:49 (1977).

5. D. F. Gill, V. K. Jenkins, R. R. Kempen and S. Ellis, The importance of pluropotential stem cells in benzene toxicity, Toxicol. 16:163 (1980).

6. K. Horigaya, M. E. Miller, E. P. Cronkite and R. T. Drew, The detection of in vivo hematotoxicity of benzene by in vitro liquide bone marrow cultures, Toxicol. 60:346 (1981).

7. G. Post, R. Snyder and G. F. Kalf, Metabolism of benzene and phenol in macrophages in vitro and the inhibition of RNA synthesis by benzene metabolites, Cell. Biol. Toxicol. 2231 (1986).

8. M. M. Le Noire et Claude, Sur une cas de purpura attribue a intoxication par le benzene, Bull. Mem. Soc. Med. Hop. Paris, 3, 1251 (1897).

9. C. Maltoni e C. Scarnato, Le prime prove sperimentali dell' azoine cancerogene del benzene, Gli. Ospedali Della Vita 4:111 (1977).

10. T. M. Dexter, C. M. Heyworth and A. D. Whetton, The role of haemopoietic cell growth factor (interleukin 3) in the development of haemopoietic cells in : "Growth factors in biology and medicine" Ciba Foundation Symposium 116", D. E. Evered and J. Nugent, ed., Pitman, London (1985).

11. M. Aksoy, K. Dinçol, T. Akgün and Ş. Erdem, Haematological effects of chronic benzene poisoning in 217 workers, Brit. J. Ind. Med. 28:296 (1971).

12. M. Aksoy, Ş. Özeriş, H. Sabuncu, Y. Inanıcı and R. Yanardağ, Exposure to benzene in Turkey between 1983 and 1985: a haematological study in 231 workers, Br. J. Ind. Med. 44:785 (1987).

13. M. Aksoy, Malignancies due to occupational exposure to benzene, Am. J. Ind. Med. 7:395 (1985).

14. G. G. Santesson, Über chronische Vergiftungen mit Steinkohlen Benzin, Vier Todes Falle, Arch. Hyg. 31:336 (1897).

15. M. Aksoy, Ş. Erdem and G. Dinçol, Leukemia in shoe-workers exposed chronically to benzene, Blood 44:837 (1974).

16. M. Aksoy, Benzene Hematotoxicity and Benzene Carcinogenecity, in: "Benzene Carcinogenecity", M. Aksoy, ed., CRC Press, Boca Raton (1988).

17. T. P. Moszcynski and J. Lisewitz, Occupational exposure to benzene, toluene and xylene and T lymphocyte function, Haematologica 17:49 (1984).

18. A. Biancacchio and V. Farmarielle, Protoporfirine libere eritrocitoire e coproporfirino urinaire nel'autossicazione subcronica sperimentale de benzol., Folia Med. (Naples) 43:558 (1961).

19. H. Kahn and V. Muzyka, The chronic effect of benzene on porphyrine metabolism, Work. Environ. Health 110:146 (1973).

20. A. Lange, R. Smolik, W. Zatonski and J. Szymanska, Serum immunoglobulin levels in workers exposed to benzene, toluene and xylene, Int. Arch. Arbeitmed. 31:37 (1973).

21. R. Smolik, A. Grzysek-Hryncewicz, A. Lange and W. Zatonski, Serum complement level in workers exposed to benzene, toluene and xylene, Int. Arch. Arbeitmed. 31:223 (1931).

22. W. Sobczyk, W. Siedecka, Z. Gejevska, W. Horyol and C. Mezer, EEG recordings in workers exposed to benzene compounds, Med. Pr. 24:273 (1973) (Polish, English Abstract).

23. A. S. Suston, E. L. Dames, J. R. Burg and R. W. Riemeyer, Percutaneous penetration of benzene in hairless mice an estimation of dermal absoption during tire-building operation Am. J. Ind. Med. 7:323 (1985).

24. H. Heimpel, H. , Aplastic Anemia, in: Trends in Haematology, Part I, 8th Meeting of European-African Div. of ISH, September 8-13, 1985, Warsaw, Poland, 181 (1985).

25. L. E. Böttiger, Epidemiology and Aetiology of Aplastic Anemia, in: "Aplastic Anemia, Pathophysiology and Approaches to Therapy", H.Heimpel, E.C.Gordon-Smith, ed., Springer Verlag, Basel (1979).

26. M. Aksoy, Ş. Erdem, G. Dinçol, I. Bakioğlu and A. Kutlar, Aplastic anemia due to chemical and drugs. A study of 108 patients, Trans. Dis. Suppl. 11:347 (1984).

27. M. Aksoy, Ş. Erdem, T. Akgün, Ö. Okur, and K. Dinçol, Osmotic fragility studies in three patients with aplastic anemia due to chronic benzene poisoning, Blut 13:85 (1966).

28. M. Aksoy, K. Dinçol, Ş. Erdem, T. Akgün and G. Dinçol, Details of blood changes in 32 patients with pancytopenia associated with long-term exposure to benzene, Br. Ind. Med. 29:56 (1972).

29. R. Truhaux, Report of the meeting of exposure on the safe use of benzene and solvents containing benzene, International Labor Office Geneva in: "Benzene: Uses, Toxic Effects, Substitutes", International Labor Office, Geneva (1968).

30. A. Baslo and M. Aksoy, Neurological abnormalities in chronic benzene poisoning. A study of six patients with aplastic anemia and two with preleukemia, Environ. Res. 27:457 (1982).

31. T. B. Mallory, E. A. Gall and W. J. Brickley, Chronic exposure to benzene, J. Ind. Hyg. Toxicol. 21:355 (1939).

32. S. Moeschlin and B. Speck, Experimental studies on the mechanism of action of benzene on the bone marrow (radiographic studies using ^{3}H-thymidine). Acta Haematol. 38:104 (1967).

33. M. Aksoy and Ş. Erdem, Follow-up study on mortality and the development of leukemia in 44 pancytopenic patients with chronic benzene exposure, Blood 52:285 (1978).

34. T. Erbengi and M. Aksoy, Electron microscopic studies on the bone marrow in four patients with chronic benzene poisoning, in : "Abstracts Lectures and Symposia", ISH European and African Div. 4th Meeting, Istanbul, September 5-9 (1977).

35. G. C. De Gruchy, "Drug Induced Blood Disorders", Blackwell Scientific, Oxford (1975).

36. G. Pollini e R. Colombi, Danno cromosamico medullare nella anemia aplastica benzolica, Med. Lav. 55:241 (1964).

37. E. C. Vigliani and A. Forni, Benzene and leukemia, Environ. Res. 11:122 (1976).

38. I. M. Tough and W. M. Court-Brown, Chromosome aberations and exposure to ambient benzene, Lancet 1:684 (1974).

39. G. Erdoğan and M. Aksoy, Cytogenetic studies in 20 patients with pancytopenia and leukemia with long-term exposure to benzene in : "European and African Division of International Society of Haematology, 3rd Meeting, London, August 24-28 1975, Abstracts 1 and 6".

40. D. Picciani, Cytogenetic study of workers exposed to benzene, Environ. Res. 19:33 (1979).

41. M. Kissling and B. Speck, Chromosome aberation in experimental benzene intoxication, Helv. Med. Acta 36:59 (1972).

42. K. Morimoto, Combined cytogenetic effects of benzene and radiation of cultured lymphocytes, Jpn. J. Ind. Hlth. 17:106 (1975).

43. P. Philip and M. K. Jensen, Benzene induced chromosome abnormalities in rat bone marrow cells, Acta Pathol. Microbiol. Scand. 78:489 (1970).

44. J. Styles and C. R. Richardson, Cytogenetic effects of benzene vapour, Mut. Res. 135:203 (1984).

45. G. L. Erexson, J. E. Wilmer and A. D. Kligerman, Sister chromatid exchange induction in human lymphocytas exposed to benzene and its metabolites in vitro, Cancer Res. 45:2411 (1985).

46. P. Delore and J. Borgomono, Leucemie aigue an cours de intoxication benzenique. Sur origine toxique de certains leucemies aigues et leurs relations avec le anemies grave. J. Med. Lyon 9:227 (1928).

47. E. Vigliani and A. Forni, Leucomogenesis Professionale, Minerva Med. 57:3952 (1966).

48. A. Goguel, A. Cavigneux e J. Bernard, Le leucemies benzeniques de la region Parisienne 1950 et 1965 Etudes de 50 observationes, Nouv. Rev. Fr. Hematol. 1:465 (1967).

49. E. M. Tareef, N. M. Kontchalovskaya and L.A. Zorina, Benzene leukemias, Acta Union Int. Contra Cancerum 19:751 (1963).

50. M. Aksoy, Ş. Erdem and G. Dinçol, Types of leukemia in chronic benzene posioning. A study in thirty-four patients, Acta Haematol. 55:65 (1976).

51. S. N. Yin , G. L. Li, F. D. Tain, Z. I. Fu, C. Yin, Y. Z. Chen, S. J. Luo, P. Z. Ye, J. Z. Zhang, G. C. Wang, X. C. Zhang, H. N. Wu and Q. C. Zhang, Leukemia in benzene workers: a retrospective cohort study, Br. J. Ind. Med. 44:124 (1987).

52. E. Browning, "Toxicity and Metabolism of Industrial Solvents" Elsevier, Amsterdam (1965).

53. E. C. Vigliani and G. Saita, Benzene and leukemia, N. Engl. J. Med. 271:872 (1964).

54. A. Forni and E. C. Vigliani, Chemical leukemogenesis in man, Ser. Haematol. 7:211 (1974).

55. M. Aksoy, unpublished data.

56. F. W. Gunz, Problems in leukemia etiology, in : "Plenary Sessions" of Scientific Conference. Proc. 13th Int. Congr. Hematology, J. F. Lehmann Verlag, München (1970).

57. F. P. Li, D. J. Marchetto and G. F. Vavter, Acute leukemia and preleukemia in eight males in a family: an X-linked disorder, Am. J. Hematol. 6:61 (1979).

58. B. MacMahon and M. A. Levy, Prenatal origin of childhood leukemia: evidence from twins, N. Engl. J. Med. 270:1802 (1964).

59. C. A. Reifschneider, Benzol: Its occurrence and prevention Natl. Safety Council, 11th Annu. Congr. Proc. Detroit. 192:249 (1922).

60. M. Aksoy and Ş. Erdem, Some problems of hemoglobin patterns in different thalassemic syndromes showing the heterogenecity of beta-thalassemia genes, Ann. N. Y. Acad. Sci. 165:13 (1969).

61. M. Aksoy, Ş. Erdem, G. Erdoğan and G. Dinçol, Acute leukemia in two generations following chronic exposure to benzene, Hum. Hered. 24:20 (1974).

62. F. Penati and E. Vigliani, Sul problema delle mielopatic aplastiche, pseudo aplastiche de benzols, Rass. Med. Ind. 9:345 (1938).

63. E. C. Vigliani and G. Saita, Benzene and leukemia, New Engl. J. Med. 271:872 (1964).

64. E. C. Vigliani, Leukemia associated with benzene exposure, Ann. N. Y. Acad. Sci. 271:143 (1976).

65. P. F. Infante, J. K. Wagoner, R. A. Rinsky and R. R. J. Young, Leukemia in benzene workers, Lancet 2:76 (1977).

66. R. A. Rinsky, R. J. Young and A. B. Smith, Leukemia in benzene workers Am. J. Ind. Med. 2:217 (1981).

67. M. Aksoy, Chronic Iymphoid leukaemia and hairy cell leukaemia due to benzene. A report of three cases. Br. J. Haematol. 67:203 (1987).

68. J. Bousser, R. Nyde e A. Fabre, Un cas d'hemopathie benzolique tres retardes a type lymphosarcome, Bull. Mem. Hop. Paris 63:1100 (1947).

69. Z. T. Wirtschafter and M. G. Bichel, Reticuloendothelial responses to benzene, Arch. Pathol. 67:146 (1960).

70. R. D. Irons, D. Wierda and R. W. Pfeifer, The immunotoxicity of benzene and its metabolites, in: "Advances in Modern Environmental Toxicology," Carcinogenecity and Toxicity of benzene, M. A. Mehlman ed., Princeton Scientific Publications, N. J. 37 (1983).

71. M. Aksoy and Ş. Erdem, K. Dinçol, T. Hepyüksel and G. Dinçol, Chronic exposure to benzene as a possibly contributary etiologic factor in Hodgkin's disease, Blut 28:293 (1974).

72. N. J. Vianna and A. Polan, Lymphomas and occupational benzene exposure, Lancet 1:1349 (1979).

73. T. Norseth, A. Andersen and J. Gildvedt, Cancer incidence in the rubber industry in Norway, Scand. J. Work Environ. Health 2:69 (1983).

74. F. R. Fraumani, N. Mantel and R. W. Miller, Cancer mortality among chemists, J. Natl. Cancer Inst. 43:1159 (1969).

75. A. Torres, M. Girait and A. Reichs, Coexistence de antecedantes benzelicos cronicos Y plasmocitoma multiple. Presentation de dos cases, Sangre 15:275 (1970).

76. M. Aksoy, Ş. Erdem, G. Dinçol, A. Kutlar, I. Bakioğlu and T. Hepyüksel, Clinical observation showing the role of some factors in the etiology of multiple myeloma, Acta Haematol. 71:116 (1984).

77. P. de Coufle, W. A. Blattner and A. Blair, Mortality among chemical workers exposed to benzene and other agents, Environ. Res. 30:16 (1983).

78. R. A. Rinsky, A. B. Smith, R. Hornung, T. G. Filloon, R. J. Young, A. H. Okun, P. J. Landrigan, Benzene and leukemia: An epidemiologic risk assesement, N. Engl. J. Med. 316:1044 (1987).

79. M. Aksoy, Lung cancer and chronic benzene poisoning, Paper presented at the Proc. Int. Workshop of Toxicilogy of Benzene, Paris, November 9-11 (1976).

EPOXIDE HYDROLASE ISOENZYMES AND THEIR INDIVIDUAL

CONTRIBUTION TO THE CONTROL OF TOXIC METABOLITES

Franz Oesch, Ludwig Schladt, Michael Knehr, Johannes Döhmer
and Helmut Thomas

Institute of Toxicology, University of Mainz
Obere Zahlbacher Straße 67, D-6500 Mainz, FRG

INTRODUCTION

Epoxides are highly strained three membered cyclic ethers which are formed *in vivo* by the microsomal cytochrome P450 dependent monooxygenases as intermediates of several important biosynthetic pathways (leukotriene A4, squalene 2,3-oxide) and as metabolites of numerous xenobiotic compounds containing olefinic or aromatic double bonds. Further transformation of these epoxides may occur by either, rearrangement to phenols, aliphatic aldehydes, or ketones; by cytochrome P450 dependent reduction to the parent compound; or by spontaneous or enzymatic conjugation to glutathione. Epoxides may also bind covalently to cellular nucleophiles, such as proteins and nucleic acids thus eliciting carcinogenic, mutagenic and other toxic effects.

Epoxide hydrolases (E.C.3.3.2.3) - formerly referred to as epoxide hydrases or epoxide hydratases - catalyze the conversion of highly reactive epoxides to more polar and hence more readily excretable 1,2-dihydrodiols by addition of water. Hence the physiological role of these enzymes appears to be primarily detoxifying. However, the dihydrodiols of certain polycyclic aromatic hydrocarbons with angular structures can be metabolized further by cytochrome P450 dependent monooxygenases to yield ultimately carcinogenic dihydrodiol epoxides, in which case epoxide hydrolase expresses toxifying action (1). Considering their involvement in drug and foreign compound metabolism epoxide hydrolases have attracted interest mostly in the fields of toxicology, carcinogenesis, nutrition and biochemistry.

HISTORICAL DEVELOPMENT

In 1950, Boyland first proposed that the hydroxylation of polycyclic aromatic hydrocarbons proceeds via an epoxide intermediate, and that the hydration of the epoxide proceeds enzymatically (2). Soon afterward the formation of epoxides from insecticides and steroids was shown. The first hydration of an epoxide (16 α, 17 α-epoxy-oestratriene-3-ol) by rat liver tissue was demonstrated 1961 (3). Five years later, evidence was provided for the hydration of epoxides of polycyclic aromatic hydrocarbons and of chlordene epoxide, followed by similar demonstrations for a large number of aromatic and aliphatic oxiranes up to the early 1970's (4,5). The first enrichment of microsomal epoxide hydrolase activity, which is often termed

Molecular Aspects of Monooxygenases and Bioactivation of Toxic Compounds
Edited by E. Arınç *et al., Plenum Press, New York,*

435

"mEH$_b$", because of its diagnostic substrate benzo(a)pyrene 4,5-oxide, was reported by Watabe and Kanehira in 1970 (6), and the first purification to apparent homogeneity from rat liver in 1975 by Bentley and Oesch (7) as well as by Lu et al. (8). In 1974, while investigating the metabolism of a juvenile hormone, Gill et al. (9) obtained strong evidence for the existence of a cytosolic epoxide hydrolase activity (cEH). However, it was not until 1982 that the first purification schemes were published for the rabbit and human liver cytosolic enzymes respectively (10,11). An additional microsomal epoxide hydrolase activity rather specific for cholesterol 5,6-oxide was observed by Watabe et al. in 1981 (12) but, although separated from mEH$_b$ (13), the enzyme mEH$_{ch}$ remains to be purified. Beginning in the early 1980s immunological studies in combination with the evaluation of substrate specificities and inhibitory properties of liver microsomes and cytosol from rabbit, mouse and man indicated the existence of an additional isoenzyme in each of the two subcellular compartments (mEH$_{TSO}$*, cEH$_{CSO}$). Each of these forms is complementary in substrate specificity to the respective nominate form and resembles closely, in physical and immunological properties, the subcellular counterpart. An epoxide hydrolase activity specific for the enzymatic hydrolysis of leukotriene A4 (LTA4) was described for the first time in 1983 in human and guinea pig plasma. This activity was also found in all investigated guinea pig tissues. Purification of LTA4 hydrolase was reported from human leukocytes and erythrocytes as well as rat neutrophiles (14).

It was observed in the late 1970s that the hepatic mEH$_b$ and cEH activities as well as protein levels are selectively increased upon administration in vivo of specific chemicals. This provided a valuable method for the evaluation of the enzymes' effect on foreign compound metabolism and toxicity as well as for the purification of these enzymes, particularly cEH, in preparative yields. Epoxide hydrolases have since been purified and characterized from a number of mammalian species (15-17). The first primary sequence was published for rabbit liver mEH$_b$ in 1984 (18). Although the first report on cDNA cloning of mEH$_b$ dates back to 1981, the first complete coding nucleotide sequence and primary structure for rat liver mEH$_b$ was not published until 1986 (19). Nucleotide and deduced amino acid sequences of human liver mEH$_b$ and LTA4 hydrolase followed in 1987 and 1988 (20,21).

MULTIPLICITY OF mEH$_b$ AND cEH

Epoxide hydrolase activity is now known to be present in all organisms tested, including bacteria, fungi, plants, insects and fish (15,16). However, activity can be classified only in higher forms of life, which have been shown to contain mEH$_b$, cEH, and mEH$_{ch}$ in all tissues investigated but with varying maximal activity depending on the species. In the livers of the most widely used experimental animals specific mEH$_b$ activity is highest in the centrilobular region and follows the sequence rabbit > guinea pig > rat > mouse, while cEH activities rank in the order mouse > hamster > rabbit > human > guinea pig > rat (15-17,22,23). Except in extrahepatic organs and human liver (in which no significant sex differences were detectable), activities of mEH$_b$ and cEH are 50-200 % higher in males than in females (16). Considerable quantitative strain differences in specific activities as well as xenobiotic inducibility of mEH$_b$ and cEH have been observed in rats and mice. These and large interindividual variations in mEH$_b$ and cEH activities of rat and particularly human samples, and the varying activity ratios of a given EH-form toward different substrates, suggested that there might be a multiplicity of mEH$_b$ and cEH forms. Although the enzymes were identified in numerous species and

---

* Abbreviations: CSO, *cis*-stilbene oxide; PNSO, p-nitrostyrene 7,8-oxide; TESO, *trans*-β-ethylstyrene oxide; TSO, *trans*-stilbene oxide.

tissues either mEH_b or cEH has been purified only from livers of mouse, rat, rabbit, rhesus monkey, and human (16,17,24), with the exception of a partial purification of cEH from mouse mammary gland (24).

Indeed, different laboratories did succeed in purifying multiple forms of mEH_b: Guengerich et al. reported the separation of multiple forms from control and induced rat liver as well as from human liver (25). A similar result was obtained by Bulleid and coworkers (26) for rat liver. More recently, three forms of microsomal epoxide hydrolase have been purified from rhesus monkey liver (27), and at least four different forms have been separated from rabbit liver in our laboratory (unpublished result).

Different amino acid compositions and substrate specificities have been reported for the materials collected from different peaks obtained during chromatography implying isozymic character of rat liver mEH_b-forms. Other authors which obtained chromatographic separation of mEH_b in several peaks have reported that these result from different detergent or phospholipid contents of one enzyme protein (26).

Conventional and elegant affinity techniques opened access to purified cEH of a number of species (17,24). However, multiplicities have been observed only for the human cytosolic enzyme (11,28), the form(s) besides the classical cEH being characterized by molecular weights of 50.000 Da compared to 58.000 Da for the classical cEH. These new form(s) display a close immunological relationship to mEH_b and substrate specificities for CSO (hence abbreviated cEH_{CSO}) and PNSO rather than for the substrates TSO or TESO which are diagnostic for the classical cEH (hence abbreviated cEH_{TSO}). There has been only one report on the purification of mEH_{TSO} from rhesus monkey (27).

GENETIC ORGANIZATION OF EPOXIDE HYDROLASES

Application of recombinant technologies has permitted insight into structural details for LTA4 hydrolase (21) and mEH_b as well as the gene organisation of the latter (19,20,29). cDNA clones for rat and human liver mEH_b have been isolated that code for a 455 amino-acid polypeptide each corresponding to molecular weights of 52,581 Da and 52,956 Da, respectively. Deduced amino-acid sequences show 81 % conservation between rat and rabbit liver mEH_b and 84 % between the rat and human liver enzymes. Opposing the indications for multiplicities from preparative work, only a single functional gene and mRNA was identified so far for human and rat mEH_b, and there was no evidence of hybridization to genes for mEH_{ch} or cEH. No apparent homologies were found with the DNA sequence of LTA4 hydrolase coding for a 610 amino acid protein. Although mEH_b is induced by a range of compounds as diverse as those inducing cytochrome P450 (15,16, 30,31), the two systems appear to be under different genetic control (30, 32). However, the result is the same: enhanced rates for transcription and the steps following this event (15,16,33). Similar sequences of events have been proposed to follow induction of cEH by industrial plasticizers and hypolipidemic drugs (17), whereas so far neither mEH_{ch} nor LTA4 hydrolase could be shown to be under marked regulatory control by induction.

FUNCTIONS OF EPOXIDE HYDROLASES

mEH_b is known to express broad substrate specificity with preference toward hydrophobic epoxides without extensive substitution. The most widely used diagnostic substrates for in vitro determination of its activity are benzo(a)pyrene 4,5- oxide and styrene oxide (31,34) whereby the former is hydrolyzed specifically by mEH_b only. Inversely, cEH metabolizes

437

epoxides in polycyclic systems slowly, but hydrates a wide variety of aliphatic or extensively substituted arene epoxides. A diagnostic in vitro substrate is *trans*-stilbene oxide. Some substrates of endogenous origin, particularly epoxides of steroids and lipids, have been identified for mEH_b, while cEH is believed to play an important role in cellular protection from side chain epoxidized sterols, squalene oxides as well as oxirane metabolites of *cis*- and *trans*-unsaturated fatty acids including arachidonic acid and epoxy-eicosanoids (15-17,35). In contrast, mEH_{ch} has a very narrow substrate specificity for physiologically arising 5,6-epoxy-cholesterol and its derivatives. LTA4 hydrolase is confined to the hydro-lysis of leukotriene A_4. No distinct physiological roles for mEH_b, mEH_{TSO}, cEH_{TSO} or cEH_{CSO} have been detected.

The hydrolysis of epoxides by mEH_b has been proposed to proceed by the nucleophilic backside attack of a hydroxide ion, which is generated from water by a single essential histidine, at the least sterically hindered epoxide carbon atom. In addition, two areas in the enzyme have been described as important for substrate-binding, enantioselectivity and regioselectivity of mEH_b (36). Based on a series of inhibition studies with chalcone oxide derivatives a similar reaction mechanism has been postulated for cEH; besides two distinct hydrophobic sites, it involves a general acid and base catalyst as well as an essential cysteine residue (37,38).

Like the conjugating activity of glutathione S-transferases, epoxide hydrolases are thought to play a detoxifying role by preventing epoxides from irreversibly binding to critical cellular macromolecules. Depending upon structural and stereochemical features of the toxin in question, epoxide hydrolase may detoxify, may not play a significant role, or may even render a compound more toxic (1). Experimental evidence supports the hypothesis that mEH_b may be closely affiliated with certain forms of cytochrome P450 in the microsomal membrane and that such association might influence the functional role of this enzyme in various pathways of bioactivation and detoxication (15,39).

mEH_b has been identified immunologically as a major portion of the cytosolic preneoplastic antigen, which is expressed at very high levels in early and late hyperplastic nodules but not in regenerating liver. The appearance of soluble mEH_b is speculated to be the consequence of enhanced endoplasmic reticulum proliferation and changes in the lipid composition under this condition (40,41).

cDNA CLONING OF RAT LIVER cEH

A cDNA expression library was constructed using purified poly-A mRNA from tiadenol-induced rat liver in the plasmid pEX1/ E. Coli POP 2136 system according to Haymerle et al. (42), and screened by a colony blot hybridization procedure with polyclonal antibodies against rat liver cEH (43). 24 positive clones were confirmed upon rescreening and the cDNA inserts of the four clones having the largest cDNA inserts ranging between 0.8 and 1.2 kb were verified by crosshybridization according to Southern and Northern blotting against three independent total RNA preparations from untreated, Aroclor 1254- and tiadenol-induced rat livers.

Upon Northern blotting, all clones recognized the same mRNA which is believed to be the coding mRNA for rat liver cEH. Additionally, clone 24 picked up a second mRNA species, somewhat smaller than the putative mRNA for cEH. This result may indicate for the first time the existence of a second isoenzyme of cEH in rat liver, which may be related to human liver cEH_{CSO} (28). However, this activity has not been allocated as yet to a distinct rat cytosolic protein, although minute CSO-hydrolyzing activities of about 0.1 nmol/min per gram liver have been described in rat liver (44). Most recently, a cytosolic hepoxilin epoxide hydrolase having a molecular weight of 53,000 Da and marginal activity towards STO was isolated from rat liver (45) and may correspond to the newly identified mRNA.

CONCLUSIONS

Two forms of human liver cEH, cEH_{CSO} and cEH_{TSO} were recently identified in our laboratory, and cEH_{CSO} as well as rat liver cEH were purified to apparent homogeneity (28,43). According to their physical, biochemical and immunological properties, and taking into account the known properties of cEH from other species, an attempt was made to assign these enzyme forms to three classes with respect to common characteristics (Table 1):

Class I cEH isoenzymes are characterized by immunological crossreactivity with either cEH-antiserum against any member of this group and diagnostic substrate specificity towards TSO, STO and, with some limitations, to CSO. Mouse, rat, rabbit and hamster liver cEH meet these requirements. Class II instead comprises guinea pig, monkey and human liver cEH_{TSO} as well as human cEH_{TESO}, which have diagnostic substrate specificity for TSO and low turnover rates with CSO (44). No crossreactivity is observed among members of this class with antiserum against cEH or mEH_b from either species of group I. Class III isoenzymes, which are represented by human liver cEH_{CSO} and cEH_{PNSO}, have lower molecular weights than the TSO-hydrolases of classes I and II, and express immunological relationship with antibodies against mEH_b of either source.

From this classification it appears that the structural and functional multiplicity of cEH increases with phylogenetic differentiation. The possible impact of this cEH differentiation on the cellular protection against reactive epoxides was investigated with differentially methylated styrene oxides and microsomes or cytosol as enzyme source (Table 2). As a result, human enzymes of either source metabolized all model substrates more efficiently than the corresponding rat liver enzymes, and the ratio of microsomal/cytosolic hydration for a certain substrate is clearly shifted towards 2-7 times lower values in humans. This finding can be explained by the presence of cEH_{CSO}, which provides mEH_b-like detoxification capacities particularly for cis-substituted epoxides such as cis-2-methyl styrene oxide, cis-1,2-dimethylstyrene oxide, and 2,2-dimethyl styrene oxide.

In conclusion, a multiplicity of human as well as rat liver cEH could be demonstrated on the enzyme level for the human, and on the mRNA level for rat liver cEH. However, due to human liver cEH_{CSO} the importance of this multiplicity for the protection of different subcellular compartments against reactive epoxides appears much more pronounced in humans compared to rats, and human liver seems to be far more efficient in the detoxification of reactive epoxides by epoxide hydrolase. Rat liver may therefore not be an appropriate model for epoxide hydrolase related risk assessment in humans.

ACKNOWLEDGEMENTS

The authors wish to thank Mrs. I. Böhm for typing the manuscript. This investigation was supported by the "Deutsche Forschungsgemeinschaft" (SFB 302). L. Schladt was the recipient of a postdoctoral fellowship from the "Deutsche Forschungsgemeinschaft", and M. Knehr was supported by a stipend from the Johannes Gutenberg-Universität Mainz (Titel 68102).

Table 1. Classification of cytosolic epoxide hydrolases from different species according to physical, biochemical and immunological properties

Class	Species	Activity[a] towards			MW	pI	Crossreactivity with antiserum against		Ref.
		TSO	CSO	STO			rat cEH$_{TSO}$	rat mEH$_b$	
I	Mouse liver	939-1500	136-180	475-622	59000	5.5	+	-	46-48
	Rat liver	173	b.d.	1567	61000	5.5	+	-	43
	Rabbit liver	0.69-0.93[b]	n.m.	1380	57000	6.0	(+)	-	10
	Hamster liver	2.10-2.20[b]	0.11-0.15[b]	n.m.	57000	n.m.	+	-	44
II	Guinea pig liver	0.87-0.99[b]	0.15-0.21[b]	n.m.	57000	7.4	(-)	-	44
	Monkey liver	0.36-0.44[b]	0.94-0.96[b]	n.m.	58000	n.m.	-	-	44
	Human liver cEH$_{TSO}$	1.02[c]	0.32[c]	b.d.	n.m.	5.7	-	-	28
	Human liver cEH$_{TESO}$	150	n.m.	22	58000	5.1-6.1	-	-	11,28
III	Human liver cEH$_{CSO}$	b.d.	145	b.d.	49000	9.2	-	+	28
	Human liver cEH$_{PNSO}$	14	n.m.	44	50000	n.m.	-	+	11

a) Activities are given in nmol × min^{-1} per mg protein of the purified protein (except footnotes b and c);

b) Activities determined from cytosol;

c) Activity after anion exchange chromatography of cytosol;

b.d.: below detection; n.m.: not measured.

Table 2: Metabolism of methylated styrene oxides by rat and human liver cytosol and microsomes

Substrate	Microsomes		Cytosol		Microsomes/Cytosol	
	Rat	Human	Rat	Human	Rat	Human
Styrene oxide	4.3[a]	40.7	0.3	5.6	14.3	7.3
trans-2-methyl styrene oxide	n.m.	86500.0	n.m.	2100.0	-	41.2
cis-2-methylstyrene oxide	0.26	5.4	0.01	0.664	26.0	8.1
trans-1,2-dimethyl styrene oxide	b.d.	0.238	b.d.	b.d.	-	-
cis-1,2-dimethyl styrene oxide	0.08	0.665	b.d.	0.034	-	19.6
2,2-dimethyl styrene oxide	0.007	0.13	0.001	0.122	7.0	1.1

a) Activities were determined at 0.5 mM substrate concentration as described in Ref. 49 and are given in
   nmol $\times$ min^{-1} $\times$ mg^{-1} protein of the microsomal or cytosolic fraction, respectively.

b.d.: below detection

n.m.: not measured

REFERENCES

1. P. Bentley, F. Oesch and H. Glatt, Dual role of epoxide hydratase in both activation and inactivation of benzo(a)pyrene, Arch. Toxicol., 39: 65-75 (1977).

2. E. Boyland, The biological significance of the metabolism of polycyclic compounds, Biochem. Soc. Symp., 5: 40-54 (1950).

3. H. Breuer and R. Knuppen, The formation and hydrolysis of $16\alpha,17\alpha$-epoxy-oestratriene-3-ol by rat liver tissue. Biochim. Biophys. Acta, 49: 620-621 (1961).

4. T. Watabe and E. W. Maynert, Role of epoxides in the metabolism of olefins, Pharmacologist, 10: 203 (1968).

5. F. Oesch, N. Kaubisch, D. M. Jerina and J. W. Daly, Hepatic epoxide hydrase: structure activity relationships for substrates and inhibitors, Biochemistry, 10: 4858-4866 (1971).

6. T. Watabe and S. Kanehira, Solubilisation of epoxide hydrolase from liver microsomes, Chem. Pharm. Bull. (Tokyo), 18: 1295-1296 (1970).

7. P. Bentley and F. Oesch, Purification of rat liver epoxide hydratase to apparent homogeneity, FEBS Lett., 59: 291-295 (1975).

8. A. Y. H. Lu, D. Ryan, D. M. Jerina, J. W. Daly and W. Levin, Liver microsomal epoxide hydrolase: solubilisation, purfication and characterization, J. Biol. Chem., 250: 8283-8288 (1975).

9. S. S. Gill, B. D. Hammock and J. E. Casida, Mammalian metabolism and environmental degradation of the juvenoid 1-(4-ethylphenoxy)- 3,7-dimethyl-6,7-epoxy-trans-2-octene and related compounds, J. Agr. Food Chem., 22: 386-395 (1974).

10. F. Waechter, M. Merdes, F. Bieri, W. Stäubli and P. Bentley, Purification and characterization of a soluble epoxide hydratase from rabbit liver. Eur. J. Biochem. 125: 457-461 (1982).

11. P. Wang, J. Meijer and F. P. Guengerich, Purification of human liver cytosolic epoxide hydrolase and comparison to the microsomal enzyme, Biochemistry, 21: 5769-5776 (1982).

12. T. Watabe, M. Kanai, M. Isobe and N. Ozawa, The hepatic microsomal biotransformation of Δ^5-steroids to $5\alpha,6\beta$-glycols via α- and β-epoxides. J. Biol. Chem. 256: 2900-2907 (1981).

13. F. Oesch, C. W. Timms, C. H. Walker, T. M. Guenthner, A. Sparrow, T. Watabe, and C. R. Wolf, Existence of multiple forms of microsomal epoxide hydrolases with radically different substrate specificities, Carcinogenesis, 5: 7-9 (1984).

14. J. McGee and F. Fitzpatrick, Enzymatic hydration of leukotriene A_4: Purification and characterization of a novel epoxide hydrolase from human erythrocytes, J. Biol. Chem., 260: 12832-12837 (1985).

15. F. P. Guengerich, Epoxide hydrolase: properties and metabolic roles, in: "Reviews in Biochemical Toxicology", Vol. 4, E. Hodgson, J. R. Bend and R. M. Philpot, eds., Elsevier Science Publishing Company, New York, 5-30 (1982).

16. R. N. Wixtrom and B. D. Hammock, Membrane-bound and soluble-fraction epoxide hydrolases: methodological aspects, in: "Biochemical Pharmacology and Toxicology", Vol. 1, D. Zakim and D. A. Vessey, eds., John Wiley and Sons, New York, 1-93 (1985).

17. J. Meijer and J. W. De Pierre, Cytosolic epoxide hydrolase, Chem.-Biol. Interactions, 64: 207-249 (1988).

18. F. S. Heinemann and J. Ozols, The covalent structure of hepatic microsomal epoxide hydrolase, J. Biol. Chem., 259: 791- 804 (1984).

19. T. D. Porter, T. W. Beck and C. B. Kasper, Complementary DNA and amino acid sequence of rat liver microsomal, xenobiotic epoxide hydrolase, Arch. Biochem. Biophys., 248: 121-129 (1986).

20. R. C. Skoda, A. Demierre, O. W. McBride, F. J. Gonzalez and U. A. Meyer, Human microsomal xenobiotic epoxide hydrolase: Complementary DNA sequence, complementary DNA-directed expression in COS-1 cells, and chromosomal localization, J. Biol. Chem., 263: 1549-1554 (1988).

21. C. D. Funk, O. Rådmark, J. Y. Fu, T. Matsumoto, H. Jörnvall, T. Shimizu, and B. Samuelsson. Molecular cloning and amino acid seqence of leukotriene A4 hydrolase, Proc. Natl. Acad. Sci. USA, 84: 6677-6681 (1987).

22. J. Seidegård and J. W. DePierre, Microsomal epoxide hydrolase: properties, regulation and function, Biochim. Biophys. Acta, 695: 251-270 (1983).

23. A. Åström, M. Eriksson, L. C. Eriksson, W. Birberg, A. Pilotti, and J. W. DePierre, Subcellular and organ distribution of cholesterol epoxide hydrolase in rat, Biochim. Biophys. Acta, 882: 359-366 (1986).

24. M. H. Silva and B. D. Hammock, Affinity purification of cytosolic epoxide hydrolase from human, rhesus monkey, baboon, rabbit, rat and mouse liver, Comp. Biochem. Physiol., 87B: 95-102 (1987).

25. F. P. Guengerich, P. Wang, M. B. Mitchell, and P. S. Mason, Rat and human liver microsomal epoxide hydrolase: purification and evidence of multiple forms, J. Biol. Chem., 254: 12248-12254 (1979).

26. N. J. Bulleid, A. B. Graham and J. A. Craft, Microsomal epoxide hydrolase of rat liver: Purification and characterization of enzyme fractions with different chromatographic characteristics, Biochem. J., 233: 607-611 (1986).

27. D. E. Moody and B. D. Hammock, Purification of microsomal epoxide hydrolase from liver of rhesus monkey: partial separation of cis- and trans-stilbene oxide hydrolase, Arch. Biochem. Biophys., 258: 156-166 (1987).

28. L. Schladt, H. Thomas, R. Hartmann and F. Oesch, Human liver cytosolic epoxide hydrolases, Eur. J. Biochem., 176: 715-723 (1988).

29. C. N. Falany, P. McQuiddy and C. B. Kasper, Structure and organization of the microsomal xenobiotic epoxide hydrolase gene, J. Biol. Chem., 262: 5924-5930 (1987).

30. P. E. Thomas, L. M. Reik, D. E. Ryan and W. Levin, Regulation of three forms of cytochrome P450 and epoxide hydrolase in rat liver microsomes, J. Biol. Chem., 256: 1044-1052 (1981).

31. F. Oesch F, Mammalian epoxide hydrases: inducible enzymes catalysing the inactivation of carcinogenic and cytotoxic metabolites derived from aromatic and olefinic compounds, Xenobiotica, 3: 305-340 (1973).

32. F. Oesch, Differential control of rat microsomal aryl hydrocarbon monooxygenase and epoxide hydratase, J. Biol. Chem., 251: 79-87 (1976).

33. J. A. Craft, N. J. Bulleid, M. R. Jackson and B. Burchell, Induction of microsomal epoxide hydrolase by nitrosamines in rat liver, Biochem. Pharmacol., 37: 297-302 (1988).

34. A. Y. H. Lu and G. T. Miwa, Molecular properties and biological functions of microsomal epoxide hydrase, Ann. Rev. Pharmacol. Toxicol., 20: 513-531 (1980).

35. P. P. Halarnkar, R.N. Wixtröm, M. H. Silva and B. D. Hammock, Catabolism of epoxy fatty esters by the purified epoxide hydrolase form mouse and human liver, Arch. Biochem. Biophys., 272: 226-236 (1989).

36. R. N. Armstrong, B. Kedzierski, W. Levin and D. M. Jerina, Enantioselectivity of microsomal epoxide hydrolase toward arene oxide substrates, J. Biol. Chem., 256: 4726-4733 (1981).

37. G. D. Prestwich, I. Lucarelli, S.-K. Park, D. N. Loury, D. E. Moddy and B. D. Hammock, Cyclopropyl oxiranes: reversible inhibitors of cytosolic and microsomal epoxide hydrolases, Arch. Biochem. Biophys., 237: 361-372 (1985).

38. G. D. Prestwich, J.-W. Kuo, S.-K. Park, D.N. Loury, and B. D. Hammock, Inhibition of epoxide metabolism by α,β-epoxyketones and isosteric analogs, Arch. Biochem. Biphys., 242: 11-15 (1985).

39. F. Oesch and J. Daly, Conversion of naphthalene to *trans*-naphthalene dihydrodiol: evidence for the presence of a coupled aryl monooxygenase-epoxide hydrase system in hepatic microsomes, Biochem. Biophys. Res. Commun., 46: 1713-1720 (1972).

40. W. Levin, A. Y. H. Lu, P. E. Thomas, D. Ryan, D. E. Kizer and M. J. Griffin, Identification of epoxide hydrase as the preneoplastic antigen in rat liver hyperplastic nodules, Proc. Natl. Acad. Sci. USA, 75: 3240-3243 (1978).

41. M. J. Griffin and K. Noda, Quantitation of epoxide hydrolase released from hyperplastic nodule and hepatoma microsomes, Cancer Res., 40: 2768-2773 (1980).

42. H. Haymerle, J. Herz, G. M. Bressan, R. Frank, and K. K. Stanley, Efficient construction of cDNA libraries in plasmid expression vectors using an adaptor strategy, Nucl. Acids Res., 14: 8615-8624 (1986).

43. L. Schladt, R. Hartmann, W. Wörner, H. Thomas and F. Oesch, Purification and characterization of rat-liver cytosolic epoxide hydrolase, Eur. J. Biochem., 176: 31-37 (1988).

44. J. Meijer, G. Lundqvist and J. W. DePierre, Comparison of the sex and subcellular distributions, catalytic and immunochemical reactivities of hepatic epoxide hydrolases in seven mammalian species, Eur. J. Biochem., 167: 269-279 (1987).

45. C. R. Pace-Asciak and W.S. Lee WS, Purification of hepoxilin epoxide hydrolase from rat liver, J. Biol. Chem., 264: 9310-9313 (1989).

46. S. S. Gill, Purification of mouse liver cytosolic epoxide hydrolase, Biochim. Biophys. Acta, 112: 763-769 (1983).

47. J. Meijer and J. W. DePierre, Properties of cytosolic epoxide hydrol-ase purified from the liver of untreated and clofibrate-treated mice. Characterization of optimal assay conditions, substrate specificity and effects of modulators on the catalytic activity, Eur. J. Bio-chem., 150: 7-16 (1985).

48. S. S. Gill, Immunological similarity of epoxide hydrolase activity in the mitochondrial and cytosolic fractions of mouse liver, Biochem. Biophys. Res. Commun., 122: 1434-1440 (1984).

49. U. Milbert, W. Wörner and F. Oesch, Characterization of rat hepatic and renal glutathione S-transferases, in: "Primary Changes and Control Factors in Carcinogenesis", T. Friedberg and F. Oesch, eds., Deutscher Fachschriften-Verlag, Wiesbaden, 14-21 (1986).

ROLE OF THE WELL-KNOWN BASIC AND THE RECENTLY DISCOVERED ACIDIC

GLUTATHIONE S-TRANSFERASES IN THE CONTROL OF GENOTOXIC METABOLITES

Franz Oesch, Ingolf Gath, Takashi Igarashi,
Hansruedi Glatt and Helmut Thomas

Institute of Toxicology, University of Mainz
Obere Zahlbacher Strasse 67, D-6500 Mainz, FRG

INTRODUCTION

Glutathione S-transferases (GSTs; E.C. 2.5.1.18) are a family of enzymes which have increasingly attracted the interest of toxicologists, pharmacologists, biochemists and clinicians since their discovery in 1961 (1). Initially, GSTs were believed to serve as intracellular transport proteins for endogenous compounds with limited solubility in water, thus acting as an intracellular equivalent to albumin in blood plasma. In this assumed capacity of reversible binding and transport of various ligands, the corresponding protein was named ligandin (2). Following the discovery of abundant GST occurrence in most forms of aerobic life including plants, and the GST-catalysed conjugation of a wide variety of electrophilic substrates with glutathione, GSTs are now generally considered to play a crucial role in the detoxification of foreign compounds such as mutagens, carcinogens and other noxious chemicals (for reviews see Refs. 3-6) by conjugation to glutathione. In addition, GSTs are believed to provide cellular protection by covalent binding of reactive electrophiles to the enzyme, which results in immobilization and inactivation of the compound in question.

The expression of multiple forms of GSTs is largely species and tissue dependent (7,8) and may be altered by inducers of drug metabolism. Also, dramatic ontogenic and organ specific changes in the expression of GSTs have been observed (9,10), and sex-related differences in the expression of specific GST isoenzymes in rat and mouse liver have been described (11,12). Particular attention has focussed recently on the high level expression of certain isoenzyme forms in mammalian tumors, their possible clinical use as diagnostic markers for neoplasia, and as factors responsible for the development of drug resistance, which is often encountered in the chemotherapy of cancer (13).

Current studies on the structure of GST genes and control of expression are expected to provide deeper insight into multiplicities, regulation, and function of GSTs (14).

Research in our laboratory has been concentrating manily on the GST-mediated detoxification of epoxides and epoxide hydrolase-resistant diol epoxides on polycyclic aromatic hydrocarbons, and the characterization of GST isoenzymes involved therein.

Molecular Aspects of Monooxygenases and Bioactivation of Toxic Compounds
Edited by E. Arınç *et al., Plenum Press, New York,*

MULTIPLICITY, NOMENCLATURE, CLASSIFICATION AND PROPERTIES OF GLUTATHIONE
S-TRANSFERASES

GSTs have been most extensively investigated in rat, mouse and human tissues, and, with the exception of microsomal GST, been found to constitute functional homo- and heterodimers (5). In the rat, for example, at least eight different subunits have been characterized, although more are known to exist, constituting 12 different isoenzymes. Several nomenclatures are currently being used, the most favoured, however, designating each subunit with an arabic numeral based on the chronological order of its characterization (Table 1).

Complete cDNA sequences and deduced primary structures are known for subunits 1, 2, 3, 4, and 7 (15-19). These data show 69 % identity in full length sequences between subunits 1 and 2, and 77 % identity between subunits 3 and 4. There is very little identity between the group comprising subunits 1 and 2, as compared with the group comprising subunits 3 and 4, and with subunit 7, indicating that these three categories belong to separate multigene families. Correspondingly, and supported by complete or partial amino acid sequences, immunological crossreactivity as well as enzymatic properties of the corresponding homodimers, the subunits given in Table 1 have been assigned to three families, Alpha (subunits 1, 2 and 8), Mu (subunits 3, 4 and 6), and Pi (subunit 7), while subunit 5 has yet to be classivied (5). So far, evidence has been obtained only for the formation of heterodimers from subunits within a multigene family with the resulting dimer reflecting the functional properties of its constituting subunits (Table 2). Comparison of orthologous isoenzymes from various species has further shown, that the classification of soluble GSTs into classes Alpha, Mu, and Pi is species-independent.

Members of a certain class express diagnostic substrate specificities, which have proven useful for the identification and classification

Table 1. Nomenclature for the cytosolic rat glutathione transferases

New nomenclature	Previous nomenclature		Alternative subunit designations
Glutathione transferase 1-1	B (ligandin)	L_2	Ya
Glutathione transferase 1-2		BL	Ya, Yc
Glutathione transferase 2-2	AA	B_2	Yc
Glutathione transferase 3-3	A	A_2	Yb1
Glutathione transferase 3-4	C	AC	Yb1, Yb2
Glutathione transferase 3-6	P		
Glutathione transferase 4-4	D	C_2	Yb2
Glutathione transferase 4-6	S		
Glutathione transferase 5-5	E		
Glutathione transferase 6-6	M_T		Yn
Glutathione transferase 7-7	P		Yp or Yf
Glutathione transferase 8-8	K		Yk

Modified and updated from Ref. 34

Table 2. Physicochemical characteristics of rat glutathione transferases

Isoenzyme	Class	Apparent subunit MW (kdalton)[a]	Subunit MW[b]	No. of amino acids per subunit[b]	Isoelectric point
1-1	Alpha	25	25 434	221	10
1-2	Alpha	25 + 28	-	-	9.9
2-2	Alpha	28	25 209	220	9.8
3-3	Mu	26.5	25 806	217	8.9
3-4	Mu	26.5	-	-	8
3-6	Mu	26.5 + 26	-	-	7.4
4-4	Mu	26.5	25 592	217	6.9
4-6	Mu	26.5 + 26	-	-	6.1
5-5	-[c]	26.5	-	-	7.3
6-6[d]	Mu	26	-	-	5.8
7-7	Pi	24	23 307	209	7.0
8-8	Alpha	24.5	-	-	6.0
Microsomal	-	17	17 237	154	10.1

Data from Ref. 5 with permission.

a Relative values estimated by sodium dodecyl sulfate/polyacrylamide gel electrophoresis.
b N-terminal methionine residue not included.
c Not yet classified.
d Recent work suggests that the major testicular enzyme, designated transferase 6-6, is a heterodimer.

of newly discovered isoenzymes (Table 3). Class Alpha GSTs and the as yet unassigned GST 5-5 for example are highly active with organic hydroperoxides such as cumene hydroperoxide, class Mu transferases conjugate *trans*-stilbene oxide with a comparatively high efficiency, while ethacrynic acid is a marker substrate for class Pi GSTs (Table 3). From the substrates listed in Table 3, 1-chloro-2,4-dinitrobenzene (CDNB) is the most commonly used to determine the overall activity of unfractionated GSTs, since it is well utilized by all rat GSTs except subunit 5, which, however, is usually present in small amounts only. The other substrates are chosen, because they are fairly specific for a particular subunit and may be selected to detect and quantify it.

CYTOTOXIC, MUTAGENIC AND CARCINOGENIC SUBSTRATES OF GLUTATHIONE S-TRANSFERASES

GST 8-8, a recently discovered Alpha class acidic isoenzyme from rat liver is highly reactive in detoxifying cytotoxic 4-hydroxyalkenals (20). The fact that among all other GSTs of rat liver investigated so far, the next most active (the near neutral Mu class isoenzyme 4-4) exhibits a specific activity towards 4-hydroxynonenal (4-HNE) which is more than 20-fold lower, suggests an important role of the mainly acidic GSTs in the detoxification of hydroxyalkenals, which are major products of endogenous lipid peroxidation (21).

Among the mutagenic and carcinogenic substrates of GSTs aflatoxin B1-8,9-oxide certainly needs mentioning. This compound is a major metabolite of aflatoxin B1 and a potent hepatocarcinogen in the rat. It is conjugated, although at the low rate of 1 nmol/min/mg, exclusively by the subunits 1 and 2 (22). The important point of this reaction, however, is its 4- to 5-fold acceleration following the induction of subunits 1 for example by ethoxyquin. Ethoxyquin is one of a wide range of chemicals which induce GSTs in the rat and include polycyclic aromatic hydrocarbons, chlorinated biphenyls, certain antioxidants and *trans*-stilbene oxide. Most of these inducers are selective for subunits 1 and 3, and therefore influence the detoxification of substrates for the corresponding enzymes (23,24).

Polycyclic aromatic hydrocarbons give rise to the metabolic formation of various electrophilic metabolites including "K-region" epoxides and "bay-region" dihydrodiol epoxides, some of which are considered powerful mutagens and initiators of carcinogenesis (25). These epoxides have been recognized as another important group of substrates primarily for near neutral and acidic GSTs, since *anti*-benzo(a)pyrene-7,8-diol-9,10-epoxide (BPDE), an ultimate carcinogen derived from benzo(a)pyrene, was shown to be metabolized by all rat GST isoenzymes investigated, and best by GST 7-7 and GST 4-4 (26,27). However, there was a considerable difference in the catalytic efficiency of 0.287 $s^{-1} \mu M^{-1}$ for GST 7-7 compared to 0.046 s^{-1} μM^{-1} for GST 4-4 as determined from isoenzymes of rat lung (27). Furthermore, rat liver GST 4-4 metabolized BPDE at a rate of 357 nmol/min/mg about 2- and 3.4-times more efficiently than GST 3-4 and GST 1-1 (26). Similar to the observations with rat GSTs, the acidic human placental GST π metabolized BPDE with an apparent Vmax of 825 nmol/min/mg more efficently than the near neutral and basic human liver GSTs μ and α-ε with apparent Vmax values of 570 nmol/min/mg and 38 nmol/min/mg, respectively (28). It should be mentioned, that the rat GSTs 4-4 and 7-7 as well as the human GST π express high selectivity for the (+)-enantiomer of BPDE which is fortunate with regard to the preferential metabolic formation of this ultimately carcinogenic intermediate. The same stereoselectivity is shown by GST 4-4 towards bay-region diol epoxides of chrysene and benz-(a)anthracene (29).

450

Table 3. Specific activities (μmol/min/mg) of rat glutathione transferases

Substrate	Class: Alpha			Mu			Pi	Unknown	Microsomal
Enzyme:	1-1	2-2	8-8	3-3	4-4	6-6[a]	7-7	5-5	(activated)
1-Chloro-2,4-dinitrobenzene	50	17	10	58	17	190	24	<0.15	30
1,2-Dichloro-4-nitrobenzene	0.04	0.04	0.12	5.3	0.18	2.85	0.048[b]	Nil	0.06
Bromosulfophthalein	<0.01	<0.01	-	0.94	0.04	-	0.01[b]	-	<0.01
Ethacrynic acid	0.08	1.24	7.0	0.08	0.62	0.057	3.84	Nil	<0.01
trans-4-Phenyl-3-buten-2-one	<0.004	<0.004	0.10	0.05	1.18	0.019	0.22	<0.001	0.001
4-Hydroxynonenal	2.6	0.67	170	2.7	6.9	-	-	-	-
Leukotriene A4[c]	0.002	0.0005	-	0.002	0.077	-	-	-	-
1,2-Epoxy-3-(p-nitrophenoxy)-propane	<0.1	<0.1	-	0.53	1.37	-	-	25.5	<0.01
trans-Stilbene oxide[d]	0.001	0.003	0.033	0.10	2.0	0.13	0.005	-	-
Benzo(a)pyrene 7,8-diol-9,10-oxide	-	0.006	0.18	0.012	0.68	-	5.5	-	-
Cumene hydroperoxide	3.1	7.9	1.10	0.35	0.72	0.19	0.048	12.5	0.8
H2O2	<0.01	<0.01	<0.01	<0.01	<0.01	<0.01	<0.01	-	<0.04
Δ5-Androstene-3,17-dione	4.2	0.36	-	0.02	0.002	-	-	-	-
p-Nitrophenyl acetate	0.79	0.20	-	1.01	0.28	0.19	-	-	-

Data from Ref. 5 with permission.

a Unpublished work by Guthenberg, C. and Mannervik, B.
b Unpublished work by Tahir, M.K. and Mannervik, B.
c Unpublished work by Örning, L., Söderström, M., Hammarström, S., and Mannervik, B.
d Unpublished work by Seidegård, J., Danielson, U.H., and Mannervik, B.
- not detected

Most strikingly, the prominent contribution of near neutral and acidic GSTs to the detoxification of polycyclic aromatic epoxides became obvious with the discovery of a weakly acidic rat liver GST termed GST-X (30). GST-X was identified as a homodimer with a subunit molecular weight of 23,500 Da, an isoelectric point of 6.9, and a particularly high turnover of 1,2-dichloro-4-nitrobenzene in comparison to other Alpha and Mu class GSTs (Table 4). This isoenzyme on a molecular basis was shown to be at least 6.7- and 5-fold more efficient in the conjugation of the mutagenic K-region epoxide benz(a)anthracene-5,6-oxide (BA-5,6-oxide) and the non-bay-region diol epoxide *r*-8,*t*-9-dihydroxy-*t*-10,11-oxy-8,9,10,11-tetrahydrobenz(a)anthracene (BA-8,9-diol-10,11-oxide) when compared with the major rat liver GSTs C (3-4), B (1-2) and A (3-3) (Table 5; Figures 1 and 2). The relative efficiency of all four investigated GSTs was similar with both substrates in spite of the structural differences in the epoxides. However, inactivation of the diol epoxide required about 1000-fold higher concentrations of GSTs than the inactivation of the K-region epoxide. An even stronger preference for K-region epoxides, as compared to diol-epoxides, was observed for microsomal epoxide hydrolase (31).

GLUTATHIONE S-TRANSFERASE X AND LOW ABUNDANCE CLASS MU ISOENZYMES

Further characterization of GST-X by fast atom bombardement mass spectroscopy of tryptic peptides revealed clear differences from GST 4-4, but also a number of structural homologies to this isoenzyme, from which a close relationship of GST-X to the Mu class GSTs could be deduced.

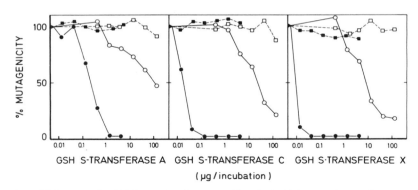

Figure 1. Effect of different forms of glutathione transferase (GST), purified from untreated rats, on the mutagenicity of BA 5,6-oxide in the presence (●) and absence (■) of glutathione (2 mM) and of BA-8,9-diol-10,11-oxide in the presence (○) and absence (□) of glutathione (2 mM). The numbers of mutants above solvent control induced by 1 μg of BA 5,6-oxide or by 3 μg of BA-8,9-diol-10,11-oxide in the presence of various amounts of the purified glutathione transferases are expressed as the percentage of the corresponding value obtained in the absence of enzyme. The absolute numbers of colonies obtained in the absence of enzyme were 74 and 85 for the solvent control, 730 and 1040 for BA 5,6-oxide, and 1520 and 1540 for BA-8,9-diol-10,11-oxide, in the presence and absence of glutathione, respectively. Triplicate incubations were performed. The variation in the numbers of colonies on replicate plates was less than 10 %.
Data from Ref. 31 with permission.

Table 4. Substrate specificity of glutathione S-transferase X compared with those of glutathione S-transferases C and B

Substrate	Specific activity (µmol/min/mg)		
	Transferase X	Transferase C (3-4)	Transferase B (1-2)
1-Chloro-2,4-dinitrobenzene	21 500	22 500 (10 000)	16 000 (11 000)
1,2-Dichloro-4-nitrobenzene	2 700	1 000 (2 000)	4 (3)
trans-4-Phenylbut-3-en-2-one	700	470 (400)	<5 (1)
Menaphthyl sulphate	0*	-	6 (4)
1,2-Epoxy-3-(p-nitrophenoxy)-propane	360	280 (0)*	- (0)*
2-(4-Nitrophenyl)-ethyl bromide	180	45	-
p-Nitrobenzyl chloride	3 300	(10 200)	(100)

Data from Ref. 30 with permission.

* A value of zero does not necessarily mean that the enzyme is incapable of catalysing the reaction, but rather that the reaction rate was not significantly greater than the spontaneous rate in trials with reasonable amounts of enzyme (> 1 mg of enzyme/assay in our experiments).

Numbers in parentheses give the corresponding literature values (35). Apparently homogeneous preparations of glutathione S-transferases X, C and B were used throughout. Assays were performed as described previously (36,37). Assays were performed in triplicate. Deviations from the mean were less than 10 %.

Table 5. Quantitative comparison of the inactivation of BA 5,6-oxide and BA-8,9-diol-10,11-oxide by different enzymes

Enzyme	Enzyme concentration in liver[a] (μg/mg tissue)	Amounts of enzyme required for a 50 % reduction in mutagenicity			
		BA 5,6-oxide		BA-8,9-diol-10,11-oxide	
		μg/incubation	mg liver equivalents	μg/incubation	mg liver equivalents
Experiment shown in Figure 1					
Glutathione transferase A (3-3)	0.5	0.2	0.4	100	200
Glutathione transferase C (3-4)	1.1	0.02[b]	0.02	20	20
Glutathione transferase X	0.25	<<0.01[b]	<<0.06	8	30
Experiment shown in Figure 2					
Glutathione transferase A (3-3)	0.5	0.11	0.2	110	200
Glutathione transferase B (1-2)	2.2	0.5	0.2	130	60
Glutathione transferase C (3-4)	1.1	0.02	0.017	30	20
Glutathione transferase X	0.25	0.003	0.011	6	20
Experiments using other enzymes[c]					
Microsomal epoxide hydrolase	0.5	0.7	1.3	Inactive(>>170)	>>300
Cytosolic epoxide hydrolase	0.16	4	30	Inactive(>> 30)	>>200
Dihydrodiol dehydrogenase	0.45	Inactive(>>3)	>>1000	70	170

Data from Ref. 31 with permission.
a Values refer to untreated, adult males of the species from which the enzyme was purified.
b Determinable only as an upper limit from the experiment.
c Data taken from Ref. 33. These experiments were performed in the same way as those using glutathione transferases.

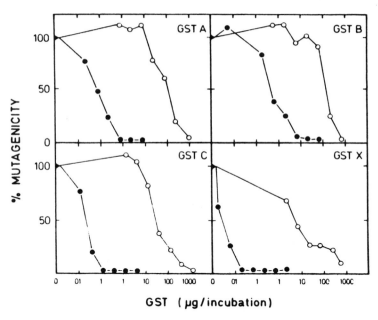

GST (µg/incubation)

Figure 2. Effect of different forms of glutathione transferases (GST), purified from Aroclor 1254-treated rats, on the mutagenicity of BA 5,6-oxide (●) and BA-8,9-diol-10,11-oxide (O). The experiment was conducted as described in the legend to Figure 1 except that the enzymes were purified from Aroclor 1254-treated rats and they were tested only in the presence of glutathione (2 mM). The absolute numbers of colonies obtained in the absence of enzyme were 109, 405, and 1270 for the solvent controls, for incubations with 1 µg BA 5,6-oxide, and for incubations with 2.5 µg BA-8,9-diol-10,11-oxide, respectively. Incubations were performed in triplicate, and the variation in the numbers of colonies on replicate plates was less than 15 %.
Data from Ref. 31 with permission.

Evidence was provided recently by chromatofocussing of S-hexylgluta-thione-Sepharose 6B purified GSTs for the presence of at least seven near neutral to acidic isoenzymes in rat liver (32). These newly discovered isoenzymes also appeared to be mainly members of the Mu class GSTs. Therefore it was assumed that they may have equal or even higher capacities for the detoxification of polycyclic aromatic epoxides or lipid peroxidation products than the so far characterized rat GSTs.

To elucidate the properties of these isoenzymes and their role in the cellular protection against toxic, mutagenic and carcinogenic inter-mediates of endogenous and foreign compounds, we disigned a method for the efficient purification particularly of acidic rat liver GSTs. This method includes chromatography of the cytosol on QA-cellulose at pH 8.5, affinity chromatography on GSH-Sepharose and chromatofocussing in the range between pH 7.0 and 4.0. Throughout the purification GST activity was monitored with CDNB and 4-HNE as substrates, the latter acting as a marker substrate specifically for acidic GSTs. Chromatofocussing resolved three major GST containing fractions (CF I - CF III) with isoelectric points at 6.8, 6.2, and 5.8 (Figure 3). CF I and CF II were further purified on hydroxylapa-tite to yield a total of four near neutral to acidic GST isoenzymes, HA I-a (Y3-Y4), HA I-b (Y5-Y5), HA II (Y1-Y1) and CF III (mainly Y2-Y2 and a trace of Y6-Y6), consisting of six different subunits which were tenta-tively termed Y1 to Y6 (Table 6; Figure 4). The apparent molecular weight of these subunits ranged between 25,500 Da and 27,000 Da.

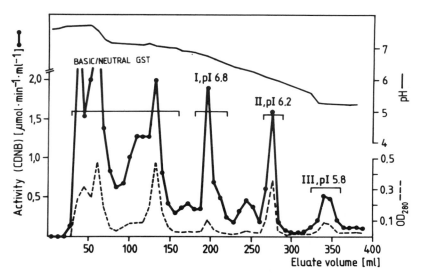

Figure 3. Chromatofocussing of GSTs in the range of pH 7-4. GSH-Sepharose 6B eluates were concentrated and dialyzed over night against 25 mM imidazole/HCl buffer (pH 7.4). Chromatofocussing was performed on a PBE 74 column (1x51 cm) according to the instructions given by the manufacturer.

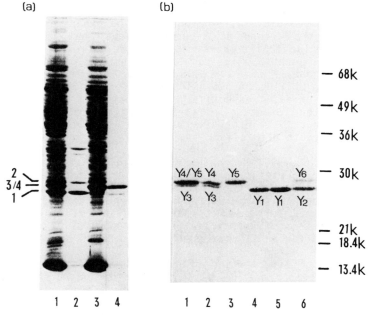

Figure 4. SDS/polyacrylamide-gel electrophoresis of crude (a) and purified acidic rat liver GSTs (b). GSTs were subjected to electrophoresis in a 12.5 % SDS/polyacrylamide gel and protein bands were visualized by Coomassie Brillant Blue R 250.

a. Lanes 1-4: Cytosol; flow through of QA-52 cellulose; QA-52 cellulose eluate; GSH affinity eluate. Subunits apparently corresponding to characterized rat liver GSTs are marked with arabic numerals (see Table 1).

b. Purified GSTs after chromatofocussing (CF) or chromatofocussing and hydroxylapatite chromatography (HA).
    Lanes 1-6: CF I; HA I-a; HA I-b; CF II; HA II; CF III.

Table 6. Purification of acidic glutathione S-transferases from rat liver

	Volume (ml)	Protein (mg)	C D N B		4 - H N E		Ratio 4-HNE/CDNB
			Specific activity (µmol/min x mg)	Total activity (µmol/min)	Specific activity (µmol/min x mg)	Total activity (µmol/min)	
Cytosol	192	2937.6	2.44	7181(100%)	1.21	3562(100%)	0.50
QA-cellulose eluate	375	1237.5	0.71	879(12.2%)	2.92	3614(101.5%)	4.11
GSH affinity eluate	158	45.8	12.07	552(7.7%)	52.41	2401(67.4%)	4.35
CF (basic/ neutral forms)	130	23.1	14.27	330(4.6%)	2.53	59(1.7%)	0.18
CF I (pI 6.8)	43.5	2.3	15.66	36(0.5%)	6.79	16(0.5%)	0.44
CF II (pI 6.2)	28	5.5	4.36	24(0.3%)	121.23	662(18.6%)	27.58
CF III (pI 5.8)	40	5.3	2.41	13(0.2%)	64.14	341(9.6%)	26.23
HA I-a	23	0.9	22.75	21(0.3%)	4.50	4.2(0.1%)	0.14
HA I-b	14	0.4	12.00	4(0.06%)	0.80	0.2(0.01%)	0.21
HA II	16.5	3.1	4.61	13.8(0.2%)	185.44	550.8(15.5%)	35.59

457

Table 7. Specific activities of the new acidic rat liver GST isoenzymes with various model substrates

Subunit (MW x 10^{-3})	HA I-a Y_3 (26.5) Y_4 (27.0)	HA I-b Y_5 (27.0)	HA II Y_1 (25.5)	CF III Y_2 (25.5) Y_6 (27.0)
Substrate				
1-Chloro-2,4-dinitro-benzene	13.08[a]	2.84	5.03	8.16
1,2-Dichloro-4-nitrobenzene	1.23	ND	ND	0.27
4-Hydroxynon-2-enal	2.32	2.57	114.36	99.29
Ethacrynic acid	0.42	0.91	3.64	3.73
Δ^5-Androsten-3.17-dione	ND	ND	0.02	0.03
trans-4-Phenyl-3-buten-2-one	0.34	0.07	ND	0.03
1,2-Epoxy-3-(p-nitro-phenoxy)propane	ND	ND	ND	ND
Cumene hydroperoxide	0.10	ND	0.23	0.19

a) Activities in $\mu mol \cdot min^{-1} \cdot mg^{-1}$.
ND: not detected.

While subunits Y1 and Y2 were shown by strong immunological crossreactivity with antiserum against subunit 8 to be members of the class Alpha GSTs, subunits Y3 to Y6 were identified by Western blotting as belonging to the class Mu subunits. Ha II (Y1-Y1), based on its isoelectric point and crossreactivity with GST 8-8, is closely related to, if not identical with, GST 8-8. CF III (Y2-Y2), however, although immunologically closely related to the Alpha class subunit 8 differs in its isoelectric point and enzymatic properties from GST 8-8 and appears to be a new isoenzyme. Subunit Y3 expressed similarity to GST subunit 6, which has been characterized so far only from rat brain and testis, and suggests the presence of a 6-Y4 heterodimer (HA I-a) in rat liver. Substrate specificities of the new isoenzymes tested with a number of model compounds are rather undiagnostic with the exception of the high turnover of 4-HNE with CF III (Table 7). In summary, we have identified and purified at least three novel acidic GSTs which appear to be minor forms representing between 0.06 % (HA I-b) and 0.3 % (HA I-a) of total rat liver CDNB-conjugating GST activity. Their efficiency in the metabolism of polycyclic aromatic epoxides is currently under investigation.

CONCLUSIONS

Glutathione S-transferases constitute a superfamily of dimeric proteins, the subunits of which may be assigned with respect to structure homologies, enzymatic, immunological and functional properties to three gene families, Alpha, Mu, and Pi. The main function of these enzymes appears to consist in the cellular protection against toxic, mutagenic and carcinogenic electrophilic metabolites arising from endogenous and exogenous compounds. Mu and Pi class isoenzymes containing the subunits 4 and 7 have proven particularly efficient in the detoxification of K-region and bay-region diol epoxides from polycyclic aromatic hydrocarbons, whereas the Alpha class GST 8-8 was by far the most effective in conjugating the highly cytotoxic and highly mutagenic lipid peroxidation product 4-HNE. These findings outline the dominant role of near neutral and acidic GSTs in the control of genotoxic metabolites. Based on these observations four acidic GSTs consisting of 6 individual subunits were purified by a newly developed method. From their close relationship to the Mu class GSTs, which are very potent in the conjugation of polycyclic aromatic epoxides it is concluded, that they, although minor forms in rat liver, may be of importance in the cellular protection against mutagenic and carcinogenic metabolites of foreign compounds.

ACKNOWLEDGEMENTS

This investigation was supported by the Deutsche Forschungsgemeinschaft (SFB 302). The kind gifts of 4-HNE (H. Esterbauer, University of Graz) and antibodies against GST 6-6 (K. Sato, Hirosaki University) are greatfully acknowledged. T.I. is the recipient of a postdoctoral fellowship from the Humboldt Foundation.

REFERENCES

1. J. Booth, E. Boyland and P. Sims, An enzyme from rat liver catalysing conjugations with glutathione. Biochem. J., 79: 516-526 (1961).

2. G. Litwack, B. Ketterer and I. M. Arias, Ligandin: a hepatic protein which binds steroids, bilirubin, carcinogens and a number of organic anions, Nature (London), 234: 466-467 (1971).

3. B. Mannervik, The isoenzyme of glutathione transferase, Adv. Enzymol. Rel. Areas Mol. Biol., 57: 357-417 (1985).

4. B. Ketterer, Protective role of glutathione and glutathione transferases in mutagenesis and carcinogenesis, Mutat. Res., 202: 343-361 (1988).

5. B. Mannervik and U. H. Danielson, Glutathione transferases-structure and catalytic activity, CRC Crit. Rev. Biochem., 23: 283-337 (1988).

6. H. Sies and B. Ketterer (eds), Glutathione conjugation: mechanisms and biological significance, Academic Press, New York (1988).

7. T. Igarashi, T. Satoh, K. Ueno and H. Kitagawa, Species difference in glutathione level and glutathione related enzyme activities in rats, mice, guinea pigs and hamsters, J. Pharm. Dyn., 6: 941-949 (1983).

8. T. Igarashi, N. Tomihari, S. Ohmori, K. Ueno, H. Kitagawa and T. Satoh, Comparison of glutathione S-transferases in mouse, guinea pig, rabbit and hamster liver cytosol to those in rat liver, Biochem. International, 13: 641-648 (1986).

9. A. A. Fryer, R. Hume and R. C. Strange, The development of glutathione S-transferase and glutathione peroxidase activities in human lung, Biochim. Biophys. Acta., 883: 448 (1986).

10. C. G. Faulder, P.A. Hirrell, R. Hume and R. C. Strange, Studies of the development of basic, neutral and acidic isoenzymes of glutathione S-transferases in human liver, adrenal, kidney and spleen, Biochem. J., 241: 221-228 (1987).

11. T. Igarashi, T. Satoh, K. Iwashita, S. Ono, K. Ueno and H. Kitagawa, Sex difference in subunit composition of hepatic glutathione S-transferase in rats, J. Biochem., 98: 117-123 (1985).

12. I. Hatayama, K. Satoh and K. Sato, Development and hormonal regulation of the major form of hepatic glutathione S-transferase in male mice, Biochem. Biophys. Res. Commun., 140: 581-588 (1986).

13. K. Sato, Glutathione transferases as markers of preneoplasia and neoplasia, Adv. Cancer Res., 52: 205-255 (1989).

14. C. A. Telakowski-Hopkins, R. G. King and C. B. Pickett, Glutathione S-transferase Ya subunit gene: identification of regulatory elements required for basal level and inducible expression, Proc. Natl. Acad. Sci. USA, 85: 1000-1004 (1988).

15. C. B. Pickett, C. A. Telakowski-Hopkins, G. J. F. Ding, L. Argenbright and A. Y. H. Lu, Rat liver glutathione S-transferases. Complete nucleotide sequence of a glutathione S-transferase mRNA and the regulation of the Ya, Yb, and Yc mRNA by 3-methylcholanthrene and phenobarbital, J. Biol. Chem., 259: 5182-5188 (1984).

16. C. A. Telakowski-Hopkins, G. S. Rothkopf and C. B. Pickett, Structural analysis of a rat liver glutathione S-transferase Ya gene, Proc. Natl. Acad. Sci. USA, 83: 9393-9397 (1986).

17. G. J. F. Ding, A. Y. H. Lu and C. B. Pickett, Rat liver glutathione S-transferases. Nucleotide sequence analysis of a Yb1 cDNA clone and prediction of the complete amino acid sequence of the Yb1 subunit, J. Biol. Chem., 260: 13268-13271 (1985).

18. G. J. F. Ding, V. D. H. Ding, J. A. Rodkey, C. D. Bennett, A. Y. H. Lu and C. B. Pickett, Rat liver glutathione S-transferases. DNA sequence analysis of a Yb2 cDNA clone and regulation of the Yb1 and Yb2 mRNAs by phenobarbital, J. Biol. Chem., 261: 7952-7957 (1986).

19. Y. Suguoka, T. Kano, A. Okuda, M. Sakai, T. Kitagawa and M. Muramatsu, Cloning and the nucleotide sequence of rat glutathione S-transferase P cDNA, Nucleic Acids Res., 13: 6049-6057 (1985).

20. H. Jensson, C. Guthenberg, P. Alin and B. Mannervik, Rat glutathione transferase 8-8, an enzyme efficiently detoxifying 4-hydroxyalk-2-enals, FEBS Lett., 203: 207-209 (1986).

21. P. Alin, H. Danielson and B. Mannervik, 4-Hydroxy-alk-2-enals are substrates for glutathione transferase, FEBS Lett., 179: 267-270 (1985).

22. B. Coles, D. J. Meyer, B. Ketterer, C. A. Stanton and R. C. Garner, Studies on the detoxication of microsomally-activated aflatoxin B1 by glutathione and glutathione S-transferases *in vitro*, Carcinogenesis, 6: 693-697 (1985).

23. T. Igarashi, N. Irokawa, S. Ono, S. Ohmori, K. Ueno and H. Kitagawa, Difference in the effects of phenobarbital and 3-methylcholanthrene treatment on subunit composition of hepatic glutathione S-transferases in male and female rats, Xenobiotica, 17: 127-137 (1987).

24. Z. Gregus, C. Madhu and C. D. Klaasen, Inducibility of glutathione S-transferases in hamsters, Cancer Lett., 44: 89-94 (1989).

25. D. R. Thakker, H. Yagi, W. Levin, A. W. Wood, A. H. Conney and D. M. Jerina, Polycyclic aromatic hydrocarbons: metabolic activation to ultimate carcinogens. In: M. W. Anders (ed.): "Bioactivation of Foreign Compounds", New York: Academic Press, pp. 177-242 (1985).

26. B. Jernström, M. Martinez, D. J. Meyer and B. Ketterer, Glutathione conjugation of the carcinogenic and mutagenic electrophile (+)-7β-8α-dihydroxy-9α,10α-oxy-7,8,9,10-tetrahydrobenzo(a)pyrene catalyzed by purified rat liver glutathione transferases, Carcinogenesis, 6: 85-89 (1985).

27. I. G. C. Robertson, H. Jensson, B. Mannervik and B. Jernström, Glutathione transferases in rat lung: the presence of transferase 7-7, highly efficient in the conjugation of glutathione with the carcinogenic (+)-7β,8α-dihydroxy-9α,10α-oxy-7,8,9,10-tetrahydrobenzo(a)pyrene, Carcinogenesis, 7: 295-299 (1986).

28. I. G. C. Robertson, C. Guthenberg, B. Mannervik and B. Jernström, Differences in stereoselectivity and catalytic efficiency of three human glutathione transferases in the conjugation of glutathione with 7β,8α-dihydroxy-9α,10 α-oxy-7,8,9,10-tetrahydrobenzo(a)pyrene, Cancer Res., 46: 2220-2224 (1986).

29. I. G. C. Robertson and B. Jernström, The enzymatic conjugation of glutathione with bay-region diol-epoxides of benzo(a)pyrene, benz-(a)anthracene and chrysene, Carcinogenesis, 7: 1633-1636 (1986).

30. T. Friedberg, U. Milbert, P. Bentley, T. M. Guenthner and F. Oesch, Purification and characterization of a new cytosolic glutathione S-transferase (glutathione S-transferase X) from rat liver, Biochem. J. 215: 617-625 (1983).

31. H. Glatt, T. Friedberg, P. L. Grover, P. Sims and F. Oesch, Inactivation of a diol epoxide and a K-region epoxide with high efficiency by glutathione transferase X, Cancer Res., 43: 5713-5717 (1983).

32. U. Milbert, Ph.D.-Thesis, University of Mainz (1986).

33. H. R. Glatt, C. S. Cooper, P. L. Grover, P. Sims, P. Bentley P, I. Merdes, F. Waechter, K. Vogel, T. M. Guenthner and F. Oesch, Inactivation of a diol-epoxide by dihydrodiol dehydrogenase, but not by two epoxide hydrolases, Science, 215: 1507-1509 (1982).

34. W. B. Jakoby, B. Ketterer and B. Mannervik, Glutathione transferases: nomenclature, Biochem. Pharmacol., 33: 2539-2540 (1984).

35. W. B. Jakoby, W. H. Habig, J. H. Keen, J. N. Ketley and M. J. Pabst, Glutathione S-transferases: catalytical aspects. In: I. M. Arias and W. B. Jakoby (eds.): "Glutathione: Metabolism and Function", New York: Raven Press, pp. 189-201 (1976).

36. W. H. Habig, M. J. Pabst and W. B. Jakoby, Glutathione S-transferases. The first enzymatic step in mercapturic acid formation, J. Biol. Chem., 249: 7130-7139 (1974).

37. B. Gilham, The reaction of aralkyl sulphate esters with glutathione catalysed by rat liver preparations, Biochem. J., 121: 667-672 (1971).

CHARACTERISATION AND REGULATION OF UDP-GLUCURONOSYLTRANSFERASES*

Brian Burchell

Department of Biochemical Medicine
Ninewells Hospital and Medical School
The University
Dundee, DD1 9SY, Scotland

INTRODUCTION

Many drugs, xenobiotics and endogenous compounds are metabolised via common two stage pathway, which was designated as phase 1 (oxidative metabolism) and phase II (conjugation) by Williams (1). Glucuronidation reactions catalysed by the UDP-glucuronosyltransferases (UDPGTs) is arguably the major conjugation system of phase II metabolism. The metabolism of phenacetin, in the example illustrated below (Fig 1), proceeds via oxidative dealkylation to produce paracetamol, a reactive and potentially toxic metabolite. Paracetamol is then rapidly glucuronidated to form a harmless water soluble metabolite which is excreted via the kidneys and urine.

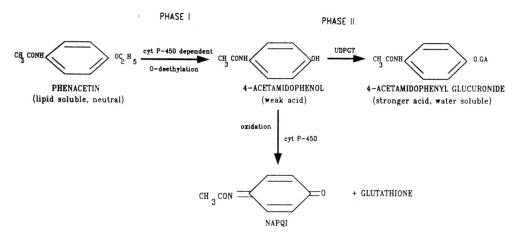

Fig. 1.   Metabolism of Phenacetin by Phase I and Phase II
          Enzyme systems

---

*This paper contains transcribed notes from a lecture given at the NATO workshop and does not attempt to comprehensively review this area of research

Molecular Aspects of Monooxygenases and Bioactivation of Toxic Compounds
Edited by E. Arınç et al., Plenum Press, New York,

The glucuronidation of paracetamol is catalysed by UDP-glucuronosyltransferase in a chemical reaction where a co-substrate UDP-glucuronic acid is required (see Dutton (2) and Fig 2).

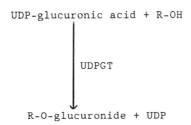

UDP-glucuronic acid + R-OH

UDPGT

R-O-glucuronide + UDP

Fig. 2. The Reaction Catalysed by UDPGT

There are many substrates accepted by the UDP-glucuronosyltransferases such that ester, ether, N-, S- and C-glucuronides can be formed (see 2,3). The UDPGTs are membrane bound enzymes mainly present in the endoplasmic reticulum and nuclear envelope of liver, kidney and intestine although low levels of activity can be detected in other tissues (4).

ASSAY OF MICROSOMAL UDPGT ACTIVITIES

The membrane bound location of UDPGTs has led to many problems with accurate assay of these enzymes. The transferases are deemed to be latent and do not express the full potential activity due to the constraints of the endoplasmic reticulum membrane. Damage of the endoplasmic reticulum by vigorous homogenisation or unwise selection of buffer systems may lead to large differences in microsomal enzyme activities measured. The classical demonstration of this problem was provided by (5), who isolated liver microsomal fractions after homogenisation in isotomic KCl (0.154M) or 0.25M sucrose 1mMEDTA, 5mMTRIS-HCl,pH74. The UDPGT activity towards 4-nitrophenol was 3-5 x higher in microsomes prepared in KCl than the UDPGT activity in microsomes prepared in sucrose. The latency of the Guinea pig liver KCl-microsomes was poor and was not further activated by detergents. This example illustrates the difficulties in comparison of UDPGT activities obtained in different laboratories. Tissue E.R. fractions must be consistently prepared in 0.25M sucrose, 1mMEDTA, 5mMTRIS-HCl,pH74. The latency or intactness of individual microsomal preparations should be independently assessed by measurement of mannose-6-phosphate hydrolysis (6), to enable comparison of different preparations obtained from different tissue specimens. Alternatively, maximum levels of UDPGT activity could be used as a basis for data comparison, after release of latent activity by treatment of microsomes with a suitable detergent such as Lubrol PX. Unfortunately, to measure maximal UDPGT activities, detergent must be titrated with microsomes to determine the maximum response (see 7). The maximal transferase activity will depend on the original intactness of the microsomes and the aglycone substrate used and thus selection of optimal levels of detergent addition should be determined in each case (see Fig 3).

Fig 3 Demonstrates that different amounts of detergent were required to determine the maximal activity of bilirubin UDPGT and 1-Naphthol UDPGT in the same preparation of hepatic microsomes. This information demonstrates the difficulties encountered in the accurate comparisons of UDPGT activities.

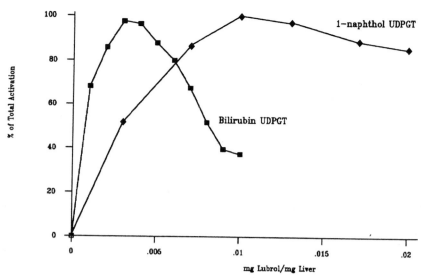

Fig. 3. Activation of human liver bilirubin and 1-Naphthol UDPGT
Activities in homogenates by titration with detergent.
Aliquots (10mg of protein) of liver homogenate were mixed with
various amounts of detergent and incubated for 30 min to $0^{\circ}C$.
Samples were then removed for assay for bilirubin and -
Napththol UDPGT by the normal assay procedures.

PURIFICATION OF UDPGTs

Purification of UDPGTs from various tissues especially human liver is
still a laborious and difficult problem. UDPGTs are membrane bound enzymes
and require 'solubilisation' by suitable non-ionic detergents prior to
purification by fairly conventional procedures (8). Unfortunately, after
solubilisation in high levels of detergent (up to 1 mg detergent/mg
microsomal protein) the transferases become very unstable. High levels of
detergent can be removed by ammonium sulphate fractionation and then the
UDPGTs can be retained in solution using conventional buffers containing
0.05% w/v detergent (9). UDPGT activities towards some hydrophobic
substrates in purified fractions can only be demonstrated after
relipidation of the preparation. For example, bilirubin UDPGT activity in
ammonium sulphate fraction of rat liver microsomes was increased 15-fold by
addition of phosphatidyl choline liposomes (10).

Separation of UDPGT isoenzymes can be achieved by ion exchange
chromatography using various DEAE-columns and chromatofocussing on Mono P
columns (Pharmacia) (see Fig 4)

Male Wistar rats (200g) were injected twice daily with clofibrate (200
mg/kg in 0.5 ml corn oil, i.p.) for 4 days, and the livers removed 18 hours
after the final injection. Microsomes were prepared from a 20% w/v
homogenate and solubilised in 0.2 M potassium phosphate buffer, pH 7.0, 5
mM 2-mercaptoethanol containing Lubrol PX such that the detergent/protein
ratio was 0.5. After centrifugation at 105,000 g for 1 hour, the
supernatant was subjected to a 20-60% ammonium sulphate fractionation. The
precipitate was dissolved in 10 mM K_2HPO_4, pH 8.0/0.005% Lubrol PX/5%

glycerol/5 mM 2-mercaptoethanol (buffer A) and dialysed against this buffer for 18 hours (100 vol x 3).

A. Dialysed material was applied to a column (70 x 2.6 cm) of DEAE cellulose (DE52 Whatman), and after eluting unbound material with 300 ml buffer A at 50 ml/hr, bilirubin UDPGT activity was eluted using a linear gradient of 0-0.3 M KCl in buffer A (Fig. 1). Selected fractions exhibiting bilirubin UDPGT activity were concentrated 25-fold by ultrafiltration (Amicon - PM10). The concentrated fractions from DEAE cellulose were applied to a column of G-25 (PD10 - Pharmacia) previously equilibrated with 75 mM TRIS acetate pH 9.3/0.05% Lubrol PX/5% Glycerol/5 mM 2-mercaptoethanol (buffer B).

B. The protein eluting in the void volume was applied to a Mono P column attached to an FPLC system (Pharmacia). After washing unbound protein with 10 ml of buffer B, Bilirubin UDPGT was eluted using a pH gradient of 9 - 6 generated by 30 ml of 10% polybuffer 96 (Pharmacia) pH 6.0 with acetic acid/0.05% Lubrol PX/5% glycerol/5 mM mercaptoethanol.Bilirubin UDPGT activity eluted at pH 7.6 and was routinely more than 90% pure as determined by S.D.S. P.A.G.E.

Therefore, UDPGTs can be purified by conventional procedures after suitable enzyme solubilisation and concentration. Further purification using CON-A-Sepharose (Pharmacia) or affinity chromatography using UDP-Sepharose (see 11). Recommendations for a successful purification procedure are listed in Table 1.

Several UDPGT isoenzymes have been purified to apparent homogeneity using these procedures and the substrate specificities of the enzymes have been determined (see Table 2). All of these UDPGTs except testosterone UDPGT are glycoproteins.

The reader will observe that certain isoenzymes are very specific and only catalyse the glucuronidation of single substrates such as bilirubin UDPGT, although xenobiotics can be designed and accepted by these enzymes (15). Other isoenzymes glucuronidate more than one substrate. Further, the glucuronidation of some substrates such as 4-nitrophenol is catalysed by several different isoenzymes (Table 2).

ANTI-UDPGT ANTIBODIES FACILITATE STUDY OF TRANSFERASE INDUCTION AND DEVELOPMENT

Induction is a process whereby the levels of an enzyme protein and hence its activity are increased by a mechanism requiring protein synthesis. These changes in transferase activity and protein levels caused by endogenous regulators or administered xenobiotics can be assessed by using antibodies raised against purified UDPGT antigens.

Antibodies raised in sheep or goats in our laboratory against purified UDPGTs have given antisera which recognise several UDPGTs, probably due to the sequence similarities of the different isoenzymes. These broad spectrum antisera, which recognise several polypeptides by Western blotting of tissue microsomal fractions have proved useful in studies of development, xenobiotic induction and genetic deficiency of UDPGTs(16). The antibodies were known to inhibit several UDPGT activities and comparison of immunostaining patterns with mobilities of purified enzymes has facilitated identification of changes of individual UDPGT levels(17) summarised in Fig 5.

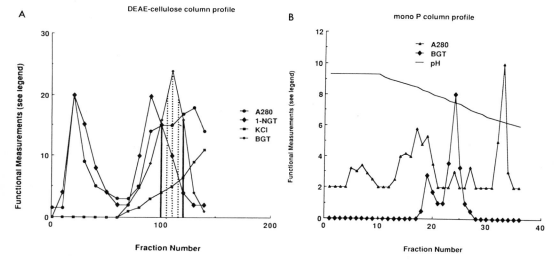

Fig. 4. Separation of UDPGT isoenzymes by DEAE cellulose and Mono P
chromatography

Table 1   Successful procedures used in the Purification of Microsomal
UDPGTS

Step Number	Procedure	Reference
1.	Isolation of Microsomal Fraction using buffered 0.25M Sucrose	9
2	Solubilisation of the UDPGTs using a non-ionic detergent	9
3	Ammonium Sulphate Fractionation	9
4.	Anion Exchange chromatography - Salt Gradient	9,10,12,13
5.	Chromatofocussing on MONO P - pH gradient elution	14
6.	UDP-Sepharose chromatography - UDPGA gradient elution	11,14

Table 2   Characteristics of UDPGT Isoenzymes Purified from Rat Liver

Isoenzyme	Subunit $(M_r)^a$	Substrate
4-Nitrophenol UDPGT	56,000	4-Nitrohenol 4-Methylumbelliferone 1-Naphthol
17ß-Hydroxysteroid UDPGT	50,000	Testosterone ß-Estradiol 1-Naphthol 4-Nitrophenol
3-α-Hydroxysteroid UDPGT (androsterone)	51,000	Androsterone Etiocholanolone Lithocholic acid
Morphine UDPGT	56,000	(-)-Morphine
Bilirubin UDPGT	54,000	Bilirubin Bilirubin monoglucuronide
Phenol UDPGT	53,000	1-Naphthol 4-Nitrophenol
5-Hydroxytryptamine UDPGT	55,000	5-Hydroxytryptamine 4-Hydroxybiphenyl 4-Methylumbelliferone
Digitoxigenin Monodigitoxide UDPGT	?	dt1 dt2
4-Hydroxybiphenyl UDPGT	?	4-Hydroxybiphenyl
Oestrone UDPGT	?	Estrone

a - Assessed by SDS-polyacrylamide gel electrophoresis

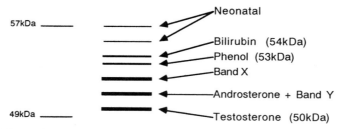

Fig. 5.   Schematic Representation of the recognition of UDPGT
enzymes by sheep anti-rat liver UDPGT antibodies
The molecular weight range of the proteins determined by
immunoblot analysis of SDS polyacrylamide gels is indicated on
the left.

Table 3   Induction of UDPGT Activities by Xenobiotics in Wistar Rat
           Liver

INDUCER	UDPGT ACTIVITIES		
	2-Aminophenol	Testostrone	Bilirubin
None	1.0	1.0	1.0
3-methylcholanthrene	4.6	0.7	1.0
Phenobarbitone	1.3	2.0	1.4
Clofibrate	0.4	0.7	1.8

The increased levels of microsomal UDPGT activities in response to
pretreatment of animals with xenobiotics can be correlated with increased
levels of individual UDPGT proteins Fig 6.

In particular, 3-Methylcholanthrene induced 1-naphthol UDP-
glucuronosyltransferase 4.5 fold and a corresponding change in the level of
transferase protein catalysing this reaction can also be observed (Fig 6).

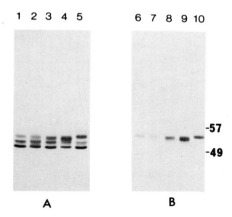

Fig. 6   Induction of UDPGT proteins by xenobiotics revealed by
          immunoblot analysis.
          Microsomes were prepared from livers of Wistar rats treated
          with various inducing agents and subjected to electrophoresis
          of SDS-polyacrylamide (7%) gels, followed by electrophoretic
          transfer to nitrocellulose. Immunostaining with anti-rat liver
          (A) or anti-rat kidney (B) UDPGT antibody. Samples were
          applied as follows: lanes 1 and 6, untreated; lanes 2 and 7,
          phenobarbital treated; lanes 3 and 8, ß-naphthoflavone treated;
          lanes 4 and 9, 3-methylcholanthrene treated; lanes 5 and 10,
          clofibrate treated. 15µg of microsomal protein was applied to
          each lane.

Therefore, using specific or broad spectrum anti-UDPGT antibodies studies of the regulation of individual UDPGT in response to exogenous or endogenous inducing agents can be undertaken.

TOPOLOGY OF UDPGTs - IMPLICATIONS FOR GLUCURONIDE SYNTHESIS AND EXCRETION IN HEPATOCYTES

UDP-glucuronosyltransferases are primarily located in hepatic endoplasmic reticulum and nuclear envelope. Fig 7 shows the route of transport of substrates and excretion of glucuronides indicating the membrane barriers and acqueous compartments which confront the passage of hydrophilic and hydrophobic compounds.

The topological location of UDPGTs is important to the understanding of restrictions of glucuronide formation and the limits imposed upon glucuronide excretion. All microsomal UDPGTs have been demonstrated to be latent enzymes and many experiments by different laboratories have indicated that the active sites of the transferases may be located in the lumen of the E.R.(18).

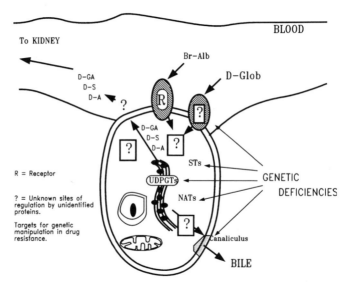

Fig. 7. A Model Hepatocyte indicating possible transporters required for the uptake of aglycones and excretion of conjugates. D, drug; GA, glucuronide; S, sulphate; A, acetylate; ST, sulphotransferase; NATs, N acetyltransferases.

The transmembrane topology of UDPGTs has been examined using proteases with antibodies(19) and complemented by computer bases analysis of DNA sequences(20,21,22). The picture that emerges is of the transferases slightly exposed on the cytoplasmic surface on the membrane joined by a transmembrane region to the majority of the protein, including the active site on the lumen of the endoplasmic reticulum (Fig 8).

These topological studies indicate that glucuronidation catalysed by UDPGTs takes place in the lumen of the endoplasmic reticulum behind a membrane barrier. These results would explain the latency of UDPGTs and their activation by detergents. The location of these enzymes within the E.R. is of great significance for the metabolic disposition of excretion of glucuronides, and these membrane barriers may be rate-limiting to the facilitated elimination of xenobiotic metabolites.

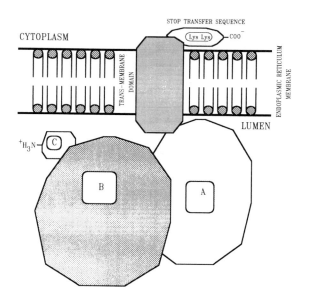

Fig. 8.   A Model of the Transverse Topology of UDPGT in the Endoplasmic Reticulum.
A - Conserved UDP-glucuronic acid binding region
B - Variable substrate binding region
C - Conserved region

CONCLUDING REMARKS

The information described here illustrates the importance of accurate assay of microsomal UDPGTs and the difficulties inherent in purification and characterisation of the membrane bound enzyme. However, the investment of time and effort to obtain these proteins provides dividends in that antibody produces can be obtained to study the regulation, topology and cellular function of UDP-glucuronosyltransferases.

REFERENCES

1.   Williams, R.T. "Detoxication Mechanisms" 2nd Ed. Chapman & Hall, London 1959.
2.   Dutton, G.J. "The Biosynthesis of Glucuronides in Glucuronic Acid", G J Dutton, ed., Academic Press, New York, pp 185-299 (1966).
3.   Dutton, G.J. "Glucuronidation of Drugs and Other Compounds", CRC Press, Boca Raton, Florida (1980).
4.   Dutton G.J. and Burchell, B. Newer aspects of glucuronidation Prog. Drug. Metab. 2: 1-70 (1977).
5.   Graham, A.B. and Wood, G.C., Factors affecting the response of microsomal UDP-glucuronyltransferase to membrane perturbants. Biochem. Biophys. Acta 311: 45-50 (1973).
6.   Arion, W.J., Measurement of Intactness of rat liver Endoplasmic Reticulum, Meth. Enzymol. 174:58 (1990).

7. Burchell, B., Coughtrie, M.W.H., Jackson, M.R., Shepherd, S.R.P., Harding, D. and Hume, R., Genetic deficiency of bilirubin glucuronidation in rats and humans, Molec. Aspects Med. 9: 429-455 (1987).

8. Burchell, B., Identification and purification of multiple forms of UDP-glucuronosyltransferase, Rev. Biochem. Toxic. 3: 1-39 (1981).

9. Burchell, B., Studies on the Purification of rat liver UDP-glucuronyltransferase Biochem. J. 161: 534-549 (1977).

10. Burchell, B., Isolation and purification of bilirubin UDP-glucuronyltransferases from rat liver, FEBS Lett. 111: 131-135 (1980).

11. Burchell, B. and Weatherill, P.J., Purification of 4-nitrophenol UDP-glucuronyltransferase from rat liver, Meth. Enzym. 77: 169-177 (1981).

12. Bock, K.W., Josting, D., Lilienblum, W. and Pfeil, H. Purification of rat liver microsomal UDP-glucuronyltransferase, Eur. J. Biochem. 98: 19-26 (1979).

13. Tukey, R.H. and Tephly, T.R. Estrone and 4-nitrophenol UDP-glucuronyltransferases (Rabbit Liver) Meth. Enzymol. 77: 177-188 (1981).

14. Falany, C.N. and Tephly, T.R., Separation purification and characterization of three isoenzymes of UDP-glucuronyltransferase from rat liver microsomes, Arch. Biochem. Biophys 227: 248-258 (1983).

15. Fournel-Gigleux S., Shepherd, S.R.P., Carre, M-C., Burchell, B., Siest, G. and Caubere, P., Novel inhibitors and substrates of bilirubin UDP-glucuronsyltransferase, Eur. J. Biochem 183: 653-659 (1989).

16. Coughtrie, M.W.H., Burchell, B., Shepherd, I.M. and Bend, J.R., Defective induction of phenol glucuronidation by 3-methylcholanthrene in Gunn rats is due to the absence of a specific UDP-glucuronosyltransferase isoenzyme, Mol. Pharmacol. 31: 585-591 (1987).

17. Coughtrie, M.W.H., Burchell, B., Leakey, J.E.A. and Hume, R., The inadequacy of perinatal glucuronidation Molec. Pharmac. 34: 729-735 (1988).

18. Hallinan, T. and De Brito, A.E.R., Topology of endoplasmic reticular enzyme systems and its possible regulatory significance. in: "Hormone and Cell Regulation", Vol 5, pp 73-95 J E Dumont and J. Nunez, eds., Elsevier, Amsterdam (1981).

19. Shepherd, S.R.P., Baird, S.J., Hallinan, T. and Burchell, B., An investigation of the transverse topology of bilirubin UDP-glucuronosyltransferase in rat hepatic endoplasmic reticulum, Biochem. J. 259: 617-620 (1989).

20. Jackson, M.R. and Burchell, B. Full length coding sequence of rat liver adrosterone UDPGT-glucuronosyltransferse cDNA and comparison with other members of this gene family, Nucl. Acids Res. 14: 779-795 (1986).

21. Mackenzie, P. Rat liver UDP-Glucuronosyltransferase cDNA. Ssequence and Expression of a cDNA encoding a phenobarbital inducible form. J. Biol. Chem 261: 6119-6125 (1986).

22. Iyanagai, T., Haniv, M., Sagawa, K. Fujii-Kuriyama, Y., Watanabe, S., Shively, J.E. and Anan, K.F., Cloning and characterisation of a cDNA encoding 3-methylcholanthrene inducible rat mRNA for UDP-glucuronosyltransferse, J Biol. Chem. 261: 15607-15614 (1986).

MOLECULAR CLONING, EXPRESSION AND GENETIC DEFICIENCIES OF UDP-

GLUCURONOSYLTRANSFERASES*

Brian Burchell

Department of Biochemical Medicine
Ninewells Hospital and Medical School
The University
Dundee, DD1 9SY, Scotland

INTRODUCTION

How do we study the specificity of human liver glucuronidation?

It is extremely difficult to routinely obtain human liver for drug-screening and other laboratory studies. Therefore, we need to devise alternative methods to study human glucuronidation.

Cloning and expression of cloned genes coding for GTs provides protein structures and allows assessment of enzymological substrate specificity for rational design of new drugs.

Inhibition kinetic analysis of UDPGT activities (1) and immunoblot analysis of human liver microsomes (2) has indicated the presence of multiple forms of UDPGTs. Therefore we have used a molecular biological approach to facilitate the study of the human UDPGTs.

MOLECULAR CLONING OF UDPGTs

The molecular cloning of rat liver UDPGTs started in the early 1980s with rather laborious techniques using enriched poly A^+ mRNA sources to attempt to distinguish individual clones from plasmid libraries. More sophisticated methods using antibody screening of bacteriophage λGT11 cDNA expression libraries were introduced following the invention of this procedure by Young and Davis (3). In 1984, both methods were successfully used to clone UDPGTs (4,5). The problem that remained was to confirm the identity of the cloned UDPGTs and reveal the function of the transferases that were encoded by the cDNAs. The identity of the UDPGTs was confirmed by hybrid selection analysis (5) and subsequently by comparison with protein sequences (see Table 1). The function encoded by the first cloned

---

*This paper is comprised of transcribed notes from a lecture given at the NATO Workshop and does not attempt to comprehensively review this field of research

Molecular Aspects of Monooxygenases and Bioactivation of Toxic Compounds
Edited by E. Arınç *et al., Plenum Press, New York,*

Table 1   Identification of Cloned UDPGTs (1986)

Methodology	Reference
Expression in COS cells in culture	7
Expression in vivo/genetic defect	6
Comparison with amino acid sequence	8, 9

UDPGT cDNAs was revealed by using the cDNAs in Northern blot analysis of mRNA preparations isolated from xenobiotic pretreated and androsterone UDPGT activity deficient rats (6). The studies revealed that one of the UDPGT cDNAs encoded a transferase specific for catalysis of androsterone glucuronidation (6). At the same time, Mackenzie reached a similar conclusion by expression of the same cDNA in cell cultures (7). Subsequently, several rat UDPGT cDNAs have been cloned and identified (see Table 2).

Table 2   Substrate Specificity of rat UDPGT encoded by cloned cDNAs

cDNA Clones	Substrates	References
pUDPGT4	androsterone	6,7
RLUG23	etiocholanolone	
	lithocholic acid	
pUDGT3	testosterone	9,10
	dihydrotestosterone	
4NPGT	4 nitrophenol	8
UDPGTK39	1-naphthol	11
	4-methylumbelliferone	
	4-nitrophenol	
	4-ethylphenol	
	4-methylphenol	
pUDGT2	4-methylumbelliferone	10
	chloramphenicol	
	4-hydroxybiphenyl	
	testosterone	
	dihydrotestosterone	
	ß-estradiol	

The rat UDPGT cDNAs and antibodies have been used to cross species and clone human liver UDPGT cDNAs (12). Computerised analysis and comparison of UDPGT cDNA sequences (after translation to amino acid sequences) has allowed a number of predictions concerning the synthesis, structure and properties of UDPGTs, which can be confirmed by biochemical analysis.

A more detailed examination of the sequence data has enabled predictions that all of the cDNAs isolated so far, with the exception of that encoding 17ß-hydroxysteroid UDPGT, possess at least one putative asparagine-linked glycosylation consensus sequence (Asn-X-Ser/Thr) in the translated protein. Treatment of purified proteins with endoglycosidases has confirmed the existence of small oligosaccharides attached to the transferases. Hydropathy profiles indicate that all UDPGTs possess hydrophobic (cleaved) signal sequences (Fig. 1).

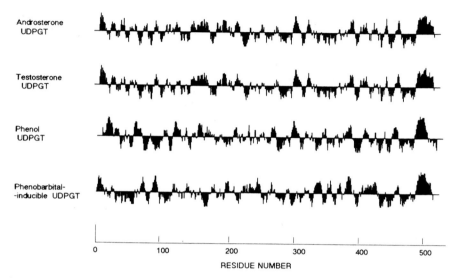

Androsterone UDPGT

Testosterone UDPGT

Phenol UDPGT

Phenobarbital-
-inducible UDPGT

0   100   200   300   400   500

RESIDUE NUMBER

Fig. 1.   Hydrophobicity profile comparison of four rat UDPGT
             protein sequences.
             The amino acid sequences were deduced from translation of
             the open reading frames of the cDNA sequences and
             analysed by computer using the Kyte and Doolittle
             programme with a 15 residue window.

Comparison of N-terminal protein sequences obtained by sequencing of purified UDPGT proteins with N-terminal sequences derived from UDPGT cDNAs has confirmed the existence of the cleaved signal sequence (see Fig. 2).

Examination of the synthesis of a UDPGT by in vitro transcription and translation has provided evidence to show that the newly synthesised protein is initially reduced in size and increased in size by the presence of dog pancreas microsomes (13). Presumably these observations were due to cleavage of the peptide signal and glycosylation of the nascent protein as predicted by sequence analysis. Indeed, inhibition of glycosylation of UDPGTs in cell cultures by addition of tunicamycin shows the reduced size of the newly synthesised transferases (11,12). Deglycosylation does not affect the activity (14) or specificity of the transferases (15).

<----------- Cleaved signal sequence ? ------------> <---- Mature UDPGT

-20 -15 -10 -5 - 1 1 5 10

Pre-Androsterone M P R K W I S A L F L L Q I S Y C F K S G H C G K V L V W P M D F

Androsterone-GT [a] G K V L V W P M D F

RLUG 38 M P G K W I S A L L L L Q I S C C F Q S G N C G K V L V W P M D F

Testosterone-GT [a] G K V L V W P M D F

Pre- Phenol GT[b] M A C L L P A A R L P A G F L F L V L W G S V L G D K L L V V P Q D G

Phenol GT[b] D K L L V V P Q D G

Fig. 2. Signal Processing or Pre-UDPGTs.
The deduced N-terminal amino acid sequence is shown above the protein N-terminal obtained by protein sequencing for three different UDPGTs. (a) see reference 9 (b) see reference 8.

Further, studies of the hydrophobicity profiles (Fig. 1) show that although there are marked differences in derived amino acid sequences between different UDPGTs, they appear to be very closely related structurally.

A highly hydrophobic sequence at the far end of the COOH-terminal region, is located between two highly charged areas. This is characteristic of the halt-transfer signals of transmembrane proteins (16). Indeed, lysine residues at positions -3, -4, -5 from the carboxylterminal are important in retention of UDPGTs in this orientation in the endoplasmic reticulum (17). The COOH-terminal half of the sequences, in general, contains the majority of the sequence homology between the different UDPGTs, which is suggestive of some conserved function - perhaps UDPGA binding or some form of conformational requirement. Further support for this concept has recently been provided by Mackenzie (18).

The major region of dissimilarity, even between enzymes showing more than 80% identity, is concentrated in the N-terminal part between residues 60-120, which presumably provides the specificity for aglycone substrate binding.

A summary of the general structural map of UDPGTs studied is shown in Fig. 3.

EXPRESSION OF UDPGT ACTIVITIES IN CELL CULTURES

Cloned UDPGT cDNA can be expressed in cell cultures to facilitate the study of the substrate specificity of the enzymes. The cDNA clones are inserted into a plasmid expression vector such as pKCRH2 (11,12). This recombinant vector containing the UDPGT cDNA is transfected into COS-7 cells expressing the large T antigen (see Fig. 4) using the calcium phosphate/glycerol shock procedure or lipofection.

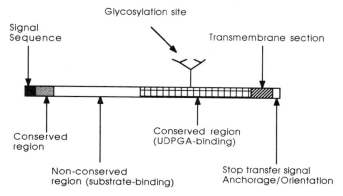

Fig. 3.   Structural Overview of UDPGT Proteins

Prominent features of the UDPGT proteins illustrated has been deduced
from studies of their synthesis, sequences and molecular biological
manipulations of the structure.

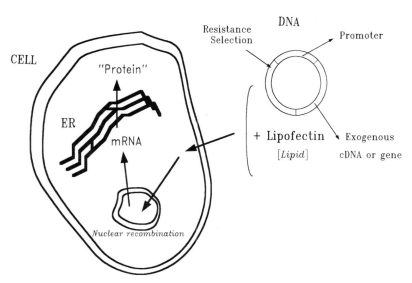

Fig. 4.   Gene Transfection and Expression in Cell Cultures
          Recombinant DNA in a vector system can be transfected into
          tissue culture cells using lipofectin, calcium phosphate/
          glycerol shock or electroporation. Promoter present in the
          vectors drive the expression of the foreign gene.

The UDPGT cDNA is transiently transcribed under the control of the SV40 promoter in the vector. The newly synthesised transferases can be detected using antibodies or by assay of UDPGT activities (11,12). Thus the substrate specificity of the enzymes can be assessed. For example, a crude analysis of the dimensions and electronic configurations of a series of naphthols by a cloned expressed human phenol UDPGT is shown in Table 3.

Table 3   Human phenol UDP-glucuronosyltransferase activity towards Naphthols

Substrate	UDP-Glucuronosyl-transferase activity (pmol/min/mg protein)		
	Vector Transfected COS-7 cells	HLUGP1 Transfected COS-7 cells	Human Liver Microsomes
1-naphthol	0.3	23.0 (1.0)	177.5 (1.00)
2-naphthol	0.2	15.3 (0.66)	210.0 (1.18)
5,6,7,8 tH 1-naphthol	N.D	4.4 (0.19)	139.0 (0.78)
R1,2,3,4 tH 1-naphthol	N.D	0.1 (0.00)	6.3 (0.03)
S1,2,3,4 tH 1-naphthol	N.D.	0.03 (0.00)	1.5 (0.01)

The data shown are the mean of measurements of at least two different cellular or microsomal preparations. The figures in parenthesis show the ratio of glucuronidation of each substrate relative to the rate of glucuronidation of 1-naphthol.

Similarly, the substrate specificity of a human UDPGT apparently encoding a transferase catalysing the glucuronidation of hyodeoxycholic acid was examined (13). Transient expression of this enzyme in COS-7 cells was poor and therefore we have obtained a clone V-79 cell line in which the H25 UDPGT cDNA encoding hyodexycholate UDPGT is stably expressed. This procedure leads to expression of the transferase protein and activity at more than 10-fold higher levels in these cell lines and therefore a better opportunity to assess the substrate specificity of the UDPGTs.

GENETIC DEFICIENCY OF UDPGTs IN RATS AND HUMANS

A Defect of Androsterone UDPGT in rats

A mutant rat strain which has a defect of androsterone UDPGT activity was reported in 1979 (19) and observed to be present in the Wistar rat colonies around the world (20,21). UDPGT activities toward several substrates - bilirubin, testosterone and 'phenols' were determined to be 'normal' i.e. comparable to control Wistar rat populations (19,21,22). Molecular biological examination of this genetic defect in LA Wistar rats revealed that the genetic lesion was caused by a gene deletion (20), preventing mRNA production (20) and synthesis of androsterone UDPGT protein (20,21,22) see Fig. 5.

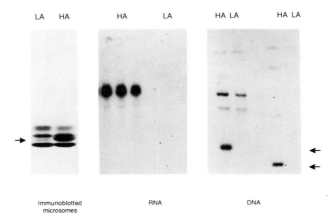

| LA | HA | | HA | LA | | HA | LA | | HA | LA |

Immunoblotted microsomes   RNA   DNA

Fig. 5.   The Molecular Basis of the Inherited Deficiency of Androsterone UDP-glucuronosyltransferase in Wistar rats A major UDP-glucuronyltransferase in rat liver (51 kDa), corresponding to androsterone glucuronidating activity, has been identified by immunoblot analysis. This isoenzyme is absent from Wistar rats exhibiting the low androsterone (LA) UDPGT activity phenotype. Northern blot analysis of total RNA from normal and androsterone glucuronidation deficient Wistar rats demonstrated that the mRNA encoding this protein was not synthesised. Differences in restriction fragment length observed on Southern blotting of genomic DNA from LA Wistar rats indicate that this inherited deficiency is the result of a deletion in the androsterone UDP-glucuronyltransferase gene.

Defects of UDPGTs in Gunn rats

This hyperbilirubinemic mutant strain of Wistar rat was originally described by Gunn (23). Gunn rats exhibit a lifelong severe unconjugated hyperbilirubinemia, characterized by plasma bilirubin levels of between 120 and 350 uM (6-15 mg/dl) which appears immediately after birth and does not respond to treatment with phenobarbitone or glutethimide (see 24). Bilirubin production rate and hepatic function tests are normal, although plasma bilirubin levels are elevated. Subsequently, it was reported that Gunn rats were unable to form bilirubin glucuronide (25,26). Bilirubin conjugates are not detected in bile by conventional methods (27). The disease is inherited as an autosomal recessive trait, and the animals frequently develop neurological problems associated with bilirubin encephalopathy (28). The Gunn rat has been considered as a model for human Crigler-Najjar Type I syndrome as a result of the total absence of bilirubin UDPGT activity (see Table 4). However, the defect in the Gunn rat is not restricted to bilirubin glucuronidation. UDPGT activity towards many other substrates, in particular planar phenols such as 2-aminophenol, 1-naphthol and 4-nitrophenol, is significantly decreased (to 5-20% of Wistar levels) whereas the glucuronidation of other compounds such as - morphine and testosterone is essentially normal (29,30,31,32). Comparisons of UDPGT activities in hepatic microsomes obtained from xenobiotic-treated

Gunn and Wistar rats shows that the defective UDPGT activities were not induced in Gunn rats. Bilirubin UDPGT activity was not detected in Gunn rats, nor was it induced to appear by treatment of the rats with the specific inducer of this UDPGT, clofibrate (31). UDPGT activity towards 2-aminophenol in Gunn rat liver is very low and was not induced by 3-methylcholanthrene (31). However, testosterone UDPGT activity in Gunn rat liver was expressed at a similar level to that in Wistar rat liver, and induced 2-fold by phenobarbital treatment (31). The cytochrome P-450 dependent monooxygenase activities, 7-ethoxycoumarin-0-deethylase and aryl hydrocarbon hydroxylase were induced some 10-fold in both Gunn and Wistar rat liver microsomes by 3-methylcholanthrene pretreatment (31).

Antibodies raised against purified UDPGTs have been used to search for the existence of UDPGT proteins in Gunn rat liver microsomes. Liver microsomes from untreated or xenobiotic-treated liver microsomes were analysed with two different antibodies following Western blotting. Bilirubin or phenol UDPGT proteins were not detected in Gunn rat microsomes, indicating that these proteins were apparently absent from microsomes and the liver (31). It was likely that the residual diminished glucuronidation of planar phenol was catalysed by at least one of the many other isoforms, eg testosterone UDPGT (33). Indeed, 1-naphthol inhibited testosterone UDPGT activity (and vice versa) in Gunn rat liver microsomes (34). More recently, we have also examined congenic strains of rat homozygous for the jaundiced locus (35) and similar results for defective levels of enzyme activities and loss of microsomal UDPGT proteins were observed, although this strain also exhibits the additional genetic lesion observed in LA rat (36). Confirmation of this work has been obtained by two independent laboratories using different polyclonal (37) or monoclonal antibodies (38).

Analysis of the phenol UDPGT gene function in Gunn rat DNA

Earlier preliminary studies have suggested that a clofibrate-inducible mRNA, which may encode bilirubin UDPGT was apparently absent from Gunn rat liver (34). More extensive examination of phenol UDPGT mRNA in Gunn rat livers has recently been performed due to the availability of a specific phenol UDPGT cDNA probe for this analysis (39). The results of examination of mRNA levels in three tissues from untreated animals and 3-methylcholanthrene-treated Gunn rats show that the phenol UDPGT mRNA was present in intestine, kidney and liver in drastically reduced amounts (20% of Wistar levels) and that these mRNA species were inducible by 3-methylcholanthrene, although to a more limited extent than in Wistar liver (37,39).

Recent cloning of the phenol UDPGT cDNA encoding a mutant Gunn rat liver enzyme has shown the existence of a point mutation in the mRNA, which may allow synthesis of a truncated protein 115 residues shorter at the COOH terminal end (37). However, the truncated protein was not detectable. Further a truncated protein was not detectable by immunoprecipitation of protein products from in vitro translation of Gunn rat liver poly A + RNA using an anti UDPGT polyclonal antibody. This latter result casts a doubt on this type of mutation, although the antibody may not recognise the truncated protein or this protein may not be made at all. The mutation proposed by Iyanagi's team (37) must be confirmed by examination of genomic DNA and the loss of the Bst N1 restriction site.

DNA isolated from Gunn and Wistar rat livers was digested with restriction endonucleases, run on agarose gels, transferred to nylon membrane and hybridized with (32P)-labelled phenol UDPGT cDNA. Comparison of the restriction fragments obtained from Gunn and Wistar rats yielded identical results and showed that the phenol UDPGT was encoded by a single unique gene (37,39).

What is the Nature of the Genetic Lesion in the Gunn Rat?

Bilirubin and phenol UDPGT proteins (and possibly others not yet detectable, e.g. UDPGT catalysing the glucuronidation of digitoxigenin monodigitoxoside) are apparently absent, accompanied by a reduction in the levels of specific mRNAs down to approx 20% of those in Wistar rats. The single unique phenol UDPGT gene appears to be present and apparently normal as judged by limited RFLP analysis.

Mutant phenol and bilirubin UDPGT genes have been observed to be cosegregated and are probably therefore closely linked on the same chromosome (40), although the two enzyme levels are differentially regulated during development (41) and by treatment of animals with xenobiotics (31).

A single genetic defect accounting for the loss of two different gene products as apparently occurs in the Gunn rat could be explained by a lesion which affects the transcription of certain UDPGT genes or a decrease in the stability of UDPGT mRNAs which prevent their translation (39). There are examples where a single genetic lesion at a regulatory site simultaneously affects transcription or translation of different proteins (42). This type of mutation could explain the existing description of this genetic deficiency. Alternatively, a mutation within the coding region of an exon of the phenol UDPGT gene coding region should be common to more than one defective gene and this may be explained by differential splicing using common exons to produce different protein products form single gene.

Human genetic deficiency of UDPGT cause hyperbilirubinemia

Gilbert's syndrome described by Gilbert and Lereboullet in 1901 is characterized by a mild chronic, unconjugated hyperbilirubinemia. This fairly benign disorder is found in approximately 2-5% of the population (28). The pathogenesis of unconjugated hyperbilirubinemia in Gilbert's syndrome is not completely understood (43). However a defective glucuronidation is apparently the major abnormality found (44). Decreased amounts of bilirubin diglucuronide and increased levels of bilirubin monoglucuronide were found in bile in parallel to the decreased transferase activity (45).

The molecular biochemistry of this clinical problem has been investigated. There are no obvious indications of impaired drug oxidation, acetylation of sulfation (46). This disease is the most interesting human genetic deficiency in terms of drug glucuronidation, because of the apparent prevalence of this problem within the population. Crigler-Najjar syndrome is a familial form of severe unconjugated hyperbilirubinemia (47) which has been observed in the Americas, Asia, Europe and Africa. Infants often develop severe neurological damage from bilirubin encephalopathy (kernicterus). More recently Arias et al (48) have categorized patients into two types based on (a) the level of unconjugated hyperbilirubinemia (Type I, greater than 300 uM or 20 mg/dl and Type II, below this division) and (b) the dramatic response of Type II patients to barbiturate or glutethimide therapy. This clear difference in response to barbiturate therapy between Type I and Type II patients suggests a fundamental difference in the molecular basis of the metabolic defect (see Table 4).

Table 4   Genetic Defects of UDPGTs Causing Hyperbilirubinaemia

Deficiency	Serum Bilirubin (uM)	Bilirubin Glucuronides in Bile	Response to Barbiturates	Inheritance
Gunn Rat	>200	None	-	Autosomal Recessive
Human				
Crigler-Najjar Syndrome (Type I)	>300	None	-	Autosomal Recessive
Crigler-Najjar Syndrome (Type II)	<300	Low levels of monoglucuronide	+	Autosomal Dominant?
Gilbert's Patients	20-60	Mainly monoglucuronide	+	?

Patients with Crigler-Najjar syndrome Type I have a complete defect in hepatic bilirubin glucuronidation. However, the genetic lesion would appear to affect other UDPGT isoenzymes as glucuronidation of other aglycones was reduced (24). In particular, there was a defect in steroid glucuronidation and in glucuronidation of N-acetyl-p-aminophenol (24) or menthol (48) although there is considerable variation in the results reported.

Recently, UDPGT activities measured in liver microsomes from Crigler-Najjar patients were determined for three aglycone substrates, after optimal activation with detergent. These enzyme activities were also determined in three 'normal' postmortem live samples from infants of the same age. Bilirubin UDPGT activity was determined by HPLC analysis of the glucuronides formed (see Fig. 6).

No UDPGT activity towards bilirubin was detectable in liver homogenates from Crigler-Najjar type I patients, whereas activity towards 1-naphthol as substrate was determined to be same value as the control. Our studies of Crigler-Najjar type II patients showed the synthesis of low levels of bilirubin monoconjugates catalysed by liver homogenates and again levels of 1-naphthol UDPGT activity were the same as controls. The results suggested that the genetic defect may affect the affinity of UDPGA binding to bilirubin UDPGT. Immunoblot analysis of Crigler-Najjar liver microsomes provided the first insights into the molecular basis if this genetic disease (34,36) The surprising feature of the immunostaining pattern of the Crigler-Najjar liver microsomes was the considerable increase in immunostaining of the 52kDa polypeptide. This UDPGT protein induction may well be related to the measured increase in the microsomal testosterone UDPGT activity (34). The Crigler-Najjar infant was treated with barbiturates which may explain the results obtained (the testosterone UDPGT isoenzyme is induced by phenobarbitone in rat liver). The hyperbilirubinemia in this single Crigler-Najjar Type I patient was certainly due to loss of bilirubin UDPGT activity, but the bilirubin UDPGT isoenzyme protein did not appear to be absent (36). The routine barbiturate treatment of severely jaundiced infants may induce certain UDPGT isoenzymes such as testosterone UDPGT. If this isoenzyme also catalyses the glucuronidation of phenols, then defects of phenol glucuronidation may be masked, because there are no suitable barbiturate-treated 'normal' controls.

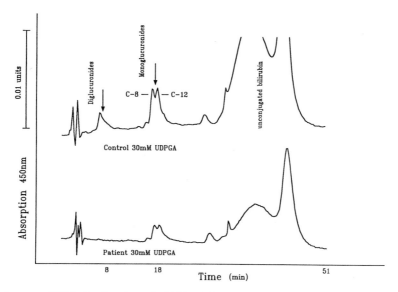

Fig. 6.   HPLC Analysis of Bilirubin Glucuronide Production By Crigler-Najjar Type II Liver Homogenate.
Top graph   Normal adult liver
Bottom graph   Crigler-Najjar Type II patient liver.
Separation was performed on a μBondapak (Waters) C18 reverse phase column using an acetonitrile/water gradient system (unpublished work).

CONCLUDING REMARKS

There is no doubt that a molecular approach to understand the biological mechanisms of glucuronidation is helping to elucidate the substrate specificity and function of individual enzymes. The availability of cloned cDNAs and gene structures will enable studies of the cellular regulation of UDPGTs during development and determination of genetic lesion associated with the hyperbilirubinemias.

Further, the antibody and cDNA probes will facilitate the study of the cellular biology of substrate entry, glucuronide formation and excretion from the endoplasmic reticulum and the liver cell. The ability to manipulate levels of transferase and permease expression in the cell will provide further information about hepatic anion transport and the pathophysiology of bile secretion.

ACKNOWLEDGEMENTS

I wish to thank the British Council for enabling me to participate in this workshop and the M.R.C. and Wellcome Trust for research support.

REFERENCES

1. Bock, K.W., Lilienblum, W. and von Bahr, C. Studies of UDP-glucuronyltransferase activities in human liver microsomes. Drug Metab. Dispos. 12: 93-97 (1984).

2. Jackson, M.R., McCarthy, L.R., Harding, D., Wilson, S., Coughtrie, M.W.H. and Burchell, B. Cloning of a human liver microsomal UDP-glucuronosyltransferase cDNA. Biochem. J. 242: 581-588 (1987).

3. Young, R.A. and Davis, R.W. Efficient isolation of genes using antibody probes. Proc. Natl. Acad. Sci. (USA) 80: 1194-1198 (1983).

4. Mackenzie, P., Gonzalez, F.J.. and Owens, I.S. Cloning and characterisation of DNA complementary to rat liver UDP-glucuronosyltransferase mRNA. J. Biol. Chem. 259: 12153-12160 (1984).

5. Jackson, M.R., McCarthy, L.R., Corser, R.B., Barr, G.C. and Burchell, B. Cloning of cDNAs coding for rat hepatic microsomal UDP-glucuronyltransferases. Gene 34: 147-153 (1985).

6. Jackson, M.R. and Burchell, B. The full length coding sequence of rat liver androsterone UDP-glucuronosyltransferase cDNA and comparison with other members of this gene family. Nucl. Acids Res. 14: 779-795 (1986).

7. Mackenzie, P. Rat liver UDP-glucuronosyltransferase. cDNA sequence and expression of a form glucuronidating 3-hydroxyandrogens. J. Biol. Chem. 261: 14112-14117 (1986).

8. Iyanagi, T., Haniu, M., Sagawa, K., Fujii-Kuriyama, Y., Watanabe, S. Shively, J.E. and Anan, K.F. Cloning and characterisation of cDNA encoding 3-methylcholanthrene-inducible rat mRNA for UDP-glucuronosyltransferase. J. Biol. Chem. 261: 15607-15614 (1986).

9. Harding, D., Wilson, S.M., Jackson, M.R., Burchell, B., Green, M.D. and Tephly, T.R. Nucleotide and deduced amino acid sequence of rat liver 17ß-hydroxy steroid UDP-glucuronosyltransferase. Nucl. Acids Res. 15: 3936 (1987).

10. Mackenzie, P. Rat liver UDP-glucuronosyltransferase. Identification of cDNA encoding two enzymes which glucuronidate testosterone, dihydrotestosterone and ß-estradiol. J. Biol. Chem. 262: 9744-9749 (1987).

11. Jackson, M.R., Fournel-Gigleux, S., Harding, D. and Burchell, B. Examination of the substrate specificity of cloned rat kidney phenol UDP-glucuronosyltransferase expressed in COS-7 cells. Molec. Pharmac. 34: 638-642 (1988).

12. Harding, D., Fournel-Gigleux, S., Jackson, M.R. and Burchell, B. Cloning and substrate specificity of a human phenol UDP-glucuronosyltransferase expressed in COS-7 cells. Proc. Natn. Acad. Sci. U.S.A. 85: 8381-8385.

13. Fournel-Gigleux, S., Jackson, M.R., Wooster, R. and Burchell, B. Expression of a human liver cDNA encoding UDP-glucuronosyltransferase catalysing the glucuronidation of hyodeoxycholic acid in cell culture. FEBS Lett. 243: 119-122 (1989).

14. Shepherd, S.R.P., Baird, S.J., Hallinan, T. and Burchell, B. An investigation of the transverse topology of bilirubin UDP-glucuronyltransferase in rat hepatic endoplasmic reticulum. Biochem. J. 259: 617-620 (1989).

15. Mackenzie, P.I. The effect of N-linked glycosylation on the substrate preferences of UDP-glucuronosyltransferases. B.B. Res. Commun. 166: 1293-1299 (1990).

16. Sabatini, D.D., Kreibich, B., Morimoto, T. and Adesnik, M. Mechanisms for the incorporation of proteins into membranes and organelles. J. Cell Biol. 92: 1-22 (1982).

17. Nilsson, T., Jackson, M. and Peterson, P.A. Short cytoplasmic sequences serve as retention signals for transmembrane proteins in the endoplasmic reticulum Cell 58: 707-718 (1989).

18. Mackenzie, P.I. Expression of Chimeric cDNAs in cell culture defines a region of UDP-glucuronosyltransferases involved in substrate selection. J. Biol. Chem. 265: 3432-3435 (1990).

19. Matsui, M., Nagai, F. and Aoyagi, S. Strain differences in rat liver UDP-glucuronyltransferase activity towards androsterone. Biochem.J. 179: 483-487 (1979).

20. Corser, R.B., Coughtrie, M.W.H., Jackson, M.R. and Burchell, B. The molecular basis of the inherited deficiency of androsterone UDP-glucuronyltransferase in Wistar rats. FEBS Lett. 213: 448-452 (1987).

21. Green, M.D., Falany, C.N., Kirkpatrick, R.B. and Tephly, T.R. Strain differences in purified rat hepatic 3a-hydroxysteroid UDP-glucuronosyltransferase Biochem. J. 230: 496-534 (1985).

22. Matsui, M. and Nagai, F. Genetic deficiency of androsterone UDP-glucuronosyltransferase activity in Wistar rats is due to the loss of enzyme protein Biochem.J. 234: 139-144 (1986).

23. Gunn, C.H. Hereditary alcoholuric jaundice in rats. J. Hered 29: 137-139 (1938).

24. Schmid, R. and McDonagh, A.F. Hyperbilirubinaemia in: "The Metabolic Basis of Inherited Disease", pp 1221-1257, J.B. Stanbury, J.B. Wyngaarden and D.S. Fredrickson, eds., McGraw Hill, New York (1978).

25. Lathe, G.H. and Walker M. An enzymatic defect in human neonatal jaundice and in Gunn's strain of jaundiced rats. Bichem. J. 67: 9P (1957).

26. Schmid, R., Axelrod, J. Hammaker, L. and Swarm, R.L. Congenital jaundice in rats due to a defect in glucuronide formation. J. Clin. Invest. 37: 1123-1130 (1958).

27. Blanckaert, N. Defective bilirubin conjugation in Crigler-Najjar disease and Gunn rats in: "Familial Hyperbilirubinaemia", pp 51-63, L. Okolicsanyi, ed., Wiley, Chichester (1981).

28. Odell, G.B. and Childs, B. Hereditary hyperbilirubinaemias in: "Progress in Medical Genetics", Vol. 4, pp 103-134, A.G. Steinberg, A.G. Bearn, A.G. Motulsky and B. Childs eds. W. B. Saunders, Philadelphia, U.S.A. (1980).

29. ackenzie, P. and Owens, I.S. Differences in UDP-glucuronosyltransferase activities in congenic inbred rats homozygous and heterozygous for the jaundiced locus. Biochem. Pharmac 32: 3777-3781 (1983).

30. Boutin, J.A., Antoine, B., Fournel, S. and Siest, G. Heterogeneity of hepatic microsomal UDP-glucuronosyltransferase activities: use and comparison of differential inductions in some mammalian species. Comp. Biochem. Physiol. 87B: 513-522 (1987).

31. Coughtrie, M.W.H., Burchell, B., Shepherd, I.M. and Bend, J.R. Defective induction of phenol glucuronidation by 3-methylcholanthrene in Gunn rats is due to the absence of a specific UDP-glucuronosyltransferase isoenzyme. Molec. Pharmacol. 31: 585-591 (1989).

32. Raza, H., Levine, W.G., Roy Chowdhury, N. and Roy Chowdhury, J. Microsomal azoreduction and glucuronidation in the metabolism of dimethylaminoazobenzene by rat liver. Xenobiotica 17: 669-677 (1987).

33. Falany, C.N. and Tephly, T.R. Separation, purification and characterization of three isoenzymes of UDP-glucuronyltransferase from rat liver microsomes. Archs Biochem. Biophys. 227: 248-258 (1983).

34. Burchell, B., Coughtrie, M.W.H., Jackson, M.R., Shepherd, S.R.P., Harding, D. and Hume R. Genetic deficiency of bilirubin glucuronidation in rats and humans. Molec. Aspects Med. 9: 429-455 (1987).

35. Leyten, R., Vroemen, J.P.A.M., Blanckaert, N. and Heirwegh, K.P.M. The congenic normal R/APfd and jaundiced R/APfd-h/j rat strains: a new animal model of hereditary non-haemolytic unconjugated hyperbilirubinaemia due to defective bilirubin conjugation. Lab. Animals 20: 335-342.

36. Coughtrie, M.W.H., Harding, D., Wilson, S., Jackson, M.R., Fournel-Gigleux, S., Hume, R. and Burchell, B. Genetic deficiency of rat and human UDP-glucuronosyltransferases in: "Molecular and Cellular Aspects of Glucuronidation" G. Siest, J. Magdalou and B. Burchell eds., John Libbey Eurotext, Montrouge,France, Inserm Vol 173 pp. 69-73 (1988).

37. Iyanagi, T., Watanabe, T. and Uchiyama, Y. The 3-methylcholanthrene inducible UDP-glucuronosyltransferase deficiency in the hyperbilirubinaemic Gunn rat is caused by a -1 Frameshift Mutation. J. Biol. Chem. 264: 21302-21307 (1989).

38. Van Es, H.H.G., Peters, W.H.M., Goldhoorn, B.G., Paul-Abrahamse, M., Oude Elferink, R.P.J. and Jansen, P.L.M. Immunochemical characterisation of UDP-glucuronosyltransferase by monoclonal antibody techniques in: "Molecular and Cellular Aspects of Glucuronidation". G.Siest, J. Magdalou and B. Burchell eds. John Libbey Eurotext, Montrouge, France, Inserm Vol 173 193-200 (1988).

39. Harding, D., Jackson, M.R., Corser, R. and Burchell, B. Phenol UDP-glucuronosyltransferase deficiency in Gunn rats: mRNA levels are considerably reduced. Biochem. Pharmac. 38: 1013-1017 (1989).

40. Naga, G., Homma, H., Tanase, H. and Matsui, M. Studies on the genetic linkage of bilirubin and androsterone UDP-glucuronosyltransferase by cross-breeding two mutant rat strains. Biochem. J. 252: 897-900 (1988).

41. Coughtrie, M.W.H., Burchell, B., Leakey, J.E.A. and Hume, R. The inadequacy of perinatal glucuronidation. Molec. Pharmacol. 34: 729-739 (1988).

42. Sala Trepat, J.M. Poiret, M., Sellem, C.H., Bessada, R., Erdos, T. and Gluecksohn-Waelsch, S. A lethal deletion on mouse chromosome 7 affects regulation of liver-cell specific functions: Posttranscriptional control of serum protein and transcriptional control of aldolase B synthesis. Proc. Natn. Acad. Sci. 82: 2442-2446 (1985).

43. Wolkoff, A.W., Roy Chowdhury, J. and Arias, I.M. Hereditary jaundice and disorder of bilirubin metabolism in: "Metabolic Basis of Inherited Disease" Fifth Edn, pp 1385-1420, J.B. Stanbury, D.S. Wyngaarden, D.S. Fredrickson, D.S. Goldstein and M.S. Brown eds. McGraw-Hill Book Company, New York (1985).

44. Fevery, J. Pathogenesis of Gilbert's syndrome. Eur. J. Clin. Invest. 11:417-418 (1981).

45. Fevery, J., Blanckaert, N., Heirwegh, K.P.M., Preaux, A.M. and Berthelot, P. Unconjugated bilirubin and an increased proportion of bilirubin monoconjugates in the bile of patients with Gilbert's syndrome and Crigler-Najjar disease. J. Clin Invest. 60: 970-979 (1977).

46. Ullrich, D., Sieg, A. Blume, R., Bock, K.W., Schroter, W. and Bircher, J. Normal pathways for glucuronidation, sulphation and oxidation of paracetamol in Gilbert's syndrome. Eur. J. Clin. Invest. 17: 237-240 (1987).

47. Crigler, J.F. and Najjar, V.A. Congenital familial non-haemolytic jaundice with kernicterus. Paediat. 10: 169-180 (1952).

48. Arias, I.M., Cartner L.M., Cohen, M., Ben Esser, J. and Levi, A.J. Chronic non-haemolytic unconjugated hyperbilirubinaemia with glucuronosyltransferase deficiency. Am. J. Med 47: 395-409 (1969).

CONTRIBUTORS

1. Muzaffer Aksoy, Scientific and Technical Research
    Council of Turkey, P.O. Box 74, 41401 Gebze, Kocaeli,
    Turkey.

2. Emel Arınç, Joint Graduate Program in Biochemistry,
    Department of Biology, Middle East Technical University,
    06531 Ankara, Turkey.

3. Claude Bonfils, INSERM U-128, Sanofi-Recherche, Rue J.
    Blayac, 34082 Montpellier, France.

4. Brian Burchell, Department of Biochemical Medicine,
    University of Dundee, Ninewells Hospital and Medical
    School, Dundee, DD1 9SY Scotland, U.K.

5. Jan-Åke Gustafsson, Department of Medical Nutrition,
    Karolinska Institute, Huddinge University Hospital
    F-69, S-14186 Huddinge, Sweeden.

6. Ernest Hodgson, Department of Toxicology, North
    Carolina State University, P.O. Box 7633, Raleigh, NC
    27695, U.S.A.

7. Eric F. Johnson, Department of Basic and Clinical
    Research, Scripps Clinic and Research Foundation,
    La Jolla, Calif. 92037, U.S.A.

8. David Kupfer, Worchester Foundation for Experimental
    Biology, Shrewsbury, MA 01545, U.S.A.

9. Anthony Y.H. Lu, Merck Sharp & Dohme Research Labs,
    Rahway, NJ 07065, U.S.A.

10. Franz Oesch, Institute of Toxicology, University of
    Mainz, Obere Zahlbacher Strasse 67, D-6500 Mainz,
    Federal Republic of Germany.

11. Richard M. Philpot, Laboratory of Cellular and Molecular
    Pharmacology, NIEHS/NIH, P.O. Box 12233, Research
    Triangle Park, NC 27709, U.S.A.

12. Walter Pyerin, Institute of Experimental Pathology,
    German Cancer Research Center, Im Neuenheimer Feld 280,
    D-6900 Heidelberg 1, Federal Republic of Germany.

13. John B. Schenkman, Department of Pharmacology,
University of Connecticut Health Center, Farmington, CT
06032, U.S.A.

14. Robert Snyder, Joint Graduate Program in Toxicology,
Rutgers College of Pharmacy, Busch Campus, Piscataway,
NJ 08855-0789, U.S.A.

15. Anton Stier, Max-Plank Institute of Biophysical
Chemistry, D-3400 Göttingen, Federal Republic of
Germany.

16. Hugo Vanden Bossche, Janssen Pharmaceutical Research
Laboratories, 2340 Beerse, Belgium.

Absorption dichroism, 98, 100,
    105–106, 109
Accessibility to antibodies, 97,
    99
Acetylation, 202
Acetylcholinestrase, 77
ADP-ribosylation, 202
Adrenal, 310, 312, 353, 356
Adrenodoxin, 312
Aflatoxin B₁, 43
Aldicarb, 18
Allelochemicals, 79
Allosteric effects, 125
Amine oxidase, 11
Aminoglutethimide, 354, 356
Aminothiols, 11
Androstenedione, 307–308, 312,
    316, 352, 354
Anisotropy decay, 104, 108,
    116–117
Antibodies to
  b5, 173
  FMO, 17, 56, 64–65, 68
  P450, 28–32, 56, 63, 85, 220,
    234, 236, 257, 261, 337
  P450 reductase, 12, 14, 135–136
  UDPG transferases, 467–468, 474,
    480, 483
Antibody selectivity, 28, 30–32
Antifungals, 345, 347–351,
    353–354, 357
Aplastic anemia, 387, 396, 403,
    416–421, 424

Arachidonic acid
  hydroxylation, 25
  metabolism, 293, 317–318

Benzene
  hydroxylation by P450, 376–380
  metabolism, 375–384, 415
  toxicity, 376, 383, 387–408,
    415–428
Benzphetamine, 7, 79–80, 117, 124,
    174–175, 180, 258–259
Benzo(a)pyrene, 36, 43, 76, 79

Bile acid synthesis, 320–321
Bone marrow, 376–377, 383, 387,
    389, 391, 394–395, 397,
    402, 404–405, 408, 415,
    417–422, 427

Candida albicans, 309, 348–349,
    351, 353, 357
Candida tropicals, 309, 351
cDNA
  of b5 reductase, 157
  of epoxide hydrolase, 437–438
  of FMO, 56, 65–67
  of glutathione S-transferase,
    448
  of P450, 8, 25–27, 32, 56–58,
    60, 63, 85, 317, 320
  of P450 reductase, 141
  of UDPG transferase, 473–475,
    478, 480
Chlorotrianisene, 333–343
Chlorpromazine, 16
Cholesterol, 305–306, 308–311,
    320, 347, 349–350, 354,
    357
  metabolism, 305, 308, 320
Clotrimazol, 345–349, 353–354
Computational prediction, 98
Congenital adrenal hyperplasia,
    315
Corneal microsomes, 317
Corticosterone, 312
Cortisol, 312, 315, 356
Covalent binding, 334, 337–339,
    342–343, 366, 368,
    382–383, 387, 396
Covalent protein modification,
    202–203
Cycloartenol, 305
Cysteamine, 11, 16
Cytochrome b5
  antibodies to, 173
  chemical modification of, 154,
    155, 161–164, 188–189,
    191–192
  electron transfer pathway, 149,

Cytochrome b5 (continued)
   electron transfer pathway
      (continued)
      151, 172–173, 176, 186,
      192, 196
   essential amino acids, 154–155,
      187
   functions of, 149–151, 185
   hydrophilic peptide, 150, 152,
      154–155, 159
   hydrophobic peptide, 149–150,
      152, 159–165
   of insects, 84
   interactions with
      b5 reductase, 149–152, 155,
         157, 187
      P–450, 7, 76, 149, 151, 155,
         171–175, 177, 179–180,
         186–189, 192–197, 314,
         317, 379–380
      P–450 reductase, 136–137, 172,
         180, 186–187, 191–192
   involvement in drug metabolism
      7, 151, 174–175, 184,
      186–188
   involvement in fatty acid
      metabolism, 76
   kinetics, 155
   of liver, 149–155, 159–165, 171,
      184
   of lung, 149, 155–156
   membrane orientation, 94,
      159–160, 164–165
   molecular weight, 150, 156, 171
   phosphorylation of, 227
   primary structure, 150, 152–153
   properties of, 150, 152, 154–155
   reactions of, 149, 151, 159, 172
   in stearyl CoA desaturation,
      149, 151, 159–161, 172–173
   topology in liposomes, 159–165,
      185
Cytochrome P450
   antibodies to, 28–32, 56, 63,
      85, 220, 234, 236, 257,
      261, 337
   chemical modifications of, 188
   cycle, 4, 6, 185–186, 196
   –dependent monooxygenase
      an overview, 1–10
   developmental changes, 236–237,
      239, 245–248, 255–264
   functional amino acids, 121–122,
      214–215
   genes, 8, 23, 26, 263, 284
   of human liver, 255–256, 258
   interactions with b5, see
      cytochrome b5

Cytochrome P450 (continued)
   interactions with estrogenic
      pesticides, 331–343
   interactions with P450
      reductase, see
      NADPH–cytochrome reductase
   isozymes phosphorylation, 200,
      208–210, 212–215, 218–223,
      227
   mechanism of oxidation, 7
   membrane topology, 40–41, 93,
      98–99, 115–126
   nomenclature, 23–24, 26, 55,
      235, 256–257, 298–299, 301
   reactions catalyzed by, 1–4
   role in biosynthetic reactions,
      305–307, 309–317, 319–321
   role in metabolic reactions,
      305, 308, 316–321
   role in prostaglandin
      metabolism, 293–303
   secondary structure, 40
Cytochrome P450 cam, 1, 40, 192,
      215, 233, 314
Cytochrome P450 14 α–demethylase
      (P450$_{14DM}$), 306–307,
      309, 311, 345, 347–349,
      351, 354, 356
Cytochrome P450 (insects)
   antibodies to, 85
   cDNA, 85
   distribution, 75–76
   functions of, 75–78
   and insect-plant interactions,
      78–79
   and insecticide resistance, 75,
      78–82
   molecular biology, 75
   molecular weight, 78, 83–84
   multiple forms, 75, 78, 83–85
   purification, 75, 82–85
   spectra, 76, 82
Cytochrome P450 (rabbit,
      microsomal)
   amino acid sequences, 24, 26–27,
      32, 40, 58, 61–62, 263
   antibodies to, 28–32, 56, 63,
      261
   in benzene metabolism, 376–380
   cDNA, 25–27, 32, 56–58, 60, 63
   developmental changes, 255,
      258–264
   enzyme activation, 43–45
   genes, 23, 26, 55
   genetic polymorphism, 24–25,
      32–33, 36
   in hydroxylation of prostanoids,
      298–300

Cytochrome P450 (rabbit,
    microsomal) (continued)
  induction of, 41–43, 55, 60–62,
      239, 250, 258, 261, 263,
      298
  interstrain differences, 36–37
  localization in kidney, 31–32
  of liver, 23–45, 55–61, 255,
      297–299
  of lung, 25, 55–62, 215, 297,
      300
  molecular biology, 25
  molecular weights, 299
  monoclonal antibodies, 29–30
  mRNA, 25–26, 36, 41–42, 56,
      59–60
  nomenclature, 23–24, 26, 55,
      256, 298–299
  regulation of, 41–45, 55, 59–63
  species differences, 24, 60–63
  substrate specificities, 36–37,
      41–43, 55
  tissue differences, 25, 55, 59,
      62–64
Cytochrome P450 (rat, microsomal)
  antibodies to, 234, 236, 257
  constitutive forms, 233,
      235–236, 247–248, 300
  developmental changes, 236–237,
      239, 245–248, 255, 257–258
  and diabetes, 237–243, 248–250
  gene expression regulation by
      TCDD, 283–291
  and growth hormone, 248–249
  in hydroxylation of prostanoids,
      299–301
  and hypertension, 244–247
  induction of, 233, 235–236, 239,
      242, 247, 250, 283–285,
      298, 337–338
  molecular weights, 238
  nomenclature of isozymes, 235,
      256–257, 301
  sex specific isozymes, 234–237,
      240–242, 244, 248–249,
      257–258, 264
  substrate specificities, 235,
      237, 239–240
Cytochrome P450 (steroidogenic)
  aromatase (P450$_{AROM}$), 307,
      316–317, 354
    cDNA, 317
  cholesterol side-chain cleavage
      (P450$_{scc}$), 307, 310–313,
      354, 356
    amino acid sequence, 313
  11–β hydroxylase (P450$_{11\beta}$),
      307, 310, 312–313

Cytochrome P450 (steroidogenic)
    (continued)
  11–β hydroxylase (P450$_{11\beta}$)
      (continued)
    amino acid sequence, 313
  17–α hydroxylase (P450$_{17\alpha}$),
      307, 313–316, 352–354, 356
    amino acid sequence, 313, 315
    deficiency, 315
  21–α hydroxylase (P450$_{21\alpha}$),
      307, 313, 315–316, 353–354
    amino acid sequence, 313
    deficiency, 315
  of microsomes, 310, 314–317
  of mitochondria, 310–314
  reactions of, 307, 310–317

DDT, 332
Delayed fluorescence, 93, 100,
    103–106
1,25-Dihydroxyvitamin D$_3$, 307
Diol epoxides, 447, 450, 452,
    454–455
Disulfides, 11

Ecdysone, 317
Econazole, 345, 348, 353–355
Electron transfer pathways, 6,
    137, 149, 151, 172–173,
    176, 185–186, 196
Epidermal microsomes, 321, 349
Epoxide hydrolase
  cDNA, 437–438
  cytosolic, 436–437
  functions of, 435, 437–439
  and inhibition by clotrimazole,
      349
  microsomal, 435–439
  molecular weight, 437
  multiple forms, 435–439
  phosphorylation of, 227
  purification, 436–437
  resistant diol epoxides, 447
Ergosterol, 305–306, 346–350, 354,
    356
ESR spectroscopy, 98, 100–101
Estradiol, 308, 312, 332, 341,
    351, 356
Estrogen receptor, 331–332,
    334–335, 341–343
Estrogenic activity, 331–333,
    335–336

Fatty acids, 77, 251, 293, 299
Fenarimol, 345
Fenthion, 18
Flavin containing monooxygenase,
    11–19, 55–71

Flavin containing monooxygenase
      (continued)
   antibody to, 17, 56, 64–65, 68
   cDNA, 56, 65, 67
   functions of, 11–13
   importance of, 12–13
   kinetic values, 15
   of liver, 13–18, 55, 64–70
   of lung, 13–16, 18, 55, 64–70
   molecular biology, 11, 18, 55
   molecular weight, 17
   organ differences, 16
   primary sequences, 66
   regulation, 64
   species differences, 15–16
   substrate specificity, 11–13,
      18, 55, 64
Fluorescence polarization, 102
FMN cleavage, 226
FMO, see flavin-containing
      monooxygenase
Fungal microsomes, 305, 348
Fungi, 305, 309, 311, 348, 350

Genetic engineering, 99
Genetic factors, 423–424
Genetic polymorphism, 24–25,
      32–33, 36
Gerbil adrenals, 312
Gibberellin synthesis, 307, 319,
      357–358
Glucocorticoid receptor, 267–276,
      284–285
Glutathione S-transferases
   acidic, 447–453, 456–458
   basic, 447–451, 455
   cDNA, 448
   in carcinogenecity, 447, 450,
      459
   in cellular protection, 447, 459
   in detoxification, 447, 450,
      452, 459
   and diol epoxides, 447, 450,
      452, 454–455
   molecular weight, 449
   in mutagenicity, 450, 452,
      454–455, 459
   nomenclature, 448
   phosphorylation of, 227–229
   and polycyclic hydrocarbons,
      447, 450, 459
   purification of, 455–457
Glycoprotein, 312
Glycosylation, 203, 475
GST, see glutathione
      S-transferases

Hair follicles, 349

Heat-shock protein, 335
Heterooligomer, 96, 99
Hexamer, 99, 119–121
HMG–CoA reductase, 305–306, 309,
      321, 350
Homooligomer, 95–96, 121
Hormone, 76, 96–97, 247–249,
      267–276, 357
Human hair, 349
Human skin, 317
Hybrid enzymes, 37–39
Hydrazines, 12
Hydroquinone, 375–383, 395–398,
      404–405, 407–408
21-Hydroxylase B genes, 316
25-Hydroxyvitamin D₃, 307, 356
Hydrophobic contacts, 95, 122

Imidazole, 309, 345, 347–348, 351,
      353–354, 356, 358
Imipramine, 16
Indomethacin, 395, 404
Insect hormones, 76
Insect cytochrome P450, see
      cytochrome P450 (insects)
Insect-plant interactions, 78–79
Insecticide resistance, 75, 77–82
Itraconazole, 345, 355–357

Juvenile hormones, 76, 436

Keratinocytes, 319
Ketoconazole, 309, 321, 345–346,
      348–349, 351–354, 356

Lanosterol, 305–307, 309, 345,
      347, 349, 351
Lauric acid, 25–26, 32, 76–77, 79
Leukemia, 387, 403, 416, 420–428
Leukotriene metabolism, 305, 308,
      318
Leydig cells, 314
Lipid peroxidation, 455, 459
LTB4, 308
Lung cancer, 428
Lung P450LgM2 phosphorylation, 215
17,20-Lyase, 314, 352–353, 356–357

Membrane
   asymmetry, 94
   contacts, 122
   topology, 93–102, 109, 115–117,
      120–128, 139, 141–142,
      470–471
Metabolic activation, 141,
      333–334, 357, 365, 367,
      370, 377–378, 383
Methoxychlor, 332–341

24-Methylenedihydrolanosterol, 306, 309, 345, 347
14-Methylfecosterol, 309
Microsomal membranes, 97, 115
Mitochondria, 310-311
Molecular biology, 11, 18, 25, 55, 141
Monoclonal antibodies, 29-30, 99, 121, 331
Monotopic proteins, 94
Muconaldehyde, 381-382, 395, 404-405
Mutagenicity, 365, 370, 372, 450, 452, 454-455, 459
Myeloperoxidase, 383

NADH-cytochrome b5 reductase, 149, 156-160
  absorption spectrum, 157
  amino acid sequence, 156-157
  cDNA, 157
  chemical modifications of, 158
  cofactors, 149, 156, 159
  electron transport chain, 149-151, 172
  essential amino acids, 157-158
  functions of, 149-151
  hydrophilic peptide, 156-159
  hydrophobic peptide, 157, 159-160
  interactions with b5, 149-152, 155, 157-159, 187
  kinetics, 158
  of liver, 156-160
  of lung, 158-159
  N-terminus of, 157, 159, 203
  membrane interactions, 159-160
  molecular weight, 159
  phosphorylation of, 227
  purification, 156, 158-159
  secondary structure, 157
  topology in liposomes, 159-160
NADPH-cytochrome P450 reductase
  antibody to, 12, 14, 135-136
  catalytic reactions, 135-136
  cDNA, 141
  chemical modifications of, 189-191
  in electron transport pathways, 6, 137, 151, 185-186
  essential amino acids, 140
  functions, 135-137, 141, 143
  from house fly, 75, 84
  hydrophilic peptide, 135-137
  hydrophobic peptide, 135, 137-138
  interactions with

NADPH-cytochrome P450 reductase (continued)
  interactions with (continued)
    b5, 136-137, 172, 180, 186-187, 191-192
    P450, 6, 41, 137, 139, 185-186, 189-191
  membrane topology, 139, 141-142
  and molecular biology, 141
  molecular weight, 135-136
Negative cooperativity, 125-128
5-Nitroimidazoles, 365-372
NMR spectroscopy, 98, 100-101
Nuarimol, 345

Obtusifoliol, 307, 345
n-Octylamine, 17, 55
Oligomers, 96-97, 99, 101, 115-116, 121
Ontogenesis, 255, 257-258
Ovary cell, 311

Pancytopenia, 387, 403, 416-421, 423, 428
Peroxidase, 405
Pesticides, 77-78, 331, 341
Phase I enzymes, 463
Phase II enzymes, 115, 228, 463
Phenol, 375-383, 395-396, 398, 405, 407-408, 415, 417, 468, 478
Phorate, 12-14, 18
Phorate sulfoxide, 12-15
Phosphatase effect, 223-227
Phosphine oxidases, 12
Phosphine sulfides, 12
Phosphines, 12
Phosphorescence, 93, 98, 100, 103, 109
Phosphorylation
  of b5, 227
  of b5 reductase, 227
  of epoxide hydrolase, 227
  of GST, 227-229
  of P450 isozymes, 121, 200, 208-210, 212-215, 218-223, 227
  of P450 reductase, 200, 209-210, 227
Phosphorus oxidase, 11
Photoselection, 93, 102, 104,-105, 115
Ping-pong mechanism, 158-159
Piperonyl butoxide, 12, 14
Plant allelochemicals, 79
Plant microsomes, 305
Polycyclic aromatic hydrocarbons, 283, 435, 447, 449-450

Polyphagy, 75
Polytopic proteins, 94
Positive cooperativity, 125–128
Posttranslational modification, 199–200
Pregnenolone, 307, 351–353
Product release, 115
Progesterone 21-hydroxylation, 29–30, 37–39
Progesterone metabolism, 308, 310
Prostacyclin, 293–294, 307, 318
Prostaglandin hydroxylation, 25–26, 295–303
Prostaglandins, 25, 251, 293–303, 307
Prostate carcinoma, 354
Protein conformation, 96
Protein kinases, 200, 203–208, 210, 213, 215–218
Protein phosphorylation, 199–203
Protein-lipid interactions, 95–96
Protein-protein interactions, 99, 185
Protomer interfaces, 123
Pyrimidine antifungals, 345, 348

Quaternary structure, 93, 95

Reactive intermediates, 11, 331, 338–342, 367, 383, 387, 405
Receptors, 267–277, 283–290, 331–332, 334–335, 341–343
Retinoic acid metabolism, 321, 349, 354, 356
Ronidazole, 365–372
  -protein adducts, 367–369
  mechanism for activation, 369–371
  mutagenicity of, 370, 372
Rotational correlation time, 107–109, 117–120
Rotational diffusion 93, 96, 102, 104, 109, 115, 120–121, 127

Saccharomyces cerevisiae, 306, 309, 311, 345, 347–348, 351
Secondary amines, 12
Semen prostaglandins, 301–303
Sister chromatid exhange, 404
Shoeworkers, 421–424
Skin, 18, 317, 349, 417–418
Skin proliferation, 319
StearylCoA desaturase, 17, 151, 159–161

Steroid hormone
  action, 267, 272
  receptor, 267–277, 284–285
    regulation, 274–277
    structure, 268–272, 285
  synthesis, 233, 311–317, 345, 351–357
Sterol synthesis, 305–307, 309, 348, 355, 357
Sulfides, 11
Sulfoxides, 13–14
Sulfur oxidase, 11
Swine dysentery, 365–366

Target inactivation, 98–99
TCDD, 24, 31, 41
TCDD receptor, 283–290
Tertiary structure, 93, 101, 120
Tertiary sulfoxides, 11
Testosterone, 187, 221, 235, 337, 242, 319, 350–351, 356
Testosterone metabolism, 308
Theca cells, 314
Thioacids, 11
Thiobenzamide, 12–13
Thiocarbamides, 11
Thiols, 11
Thioridazine, 15
Thromboxane, 318
Triarimol, 345–347
Tyrosine kinases, 203

UDP-Glucuronosyltransferase,
  antibody to, 467–468, 474, 480, 483
  cDNA cloning, 473–475, 478, 480
  DNA sequencing, 470
  in drug metabolism, 463–464, 468–469, 482
  enzyme assay, 464–465
  genetic deficiencies of, 478–482
  genetic lesion, 481
  glycosylation, 475
  hyperbilirubinemia, 481–482
  induction of, 467, 469
  isoenzymes, 465–466, 468–469, 474–476, 483
  membrane topology, 470–471
  molecular cloning of, 473–475, 478, 480
  protein sequences, 473, 475–476
  purification, 465–467
  substrate specificity, 469, 474, 478–479
  synthesis, 473–475, 478–479
Uniaxial rotation, 107–109
Ustilago maydis, 345

Vitamin A, 305, 308, 321
Vitamin D₃, 307, 318
Vectorial labelling, 98

Wobbling rotation, 107–108

X-ray diffraction, 98, 100, 157,
    163
Xenobiotics, 11, 13, 77, 236, 247,
    321

Yeast, 305, 309, 348–351

DATE DUE

MAY 2 7 1998			